PRACTICAL COOKERY

FOR THE LEVEL 3 ADVANCED TECHNICAL DIPLOMA IN PROFESSIONAL COOKERY

DAVID FOSKETT

NEIL RIPPINGTON

STEVE THORPE

PATRICIA PASKINS

HODDER
EDUCATION
AN HACHETTE UK COMPANY

City & Guilds have endorsed this product as a supporting resource for the hospitality and catering sector. However, this endorsement does not guarantee exact coverage of any City & Guilds qualification and is not directly linked to any City & Guilds assessment.

Although every effort has been made to ensure that website addresses are correct at time of going to press, Hodder Education cannot be held responsible for the content of any website mentioned in this book. It is sometimes possible to find a relocated web page by typing in the address of the home page for a website in the URL window of your browser.

Hachette UK's policy is to use papers that are natural, renewable and recyclable products and made from wood grown in sustainable forests. The logging and manufacturing processes are expected to conform to the environmental regulations of the country of origin.

Orders: please contact Bookpoint Ltd, 130 Park Drive, Milton Park, Abingdon, Oxon OX14 4SE. Telephone: (44) 01235 827720. Fax: (44) 01235 400454. Lines are open 9.00–5.00, Monday to Saturday, with a 24-hour message answering service. You can also order through our website: www.hoddereducation.co.uk

ISBN: 978 1 5104 0185 3

© David Foskett, Neil Rippington, Steve Thorpe and Patricia Paskins, 2017

First published in 2017 by
Hodder Education,
An Hachette UK Company
Carmelite House
50 Victoria Embankment
London EC4Y 0DZ

www.hoddereducation.co.uk

Impression number   10 9 8 7 6 5 4 3 2

Year   2021 2020

Cover photo © Andrew Callaghan

Typeset in Myriad Pro Light 10/12 pts by Aptara Inc.

Printed in India

A catalogue record for this title is available from the British Library

Contents

* The questions in the assessment strategy pages are aimed at giving learners a chance to practice their knowledge and prepare for their end assessment by answering questions in the style of the summative tests. Although this textbook is closely matched to the City & Guilds qualification and has been endorsed by City & Guilds as an excellent teaching and learning resource, these practice questions have been written by the publisher, and have not been designed, reviewed or approved by City & Guilds,. They in no way represent sample exam questions from City and Guilds, rather they are practice questions, loosely in the style of the formative assessments required by the DFE for all awarding organisations end tests.

Picture credits: p10 © Hallgerd – Shuttersock; p13 © WavebreakMediaMicro – Fotolia; p14 © NHS; p16 © Suthisa – Shutterstock; p19 *l* © Oleg Kozlov – Fotolia, *c* © Bogdan Dumitru – Fotolia, *r* © kwanchalchaludom – Fotolia; p26 *t* Crown Copyright, *b* © Africa Studio – Fotolia; p27 *t* © ranplett/ iStockphoto.com, *b* © Gail Philpott / Alamy; p28 © JPC-PROD – Shutterstock; p29 *tl* © multiart – Shutterstock, *tr* © eyewave – Fotolia, *b* © Gregory Gerber – Shutterstock; p32 © vicgmyr – Shutterstock; p43 © Kitchen CUT Ltd.; p139 © British Potato Council; p146 *l* © Sam Bailey, *r* © Lilyana Vynogradova – Fotolia; p155 © Africa Studio – Fotolia; p332 © Alexandra – Fotolia; p388 *a* © Meawpong – Fotolia, *b* © joanna wnuk – Fotolia, *c* © ftlaudgirl – Fotolia, *d* © Gray wall studio – Fotolia; p391 © Getty Images/iStockphoto/Thinkstock; p392 © kuvona – Fotolia; p393 *l* © B. and E. Dudzinscy – Fotolia; p445 © baibaz – Fotolia; p448 © WavebreakmediaMicro – Fotolia.

Except where stated above, photographs are by Andrew Callaghan. Thanks to University College Birmingham and Jack Jordan for appearing in the cover photo. Crown copyright material is licensed under the Open Government Licence v1.0.

301 Legal and social responsibilities in the professional kitchen

Learning outcomes

In this unit you will be able to:

1. Understand the importance of keeping food safe, including:
   - knowing the legislation relating to food safety in the professional kitchen
   - understanding how individuals can take personal responsibility for food safety in the professional kitchen
   - understanding the personal hygiene practices that should be followed within a professional kitchen to ensure food safety
   - understanding the importance of good personal hygiene and practice while preparing and cooking dishes
   - understanding the requirements for cleaning equipment and work areas within a professional kitchen
   - understanding how work flow and kitchen design can support food safety in a professional kitchen.

2. Understand how to maintain a healthy and safe professional kitchen, including:
   - knowing legislation relating to health and safety in the workplace
   - understanding the importance of health and safety within the professional kitchen
   - being able to identify and determine risks associated with hazards commonly found in the professional kitchen
   - understanding how control measures are used to minimise risk in the professional kitchen.

3. Understand how to design special and balanced menus, including:
   - understanding the principles of balanced diets
   - understanding alternatives needed to adjust and improve dishes to promote a healthy diet
   - understanding the differences between food preferences and special dietary requirements and the factors affecting diets
   - understanding how to adapt dishes to meet customer food preferences and special dietary requirements.

Understand the importance of keeping food safe

Food safety means putting in place all of the measures needed to make sure that food and drinks are suitable, safe and wholesome through all the processes of food provision, from production of food, selecting suppliers and delivery, right through to serving the food to the customer. Everyone consuming food and drinks prepared for them has the right and expectation to be served safe food and drinks that will not cause them illness or harm them in any way.

Eating 'contaminated' food can result in **food poisoning**, causing harm, illness and in some cases even death. Food poisoning describes a range of illnesses of the digestive system, and is usually caused by eating food that has become contaminated with bacteria and/or the **toxins** they may produce. Illness may also be caused by other contaminants and poisons being introduced to food (for example, by eating poisonous fish, plants or foods contaminated with chemicals, metal deposits or physical contaminants), or by an allergic person consuming food containing an allergen. Food poisoning may also be caused by organisms such as **viruses**. (For more information on food safety hazards within the professional kitchen see Topic 1.2.)

The main symptoms of food poisoning are often similar and typically may include:

- nausea
- vomiting
- diarrhoea
- dehydration
- abdominal pain
- fever and headache.

The number of reported cases of food poisoning in the UK each year remains unacceptably high, with around 500,000 cases of food poisoning a year from known **pathogens**. According to the Food Standards Agency, this figure would more than double if it included food poisoning cases from unknown pathogens. Also, as a large number of food poisoning cases are never reported, no one really knows the actual number.

Food poisoning can be an unpleasant illness for anyone but in some groups of people who are referred to as 'high risk' it can be very serious or even fatal. These high-risk groups include:

- babies and the very young
- elderly people
- pregnant women
- those who are already unwell or recovering from an illness (such as those with a weakened **immune system**).

KEY TERMS

Food poisoning: illnesses of the digestive system, usually caused by consuming food or drink that has become contaminated with viruses, bacteria and/or toxins.

Toxin: a poison produced by some bacteria as they multiply in food or as they die in the human body.

Virus: micro-organism even smaller than bacteria. It does not multiply in food but can enter the body via food where it then invades living cells.

Pathogens: micro-organisms such as bacteria, fungi and viruses that can cause disease.

Immune system: a system of structures and processes in the human body which defend against harmful substances and maintain good health. Occasionally the immune system in some people recognises ordinary food as harmful and a reaction occurs.

KEY TERM

Environmental health officer (EHO): a person employed by the local authority to advise upon, inspect and enforce food safety legislation in their area. An EHO is now sometimes called an environmental health practitioner (EHP).

TAKE IT FURTHER

You can find further information on Food Standards Agency guidance and regulations online at www.food.gov.uk/business-industry/guidancenotes.

1.1 Legal requirements for food safety

Food safety standards are protected by a range of legislation and enforced by **environmental health officers** (sometimes known as environmental health practitioners).

Food Safety Act 1990

The Food Safety Act 1990 provides the framework for all food safety law in Britain. It covers all types of food business in England, Wales and Scotland, and there is similar legislation in Northern Ireland. The Act covers all food activities from food production right through to distribution, retail, catering and any other form of food activity.

The Act places the following main responsibilities for all food businesses:

● to ensure that nothing is included in food, removed from food or food treated in any way making it damaging to the health of people eating it
● to ensure that any food served or sold is of the nature, substance or quality expected by consumers
● to ensure that the food is labelled, advertised and presented in a way that is not in any way false or misleading.

The Act gives the government powers to make regulations on matters of detail. The Food Standards Agency is the principal government department responsible for preparing specific regulations under the Act.

Food Hygiene Regulations 2006

The **Food Hygiene Regulations 2006** were introduced throughout the UK on 1 January 2006 to bring UK legislation in line with European legislation. The regulations update, consolidate and simplify EU food hygiene legislation.

The Regulations aim to apply effective and proportionate controls throughout the food chain, from primary production to sale or supply to the consumer (from 'farm to fork'), and clarify that it is the primary responsibility of food business operators to produce safe food. They focus on the necessary controls for public health protection.

The regulations require:
● all food businesses to be registered with the relevant authority (which authority will depend on the type of business)
● food business operators to put in place, implement and maintain permanent procedures, based on **Hazard Analysis Critical Control Point (HACCP)** principles. (For more information on food safety management systems and reporting and recording, see Topic 1.2.)
● that the appropriate level of public health protection is in place
● food businesses to apply the legislation flexibly and appropriately according to the nature of the business.

Food Information to Consumers Regulation 1169/2011

The Food Information to Consumers Regulation 1169/2011 is European legislation relating to food labelling. The Food Information Regulations 2014 enable local authorities to enforce this legislation in the UK. Much of what appears in these regulations was already detailed in Food Labelling Regulations 1996. There is no change in the principles of providing safe food, which is honestly described and not misleading. In addition, foods cannot be labelled to state that they will prevent, treat or cure an illness.

The following information is required in English on all pre-packed food labelling:

- a true name or description of the food
- the ingredients it contains, in descending weight order
- how it should be handled, stored, prepared or cooked
- who manufactured, packed or imported it
- origin information (if its absence would mislead the consumer)
- specific information declaring whether the food is irradiated or contains genetically modified material or aspartame, high caffeine, sweeteners, packaging gases, etc.
- net quantity in metric measure (grams, kilograms, litres, etc.)
- alcoholic strength where there is more than 1.2% alcohol by volume
- allergenic ingredients identified on the label in an easy-to-understand format.

Allergens

If food is non-pre-packed (such as in a restaurant or café) the information about allergens can be supplied on the menu, on menu screens, boards or provided verbally by an appropriate member of staff as well as in other formats. It must be clear, easily visible and legible. If verbal information is to be provided, it is necessary to make it clear to customers that staff can provide this information on request.

Allergen information must be specific to the actual food, complete and accurate. Refusal to give information, inaccurate or incomplete information about allergenic ingredients used in foods offered for sale would be a breach of the EU Food Information to Consumers Regulation.

1.2 How individuals can take personal responsibility for food safety

Keeping food safe so it does not cause harm to the consumer is the personal responsibility and a legal requirement for *everyone* involved in providing, handling and selling food.

Food delivery and storage

For food to remain in best condition and be safe to eat, it is essential to follow correct delivery and storage procedures.

Food must be delivered in suitable packaging, within the required **use-by** or **best before dates** and at the correct temperature.

- Use-by dates appear on perishable foods with a short life and will need refrigeration. Legally the food must be used by or before that date and not stored or offered for sale after the date. It is also an offence to change the date on the food.
- Best before dates apply to foods that are expected to have a longer life, for example, dry products or canned food. A best before date advises that food should be at its best before this date; using after the date is legal but not advised.

Only approved suppliers who can assure that food is delivered in the best condition must be used. As part of the food safety management system a record must be kept of all food suppliers used by a business. This is called **traceability**.

- All deliveries should be checked for quality and quantity then moved to the appropriate storage area as soon as possible (chilled or frozen food should be stored within 15 minutes of delivery to prevent the multiplication of spoilage organisms and pathogenic bacteria).
- Temperatures of food deliveries should be checked using a food probe – chilled food should be below 5 °C; frozen foods should be at or below −18 °C. If foods are above these temperatures on delivery speak to a head chef or supervisor before accepting them. Many suppliers will now provide a print out of delivery temperatures – keep these with other temperature records.
- Dry goods should be in undamaged packaging, well within best before dates, be completely dry and in perfect condition.
- Remove food items from outer boxes before placing the products in the refrigerator, freezer or dry store. Remove outer packaging carefully, especially from fruit and vegetables, looking out for any possible pests that may have found their way in.
- **Cleaning** of all food storage areas using **detergent** must be done thoroughly and regularly and the procedure for this must be recorded on a **cleaning schedule** that is available to everyone working in the kitchen.

KEY TERMS

Use-by date: the date by which food must be used. It must not be stored or offered for sale after this date.

Best before date: date coding appearing on packaged foods that are stored at room temperature and are an indication of quality. Use of the food within the date is not legally binding but it is bad practice to use foods that have exceeded this date

Traceability: the ability to track food through the stages of production, processing and distribution.

Cleaning: the removal of dirt and grease, usually with the assistance of hot water and detergent.

Detergent: a substance that removes grease and dirt and holds them in suspension in water. It may be in the form of liquid, powder, gel or foam. Detergent will not kill pathogens.

Cleaning schedule: planned and recorded cleaning of all areas and equipment.

Table 1.1 Storage instructions for different food types

Food type	Storage temperature	Storage instructions
Raw meat and poultry	Refrigerator running at between 0 °C and 2 °C (multi-use refrigerators should be running at below 5 °C)	Store in a refrigerator used just for meat and poultry to avoid drip contamination. If not already packaged, place in trays, cover well with cling film and label. If storing in a multi-use refrigerator, cover, label and place at the bottom of the refrigerator well away from other items.
Fish	Refrigerator running at 1–2 °C (multi-use refrigerators should be running at below 5 °C)	A specific fish refrigerator is preferable. Remove fresh fish from ice containers and place on trays, cover well with cling film and label. If it is necessary to store fish in a multi-use refrigerator, make sure it is well covered, labelled and placed at the bottom of the refrigerator well away from other items. Remember that odours from fish can get into other items such as milk or eggs.
Dairy products/ eggs	Pasteurised milk and cream, eggs and cheese should be stored below 5 °C. Sterilised or UHT milk can be kept in the dry store. After delivery, eggs should be stored at a constant temperature (a refrigerator is the best place to store them).	Pasteurised milk and cream, eggs and cheese should be stored in their original containers. For sterilised or UHT milk follow the storage instructions on the label. To avoid **cross-contamination**, prevent eggs from touching other items in the refrigerator.
Frozen foods	In a freezer running at −18 °C or below	Separate raw foods from ready-to-eat foods and never allow food to be re-frozen once it has defrosted.
Fruit, vegetables and salad items	Dependent on type; refrigerated items should be stored at around 8 °C to avoid chill damage.	Will vary according to type, e.g. sacks of potatoes, root vegetables and some fruit can be stored in a cool, well ventilated store room, but salad items, green vegetables, soft fruit and tropical fruit should ideally be kept in refrigerated storage.
Dry goods	A cool, well-ventilated dry store area.	Items such as rice, dried pasta, sugar, flour and grains should be kept in clean, covered containers on wheels, or in smaller sealed containers on shelves to stop pests getting into them. Well-managed **stock rotation** is essential. Retain packaging information as this may include essential allergy advice.
Canned products	**Ambient** (room) temperature	Stock rotation is essential. Canned food will carry best before dates and it is not advisable to use after this date. 'Blown' (swollen) cans must never be used and do not use badly dented or rusty cans. Once opened transfer any unused canned food to a clean bowl, cover and label it and store in the refrigerator for up to two days.
Cooked foods	Below 5 °C	For example, pies, pâté, cream cakes, desserts and savoury flans. They will usually be high-risk foods so correct storage is essential. For specific storage instructions, see the labelling on the individual items, but generally, keep items below 5 °C. Store carefully wrapped and labelled and well away from and above raw foods to avoid any cross-contamination.

Stock rotation – first in, first out (FIFO)

This term is used to describe stock rotation and is applied to all categories of food. It simply means that foods already in storage are used before new deliveries, providing the food is still within the required dates.

KEY TERMS

Cross-contamination: contaminants such as pathogenic bacteria transferred from one place to another. This is frequently from raw food to cooked/high risk food.

Stock rotation: managing stock by using older items before newer items, provided the older items are in sound condition and are still within use-by or best before dates.

Ambient storage: storing food at room temperature.

First in, first out (FIFO): a method of stock rotation that means that foods already in storage are used before new deliveries.

Storing food in a multi-use refrigerator

- Keep the refrigerator running between 1°C and 4°C.
- All food must be covered and labelled with name of the item and the date.
- Always store raw food at the bottom of the refrigerator with other items above.
- Keep high-risk foods well away from raw foods.
- Never overload the refrigerator – to operate properly cold air must be allowed to circulate between items.
- Wrap strong-smelling foods very well as the smell (and taste) can transfer to other foods, such as milk.
- Record the temperature at which the refrigerator is operating. Do this at least once a day and keep the refrigerator temperatures with other kitchen records. If you find that a refrigerator or freezer is running above the required temperature, check to make sure that the door has not been left open and that the door has not been opened several times in the past few minutes. If it is still showing temperatures above requirements report it immediately to whoever is in charge of the kitchen.

Any food that is to be chilled or frozen must be well wrapped or placed in a suitable container with a lid. Items may also be vacuum-packed. Make sure that all food is labelled and dated before chilling or freezing. Using colour-coded date labels or 'day dots' make it easier to recognise how long particular items have been stored and which items need to be discarded.

Clean refrigerators regularly:

- Remove the food to another refrigerator if possible.
- Clean according to the cleaning schedule using a recommended **sanitiser** (a solution of bicarbonate of soda and water is also good).

- Remember to empty and clean any drip trays and clean door seals thoroughly.
- Rinse then dry with kitchen paper.
- Make sure the refrigerator front and handle is cleaned and disinfected to avoid any cross-contamination.
- Ensure the refrigerator is down to the required temperature (1–4°C) before replacing the food in the proper positions. Check the dates and condition of all the food before replacing.

ACTIVITY

You have been asked to put away a food delivery in a multi-use refrigerator. Indicate below on which shelf you would position the different food items delivered. The items in the delivery are: raw chicken, cooked ham, cream, salmon fillets, cooked vegetable quiche, eggs, cheese, sponge cakes filled with cream, pâté, raw prawns, fresh pasta, rump steak, yoghurt, milk, sausages, butter and frozen chicken drumsticks that need to be defrosted for use tomorrow.

Shelf 1	
Shelf 2	
Shelf 3	
Shelf 4	

Preparation of food

Monitor the time that food, especially **high-risk food**, spends at kitchen temperatures and keep this to a minimum. When preparing large amounts, do so in batches, keeping the majority of the food refrigerated until it is needed. High-risk food out of refrigeration for two hours or more must be thrown away.

Work hygienically when preparing food: wash hands frequently and between tasks; make sure areas and equipment are thoroughly cleaned and sanitised at the start and end of the day, and between each task in line with cleaning schedules. Small equipment can be disinfected by putting it through a dishwasher; large equipment that cannot be moved is often cleaned with steam cleaners.

Use the correct colour-coded equipment to keep different types of food separate. As well as colour-coded chopping boards some kitchens also provide colour-coded knives, cloths, cleaning equipment, storage trays, bowls and even colour-coded uniforms.

Colour-coded chopping boards help to keep different types of food separate

Worktops and chopping boards will come into contact with the food being prepared so need special attention. Make sure that chopping boards are in good condition as cracks and splits could hold onto bacteria and this could be transferred to food. Any damaged equipment should be removed from use and reported to the person in charge of the kitchen.

Wear disposable gloves when handling high-risk foods and food such as raw chicken that could contaminate hands, other foods, equipment and areas. High-risk food is usually ready-to-eat food that will require no further cooking before it is eaten and includes:

● soups, stocks, sauces and gravies
● eggs and egg products
● milk, cream and their products
● cooked meat, fish or meat and fish products
● any foods that need to be handled or reheated.

KEY TERMS

Sanitiser: a chemical with detergent and disinfecting properties; it breaks down dirt and grease and controls bacteria.

High-risk food: usually ready-to-eat food that will require no further cooking before it is eaten.

Cooking food

Cooking food to a core (centre) temperature of 75 °C for two minutes will kill most bacteria. This temperature is especially important where large amounts are being cooked, or the consumers are in the high-risk categories. Some popular dishes on hotel and restaurant menus may be cooked to a lower temperature than this according to individual dish and customer requirements. Lower temperatures can be used when a whole piece of meat, such as a steak or a whole piece of fish, is cooked. However, always cook to the higher recommended temperature where meat has been boned, rolled or minced, or where food is part of a made up dish such as a fish pie.

Reheating

When reheating previously cooked food, reheat to 75 °C+ (the recommendation is 82 °C in Scotland.) The temperature in the centre of the food must be maintained for at least two minutes and food must only be re-heated once.

Holding for service and serving food

Cooked food being held for service must be kept above 63 °C (or below 5 °C for cold food). These temperatures also apply when serving or transporting food. When serving, the food is high risk so the highest standards of personal, food and equipment hygiene must be maintained. Handle food a little as possible and use disposable gloves, ladles, slices, serving spoons and tongs as appropriate.

Cold food dishes prepared in advance should be covered with cling film and refrigerated at 1–5 °C to keep them safe and stop them drying out. Any food out of temperature control for two hours or more must be thrown away.

Chilling

If food is being cooled or chilled to serve cold or for reheating at a later time, it must be protected from contamination and cooled quickly to 8 °C within 90 minutes. This will help prevent multiplication of any bacteria that may be present and avoids possible problems with **spore** formation. The best way to cool food is in a blast chiller, but if this is not available place the food in the coolest part of the kitchen, making sure it is protected from contamination while it is cooling.

The running temperature of refrigerators, freezers and chill cabinets should be checked and recorded at least once a day. Refrigerators and chill cabinets should be below 5 °C and freezers below –18 °C. Any piece of equipment observed running above these temperatures should be reported to the head chef or supervisor.

Identifying and controlling food safety hazards

Contamination of food means that there is something present in the food that harms the quality of the food, changes its taste, could cause harm or illness or could cause an allergic reaction. There are four main ways that food can become contaminated and all must be avoided.

Biological hazards

Biological contamination is the most dangerous type of contamination and could occur from **pathogenic bacteria** and the toxins they may produce. It is also caused by viruses, yeasts, moulds, spoilage bacteria and enzymes. These are all around us – in the environment, on raw food, on humans, animals, birds and insects.

Although all kinds of contamination of food must be taken seriously and preventative steps taken to avoid food contamination, pathogenic bacteria remains the greatest concern and is responsible for a large proportion of food poisoning cases. Pathogenic bacteria may be present in food and multiply to dangerous levels, but are not visible except with a microscope. Contaminated food may also look, smell and taste as it should, and so preventing contamination is of the greatest importance. In the right conditions (the presence of food, warmth and moisture and given sufficient time) food poisoning bacteria can multiply every 10–20 minutes by dividing in half; this is called **binary fission**.

Bacteria can multiply at the temperature range between 5 °C and 63 °C. This is referred to as the **danger zone**.

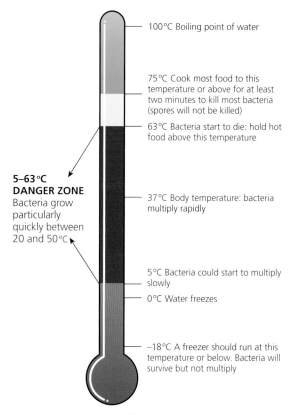

100 °C Boiling point of water

75 °C Cook most food to this temperature or above for at least two minutes to kill most bacteria (spores will not be killed)

63 °C Bacteria start to die: hold hot food above this temperature

5–63 °C DANGER ZONE Bacteria grow particularly quickly between 20 and 50 °C

37 °C Body temperature: bacteria multiply rapidly

5 °C Bacteria could start to multiply slowly

0 °C Water freezes

−18 °C A freezer should run at this temperature or below. Bacteria will survive but not multiply

Bacteria can be controlled by ensuring food is kept at controlled temperatures

To avoid biological contamination store food correctly at the recommended temperatures. Observe personal hygiene requirements and especially thorough hand washing. Keep all areas and equipment clean and sanitised, and report any illness you may have before starting work.

KEY TERMS

Spore: a resistant, resting phase for some bacteria, when they form protection round the essential part of the cell that can then survive boiling, freezing and disinfection.

Contamination: when something is present in food that harms its quality, changes its taste or could cause harm, illness or an allergic reaction.

Biological contamination: contamination caused by living organisms (for example, bacteria, toxins, viruses, yeasts, moulds and enzymes).

Pathogenic bacteria: bacteria that can cause illness by infection or by the toxins they produce.

Binary fission: the process by which bacteria divide in half and multiply.

Danger zone: the temperature range between 5 °C and 63 °C at which bacteria can multiply.

Table 1.2 Food poisoning bacteria

Major food poisoning pathogens	Source of bacteria	Preferred temperature for growth	Illness onset time	Symptoms	Can it form spores?
Salmonella This used to be main cause of food poisoning in the UK (now campylobacter). There are many different types of salmonella. Poisoning can be passed on through **healthy carriers** (someone carrying salmonella but not showing any signs of illness).	Raw meat/poultry, raw egg, intestines and excreta of humans and animals Sewage/untreated water Pests and domestic pets Food sources include raw meat and poultry, raw eggs, untreated milk and shellfish	7–45 °C	12–36 hours	Stomach pain, diarrhoea, vomiting, fever (1–7 days)	No
Clostridium perfringens Food poisoning incidents from this organism have occurred when large amounts of meat are brought up to cooking temperatures slowly then allowed to cool slowly for later use, or if meat does not get hot in the centre. Poultry where the cavity has been stuffed has also caused problems because the middle does not get hot enough to kill the bacteria. All these examples can lead to the formation of spores. Spores are very resistant to any further cooking and allow the survival of bacteria in conditions that would usually kill them.	Animal and human intestines and excreta, soil and sewage, insects, raw meat and poultry, unwashed vegetables and salads	15–50 °C	12–18 hours	Stomach pain and diarrhoea (12–48 hours); vomiting is rare	Yes
Staphylococcus aureus Produces a toxin in food; is heat-resistant and very difficult to destroy. To avoid food poisoning from this organism, food handlers need to maintain very high standards of personal hygiene.	Humans – mouth, nose, throat, hair, scalp, skin, boils, spots, cuts, burns, etc.	7–45 °C	1–7 hours	Stomach pain and vomiting; flu-like symptoms, maybe some diarrhoea; lowered body temperature (6–24 hours)	No
Clostridium botulinum A spore-forming organism also producing toxins but fortunately is fairly rare in the UK. Symptoms can be very serious, even fatal. It multiplies in conditions where there is very little or no oxygen so is of concern to canning industries and where food is vacuum-packed.	Soil, fish intestines, dirty vegetables, some animals	7–48 °C	2 hours – 8 days (usually 12–36 hours)	Difficulty with speech, breathing and swallowing; double vision, nerve paralysis; death	Yes
Bacillus cereus Produces spores and two different types of toxin. One toxin is produced in food as the organisms multiply or expire and the other in the human intestine as the organisms expire. *Bacillus cereus* can survive whether oxygen is present or not and is more difficult to destroy when fats and oils are present. It is often associated with cooking rice in large quantities, cooling it too slowly and then reheating. The spores are not destroyed and further bacterial multiplication can then take place.	Cereal crops, especially rice, spices Dust and soil (unwashed vegetables)	3–40 °C	1–5 hours; 8–16 hours	Vomiting, abdominal pain, maybe some diarrhoea (12–24 hours); stomach pain, diarrhoea, some vomiting (1–2 days)	Yes

Food-borne illness

Different organisms from those described in Table 1.2 cause what is described as food-borne illness. These do not multiply in food, but use food to get into the human gut, where they then multiply and cause a range of illnesses, some of them very serious. They include those described in Table 1.3.

Table 1.3 Food-borne illnesses

Bacteria or virus	Sources	Preferred temperature for growth	Illness onset time	Symptoms	Can it form spores?
Campylobacter jujuni This now causes more food-related illness than any other organism. One of the reasons thought to contribute to this is the increase in consumption of fresh chicken which is a significant source of this organism.	Raw poultry/meat Untreated milk or water Sewage Pets and pests, birds and insects	28–46 °C	2–5 days	Headache, fever, bloody diarrhoea, abdominal pain (mimics appendicitis)	No
E. coli 0157VTEC There are many strains of *E-coli* but significant problems have been caused in recent years by *E. coli 0157VTEC*. Symptoms can be very serious and can even be fatal.	Intestines and excreta of cattle and humans Untreated water and sewage Untreated milk Raw meat, under-cooked mince Unwashed salad items and dirty vegetables	4–45 °C	1–8 days (usually 3–4 days)	Stomach pain, fever, bloody diarrhoea, nausea Has caused kidney failure and death	No
Listeria monocytogenes This organism is of concern because it can multiply (very slowly) at refrigerator temperatures, i.e. below 5 °C. It is also of concern because of the serious outcomes that poisoning from this organism can cause.	Pâté, soft cheeses made from unpasteurised milk, raw vegetables and prepared salads **Cook/chill meals**	0–45 °C	1 day – 3 months	Meningitis, **septicaemia**, flu-like symptoms, stillbirths	No
Norovirus An air-borne virus, widely present in the environment and highly contagious. Passed from person to person. Does not grow in food – viruses only grow in living cells.	Can survive on surfaces, equipment and cloths for several hours	N/A	24–48 hours	Severe vomiting and diarrhoea	No
Typhoid/paratyphoid People who have suffered from this can become long-term carriers, which means they could pass the organism on to others through food (food handlers need six negative faecal samples before returning to work after illness).	Sewage, untreated water Also dirty fruit/vegetables	N/A	8–14 days	Fever, nausea, enlarged spleen, red spots on abdomen. Severe diarrhoea. Some fatalities	No
Bacillary dysentery (*shigella sonnei* and *shigella flexneri*)	Carriers Sewage Contaminated water	N/A	12 hours – 7 days	Abdominal pain Fever, nausea, vomiting, diarrhoea	No

Viruses are even smaller than bacteria and can only be seen with a powerful microscope. They multiply on living cells, not on food, though they may be transferred into the body on food or drinks and may live for a short time on hard surfaces such as kitchen equipment. Viruses can be air borne and water borne and are easily passed from person to person.

Toxins (poisons) can be produced by some bacteria as they multiply in food. They are heat resistant and may not be killed by the normal cooking processes that kill bacteria so remain in the food and can cause illness.

Some bacteria produce toxins as they die, usually in the intestines of the person who has eaten the food.

Spores can be formed by some bacteria in conditions such as a rise in temperature or in the presence of chemicals such as **disinfectant**. A spore forms a protective 'shell' inside the bacteria, protecting the essential parts from the high temperatures of normal cooking, disinfection, dehydration and so on. Once spores are formed the rest of the cell disperses and cannot divide and multiply as before but simply survives until conditions improve, for example, high temperatures drop to a level where the cell can re-form and multiplication can start again. Prolonged cooking times and/or very high temperatures are needed to kill spores. Time is very important in preventing the formation of spores. Large amounts of food such as meat for stewing, when brought slowly to cooking temperature, allows time for spores to form, which are then very difficult to kill. Bring food up to cooking temperature quickly and cool food quickly.

> **KEY TERMS**
>
> **Septicaemia:** blood poisoning; it occurs when an infection in the bloodstream causes the body's immune system to begin attacking the body itself.
>
> **Healthy carrier:** someone carrying salmonella in their intestine without showing any signs of illness.
>
> **Cook/chill meals:** pre-cooked foods that are rapidly chilled and packaged then held at chiller temperatures before being reheated for use.
>
> **Disinfectant:** destroys pathogenic bacteria, bringing it to safe level.

Food spoilage

Food spoilage means that food has spoiled or 'gone off'. Unlike contamination with pathogens it can usually be detected by sight, smell, taste or texture. Signs of food spoilage include:

- mould
- slimy, over-wet or over-dry food
- sour smell or taste
- discoloured and wrinkled food or other texture changes.

It is caused by natural breakdown of the food by spoilage organisms such as **spoilage bacteria**, enzymes, moulds and yeasts, which in some cases may not be harmful themselves but cause the food to deteriorate. Spoilage may also be caused by poor storage, poor handling or by contamination of the food.

Food spoilage can account for a significant amount of unnecessary waste in a business and if food stock is being managed and stored properly it should not happen.

Any food that has spoiled or is out of date must be reported to the supervisor or line manager then disposed of appropriately.

Chemical hazards

Chemical contamination can occur when chemical items get into food accidentally; these can make the consumer suffer discomfort or become ill. The kinds of chemical that can get into food could be cleaning fluids, disinfectants, machine oil, degreasers and pesticides. Problems can also occur when there are chemical reactions between such things as metal containers and acidic foods.

To avoid chemical contamination keep chemicals well away from open food. When cleaning or disinfecting use chemicals according to manufacturer and data sheet instructions. Keep all items in their original containers and do not use containers for anything else.

Physical hazards

Physical contamination is caused when something gets into food that should not be there and may be one or more of a wide range of items, such as glass, nuts, bolts and oil from machinery, flaking paint, pen tops, threads from worn clothing, buttons, blue plasters, hair or insects.

Physical contamination can be avoided by working with care and keeping work areas clean and free of any items that could cause contamination. High standards of personal hygiene and staff uniforms will also help to prevent physical contamination.

Keep work areas clean and free of items that may cause a physical hazard

Allergenic hazards

An **allergy** is when a person's immune system reacts to certain foods. Allergens can cause a range of reactions, including swelling, itching, rashes and breathlessness; they may even cause **anaphylactic shock** – a severe reaction, in which swelling of the throat and mouth blocks the airways.

Food intolerance is different to an allergy and does not involve the immune system, but there may still be a reaction to some foods and some of these reactions can be serious.

Foods usually associated with allergies and intolerances are nuts, dairy products, wheat-based products, gluten, eggs, fish, sesame seeds, preservatives and shellfish. Some people may also have an allergy to some fruits, vegetables, plants and mushrooms.

From 2014, provision of certain food allergy information became **mandatory**, and now all businesses offering food to customers need to provide this information. Under EU Regulation FIR 1169/201, specific information on 14 allergen ingredients must be available, irrespective of whether the food is packaged or chosen from a menu. (Further information on this legislation and the associated Food Information Regulations 2014 can be found in Topic 1.1.).

Avoid **allergenic reactions** by:
- remaining aware of possible allergens in the food that you prepare and cook
- providing accurate information for customers
- avoiding allergen cross-contamination, for example, by cleaning areas thoroughly between preparation of different food types
- using specific equipment for allergic customers
- taking care that items such as sauces and garnishes do not contain allergens
- retaining food packaging for allergen information.

If you are aware that food has become contaminated or become unsafe in any way during its storage, preparation, cooking or preparation for service, it is essential that you do not serve the food to a customer. Always report your concerns to someone in charge and remain aware of how you should report any problems and concerns.

More information on adapting dishes for customers with food allergies or intolerances can be found in Topic 3.2.

KEY TERMS

Food spoilage: foods spoiled by the action of bacteria, moulds yeasts or enzymes. The food may smell or taste unpleasant, be sticky, slimy, dry, wrinkled or discoloured. Food spoilage is usually detectable by sight, smell or touch.

Spoilage bacteria: cause food to change and spoil, for example, develop a bad smell or go slimy.

Chemical contamination: contamination by chemical compounds used for a variety of purposes such as cleaning and disinfection.

Physical contamination: when something gets into foods that should not be there (e.g. glass, paint, plasters, hair or insects).

Allergy: a reaction by the immune system to certain foods or ingredients; also referred to as an **allergenic reaction.**

Anaphylactic shock: a severe, potentially life-threatening reaction to food, which causes swelling of the throat and mouth and blocks airways.

Food intolerance: does not involve the immune system, but does cause a reaction to some foods.

Mandatory: something that must be done, for example rubber gloves must be worn when handling certain chemicals.

Safe food handling procedures and working practices

All food handlers are required to contribute to food safety by observing and taking part in all of the policies and procedures put in place at their workplace. These may include:
- observing personal hygiene, hand washing and correct uniform standards
- recording temperatures of food storage, cooking, holding and serving
- recording cleaning and disinfection procedures as required by cleaning schedules
- completing records such as opening and closing checks
- following procedures for safe food handling and working practices
- observing workplace procedures for reporting any problems that could affect food safety including the reporting of any illness
- reporting any concerns or problems that may affect food safety
- taking part in any food safety training and qualifications required by the employer.

Controlling cross-contamination

Cross-contamination is when bacteria or other contaminants are transferred from contaminated food (usually raw food), equipment or surfaces to ready-to-eat food. It is the cause of significant amounts of food poisoning and care must be taken to avoid it. Cross-contamination could be caused by:
- foods touching such as raw meat and cooked meat
- raw meat or poultry dripping onto high-risk foods

- soil from dirty vegetables coming into contact with high-risk foods
- dirty cloths or dirty equipment
- equipment such as chopping boards and knives used for raw then cooked food without proper cleaning or disinfection
- hands touching raw then cooked food without being washed between tasks
- pests spreading bacteria from their own bodies, urine and droppings around the kitchen.

Allergen cross-contamination can occur by allergen foods touching, dripping onto non-allergen food or from surfaces and equipment.

Cross-contamination can be avoided by always following hygienic working practices and storing, preparing and cooking food safely. This is the responsibility of all food handlers. Good personal hygiene practices by staff, especially frequent and efficient hand washing, are very important in controlling cross contamination. Good personal hygiene practices are discussed in Topic 1.3.

Reporting and recording systems

Food safety legislation requires that every food business has a recorded food safety management system, which proves how they manage all aspects of food safety. All food safety management systems must be based on **HACCP**.

Hazard Analysis Critical Control Point (HACCP)

This is a system that identifies the hazards that could occur at critical points or stages in any process and the safeguards in place to stop potential hazards from causing harm. The system must provide a documented and regularly updated record of the policies and procedures in place and of stages all food will go through right up until the time it is eaten. Full written or electronic records must be available for inspection as part of the system.

When environmental health officers or practitioners inspect a food business they will check that food safety management systems are in place and are working effectively.

Achieving the correct temperatures and recording this is an important part of food safety and businesses must not only ensure that required temperatures are achieved but the temperature records are made available for inspection. Most food businesses record refrigerator, chill cabinets and freezer temperatures at least once a day. Food being kept hot for service must not fall below 63 °C so must be tested with a temperature probe and the temperature recorded. It is also good practice to probe and record temperatures of food that is cooked to order.

Systems are now available to automatically log temperatures of all refrigerators, freezers and display cabinets in a business. Temperatures are recorded and sent to a central computer several times a day. These can then be printed or stored electronically as part of **due diligence** record keeping. Units not running at correct temperatures will be highlighted.

> **KEY TERM**
>
> **Due diligence:** written and recorded proof that a business took all reasonable precautions to avoid food safety problems and food poisoning.

Ensuring compliance with policies and procedures

Ensuring food offered for service is safe

As well as an employer's legal responsibility to produce safe, hygienically prepared and cooked food, employees have their own specific responsibilities under food safety legislation. It is the employee's responsibility to:

- partake in any food safety training provided by the employer and relevant to the work they are doing so that they understand the principles of food safety and how to avoid causing food poisoning
- comply with the safe, hygienic working practices as established by their employer and under food safety legislation, including working in a way that does not endanger or contaminate food and never serving food they know is suspect or contaminated
- report to someone in charge anything that may have an effect on food safety, such as equipment running at the wrong temperature, and co-operate with all food safety procedures the employer puts in place
- report any substandard food deliveries, out-of-date or contaminated food
- immediately report any sightings of pests or signs that pests have been present
- correctly and accurately complete food safety procedures such as recording running temperatures of equipment and temperatures of food as allocated by an employer, manager or supervisor
- report any illness, especially if stomach related, to a supervisor before starting work. After suffering such an illness they must not return to work until 48 hours after the last symptom
- maintain high standards of personal hygiene, wear clean, suitable kitchen clothing and adopt hygienic working practices.

1.3 Importance of being clean and hygienic

Because humans are a source of food poisoning bacteria,
especially staphylococcus, it is very important (and a legal
requirement) that all food handlers take care with personal
hygiene and adopt good practices when working with food.

Personal hygiene practices

Food handlers must:
- arrive at work clean (daily bath or shower) and with
  clean hair
- wash hands thoroughly and frequently (see below)
- wear clean kitchen clothing used only in the kitchen. This
  should completely cover any personal clothing
- keep hair well contained in a suitable hat or net
- keep nails short and clean, not bitten and not wear nail
  varnish or false nails
- not wear jewellery or a watch when handling food. A
  plain wedding band is permissible but could still trap
  bacteria. Jewellery can also fall into food
- avoid wearing cosmetics and strong perfumes
- not smoke in food preparation areas (bacteria from
  touching mouth area could get into food as could ash or
  smoke, and smoking can cause coughing). At break times
  food handlers must not smoke wearing kitchen clothing
- not eat food, sweets or chew gum when handling food.
  Avoid scratching the skin and spitting should never occur
  in a food area, as all may transfer bacteria to food
- wash, dry, then cover any cuts, burns or grazes with a
  blue 'food standard' waterproof dressing, then, wash
  hands
- report any illness to the supervisor as soon as possible
  and before going near any food (see below).

Uniform and protective clothing

Protective clothing is worn in the kitchen to protect food
from contamination as well as shielding the wearer from
heat, burns and splashes and to provide protection for
those with skin problems.

All items of clothing should be hard wearing, light in colour,
fit well, be in good repair and suitable for the tasks being
completed. Kitchen protective clothing usually consists of:
- chef's jacket
- trousers
- apron
- hat or hair net
- safety shoes.

Kitchen clothing itself could contaminate food. If it gets
dirty or stained change it for clean clothing.

**Protective clothing helps to protect food from
contamination**

Hand washing

Hands are constantly in use in the kitchen and will
be touching numerous materials, foods, surfaces and
equipment. Contamination from hands can happen very
easily and great care must be taken with hand washing to
avoid this. A basin with hot and cold water and preferably
mixer taps must be provided just for hand washing.

Hands should always be washed:
- when you enter the kitchen, before starting work and
  handling any food
- after a break (particularly after using the toilet)
- after smoking or eating
- between different tasks but especially after handling raw
  and before handling cooked or high risk food
- if you touch your hair, nose or mouth and face or use a
  tissue for a sneeze or cough
- after you apply or change a dressing on a cut or burn
- after using cleaning materials, cleaning preparation areas,
  equipment or contaminated surfaces
- after handling kitchen waste, external food packaging,
  money or dirty vegetables.

Hand-washing technique with soap and water

1. Wet hands with water

2. Apply enough soap to cover all hand surfaces

3. Rub hands palm to palm

4. Rub back of each hand with palm of other hand with fingers interlaced

5. Rub palm to palm with fingers interlaced

6. Rub with back of fingers to opposing palms with fingers interlocked

7. Rub each thumb clasped in opposite hand using a rotational movement

8. Rub tips of fingers in opposite palm in a circular motion

9. Rub each wrist with opposite hand

10. Rinse hands with water

11. Use elbow to turn off tap

12. Dry thoroughly with a single-use towel

13. Hand washing should take 15–30 seconds

clean your hands campaign

National Patient Safety Agency NHS

© Crown copyright 2007 283373 1p 1k Sep07

Adapted from World Health Organization *Guidelines on Hand Hygiene in Health Care*

Wash hands correctly to reduce the risk of contamination

Illness reporting

It is essential to report any illness to the supervisor as soon as possible, and before going near any food. There will be a procedure for doing this in your workplace – make sure you understand what it is and follow the procedure appropriately. This is a legal requirement.

Reportable illnesses include:

- diarrhoea and/or vomiting, nausea and stomach pain; you may not go back to handling food until 48 hours after the last symptom has passed
- infected (**septic**) cuts, burns or spots. Septic cuts have become infected with bacteria and are often white or yellow in appearance.
- eye or ear infections
- cold or flu symptoms including sore throat
- cuts, burns, spots or other injuries especially where they have become septic
- skin problems such as dermatitis
- also report illness you had when on holiday and family members or friends you have contact with who have the above symptoms, especially where they are stomach related.

Some employers require a fitness to work certificate or other medical clearance before the employee returns to working with food.

Dealing with cuts, wounds and injuries

If you are cut, burnt, wounded or injured at work assess or get someone else to assess how serious the injury is. If considered necessary seek medical help.

For minor injury run the affected part under cold running water, preferably for 10 minutes. Dry carefully with paper towels and apply a waterproof dressing

The standard bright blue plasters used in kitchen areas are easily visible if they fall into food and each one contains a thin metal strip, which would be identified by metal detectors in food manufacture. As well as protecting the wearer, a plaster provides a barrier between the wearer and food, so reduces bacterial contamination.

Disposable plastic gloves may also be worn to reduce contamination and to demonstrate good practice, but make sure these are changed regularly, especially between tasks.

All injuries, no matter how minor, must be reported to the appropriate person and recorded according to workplace procedure.

Importance of good personal hygiene practice while preparing and cooking food

Good practices and personal hygiene of food handlers is absolutely essential to ensure the safety of food. There is always a risk that bad practices can cause contamination of food and lead to possible food poisoning. For example, food that is not stored properly can support the multiplication of harmful bacteria, which can then contaminate other foods, equipment and surfaces by cross contamination. A food handler not washing their hands properly, who is unwell or wearing dirty kitchen clothing could also be the cause of cross contamination. Food left too long at kitchen temperatures can support bacterial multiplication; food cooked incorrectly can allow bacteria to survive, all of which could be the cause of food poisoning.

Adopting professional and safe working practices along with excellent personal hygiene standards and clean smart presentation will increase self-esteem and pride in the job, as well as inspiring and providing a good example and a role model for others in the team. It contributes to customer satisfaction and returning customers, who will make recommendations to others, leading to a strong and successful business.

A food business not observing good practices in storing, preparing, cooking and serving food can result in unhygienic and disorganised work areas, cross contamination and unsafe food being served to customers. Bad practice can soon result in dissatisfied customers and a poor business reputation, which in turn could cause difficulties for the continuation of the business and even business closure. However, the most serious risk from bad practice is the risk of harm and illness that could be caused to customers.

The reputation of a business is crucial to its success and no longer is it just personal recommendation and word of mouth that drives this. Entries on social media mean that customer experience, both good and bad, can be seen worldwide within seconds.

1.4 Importance of keeping work areas clean and hygienic

Clean food areas play an essential part in the production of safe food. All food handlers are responsible for the cleanliness and cleaning of their working areas.

Clean premises, work areas and equipment are important to:
- control the bacteria that cause food poisoning
- reduce the possibility of physical and chemical contamination
- reduce accidents such as slips, trips and falls
- create a positive image for customers, visitors and employees
- comply with the law
- avoid attracting pests to the kitchen.

To ensure food areas are clean and hygienic at all times, clean as you go and do not allow waste to build up; clean up any spills straight away.

It is a requirement to plan, record and check all cleaning in the form of a planned cleaning schedule, which will form part of the food safety management system. All food handlers must have an awareness of the requirements of the cleaning schedule and make sure that these are carried out.

A cleaning schedule needs to include the following information:
- What is to be cleaned.
- Who should do it (name if possible).
- When it is to be done, for example the time of day.
- Materials to be used, including chemicals and their correct dilution, cleaning equipment and the protective clothing/equipment to be used.
- Safety precautions that must be taken, for example, use of wet floor signs.
- Signatures of the cleaner and the supervisor checking the work along with the date and time.

If there is any problem that prevents correct cleaning procedures being completed, such as equipment not working properly or shortage of cleaning chemicals, report this to someone in charge of the kitchen immediately.

Cleaning products and equipment

The type and amount of cleaning equipment needed in a kitchen area will vary according to the type and size of kitchen and the work being carried out.

All kitchens will have the following chemicals, which are used for different tasks.
- Detergent will remove grease and dirt and hold them in suspension in water. It may be in the form of a liquid, powder, gel or foam and usually needs dilution and adding to water. Detergent does not kill bacteria, though if mixed with very hot water (which is how detergents work best) this may help to do so. It will clean and degrease surfaces so disinfectant can work properly.
- Disinfectant will destroy bacteria when used properly. Make sure that you only use a disinfectant intended for kitchen use, which will not leave behind a smell or taste. Disinfectant must be left on a cleaned, grease-free surface for the required amount of time (contact time) to be effective and works best with cool water. Heat may also be used to disinfect, for example, by using steam cleaners or the hot rinse cycle of a dishwasher.
- Sanitiser cleans and disinfects and usually comes in spray form. Sanitiser is very useful for frequently used work surfaces and equipment, especially between tasks and also for hand- contact surfaces such as refrigerator handles. Sanitiser may also need contact time.
- **Steriliser** can be used after cleaning to make a surface or piece of equipment bacteria free.

As well as the chemicals listed above, cleaning equipment includes cloths and/or kitchen paper, brooms, mops and mop buckets, dustpans and brushes, wire brushes and soft bristle brushes, scrapers, scouring pads and personal protection safety items such as rubber gloves, goggles, masks and waterproof aprons.

Larger equipment may include mechanical floor scrubbers, mechanical floor mops with vacuum water extraction, steam jet cleaners with various heads and attachments, degreasing and extractor cleaning equipment and others designed for specific cleaning purposes.

The type of cleaning equipment available in a kitchen will vary according to its size and the type of work carried out

Cleaning and disinfection process

Some kitchen areas such as floors and walls will need planned and thorough cleaning, but some items, especially in areas where high-risk foods are handled, need both cleaning and **disinfection**. These are:

- all food contact surfaces, such as chopping boards, bowls, spoons, whisks, etc.
- all hand contact surfaces, such as refrigerator handles and door handles
- cleaning materials and equipment, such as mops, buckets, cloths, hand wash basins.

For kitchen surfaces one of the cleaning methods shown in Table 1.4 are recommended.

Table 1.4 Cleaning methods

6-stage cleaning	4-stage cleaning
1. Remove debris and loose particles	1. Remove debris and loose particles
2. Main clean to remove soiling grease	2. Main clean using hot water (and detergent if very soiled or greasy) and sanitiser
3. Rinse using clean hot water and cloth to remove detergent	3. Rinse using clean hot water and cloth if recommended on instructions
4. Apply disinfectant. Leave for contact time recommended on the container	4. Allow to air dry or use kitchen paper
5. Rinse off the disinfectant if recommended	
6. Allow to air dry or use kitchen paper	

Dishwashing

The most efficient and hygienic method of cleaning dishes and crockery is to use a dishwasher, as this will clean and disinfect items that will then air dry, removing the need for cloths. The dishwasher can also be used to clean and disinfect small equipment such as chopping boards.

Stages in machine dishwashing:
- Remove waste food.
- Pre-rinse or spray.
- Load onto the appropriate racks with a space between each item.
- The wash cycle runs at approximately 50–60 °C using a detergent. The rinse cycle runs at 82 °C or above which disinfects items ready for air-drying.

If items need to be washed by hand:
- Scrape or rinse off residual food.
- Wash items in a sink of hot water; the temperature should be 50–60 °C, which means rubber gloves will need to be worn. Use a dishwashing brush rather than a cloth – the brush will help to loosen food particles and is not such a good breeding ground for bacteria as a cloth.
- Rinse in very hot water. This will help to kill any remaining bacteria.
- Allow to air dry (do not use tea towels).

Cleaning large kitchen equipment

Large equipment such as large mixing machines and ovens cannot be moved so need to be cleaned where they are. This is called **clean in place** and each item will have a specific method outlined on the cleaning schedule. Sometimes steam cleaning methods are used that also disinfect the items.

ACTIVITY

Tick the items below that you think should be both *cleaned* and *disinfected*.

Item	Clean only	Clean/disinfect	Item	Clean only	Clean/disinfect
Cutlery			Staff lockers		
Milk cartons for recycling			Red chopping boards		
Delivery area floor			Fridge door handle		
Grater			Inside of deep fryer		
Food containers from a hot counter			Dishcloths		
Nailbrushes			Door frames		

KEY TERMS

Disinfection: action to bring micro-organisms to a safe level. This can be done with chemical disinfectants or heat.

Clean in place: cleaning items where they are rather than moving them to a sink. This is used for large equipment such as mixing machines.

Use of chemicals for cleaning

When using chemicals:

- Never use any chemical or piece of equipment without instruction and training and follow instructions for safe chemical use including dilution, mixing and safe disposal.
- Detergent or disinfectant only stays active for an hour or two so long soaking of cloths and mops is not recommended. Wash/disinfect the items, squeeze out, then allow to air dry.
- Do not store cleaning chemicals in food preparation and cooking areas; instead use separate, lockable storage. Store chemicals in their original containers with the instruction label visible.
- Do not use sprays such as sanitiser near open food and take care not to get it into eyes or on skin.
- Kitchen areas will display safety data sheets with information on chemicals, how they are to be used and how to deal with spillages and accidents. Make sure you are familiar with this information.
- If cleaning materials or chemicals are spilled, warn others in the area, put up a wet floor sign, wear rubber gloves and a mask if needed, soak up the excess with kitchen paper then clean the area.
- If cleaning materials are spilled onto the skin wash off with cold water and dry. If there is a skin reaction seek first aid immediately and report the incident.

More information on the **COSHH (Control of Substances Hazardous to Health) Regulations**, which cover use of chemicals in the workplace, can be found in Topic 2.1.

KEY TERM

COSHH Regulations: legal requirement for employers to control substances that are hazardous to health to prevent any possible injury to those using the substances

Disposal of waste

Kitchen waste should be placed in waste containers (bins) with lids. Ideally these should be foot operated and lined with a strong bin liner. Waste bins need to be strong, easy to clean, pest proof and kept in good, sound condition, away from direct sunlight. They must be emptied regularly to avoid cross contamination and odour and never left un-emptied in the kitchen overnight as this could result in the multiplication of bacteria and attracting pests.

Outside waste areas should not be too close to the kitchen door and waste bins should be on wheels so they can be easily moved, and allow for cleaning underneath. They should have closely fitting lids and be situated on a hard standing area that can be easily cleaned and hosed.

Pests may be attracted to waste areas so not allowing a build-up of waste and regular cleaning and disinfection will help to discourage this. Staff need to remain aware of possible signs of pests and report any suspicion of a pest presence immediately (see below).

Recycling is now everyday practice in kitchens and staff need to become familiar with the separation of different waste items ready for collection. These may be bottles, cans, waste food, paper and plastic items.

How kitchen workflow and design can support food safety

Suitable buildings with well-planned fittings, layout and equipment are essential for all food areas. It is a legal requirement that food premises are designed with good food safety as a priority, ensuring that there is the correct equipment to keep food at the required temperatures. Design and choice of equipment must allow for ease of cleaning and prevention of pest entry.

Certain essentials must be in place if a building is to be used for food production. There must be:

- electricity supplies, and preferably gas supplies
- drinking water and good drainage
- suitable road access for deliveries and refuse collection
- no risk of contamination from surrounding areas and buildings, for example, chemicals, smoke, odours or dust.

Other factors to be considered in kitchen design:

- There must be adequate storage areas – proper and sufficient refrigerated storage is especially important.
- Suitably positioned hand washing and drying facilities that are only used for this purpose must be provided.
- Cleaning and disinfection must be planned with separate storage for cleaning materials and chemicals and all areas must be designed to allow for thorough cleaning, disinfection and pest control.
- Personal hygiene facilities must be provided for staff as well as changing facilities and storage for personal clothing and belongings.
- First aid equipment must be provided for staff and any accidents must be recorded.

Workflow

When planning food premises, separating areas for different foods, storage, processes and service areas will help to reduce the risk of contamination and assist efficient working and effective cleaning. A planned **linear workflow** should be in place. For example:

Delivery → Storage → Preparation → Cooking → Hot holding → Serving.

This means there will be no cross-over of activities that could result in cross contamination.

'Dirty areas' that involve preparation or storage of raw foods or washing of dirty vegetables need to be kept separate from the 'clean areas' where cold preparations, finishing and serving takes place. The direction of ventilation and cleaning processes should move from clean areas to dirty areas. If separate areas for raw and high risk foods are not possible, keep them well away from each other, making sure that working areas are thoroughly cleaned and disinfected between tasks.

> **KEY TERM**
>
> **Linear workflow:** a flow of work that allows the processing of food to be moved smoothly in one direction through the premises from the point of delivery to the point of sale or service.

Surfaces and equipment

Installing the right equipment, which is properly placed and used correctly, will keep food in best condition and avoid any cross contamination. Equipment and surfaces need to be smooth, impervious (not absorb liquids) and must be non-corrosive, non-toxic and must not crack, chip or flake. This will make areas and equipment easier to clean effectively, as well as helping to avoid contamination of food.

All of the following must also be considered in the planning and layout of premises intended for food use:

- lighting and ventilation (natural and artificial)
- drainage
- floors
- walls
- ceilings
- windows/doors.

All surfaces and equipment must be suitable for the required purpose, in good repair and easy to clean as well as disinfect where necessary.

All food premises, fittings and equipment must be kept in good repair to ensure food safety. For example, cracked surfaces or chipped equipment could support the multiplication of bacteria and a refrigerator running at the wrong temperature may also allow bacteria to multiply in food. If you notice anything is damaged, broken or faulty report it to a supervisor immediately. You may have specific reporting forms to do this.

Pests

Pests can be a serious cause of contamination and disease in premises. Pest prevention is a legal requirement and evidence of pests in a food business can result in legal action, loss of reputation, reduced profit, poor staff morale and closure of the business. Any suspicion that pests may be present must be reported to whoever is in charge of the kitchen immediately.

Common pests that cause food contamination

Some of the common pests that may cause problems in food areas are shown in Table 1.5.

Table 1.5 Signs of pest presence

Pest	Signs that they are present
Rats and mice	Sightings of rodents, droppings, unpleasant smell, gnawed wires and/or packaging, greasy marks on lower walls, damaged food stock and spillages, paw prints, holes in frames, skirting boards, etc.
Flies and wasps	Sight/sound of flies and wasps, dead insects, maggots.
Cockroaches	Sighting (dead or alive) usually at night, pupae, egg cases, dust piles in corners and small spaces, unpleasant smell.
Ants	Sightings and can be present in food. The tiny, pale-coloured pharaoh ants are very difficult to spot but can still be the source of a variety of pathogens.
Weevils	Sightings of weevils in stored products such as flour and cornflour. These are very difficult to see – tiny black insects moving in flour.
Birds	Sighting and droppings in outside storage areas and around refuse.
Domestic pets and wild cats	These must be all kept out of food areas as they carry pathogens on fur, whiskers, saliva, urine, etc.

Measures to keep pests out of the building are important and need to be planned as part of food safety management. Some possible measures are described below.

● Good environmental design – blocking entry for pests (for example, no holes around pipe work, avoid all gaps and cavities where pests could get in, use sealed drain covers and appropriate proofing).
● Repair any damage to the building or fixtures and fittings quickly.
● Use screening or netting on windows and doors.
● Doors should be self-closing, with metal kick plates and bristle strips along the bottom.
● Check deliveries and packaging for pests.
● Place physical or chemical baits and traps in relevant places (a pest contractor will do this).
● Use electronic fly killers (EFKs) to eliminate flying insects such as flies, bluebottles, mosquitos and wasps – these insects can enter through very small spaces or from other parts of the building and can spread bacteria around the

kitchen and onto open food. EFKs use ultraviolet light to attract the insects that are then electrocuted on charged wires and fall into catch trays. Do not position the EFK directly above food preparation areas.

● Keep food in sealed containers and ensure no open food is left out uncovered.
● Check food stores regularly for signs of pests.
● Do not allow a build-up of waste in the kitchen.
● Do not keep outside waste too close to the kitchen, and containers need to be emptied regularly. Keep the area clean and tidy.
● Arrange for professional and organised pest management control, surveys and reports.

Pest control contractors can offer advice on keeping pests out and organise eradication systems for those that do get in. A good pest control contractor can also complete formal audits and supply certificates proving pest control management is in place.

TAKE IT FURTHER
Visit the following websites to find out more information on food safety:
● Food Standards Agency **www.food.gov.uk**
● Chartered Institute of Environmental Health: **www.cieh.org**
● Highfield: **www.highfield.co.uk**
● Foodlink: **www.foodlink.org.uk**
● Food Law Reading: **www.foodlaw.rdg.ac.uk**
● Royal Institute of Public Health and Hygiene: **www.rsph.org.uk**

TEST YOURSELF

1 List the groups of people considered to be 'high risk' if they contract food poisoning.

2 Name **three** items of legislation relating to food safety.

3 What are the conditions/temperatures you would recommend for storing the following foods:
   a) dairy food
   b) frozen fish
   c) fresh meat
   d) soft fruit
   e) flour.

4 Food held for service should be kept above 63°C or below 5°C.
   a) Why is this important?
   b) What is the name given to this range of temperatures?

5 Give **four** examples of foods that could cause an allergic reaction in some people.

6 There are certain illnesses that a food handler must report to their supervisor before being involved with food. What are they?

7 Give **four** examples of how you can avoid cross-contamination in a kitchen.

8 When cleaning what do each of the following chemicals do?
   a) Detergent
   b) Sanitiser
   c) Disinfectant

9 Pests should be kept out of food premises.
   a) Why is this important?
   b) Suggest three ways that you could keep pests out.
   c) How can flying insects be dealt with in a food area?

10 Your restaurant is going to run an outside catering event in a marquee. The food will be cold and prepared in the main restaurant. Consider what arrangements need to be made for:
   - hygienic transport of food
   - storage of food at the venue
   - safe service of food including how long food can be out of refrigeration
   - checking that agency staff are suitably knowledgeable in food safety matters
   - staff hygiene facilities
   - cleaning of equipment and dishwashing.

Understand how to maintain a healthy and safe professional kitchen

2.1 Importance of health and safety in the professional kitchen

Professional kitchens can potentially be dangerous places to work if not properly managed. It is a legal requirement for every establishment to identify risks and to introduce necessary measures to minimise risks and possible workplace accidents and injury, to ensure that the workplace is as safe as it can possibly be.

It is of the greatest importance that good practices are developed that make everyone working in the area aware of the importance of maintaining high standards of health and safety. Making health and safety part of daily working practices, planning, feedback, handover and meetings will help to ensure safer working practices and a safe workplace.

There is a significant amount of legislation in place to ensure that business owners comply with what is necessary to avoid employees being injured or made unwell in any way by their workplace or the work they complete.

Health and Safety at Work Act

The Health and Safety at Work Act 1974 (which was updated in 1994) imposes a general duty on an employer 'to ensure so far as is reasonably practicable, the health, safety and welfare at work of all employees'. It is largely about establishing and maintaining good health and safety practice in the workplace. It requires that employers comply with all parts of the Act, including completing **risk assessments** of all areas and procedures and introducing safe ways of working.

Employers will need:
- a health and safety policy that includes risk assessments (unless there are fewer than five employees)
- to keep premises and equipment in safe working order and provide/maintain personal protective equipment for employees.

Employees must:
- take reasonable care for the health and safety of themselves and of other persons who may be affected by their actions or omissions at work

- co-operate with the employer so far as is necessary to meet or comply with any requirement concerning health and safety and not interfere with, or misuse, anything provided in the interests of health, safety or welfare.

The Act is particularly relevant to the hospitality and catering industries, where many hazards are a naturally occurring part of everyday work. Working areas themselves, machinery and tools can all present dangers. Where so many hazards can be present in working areas it is of particular importance to make sure that staff take part in regular safety training and that risk assessments address any dangers and regularly update the controls in place.

Control of Substances Hazardous to Health (COSHH) Regulations

Under COSHH, risk assessments must be completed by employers for all hazardous chemicals and substances that employees may be exposed to at work. This includes assessment of their safe use and proper disposal.

A range of chemicals will be used in kitchens, and each chemical must be given a risk rating that is recorded. Employees must be given relevant information and training on the chemicals they will be using. Data sheets or cards with information about the chemicals, how they must be used and what to do if an accident occurs with the chemical must be kept in an easily accessible place so staff can refer to them when needed. Measures must be taken to ensure that all staff understand these, with consideration given to those who do not speak English as their first language or who have other possible barriers to understanding the information.

Everyone must be trained in the safe handling and use of chemicals and proper dilution procedures where needed and must wear protective equipment – goggles, gloves and face masks – as appropriate. Eye goggles should be worn when using oven cleaners, gloves when hands may come into contact with any chemical cleaner and face masks when using grease-cutting and oven degreasers. Extra protective aprons or overalls may be needed. It is essential that staff are trained to take precautions and not to take unnecessary risks.

Reporting of Injuries, Diseases and Dangerous Occurrences Regulations (RIDDOR)

These regulations require employers and/or the person with responsibility for health and safety within a workplace to report and keep records of any:

- work-related fatal accidents
- work-related disease
- accidents and injury resulting in the employee being off work for three days or more
- dangerous workplace events (including 'near miss' occurrences)
- major injuries, loss of limbs or eyesight.

Accidents and incidents under **RIDDOR** need to be reported. This can be done by contacting HSE by telephone: 0845 300 9923; or by completing the RIDDOR form on their website: **www.hse.gov.uk/riddor**.

Management of Health and Safety at Work Regulations

These regulations state that where an employer has five or more employees there must be a written health and safety policy in place that is issued to every member of staff. This will outline the responsibilities of the employer and employee in relation to health and safety. It states that health and safety information must be provided for all employees.

Manual Handling Operations Regulations

These regulations are in place to protect employees from injury or accident when required to lift or move heavy or awkwardly shaped items. A risk assessment must be completed, employees trained in correct manual handling techniques and lifting or moving equipment provided where appropriate. Handling of objects can be more difficult in hospitality areas because they could be very hot, frozen or sharp.

Personal Protective Equipment at Work Regulations

These regulations require employers to assess the need for and to provide suitable personal protective equipment (PPE) and clothing at work. In a commercial kitchen environment these may include chefs' uniforms and items for cleaning tasks such as rubber gloves, goggles and masks. Employers must keep these items clean, in good condition and provide suitable storage for them. Employees must use them correctly and report any defects or shortages.

> ### KEY TERMS
>
> **Risk assessment:** the process of identifying and evaluating hazards and risks and putting measures in place to control them.
>
> **RIDDOR:** Reporting of Injuries, Diseases and Dangerous Occurrences Regulations 2013 – all injuries, diseases and dangers occurrences happening in the workplace or because of work carried out on behalf of the employer must be reported to the Health and Safety Executive. This is a legal requirement and it is the employer's responsibility to make any such reports.

2.2 Controlling hazards in the professional kitchen

When we assess health and safety we consider **hazards**. A hazard is anything that can cause harm. The following aspects of the kitchen environment have the potential for hazards:

- Equipment – knives, liquidisers, food processors, mixers, mincers and a range of other kitchen equipment.
- Substances – cleaning chemicals, detergents, sanitisers, disinfectants, degreasers and other substances.
- Work methods – dangerous or careless working, for example, carrying knives and equipment incorrectly, not using equipment properly, not following workplace safety rules, becoming distracted or not concentrating, not wearing correct PPE, trying to work too quickly due to pressure/stress.
- Work areas – badly designed work areas, inadequate cleaning standards (for example, spillages not cleaned up), overcrowded work areas, insufficient work space, uncomfortable work conditions due to extreme heat, cold or insufficient lighting or ventilation.

These hazards could lead to the following common accidents in kitchens:

- slipping on a wet or greasy floor
- walking into, tripping or falling over objects
- lifting objects in the wrong way and lifting heavy or awkward objects
- being exposed to hot items or dangerous substances such as steam or oven-cleaning chemicals
- being hurt by moving objects, such as falling stacked boxes
- cuts from knives or other cutting equipment
- entrapment in machines, such as large mixing machines
- injuries caused by machines, such as vegetable cutting machines, liquidisers, mincers
- fires and explosions
- electric shocks from equipment.

A **risk** is the chance of somebody being harmed by a hazard present in the working area. There may be a high risk, medium risk or a low risk of harm occurring. For example, using a stick blender is a hazard but what is the likelihood of it causing harm? If it is not used properly and the hands are placed near the moving blades there could be a serious injury. If the blender is not kept well below the surface of hot soup it will splash out and could burn the hands and face. However, with proper training and supervision the blender can be used safely and no injury will occur. This risk is medium.

ACTIVITY

1 Give examples of **three** more risks there may be in using different pieces of kitchen equipment.

2 Describe the injuries that could possibly occur in using each piece of equipment.

3 Decide whether each of your examples is a low, medium or high risk.

Risk assessment

As part of managing health and safety in a business it is required by law that an employer control the potential risks in the workplace (this is detailed in the Management of Health and Safety at Work Regulations). All workplaces with five or more employees must have a written health and safety policy that is given to all employees. To do this an employer must have conducted a full risk assessment of the premises and procedures, including consideration of anything with the potential to cause harm. They must then put the necessary **control measures** in place to keep employees safe and keep records of this. If there are fewer than five employees a full written risk assessment is not required, but the employer still has a duty of care for the health and safety of employees at work.

Table 1.6 Risk assessment record

Hazard	Risk level before control measures	Current control measures	Risk level with controls	Additional control required (action, person responsible and date)	Date of staff training and any actions taken
Electricity	Medium	Use and maintain electrical equipment according to manufacturer's instructions. Switch off from supply when not in use. Do not operate electrical switches with wet hands. Report electrical faults immediately. Kitchen signs in appropriate places.	Low	Supervision at all times of employees under 18 years old. Initial training on equipment use for all staff and additional training if under 18. Refresher training every six months. Testing of new equipment. Yearly testing of existing equipment.	
Hot liquids (e.g. water, oil)	High	All staff to be made aware of the danger of hot containers. Do not overfill pans with water. When lifting the lid, always use the handle and stand to the side to allow steam to escape. Avoid drips, especially of oil. Avoid condensation settling on the floor which would then become wet. Warning signs in kitchen.	Medium	Supervision at all times of employees under 18 years old. Initial training and induction covering hot liquids and their dangers for all staff and additional training if under 18.	
Hot surfaces	Medium	If a surface becomes hot during use, precautions must be taken to prevent burns. Place pans in a position that minimises the risk of contact. Do not reach over them. Warning signs in kitchen.	Low	Initial staff training and induction as for hot liquids. Additional training if under 18.	
Knives	Medium	Ensure that knives are kept sharp and have secure handles. Keep handles dry and free from grease. Ensure that the correct size and type of knife is used. Do not leave knives lying on work surfaces or tables. When carrying knives, always do so with the point downwards.	Low	Initial staff training and induction. Additional training if under 18. Knife safety awareness training for all staff. First aid training for key staff.	

Managing risk need not be a complicated procedure and guidance, including useful templates, is available on the **Health and Safety Executive** website: **www.hse.gov.uk.**

There are many control measures that can be put in place to avoid accidents in kitchens and minimise the risk from potential hazards.

- The working areas should be well designed for the tasks being completed. They must be well maintained, with good standards of housekeeping and cleaning in place and clear visible signage as appropriate.
- All staff must be trained in safe working practices and this must be monitored, including the correct use of the necessary PPE. Staff must arrive at work ready physically and mentally for the work ahead.
- The majority of accidents in kitchens are caused by falling, slipping or tripping. Therefore, floors should be even, with no unexpected steps or gradients. Floor coverings should have an anti-slip finish and be kept clean and grease free. Any spillage must be cleaned immediately, verbal warnings given and warning notices put in place and a member of staff should stand guard until the hazard is cleared.
- Corridors and walkways must be kept clear so that people do not trip over objects that are in the way. People carrying trays and containers may have a restricted view, may not see items on the floor and so may trip over them. This could be made worse if the person falls onto a hot stove or if the item they are carrying is hot; falls can have severe consequences.
- There should be good, efficient lighting so employees can see clearly where they are going and any possible hazards.
- There should be adequate ventilation so condensation is avoided and excess heat and fumes can escape. Condensation can also result in wet, slippery floors.
- Kitchens should not be overcrowded. There should be enough space for people to move around work areas freely.
- Employees must follow the rules and safe working practices put in place. They must work in an organised way, cleaning as they go.

- Employees must move around the kitchen with care, avoiding backtracking or crossing over to avoid collisions with others. No matter how busy, never run in a kitchen.

ACTIVITY

What are the ways you inform and warn staff about staying safe in a busy kitchen?

Below is a personal safety checklist to help you stay safe:

- Always wear the correct protective clothing and any other recommended items.
- Never work under the influence of alcohol or drugs.
- Keep hair short and covered by a hat, or tied back and in a net if long.
- Do not wear jewellery or a watch at work as these can be caught in machinery.
- Always walk with care and use the intended walkways; never take short cuts.
- Look out for and follow the information on all warning notices and safety signs.
- Always use the correct lifting techniques and use trolleys or other appropriate lifting equipment if available.
- If in doubt about any procedure including the use of equipment or machinery, do not use it. Ask for help and proper training.

KEY TERMS

Hazard: any area, activity or procedure with the potential to cause harm.

Risk: the possibility that someone could be harmed by the hazard.

Control measure: measures put in place to minimise risk.

Health and Safety Executive: the national independent authority for work-related health, safety and illness. It acts to reduce work-related death and serious injury across UK workplaces.

TEST YOURSELF

1 Under the Health and Safety at Work Act 1974 what must an employer do to ensure that staff are safe at work?

2 What are COSHH regulations in place to protect?

3 What are the personal protective equipment items likely to be used by staff in a kitchen?

4 List **five** common causes of accidents in a kitchen.

5 Suggest **three** ways that accidents can be avoided in the kitchen.

6 There are a number of hazards in a professional kitchen that could result in injury if not managed correctly.
   a) Make a list of hazards that may be present in a busy professional kitchen and for each suggest a control to minimise risk.
   b) How could you make kitchen staff aware of the safety controls that are in place?

Understand how to design special and balanced menus

3.1 Understand the principles of balanced diets

Public Health England have published statistics stating that 61.7 per cent of adults are overweight or obese (65.3 per cent of men and 58.1 per cent of women). Levels of obesity have increased year on year over the past 20 years. However, healthy eating is not just about reducing the levels of obesity. Medical research suggests that a third of all cancers are caused by poor diet, including cancers of the bowel, stomach and lungs, as well as to high blood pressure, **diabetes**, **osteoporosis** and tooth decay.

Eating a balanced, nutritional diet combined with regular exercise can help to protect from these diseases as well as lowering cholesterol and reducing obesity. With increasing numbers of people eating outside of their homes, those producing food are in a strong position to influence customers in making healthy food choices. It is therefore important that those producing food are aware of the different nutrients and their importance when preparing, cooking and serving dishes as part of a balanced diet to ensure that nutritional value is maximised.

Many individuals and groups of people follow special diets or have allergies or food intolerances and caterers must be aware of the main features of these dietary needs and the cause and effects of not following the requirements properly.

No food needs to be considered 'good' or 'bad' – it is the overall balance of foods in the diet that is important. There is no perfect food that provides all the nutrients needed for the body – different foods have differing nutritional content so provide for the body in different ways. Because of this it is essential to eat a variety of foods to provide all the nutrients needed for a balanced diet; this also makes meals more interesting.

KEY TERMS

Diabetes: a disease in which the body produces insufficient or no insulin so cannot regulate the amount of sugar in the blood. If left untreated it can lead to increased risk of heart disease, blindness, kidney problems and fatigue. Type 1 diabetes tends to occur in people under 40 and in children; Type 2 diabetes is more common in overweight and older people and can sometimes be controlled by diet alone.

Osteoporosis: a disease in which the density or thickness of the bones breaks down, putting them at greater risk of fracture. Exercise and good nutrition can reduce the risk of developing osteoporosis.

Current nutritional guidelines

The Food Standards Agency has produced some easy-to-follow guidelines on good nutrition and healthy eating. The main points are:

- eat the right amount of food for how active you are
- eat a range of foods to make sure you are getting a balanced diet.

A healthy balanced diet contains a variety of types of food, including:

- plenty of fruit and vegetables
- a variety of starchy foods such as wholemeal bread and wholegrain cereals
- some protein-rich foods such as meat, fish, eggs and lentils
- some dairy foods.

The Eatwell Guide

The Eatwell Guide sets out the government's recommendations on eating healthily and achieving a balanced diet. It shows the different types of foods and drinks we should consume, and in what proportions, to have a well-balanced and healthy diet.

Eight tips for eating well

The Food Standards Agency has also put together eight tips for eating well:

1 Base your meals on starchy foods.
2 Eat lots of fruit and vegetables.
3 Eat more fish, including a portion of oily fish each week.
4 Cut down on saturated fat and sugar.
5 Try to eat less salt – no more than six grams daily for adults.
6 Get active and try to be a healthy weight.
7 Drink plenty of water.
8 Don't skip breakfast.

TAKE IT FURTHER

There is more detailed information and an explanation of each of these points on the Food Standards Agency website: **www.food.gov.uk**. The information is also available from them in booklet form.

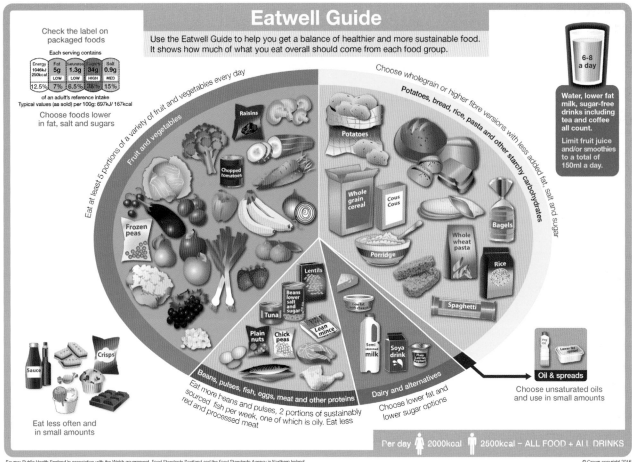

Eatwell Guide

Five a day

Fruit and vegetables are part of a healthy, balanced diet and can help us stay healthy. It's important to eat enough of them. The 'Five a day' campaign is based on advice from the World Health Organization, which recommends eating a minimum of 400 grams of fruit and vegetables a day to lower the risk of serious health problems, such as heart disease, stroke and some cancers. Evidence shows there are significant health benefits to getting at least five 80-gram portions of a variety of fruit and vegetables every day.

Aim to eat at least five portions of fruit and vegetable each day

ACTIVITY

Design a five-day menu for a residential conference centre providing three meals a day, to show how you would incorporate at least five portions of fruit and vegetables each day. Provide as much variety as possible.

Sources of nutrients

Nutrients in food help our bodies to carry out essential everyday activities and to heal themselves if they are injured. Different nutrients perform different functions and can be found in different food sources.

Carbohydrates

Carbohydrates are needed to provide energy. They are made by plants then either used by the plants as an energy source, or eaten by animals or humans for energy or as dietary fibre.

The three main types of carbohydrate are sugars, starches and fibre.

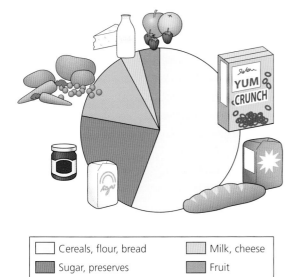

Legend:
- ☐ Cereals, flour, bread
- ▨ Milk, cheese
- ■ Sugar, preserves
- ▥ Fruit
- ▨ Potatoes, vegetables

Sources of carbohydrate in the average diet

Sugars

Sugars are the simplest form of carbohydrate; other forms of carbohydrate may be converted to sugars when they are digested. Types of sugar include:

- glucose – found in the blood of animals, in fruit and in honey.
- fructose – found in fruit, honey and cane sugar
- sucrose – found in beet and cane sugar
- lactose – found in milk
- maltose – found in cereal grains and also used in beer making.

Starches

Starches are more complex than sugars and are converted to sugars in the digestive process. Starches are present in many foods including:

- pasta
- cereals
- bread and other bread-based products (e.g. biscuits and cakes)
- wholegrains such as barley, couscous, barley and tapioca
- ground grains such as flour, cornflour, ground rice and arrowroot
- vegetables such as potatoes, parsnips, peas and beans
- pulses such as dried lentils and beans and peas
- unripe fruit such as bananas, apples and pears.

Foods high in carbohydrates

Fibre

Dietary fibre is a very important form of starch. Unlike other carbohydrates, dietary fibre cannot be digested and does not provide energy to the body. However, dietary fibre is essential for a healthy diet because it:

- helps to remove toxins and waste from the body and maintain healthy bowel movement
- helps to control the digestion and processing of nutrients
- adds bulk to the diet, which curbs the feelings of hunger so can be useful as part of a weight reduction programme.

Fibre is found in:
- fruits, vegetables and salad crops
- wholemeal or wholegrain bread
- wholegrain cereals and oats
- wholemeal pasta
- wholegrain rice
- pulses such as peas beans and lentils.

Fruit and vegetables are a source of fibre

Protein

Protein is needed for carrying out the millions of bodily functions essential to staying alive and completing human activities. Protein is essential for growth in children and is needed for repairing and growth of body cells and tissues. Protein can also provide some heat and energy and any protein not used for repair or growth of cells is converted into carbohydrate or fat and so can be used for energy.

There are two basic kinds of protein:

- Animal protein – found in meat, game, poultry, fish, eggs, milk and cheese.
- Vegetable protein – found in vegetables, seeds, pulses, peas, beans, nuts and wheat.

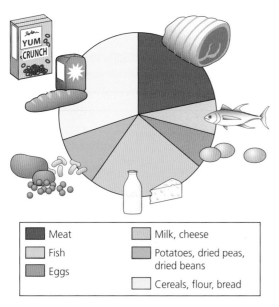

Meat	Milk, cheese
Fish	Potatoes, dried peas, dried beans
Eggs	Cereals, flour, bread

Sources of protein in the average diet

Protein is made up of chemicals known as **amino acids**, and ideally these amino acids need to be obtained from both animal and vegetable proteins. The protein in cheese is different to the protein in meat because the amino acids are different. Some amino acids are essential to the body and must be included in a balanced diet. Table 1.7 shows the proportion of protein in different animal and plant foods.

KEY TERM

Amino acid: the structural units of protein.

Table 1.7 Proportion of protein in some common foods

Animal foods	Protein (%)	Plant foods	Protein (%)
Cheddar cheese	25	Soya flour (low fat)	45
Bacon (lean)	20	Soya flour (full fat)	37
Beef (lean)	20	Peanuts	24
Cod	17	Wholemeal bread	9
Herring	17	White bread	8
Eggs	12	Rice	7
Beef (fat)	8	Peas (fresh)	6
Milk	3	Potatoes (old)	2
Cream cheese	3	Bananas	1
Butter	< 1	Apples	< 1

Fats

Fats are naturally present in many foods and are an essential part of the diet. The main functions of fats are to protect the body and as a source of energy. Fats form an insulating layer under the skin, which helps to protect vital organs and keep the body warm. Fat is needed to build cell membranes in the body and also provides a source of the fat soluble vitamins A, D, E and K.

Fats, mainly of animal source, are solid at room temperature. These include butter, dripping (from beef), suet, lard (from pork), cheese, cream, bacon and fatty meat and oily fish.

Sources of fat

Oils are usually from a plant source and are liquid at room temperature. These include the various oils such as olive, sunflower, rape seed, palm, coconut, vegetable, nut oils and soya oils.

Saturated and unsaturated fats/oils are not the same in their chemical structure and differ because of how many hydrogen atoms occur in a molecule, compared with carbon atoms.

A fatty acid molecule is one of the building blocks of fat, and the higher the proportion of hydrogen than carbon it has, the more saturated it is. Saturated fats are usually solid at room temperature and include those listed as fats above. Saturated fats are also frequently found in processed foods.

When there are fewer hydrogen atoms than carbon in the molecule, the product will be a liquid oil as listed above, as well as oils occurring in certain fish, nuts, seeds, avocado and olives.

A diet high in **saturated fat** is thought to increase the risk of heart disease as well as contributing to other poor health issues.

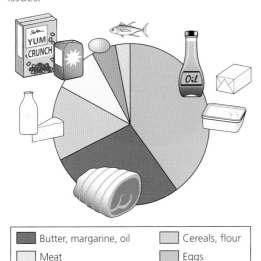

■ Butter, margarine, oil	▨ Cereals, flour
☐ Meat	▨ Eggs
▨ Milk, cheese	☐ Fish

Sources of fat in the average diet

Sources of oils in the average diet

Margarine and spreads are made from oils, which are likely to be liquid at room temperature. A process was developed where oils were made solid by passing hydrogen through the oil in controlled conditions. Margarines made in this way are said to contain hydrogenated fat. There have been serious health concerns about this process because of the production of **trans fats**, which can be a contributing factor

in heart disease. For this reason, hydrogenated fats are being used less and different processes have been developed by the margarine industry. Some tropical oils, such as palm oil and coconut oil, are naturally semi-solid and do not require hydrogenation.

Eating too much fat has negative effects on the body and can lead to:

- being overweight (obesity)
- high levels of cholesterol, which can block the blood vessels (arteries) in the heart
- heart disease
- higher risk of certain cancers and stroke
- halitosis (bad breath)
- type 2 diabetes (type 1 diabetes is not diet-related and not related to being overweight).

Table 1.8 shows the percentage of saturated in the different types of food in an average diet.

Table 1.8 Percentage of saturated fat in food in an average diet

Type of food	Saturated fat (%)
Milk, cheese, cream	16.0
Meat and meat products	25.2
This splits down into:	4.1
Beef	3.5
Lamb	5.8
Pork, bacon and ham	2.7
Sausage	9.1
Other meat products (e.g. burgers, faggots, pate)	
Other oils and fats (e.g. olive oil, margarine, sunflower oil)	30.0
Other sources, including eggs, fish and poultry	7.4
Biscuits and cakes	11.4

Vitamins

Vitamins are chemical components found in very small amounts in a variety of foods and are vital for life and for maintaining good health. Vitamins are essential for many bodily functions such as growth and protection from disease.

Vitamins can be **fat soluble** (A, D, E and K) so available from some foods containing fats and oils, or **water soluble**, especially C and the B group. Care must be taken with foods containing water soluble vitamins not to soak them in water or cook in too much water or the vitamins could be lost.

Table 1.9 describes some of the main vitamins, their functions and sources.

Table 1.9 Vitamins, their functions and sources

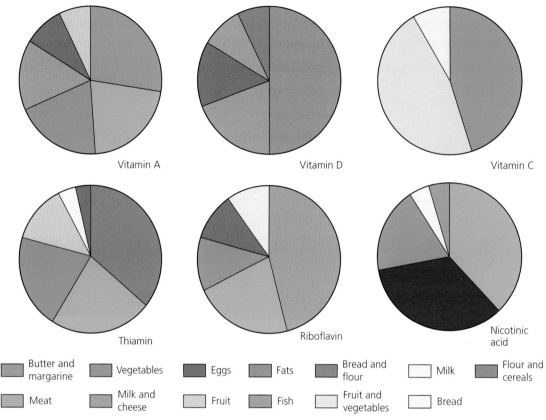

Main sources of vitamins A, C and D, thiamine, riboflavin and nicotinic acid in the average diet

Table 1.9 Vitamins, their functions and sources

Vitamin	Functions	Foods in which they are found
A	Helps children to grow. Helps the body to resist infection. Helps to prevent night blindness.	Fatty foods, dark green vegetables, halibut and cod liver oil, butter and margarine, watercress, herrings, carrots, liver and kidneys, spinach, tomatoes, apricots. Fish liver oils are the best source.
B (all types)	Helps to keep the nervous system in good condition. Enables the body to obtain energy from carbohydrates. Encourages the body to grow.	
B1 (thiamin)	Helps the body to produce energy. Necessary for the brain, heart and nervous system to function properly.	Yeast, bacon, oatmeal, peas, wholemeal bread
B2 (riboflavin)	Helps with growth. Necessary for healthy skin, nails and hair.	Yeast, liver, meat extract, cheese, egg
B (niacin or nicotinic acid)	Vital for normal brain function. Improves the health of the skin, the circulation and the digestive system.	Meat extract, brewer's yeast, liver, kidney, beef
C (ascorbic acid)	Needed for child growth. Helps cuts and broken bones to heal. Prevents gum and mouth infections.	Potatoes, blackcurrants, green vegetables, lemons, grapefruit, bananas, strawberries, oranges, tomatoes, fruit juices

Vitamin	Functions	Foods in which they are found
D	Controls the way our bodies use calcium.	Sunlight helps the body to produce vitamin D.
	Necessary for healthy bones and teeth.	Fish liver oils, oily fish, egg yolk, margarine and dairy produce.
E	An antioxidant.	Nuts, seeds, vegetables, oils, wheat germ and whole grains
	Helps the body control toxins.	
	Helps the body against infection.	
K	Needed for blood clotting, helps wounds heal.	Green leafy vegetables, vegetable oils, grains

Minerals

There are 19 minerals needed in very small quantities in order for the body to function properly.

Minerals are needed to build bones and teeth, for a variety of bodily functions and to control levels of fluid in the body. Table 1.10 shows the most common minerals needed by the body.

Table 1.10 Minerals, their functions and sources

Mineral	Functions	Foods in which they are found
Calcium	Builds bones and teeth.	Milk and milk products, green vegetables, the bones of canned oily fish (e.g. sardines), drinking water, wholemeal bread and white bread (if calcium has been added)
	Helps blood to clot.	
	Helps muscles to work.	
	(Vitamin D is also needed in order for the bodies to use calcium effectively).	
Iron	To build haemoglobin – a substance in the blood that transports oxygen and carbon dioxide around the body.	Lean meat, wholemeal flour, offal, green vegetables, egg yolk, fish
	Iron is better absorbed if vitamin C is also present.	Our bodies absorb iron more easily from meat and offal.
Folic acid	Helps protect against stroke.	Dark green vegetables such as broccoli and spinach
Phosphorous	Builds bones and teeth (together with calcium and vitamin D).	Liver, bread, cheese, eggs, kidney, fish
	Controls the structure of brain cells.	
Sodium	Regulates the amount of water in the body.	Many foods are cooked with salt (sodium chloride), have salt added to them (e.g. bacon and cheese) or already contain salt (e.g. meat, eggs, fish).
	Helps muscles and nerves to function.	
	If too much salt is eaten some of it is excreted in urine.	
	The kidneys control how much salt is lost in urine.	
	We also lose sodium when we sweat (but our bodies cannot control how much salt we lose in this way).	
Iodine	Enables the thyroid gland to function properly (produces hormones that control our growth).	Sea food, iodised salt, drinking water obtained near the sea, vegetables grown near the sea
	If the thyroid is not working properly it can make us underweight or overweight.	
Potassium	Balances the fluids in our bodies.	Bananas, avocados, citrus fruits and green leafy vegetables
	Helps to maintain a normal heart rate.	
	Helps nerves and muscles to function.	

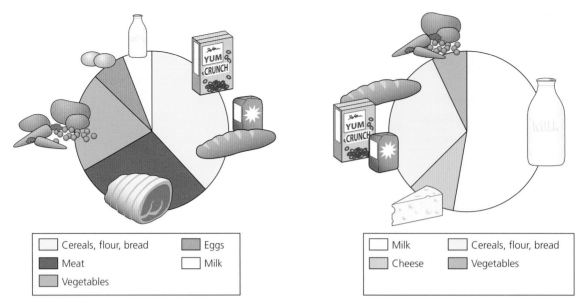

Cereals, flour, bread   Eggs

Meat   Milk

Vegetables

Milk   Cereals, flour, bread

Cheese   Vegetables

Main sources of iron (left) and calcium (right) in the average diet

Water

Consumption of water is vital to life and to remain healthy. Water is constantly lost from the body because of a range of bodily functions as well as through urine and sweat. It is recommended we drink eight glasses of water a day.

Water:

● regulates body temperature – sweat evaporates on the skin causing cooling
● helps to remove waste products from the body. If these waste products were not removed they could release poisons causing illness and organ damage
● helps the body to absorb nutrients, vitamins and minerals and assists in the digestion process
● acts as a lubricant, assisting in keeping eyes and joints working healthily
● prevents the body from becoming dehydrated.

Water is vital for good health

Sources of water include:

● drinks of all kinds
● foods such as fruit, vegetables, meat and eggs
● fibre.

Table 1.11 shows how a lack of nutrients can affect the body and Table 1.12 shows the effect of consuming too much of some nutrients.

Table 1.11 Effects of a diet lacking in different nutrients

Nutrient	Effect of having too little or none in your diet
Carbohydrate	Lack of energy; weight loss; low immune system
Fat	Lack of energy; weight loss; low immune system
Protein	Water retention; muscle wastage; hair loss
Fibre	Bowel disorders; bowel cancer; constipation
Vitamin A	Sight problems; hydration problems
Vitamin B1	Nervous disorders
Vitamin B2	Growth disorders; skin disorders
Vitamin B6	Anaemia; blood disorders
Vitamin B12	Anaemia; blood disorders; possible mental problems
Vitamin C	**Scurvy**; tiredness; blood loss if injured (as blood does not clot properly); bruising
Vitamin D	**Rickets**
Vitamin E	Mal-absorption and low immune system
Vitamin K	Slow healing of wounds
Niacin	Possible mental problems; depression; diarrhoea
Iron	Tiredness; lack of energy and strength
Calcium	Rickets
Potassium	High blood pressure
Magnesium	Slower recovery from injury or illness
Folic acid	Blood disorders

Table 1.12 Effects of having too much of some components in your diet

Component	Effect of having too much in your diet
Fat	Obesity; heart disease and heart attacks; high blood pressure
Carbohydrate	Obesity; tooth decay; diabetes (caused by too much sugar)
Salt	High blood pressure; stroke

KEY TERMS

Scurvy: a disease that can cause bleeding gums and other symptoms.

Rickets: a disease of the bones

Adjusting and improving dishes to promote a healthy diet

Providing balanced meals for customers has a wider benefit than simply supporting a healthy lifestyle. Providing the right foods as part of a balanced diet can help with natural reduction of cholesterol and obesity, reducing the risk of heart disease. Offering healthy eating choices on menus will help customers to select foods that are better for their long-term health as well as assisting them to make informed choices.

When developing dishes for menus the chef needs to be aware of the nutritional value of the ingredients used, since reducing some nutrients for health reasons could have other health implications. For example, reducing milk and dairy products in a dish to make it lower in fat may also reduce calcium, which helps to develop strong bones and teeth and prevents rickets. It would therefore not be appropriate to make this change in a menu for children. Meat and offal are good sources of iron so those following a vegetarian diet need to find alternative sources of iron to reduce the risk of anaemia.

There are a number of ways that chefs can contribute to providing healthy eating choices for their customers, and some suggestions are given below.

- Include wholemeal pasta and rice dishes on menus but reduce the fat and salt in pasta sauces and reduce the butter and cheese content in risotto.
- Reduce salt, fat and sugar content in all dishes. Experiment with how much can be reduced without losing the quality and flavour of the dish.
- Use healthier cooking methods such as grilling, steaming or poaching. For pan frying, try spray oil.
- Avoid adding butter and salt to cooked vegetables and grilled food.
- Use herbs, spices and lemon juice to add flavour rather than salt.
- Provide simple nutritional information on menus to allow customers to make informed choices.
- Increase the vegetable or fruit content of menu items. For example, a pasta dish with chopped vegetables in the sauce.
- Use wholemeal bread and rolls.
- Introduce more healthy fruit-based desserts

3.2 Understand how to design dishes for special diets and balanced menus

Individuals may adopt a particular diet or style of eating for a number of reasons, and this may change according to their needs. Growing children, changes in lifestyle, the state of their health, their occupation and ageing are all contributing factors. Food preparation staff need to be aware of these factors when developing menus and providing food for customers. Some diet choices are **food preferences** – people may choose to eat certain foods because they like or prefer them to other foods, or because they form part of their lifestyle choice. Other food choices are **special dietary requirements** – people eat or do not eat certain foods for medical, religious or cultural reasons, or because they have an allergy or intolerance to them.

Healthy eating and balanced diets

There is more information available than ever about healthy eating and balanced diets. This is promoted through the media, in schools, colleges, health centres, hospitals, day centres and anywhere where there are groups of people likely to benefit from such information. Increasingly the eating advice for healthy people is to eat a balanced diet and many people will make food choices based on these guidelines. There is also greater awareness about the health risks of eating unbalanced diets, and diets that are high in fat, sugar and salt.

ACTIVITY

A favourite item on a school menu is deep-fried bread crumbed chicken drumsticks with chips. Suggest a healthier alternative that would still appeal to children. What would you call the new menu item? How would you prepare and cook this?

Health and lifestyle preferences

Individuals may choose to base their style of eating around benefits to their health or their work and lifestyle preferences. Peoples' lifestyles vary considerably and the food they and their families eat will depend on many factors:

- Working hours and working patterns: people working long hours, in shifts or at night makes eating or the provision of meals at traditional times difficult or impossible.
- Busy lifestyles of different family members completing activities at different times, so requiring food at times convenient for them and often on the move.
- A reduction of available cooking time and actual cooking skills, so there is more reliance on convenience foods and ready cooked food.
- Increased leisure time means more disposable income for some people with a much greater emphasis in eating out and socialising over food.
- More meals are eaten in workplaces or near to them and in educational establishments. A primary school child may attend breakfast club, have a cooked lunch and then tea at an after school club. Popular food outlets and coffee shops provide a variety of breakfast items, which have become increasingly popular.
- Eating styles are frequently governed by trends originating from advertising, social media, foreign travel and many other factors.
- A desire for more healthy lifestyles or food to support high activity or sports increasingly drive food choices.

With increasing amounts of food types, styles and availability for eating out or at home, selection of food for a particular lifestyle has become easier.

Vegetarian and other non-meat diets

People may choose to adopt a vegetarian diet for ethical, cultural or religious reasons, or as a lifestyle choice. Generally, vegetarians avoid meat or fish of any kind or dishes made with or containing animal products. In addition to the general vegetarian definition there are:

- ovo-vegetarians – do eat eggs but not milk or any milk products
- lacto-vegetarians – consume milk and milk products but not eggs
- vegans – do not eat any food of animal origin, including honey, dairy products, eggs and egg products
- fruitarians or fructarians – eat only fruit, nuts and berries
- 'demi-vegetarians' – occasionally may eat fish and/or meat.

Allergies

As discussed in Topic 1.2 of this chapter, a food allergy occurs when an individual's immune system responds to certain foods or ingredients as harmful and rejects them, causing an allergic reaction to take place. The allergic reaction that some people have to certain food items can be serious and even fatal. Reactions include:

- tingling or itching or burning sensation in the in the mouth or throat
- an itchy red rash on the skin with possible raised areas
- swelling of the face, mouth or other areas
- difficulty in swallowing
- wheezing or shortness of breath
- feeling dizzy and lightheaded
- nausea or vomiting
- abdominal pain or diarrhoea
- hay fever-like symptoms, such as sneezing or itchy eyes.

The symptoms of a severe allergic reaction can be sudden and may worsen very quickly. Anaphylaxis or anaphylactic shock may be linked to people with a nut allergy (especially peanuts), but nuts are not the only food that could cause a reaction. Initial symptoms of anaphylaxis may be the same as those listed above, but especially swelling in the mouth, nose and throat. This can lead to difficulty in breathing, sudden fear and anxiety, rapid heartbeat and a sudden drop in blood pressure, causing light-headedness, confusion and unconsciousness. Anaphylaxis is a medical emergency – it can be fatal so it is essential to seek immediate medical attention.

The Anaphylaxis Campaign has warned caterers to be on the alert for foods containing flour made from lupin seeds, as this can cause an allergic reaction similar to peanuts. Lupin flour is used as an ingredient in France, Holland, Portugal and Italy because of its nutty taste, attractive yellow colour and because it is guaranteed GM free. Check ingredient lists for use of lupin.

Although there are a very large number of allergens or ingredients causing reactions because of food intolerance, **14 major allergens** have been identified in the Food Information Regulations 2014 (see Topic 1.1 for more information on this legislation). The 14 allergens that have been identified under the EU regulations are:

- crustaceans
- eggs
- fish
- peanuts
- soybeans
- milk
- cereals containing gluten such as, wheat, rye, barley, oats and related grains
- nuts: almond, hazelnut, walnut, cashew, pecan nut, Brazil nut, pistachio nut and macadamia nut

- celery and celeriac
- mustard
- sesame
- lupin
- molluscs.

- sulphur dioxide or sulphites (a preservative that may be used in wines, fruit juices, vinegar, prepared fruit and other items)

Information about these items must be given to the consumer. This is likely to be on a label of packaged foods, or in establishments such as restaurants could be verbal or on menus, but this must be accurate. Dealing with allergies and allergens must now be included in HACCP-based food safety management systems. Authorised officers now have similar powers to take action against breaches in the legislation as they have with other food safety non-compliance.

Food preparation staff need to be especially careful when providing food for customers with allergies and other dietary needs. Even tiny traces of an allergen can trigger a serious allergic reaction in some people, so useful and accurate information must be given to all customers.

- Be receptive to people with food allergies and intolerances. Find out what your workplace's policy is for providing information and follow this policy. Follow the agreed safe procedures for preparing and cooking food for these customers. If you are not confident that a 'safe' meal can be provided for an allergy sufferer then they must be informed that you cannot serve them.
- Make sure that you have had proper training in how to deal with food for allergic customers and those with special dietary needs. Those who have not been trained should not be involved with provision of the food. Service staff also need thorough training.
- Check labels and recipes carefully. Packaged food and ingredients will have the 14 main allergens highlighted so save food packaging. However, be aware that customers may inform you of an allergy or special dietary need that is not included in the 14 identified. Many kitchens now have charts of the allergens present in each of their dishes. Be aware of these and check them carefully.
- Be aware of cross-contamination. This is just as important for allergies and special diets as it is for food safety. Have separate preparation areas and separate equipment. A pack of colour coded equipment (purple) is now available for preparing food for allergic customers. Remain aware of cross contamination that could occur in fridges, with hands and with equipment.
- Take care with cooking processes to avoid contamination. For example, cooking equipment that has not been cleaned properly between use such as frying pans,

griddles and grill bars could contaminate food for the allergic customer. Oil in a deep fat fryer could also contaminate different foods.

- Cleaning surfaces and equipment very thoroughly between use will help to prevent any cross contamination. Thorough and frequent hand washing is also essential; consider too wearing disposable gloves when preparing certain foods.
- Take care with garnishes on dishes as these could cause an allergen problem either in themselves or the way they have been prepared.

TAKE IT FURTHER

The 14 allergens are described further in a chart available from the Food Standards Agency. A printer-friendly version is available on their website and could be a very useful source of information for a food business and their customers: **www.food.gov.uk/science/allergy-intolerance**.

Other useful sources of information:
- NHS: **www.nhs.uk/conditions/allergies**
- Alletess Medical Laboratory: **www.foodallergy.com**

You may wish to sign up to Food Standards Agency allergy alerts: **www.food.gov.uk/email**

Twitter: follow @foodgov, see **http://food.gov.uk/twitter** or search for #AllergyAlert.

Facebook: **www.food.gov.uk/facebook**

ACTIVITY

Produce a chart listing ten items that may appear on a popular restaurant salad bar. Across the top list the allergens covered by legislation. Mark the allergens that may be in each salad type.

As well as providing accurate information to customers about possible allergens in the food being supplied, it is essential to ensure that:

- all staff dealing with food are well informed about possible allergens in food and where they may be on the menu
- everyone is aware of the possibility of hidden allergens, especially in foods with a number of ingredients such as complex salads, soups and sauces
- any possible cross contamination is avoided, for example, from food equipment or where different foods are presented together as they would be in a salad bar.

Food intolerance

Food intolerance is different from an allergy because it does not involve the immune system. Nevertheless, the symptoms can be significant, causing illness, reactions and discomfort.

Food intolerance is much more common than food allergy and the reaction may be slower or delayed by many hours after eating the food. Symptoms could last for many hours, days or even longer. The intolerance could be to several foods, a group of foods or a single ingredient.

The symptoms of food intolerances can vary and include fatigue, joint pains, darkness under the eyes, sweats, diarrhoea, vomiting, bloating, irritable bowel, skin symptoms such as rashes, eczema, and other chronic conditions.

Coeliac disease (intolerance to gluten)

Gluten is a mixture of proteins found in some cereals such as wheat, rye barley and can include oats. If someone has a gluten intolerance, the presence of gluten causes the immune system to produce antibodies that attack the linings of the intestines. This results in severe inflammation of the gastro-intestinal tract, pain and diarrhoea, and malnutrition due to an inability to absorb nutrients.

People with gluten intolerance should avoid all products made from wheat, barley or rye, including bread; always check the label on all commercial products.

Potatoes, rice and flours made from potatoes and rice, cornflour, fresh fruit and vegetables, meat, fish eggs and most dairy products should be suitable for those with an intolerance to gluten.

Lactose intolerance

Lactose is a natural sugar found in milk and other dairy products. Lactose intolerance is a digestive problem where the body cannot digest lactose. A person with lactose intolerance should avoid milk, dairy products and foods with these as an ingredient. As an alternative they should consume non-dairy foods and those where dairy products have not been included.

Medical diets

Some people need to follow a particular diet for medical reasons. The table below includes some examples of medical diets, the foods people following these diets need to avoid and those that are permitted.

Table 1.13 Medical diets

Type of diet	Description	Foods to avoid	Permitted foods
Low cholesterol	High levels of cholesterol circulating in the bloodstream are associated with an increased risk of cardiovascular (heart) disease. People following a low cholesterol diet need to reduce the amount of cholesterol in their bloodstream.	Liver, kidney, egg yolks, fatty meats, bacon, ham, pâté, fried foods, pastry, cream, full-fat milk, butter, margarine, full-fat yoghurt, chocolate, cheeses, salad dressings, biscuits, cakes	Lean meat, poultry and fish grilled or poached, fresh fruit and vegetables, low-fat milk, low-fat yoghurt, porridge, muesli Low fat spreads can be a useful replacement for butter.
Diabetic	Diabetes is the lack of or inability to use insulin in the body, which means that the amount of sugar going into the blood is not controlled. Diabetes may be controlled by diet alone (controlling sugar intake), by tablets or insulin injections or by a combination of these. Diabetics must follow their individual medical advice.	Generally cannot have foods or drinks containing sugar. Any sugars must be carefully monitored and balanced, especially if injecting insulin. Sugar-free sweeteners can be useful. Failure to control diabetes can lead to comas and long-term problems such as increased risk of cardiovascular (heart) disease, blindness and kidney problems.	Potatoes, rice and flours made from potatoes and rice; also cornflour, fresh fruit and vegetables wholemeal bread and grains, pasta, pulses. Meat, fish eggs, most unsweetened dairy products.
Low fat	Individuals may require a low fat diet for a number of reasons – it may be a requirement of certain medical treatments or for weight loss reasons.	Any food that contains significant amounts of fat, or has been fried or roasted. High-fat sauces and desserts including chocolate and cream. Pastries, cakes and biscuits.	Lean meat, poultry and fish grilled or poached, fresh fruit and vegetables, low-fat milk, low-fat yoghurt, porridge, muesli. Use low fat spreads instead of butter and low fat salad dressings.

Type of diet	Description	Foods to avoid	Permitted foods
Low residue	This diet is followed by people with digestive tract and inflamed bowel problems including Crohn's disease.	Wholemeal bread, skins and seeds of fruit and vegetables, brown rice and pasta, nuts, dried fruits, fried and fatty food.	Fruit and vegetables without seeds skins and pips. Smooth fruit juices, white rice bread and pasta. Lean meat, poultry and fish.
Milk/dairy-free	Those with lactose intolerance or a milk allergy (often an allergy to the proteins in milk). Other medical conditions may require a dairy-free diet.	Milk, butter, cheese, yoghurt and any pre-prepared foods that include milk products (check label especially for addition of dried milk). Margarine/low fat spread unless it specifically states it is dairy free).	Meat, poultry and fish but check that sauces are dairy free. Fruits and vegetables with no added cream or butter. Check all drinks are dairy free.
Low salt	Too much salt can raise blood pressure, which means increased risk of health problems such as heart disease and stroke. Eating less salt is recommended for everyone and a low salt diet is prescribed for some medical conditions.	Foods and dishes that have had salt added in cooking or processing (including smoked and cured fishes and meats, and hard cheeses) or that contain monosodium glutamate.	Meat, poultry and fish but check that salt has not been added in cooking and that sauces are salt free. Fruits and vegetables with no added salt. Salt free dairy products. Read the labels of packaged food for salt content.
Pregnant women	The diet must provide enough energy and nutrients for the unborn baby to grow and develop, as well as keeping the mother fit and healthy.	Plenty of fruit and vegetables – at least five portions of a variety of fruit and vegetables a day. Starchy foods, such as bread, pasta, rice and potatoes. Protein, such as lean meat and chicken, fish, preferably two servings of fish a week, including one of oily fish), eggs and pulses (such as beans, peas and lentils – these are also good sources of iron). Include fibre, found in wholegrain bread, wholegrain cereals, pasta, brown rice, pulses, and fruit and vegetables. Some dairy foods such as milk, cheese and yoghurt, which contain calcium are also good sources of protein.	Untreated milk and its products; soft cheeses, such as Camembert, Brie or others that have a similar rind; soft blue cheese; paté of any type; undercooked meat; raw eggs and food containing raw or partially cooked eggs.

Religious influences on diet

Many religions have rules relating to foods that can and cannot be eaten.

- Islamic or Muslim – do not eat pork, meat that is not halal (slaughtered according to custom), shellfish and alcohol (even when used in cooking).
- Hindu – do not eat meat, fish or eggs (orthodox Hindus are usually strict vegetarians); less strict Hindus may eat lamb, poultry and fish but definitely not beef as cattle have a deep religious meaning. Milk, however, is highly regarded.
- Sikh – beef, pork, lamb, poultry and fish may be acceptable to Sikh men; Sikh women may avoid all meat.
- Jewish – do not eat pork, pork products, shellfish and eels, meat and milk served at the same time or cooked together. Strict Jews eat only kosher meat – slaughtered and butchered according to custom; milk and milk products are usually avoided at lunch and dinner (but acceptable at breakfast).
- Rastafarian – avoid all processed foods, pork, fish without fins (eels), alcohol, coffee, tea.

- Mormonism – prohibits alcohol and caffeine (in coffee, tea, chocolate, etc.).
- Jainism – observes rules for the protection of all life forms. Strict Jains don't eat meat, poultry, fish or eggs and sometimes milk; they may avoid eating root vegetables as the whole plant is killed when the root is dug up.

TAKE IT FURTHER

Find a restaurant menu or download one of your choice. Select three courses from the menu for the following customers:
- A diabetic
- A person on a low fat diet
- Someone who is allergic to eggs
- A vegan

TEST YOURSELF

1   What are the **six** main nutrient groups of food?

2   Why is dietary fibre needed in the body when it cannot be digested?

3   Why is vitamin C (ascorbic acid) needed in the body? List **three** foods rich in Vitamin C.

4   Suggest **five** ways that chefs can make the food they cook healthier.

5   What is the difference between an allergy and food intolerance?

6   What is coeliac disease? List the foods and ingredients that must be eliminated if cooking for someone with coeliac disease.

7   Which religious/cultural diets forbid the eating of pork?

8   What are the ways that a restaurant can inform customers of allergens that may be present in certain menu items?

9   What is anaphylactic shock and what are the symptoms?

302 Financial control in the professional kitchen

Learning outcomes

In this unit you will be able to:

1 Calculate dish costs and selling prices, including:
   - being able to calculate ingredients' costs based on units of purchase and the amount required for recipes
   - being able to calculate dish costs based on total ingredients' costs and recipe yield
   - being able to calculate the selling price of dishes based on gross profit margin targets as percentages
   - being able to apply and remove VAT at various rates
   - understand the factors that contribute to the production of net profit
   - being able to express net profit as a percentage of sales.
2 Monitor the financial performance of a professional kitchen, including:
   - being able to calculate the cost of wastage and loss of ingredients as a result of preparation, cooking and presenting
   - being able to re-calculate the cost of ingredients after preparation, cooking and presenting
   - understanding how to manage yield

   - understanding the purpose of sales analysis and how to analyse sales
   - understanding how to plan menus
   - understanding how the management of kitchen resources can affect profitability.
3 Understand financial management within professional kitchens, including:
   - understanding the importance of purchasing decisions in controlling costs
   - knowing the stages of the purchasing cycle
   - knowing the documents used within the purchasing cycle
   - understanding financial data on different departments of a food operating business
   - knowing how to prepare an income statement (profit and loss account) from financial data available, calculating overall business profit/loss
   - knowing that the purpose of a balance sheet is to demonstrate the financial health of a business at any given time
   - understanding how to interpret a balance sheet.

Calculate dish costs and selling price

1.1 Dish costing

Food costs and dish costing are important because it is essential to know the exact cost of each process and every item produced. Understanding these costs is important because:

- it tells you the net profit made by each section of the organisation and shows the cost of each meal produced
- it will reveal possible ways to economise and can result in a more effective use of stores, labour and materials
- costing provides the information necessary to develop a good pricing policy
- cost records help to provide speedy quotations for special functions, such as parties and wedding receptions
- it enables the caterer to keep to a budget. If food costing is controlled accurately, the cost of particular items on the menu and the total expenditure on food over a given period can be worked out.

Understanding food costs helps to control costs, prices and profits. An efficient food cost system will show up any bad buying and inefficient storage and should help to prevent waste and pilfering. This can help the chef to run an efficient business and give the customer value for money.

Factors in controlling food costs and profit
Sourcing and purchasing of food commodities

The menu dictates what a food operation needs to purchase, and based on this the buyer searches for a market that can supply these requirements. Once the right market is found, the buyer must investigate the various products available. The buyer makes decisions regarding quality, amounts, price and what will satisfy the customers but also what will make a profit. A buyer must have knowledge of the internal organisation of the company and be able to obtain the products needed at a competitive price. They must understand how these items are going to be used in the production operations (i.e. how they are going to be prepared and cooked), to make sure that the right item is purchased. Sometimes the item required may not have to be of prime quality – tomatoes for use in soups and sauces, for instance.

It is important to accurately cost dishes, so you should know the cost of all of the ingredients you are using in your dishes.

Example of dish costing

Name of dish: Navarin of lamb

Number of portions: 20

Ingredients	Amount	Cost per unit	Total cost £ p
Stewing lamb (shoulder)	2.5 kg	£6.50 per kg	£16.25
Carrots	500 g	60p per kg	£0.30
Onions	500 g	65p per kg	£0.32
Oil	75 ml	£1/litre	£0.08
Garlic clove	5 cloves	30p per bulb	£0.30
Flour (white)	120 g	£1.50 per kg	£0.18
Tomato paste	70 g	50p per 100 g	£0.35
Brown stock	2.5 litres	50p per litre	£1.25
Bouquet garni			£0.30
		Total	£19.33

Cost of 1 portion = £19.33 ÷ 20 = £0.97 per portion.

Elements of cost

Costing dishes is a very important process for a chef and allows the selling price to be established. To calculate the total cost of any one item or meal provided, it is necessary to analyse the total expenditure under several headings. The total cost of each item consists of the three main elements described below.

Food and materials costs

These are known as **variable costs** because the level will vary according to the volume of business. Variable costs include food costs, which may vary on a daily or weekly basis. It is important not to have excess stock in storage as it ties up money and over-stocking can result in food wastage. **Fixed costs** include labour and overheads – these are charges that do not change according to the volume of business.

Labour costs

In an operation that uses part-time or extra staff for special occasions, the money paid to these staff also comes under variable costs; by comparison, salaries and wages paid regularly to permanent staff are fixed costs.

Labour costs in the majority of operations fall into two categories:

1 Direct labour costs – the salaries and wages paid to staff such as chefs, waiters, bar staff, housekeepers and chambermaids, which can be allocated to income from food, drink and accommodation sales.

2 Indirect labour costs – includes salaries and wages paid, for example, to managers, office staff and maintenance staff who work for all departments in the establishment (so their labour cost should be charged to all departments).

Overheads

Overheads consist of rent, rates, cleaning materials, heating, lighting and equipment, maintenance, gas, electricity and sundry expenses, insurance and marketing costs.

KEY TERMS

Variable costs: costs that vary according to the volume of business; includes food and materials costs.

Fixed costs: regular charges, such as labour and overheads that do not vary according to the volume of business.

Overheads: expenses associated with operating the business, such as rent, rates, heating, lighting, electricity, gas, maintenance and equipment.

Gross profit

Gross profit (or kitchen profit) is the difference between the cost of an item and the price it is sold at.

Gross profit = selling price – food cost

For example, if food costs are £2 and the selling price is £4:

Gross profit = £4 – £2 = £2

Net profit

Net profit is the difference between the selling price of the food (sales) and total cost of the product (this includes food, labour and overheads).

Net profit = selling price – total cost

For example, if the selling price is £5 and the total cost is £4:

Net profit = £5 – £4 = £1

WORKED EXAMPLE

Food sales for 1 week = £25,000

Food costs for 1 week = £12,000

Labour and overheads for 1 week = £9,000

Total costs for 1 week = food costs + labour and overheads = £12,000 + £9,000 = £21,000

To work out the gross profit:

Gross profit = food sales – food cost = £25,000 – £12,000 = **£13,000**

To work out the net profit:

Net profit = food sales – total costs = £25,000 – £21,000 = **£4,000**

ACTIVITY

Food costs for one week at the restaurant you work in are £15,000. In the same week, the restaurant spends £12,000 on labour and overheads. Food sales for the week total £30,000.

1 Calculate the **gross profit** the restaurant makes for the week.

2 Calculate the **net profit** the restaurant makes for the week.

Profit is always expressed as a percentage:

$$\% \text{ net profit} = \frac{net\ profit}{sales} \times 100$$

WORKED EXAMPLE

Using the same example as above, the percentage net profit for the week is:

$$\frac{net\ profit}{sales} \times 100$$

$$\frac{4,000}{25,000} \times 100 = 16\%$$

KEY TERMS

Gross profit: the difference between the cost of an item and the price at which it is sold.

Net profit: the difference between the selling price of an item and the total cost of the product (this includes food, labour and overheads).

It is usual to express each element of cost as a percentage of the selling price. This enables the caterer to control profits.

Table 2.1 shows how each element of cost can be shown as a percentage of the sales.

Table 2.1 Elements of cost shown as percentage of sales

	Costs	Percentage of sales
Food cost	£12,000	$\frac{12,000}{25,000} \times 100 = 48\%$
Labour	£6,000	$\frac{6,000}{25,000} \times 100 = 24\%$
Overheads	£3,000	$\frac{3,000}{25,000} \times 100 = 12\%$
Total costs	£21,000	$\frac{21,000}{25,000} \times 100 = 84\%$
Sales	£25,000	
Net profit (sales – total costs)	£4,000	$\frac{4,000}{25,000} \times 100 = 16\%$

1.2 Selling prices

Calculating the selling price

If the selling price of a dish is expressed as 100 percent (the total amount received from its sale), it can be broken down into the amount of money spent on food items and the gross profit. This can be expressed in percentages. Often, caterers need to ensure that an agreed gross profit is achieved. The selling price needed to achieve a specific gross profit percentage can be calculated by dividing the total food cost by the food cost as a percentage of the sale and multiplying by 100.

Food costs as a percentage of the sale = 100 – gross profit %

WORKED EXAMPLE

If food costs are £3.50, to calculate the selling price and make sure a 65% gross profit is achieved, the following calculations can be used.

First, calculate food costs as a percentage of the sale:

Food costs as a percentage of the sale =

100 – gross profit % = 100 – 65 = 35%

This is shown in the pie chart below.

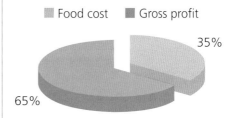

The selling price of the dish is 100%: therefore, if gross profit is 65%, the food cost as a percentage of the sale is 35%.

To calculate the selling price:

This can also be presented in monetary terms, as shown in the following diagram:

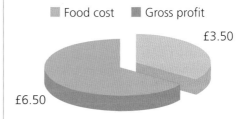

Here, the total food cost is £3.50 and the selling price is £10; the gross profit is therefore £6.50.

ACTIVITY

Food costs for a dish are £10.50. What should be the selling price of the dish to ensure you achieve 65% gross profit?

Raising the required gross profit reduces the food cost as a percentage of the selling price. For example, if the gross profit requirement was raised to 75 per cent, this would reduce the food cost as a percentage of the selling price to 25 pe rcent (100% – 75%), as shown in the diagram below.

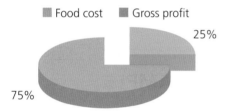

WORKED EXAMPLE

If the food cost is £3.50 and the required gross profit is 75%:

To achieve a 75% gross profit, with a £3.50 food cost, the selling price would need to be £14.00.

The percentages still add up to 100%, but the proportion spent on food is now smaller in terms of the selling price (food costs are now 25% of the selling price) because the percentage gross profit is higher. The diagram below illustrates this.

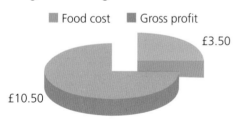

To check that this is correct, the following calculation can be applied:

(brings up to 25%)

… and

(brings up to 75%)

£10.50 (75%) + £3.50 (25%) = £14.00 (100%)

ACTIVITY

If a dish costs £3.00 to produce, what should the selling price be to achieve a 70 per cent profit on sales?

1.3 Net prices

Average spend per customer

The average amount spent by each customer is determined by dividing the total sales by the number of customers.

WORKED EXAMPLE

If total sales are £25,000 and the restaurant serves 1,000 meals, then the average amount spent by each customer is £25

If the percentage composition of sales for a month is known, the average price of a meal for that period can be further analysed.

WORKED EXAMPLE

Average price of a meal = £25.00 = 100%

£0.25 = 1%

Remember, the costs as a percentage of sales were as follows:

	Costs	Percentage of sales
Sales	£25,000	
Food cost	£12,000	48%
Labour	£6,000	24%
Overheads	£3,000	12%
Net profit (sales – total costs)	£4,000	16%

This means that the customer's contribution towards:

Food cost = £0.25 × 48 (food cost as a percentage of sales) = £12.00

Labour = £0.25 × 24 (labour as a percentage of sales) = £6.00

Overheads = £0.25 × 12 (overheads as a percentage of sales) = £3.00

Net profit (net profit as a percentage of sales) = £0.25 × 16 = £4.00

Cost price of a dish

A rule that can be applied to calculate the food cost price of a dish is as follows:

Let the cost price of the dish equal 40 per cent and fix the selling price at 100 per cent.

WORKED EXAMPLE

If the dish was sold at £10:

Selling price (100%) = £10

$$\text{Cost of dish } (40\%) = \frac{10}{100} \times 40 = £4$$

Selling the dish at £10 and making 60 per cent gross profit above the cost price, would be known as 40 per cent food cost.

WORKED EXAMPLE

If entrecote steak costs £10.00 per kg, what would be the selling price at 40 per cent food cost of a 250 g sirloin steak?

First, you need to work out how much 250 g would cost.

There are 1,000 g in 1 kg, so 250 g = 0.25 kg (250 ÷ 1,000).

$$\frac{1\text{kg}}{4} = 0.25 \text{ kg}$$

So, if 1 kg costs £10, the cost of 0.25 kg (or 250 g)

$$\frac{£10}{4} = £2.50$$

250 g entrecote steak at £10.00 per kg = £2.50

If the selling price is fixed at 40% food cost:

$$\text{Selling price} = \frac{2.50 \text{ (food cost)} \times 100 \text{ (fixed selling price\%)}}{40 \text{ (fixed cost\%)}} = £6.25$$

The selling price of 250 g of sirloin steak at 40 per cent food cost is £6.25.

Today, many chefs use computer software to work out the cost of dishes, the gross profit and the selling price. One such system commonly used is Kitchen Cut, which is able to assist the chef in many aspects of financial control.

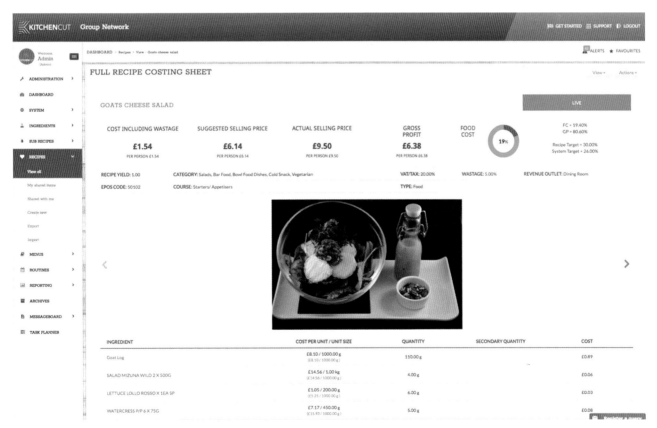

Costing sheet using Kitchen Cut

Monitor the financial performance of a professional kitchen

2.1 Calculating wastage

In order to control food wastage it is crucial to pay particular attention to the following:

- Purchasing – use detailed purchasing specifications, with no over ordering.
- Operate correct storage procedures and stock control.
- Design the menu taking into account customer requirements and sales mix.
- Adhere to correct preparation and cooking methods. Use standardised recipes which give the exact yields required.
- Use accurate, consistent portion control.
- Where possible use accurate forecasting systems and procedures. Use accurate and reliable data, for example, sales histories.

Chefs must monitor waste, check storage systems, stock rotation and sell by dates. They need to know how much is thrown away at the end of each day, each week and month, and set targets to reduce waste. Check preparation methods and that the exact yields are being obtained. Incorrect preparation, cooking and presentation methods can result in a waste of ingredients. Standard purchasing specifications and standardised recipes help to reduce waste, if written and followed correctly.

Check the plate waste (how much customers are leaving on their plates). If this is excessive then reduce the portion sizes. Every aspect of the control cycle must address waste and the cost of waste must be calculated.

2.2 Yield management and portion control

Standard purchasing specification

Standard purchasing specifications are documents that are drawn up for every commodity, describing exactly what is required and expected. They assist with the formulation of standardised recipes. A determined specification is drawn up which, once approved, will be referred to every time the item is delivered. It is a statement of various criteria related to quality, grade, weight, size and method of preparation, if required (such as washed and selected potatoes for baking). Other information given may be variety, maturity, age,

colour, shape and so on. A copy of the standard specification is often given to the supplier and the storekeeper, who are then very clear as to what is needed. These specifications assist in costing and control procedures.

Commodities that can be specified:

- Grown (primary) – butcher's meat; fresh fish; fresh fruit and vegetables; milk and eggs.
- Manufactured (secondary) – bakery goods; dairy products.
- Processed (tertiary) – frozen foods including meat, fish, fruit and vegetables; dried goods; canned goods.

Any food product can have a specification attached to it. However, the primary specifications focus on raw materials, ensuring the quality of these commodities. Without quality at this level, a secondary or tertiary specification is useless. For example, to specify a frozen apple pie, this product would use:

- a primary specification for the apple
- a secondary specification for the pastry
- a tertiary specification for the process (freezing).

However, regardless of how good the secondary or tertiary specifications are, if the apples used in the beginning are not of a very high quality, the whole product will not be of a good quality.

Example of a standard purchasing specification

- Commodity: round tomatoes
- Size: 50 grams 47–57 millimetres diameter
- Quality: firm, well formed, good red colour, with stalk attached
- Origin: Dutch, available March–November
- Class/grade: Super class A
- Weight: 6 kilograms net per box
- Count: 90–100 per box
- Quote: per box/tray
- Packaging: loose in wooden tray, covered in plastic
- Delivery: day following order
- Storage: temperature 10–13 °C at a relative humidity of 75–80 per cent
- Note: avoid storage with cucumbers and aubergines

For most perishable items, rather than entering into a long-term contract, a daily, weekly or monthly quotation system is more common. This is essentially a short-term contract reviewed regularly to ensure that a competitive situation is maintained.

Portion control

Portion control means controlling the size or quantity of food to be served to each customer. The amount of food allowed depends on the considerations described below.

Type of customer or establishment

Customer requirements differ across a range of situations. For example, the number of calories consumed by people on a daily basis will differ according to factors such as age, gender, type of work, their metabolism and personal preferences. In a restaurant offering a three-course table d'hôte menu for £25, including salmon as a main course, the size of the portion is more likely to be smaller than in a restaurant charging £17.50 for the salmon on an à la carte menu. The supplier of the salmon may also be supplying other outlets with the same salmon but in quite different circumstances – for example, in a staff restaurant or a care or residential home where different costing and pricing models are used.

Quality of food

Better-quality food usually yields a greater number of portions – low-quality stewing beef often needs so much trimming that it is difficult to get six portions to the kilo, and the time and labour involved also costs money. On the other hand, good-quality stewing beef will often give eight portions to the kilo, with much less time and labour required for preparation, as well as increased customer satisfaction

Buying price of the food

The buying price should correspond to the quality of the food if the person responsible for buying has bought wisely. A good buyer will ensure that the price paid for any item of food is equivalent to the quality – in other words, a higher price should mean better quality and a good yield, and an established portion control. If, on the other hand, an inefficient buyer has paid a high price for food of indifferent quality then it will be difficult to get the desired number of portions. The selling price necessary to make the required profit is also likely to be too high and customer satisfaction can be negatively affected. In some cases, it is possible to purchase foods that are already portioned to a tight specification. For example, cuts of meat, poultry and fish can be specified and prepared precisely so that the product is ready to go without further preparation. However, this is usually at a premium price, which has to be considered carefully when costing and pricing.

Portion control should be closely linked to the buying of the food; without a good knowledge of the food purchased it is difficult to state fairly how many portions should be obtained from it. To evolve a sound system of portion control, each type of establishment needs to take into account individual considerations. However, a golden rule for all should be 'a fair portion for a fair price'.

There are certain items of equipment that can assist in maintaining control of the size of portions:

- scoops, for ice cream and sorbets
- ladles, for soups and sauces
- fruit juice glasses (75–150 millilitres)
- soup plates or bowls (14, 16, 17, 18 centimetres)
- individual pie dishes, pudding basins, moulds and coupes
- standardised equipment, such as gastronorm trays
- pasta dishes made in specific trays, e.g. yielding 20 portions.

The following example demonstrates how portion control can save money.

If as little as 0.007 litres of milk was being lost per cup by spillage, a business selling 32,000 cups a year would lose 224 litres of milk. This adds up to a loss of hundreds of pounds per year.

When an extra penny's worth of meat is served on each plate it mounts up to a loss of £3,650 over the year when 1,000 meals are served daily.

Standardised recipes

The standard recipe is a written formula for producing a food item of a specified quality and quantity for use in a particular establishment. It should show the precise quantities and qualities of the ingredients, together with the sequence of preparation and service. It also enables the establishment to have greater control over cost and quantity, portion control and yield.

The objective of the standard recipe is to predetermine the following:

- the quantities and qualities of ingredients to be used, stating the purchase specification
- the yield obtainable from a recipe
- the food cost per portion
- the nutritional value of a particular dish or product
- equipment requirements.

This facilitates menu planning, purchasing and internal requisitioning, food preparation and production and portion control. It will also assist new staff in the preparation and production of standard products.

2.3 Sales analysis and menu planning

Considerations when planning menus

When developing a menu it is important to establish the essential and social needs of the customer, to accurately predict what the customer is likely to buy and how much he or she is going to spend. Customer satisfaction is all important: remember who pays the bill. Meeting the needs of the customer has to be successfully balanced with portioning and costing the product in order to keep within company profitability policy.

The following list summarises the essential considerations when planning and developing menus.

- Competition – be aware of any competition in the locality, including prices and quality. It may be wiser to produce a menu that is quite different to those of competitors.
- Location – study the area in which your establishment is situated and the potential target market of customers.
- Type and size of establishment – pub, school, hospital, restaurant, etc.
- Customer profile – different kinds of people have differing likes and dislikes. Analyse the type of people you are planning to cater for (e.g. office workers in the city requiring quick service).
- Decide the range of dishes to be offered and the pricing structure – price each dish separately? Or offer set two- or three-course menus? Or a combination of both?
- Number and sequence of courses.
- Estimated customer spend per head – whatever the level of catering, a golden rule should be 'offer value for money'.
- Time of day – breakfast, brunch, lunch, tea, high tea, dinner, supper, snack, special function, etc.
- Time of year – be aware of seasonality and where possible use ingredients in season. Certain dishes that are acceptable in summer may not be so in winter; foods in season are usually in good supply and reasonable in price.
- Cost factor – crucial if an establishment is to be profitable. Costing is essential for the success of compiling any menu. Modern computer techniques can analyse costs swiftly and on a daily basis.
- Space and equipment in the kitchens will influence the composition of the menu (e.g. avoiding overloading of the deep-frying pan, salamanders and steamers).
- Number and capability of staff – overstretched staff can easily reduce the standard of production envisaged. Also consider the skill level of both kitchen and service staff.
- Availability of supplies and reliability of supplier, seasonal foods and storage space.
- Modern trends in food fashions – these should be considered, alongside popular traditional dishes.
- Special requirements – consider special diets such as kosher, Muslim and vegetarian.
- Special days – Christmas, Hogmanay, Shrove Tuesday, Eid, Chinese New Year, etc.
- Food allergies – provide useful information on the menu and make sure staff are informed about the ingredients of each dish.
- Outdoor catering – are there opportunities for outdoor catering or takeaway food?
- Use menu language that customers understand.
- Strive for a sensible nutritional balance.
- There should be no unnecessary repetition of ingredients, flavours or colours from dish to dish.
- Be aware of the Trade Descriptions Act 1968 and the Consumer Rights Act 2015.

Menu engineering (sales mix)

One approach to sales analysis that has gained in popularity is 'menu engineering', also known as 'sales mix'. This is a technique of menu analysis that uses two key factors of performance in the sales of individual menu items: the popularity and the gross profit contribution of each item. The analysis results in each menu item being assigned to one of four categories, as follows.

1 Items of high popularity and high cash gross profit contributions. These are known as Stars.
2 Items of high popularity but with low cash gross profit contribution. These are known as Plough horses.
3 Items of low popularity but with high cash gross profit contributions. These are known as Puzzles.
4 Items of low popularity and low cash gross profit contribution. These are the worst items on the menu and are known as Dogs.

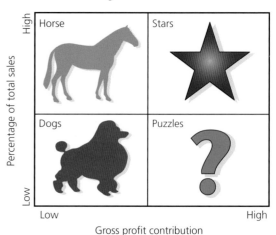

Menu engineering matrix (based on Kasavana and Smith, 1982)

Chefs and food and beverage managers operating in a competitive environment require knowledge of menu engineering in order to maximise business potential. The advantage of this approach is that it provides a simple way of graphically indicating the relative cash contribution position of individual items on a matrix, as shown in the diagram.

There are a variety of computer-based packages that will automatically generate this categorisation, usually using data from electronic point of sale (EPOS) control systems.

In order to determine the position of an item on the matrix, two things need to be calculated. These are:

- the cash gross profit
- the sales percentage category.

The cash gross profit category for any menu item is calculated by reference to the weighted average cash gross profit. Menu items with a cash gross profit that is the same as or higher than the average are classified as high. Those with lower than the average are classified as low cash gross profit items. The average also provides the axis separating Plough horses and Dogs from Stars and Puzzles.

The sales percentage category for an item is determined in relation to the menu average, taking into account an additional factor. With a menu consisting of ten items, one might expect – all other things being equal – that each item would account for 10 per cent of the menu mix. Any item that reached at least 10 per cent of the total menu items sold would therefore be classified as enjoying high popularity. Similarly, any item that did not achieve the rightful share of 10 per cent would be categorised as having a low popularity. With this approach, half the menu items would tend to be shown as being below average in terms of their popularity. This would potentially result in frequent revision of the composition of the menu. It is for this reason that Kasavana and Smith (1982) have recommended the use of a 70 per cent formula. Under this approach, all items that reach at least 70 per cent of their rightful share of the menu mix are categorised as enjoying high popularity. For example, where a menu consists of, say, 20 items, any item that reached 3.5 per cent or more of the menu mix (70 per cent of 5 per cent) would be regarded as enjoying high popularity. While there is no convincing theoretical support for choosing the 70 per cent figure rather than some other percentage, common sense and experience tend to suggest that there is some merit in this approach.

Interpreting the categories

There is a different basic strategy that can be considered for items that fall into each of the four categories of the matrix, as follows.

- Stars – these are the most popular items, which may be able to yield even higher gross profit contributions by careful price increases or through cost reduction. High visibility is maintained on the menu and standards for these dishes should be strictly controlled.
- Plough horses – again, these are solid sellers, which may also be able to yield greater cash profit contributions through marginal cost reduction. Lower menu visibility than Stars is usually recommended.
- Puzzles – these are exactly that – puzzles. Items such as flambé dishes or a particular speciality can attract customers, even though the sales of these items may be low. Depending on the particular item, different strategies might be considered, ranging from accepting the current position because of the added attraction that they provide, to increasing the price further.
- Dogs – these are the worst items on a menu and the first reaction is to remove them. An alternative, however, is to consider adding them to another item as part of a special deal. For instance, adding them in a meal package to a Star may have the effect of lifting the sales of the Dog item and may provide a relatively low-cost way of adding special promotions to the menu.

The menu engineering methodology is designed to categorise dishes into good and poor performers. For dishes with high popularity and high contribution (Stars):

- do nothing
- modify price slightly – up or down
- promote through personal selling or menu positioning.

For dishes with high popularity and low contribution (Plough horses):

- do nothing
- increase price
- reduce dish cost – modify recipe by using cheaper commodities or reducing the portion size.

For dishes with low popularity and high contributions (Puzzles):

- do nothing
- reduce price
- rename dish
- reposition dish on menu
- promote through personal selling
- remove from menu.

For dishes with low popularity and low contribution (Dogs):

- do nothing (although this would be a waste of stock, labour and effort)
- replace dish
- redesign dish
- remove dish from menu.

Benefits of menu engineering

Although there are some difficulties, the benefits of the menu engineering approach include:

- planning for continuous control of cash gross profit
- giving prominence to and controlling the determinants of menu profitability, i.e. the number of items sold, the

cash gross profit per item and the overall compositon of the menu

- application of an analytical approach, recognising that menu items belong to distinctly dissimilar groups, which have different characteristics and which require different handling in the context of cash gross profit control.

PLU No.	Menu item	Selling price	Selling ex VAT	Cost price	Cost total	Selling total ex. VAT	Qty sold	Cost %	GP%	Category identified	Under av. cost %
						Menu engineering					
	Starter										
123	Soup	£5.00	£4.17	£1.00	£11.00	£45.83	11	24.00%	76.00%		yes
124	Melon	£4.50	£3.75	£1.20	£30.00	£93.75	25	32.00%	68.00%	STAR	yes
125	Chilli prawns	£6.50	£5.42	£1.80	£21.60	£65.00	12	33.23%	66.77%		yes
126	Pâté	£3.80	£3.17	£0.90	£1.80	£6.33	2	28.42%	71.58%		yes
127	Mushrooms	£8.00	£6.67	£3.50	£63.00	£120.00	18	52.50%	47.50%		no
128	Bisque	£5.20	£4.33	£2.20	£121.00	£238.33	55	50.77%	49.23%		no
				Total	£248.40	£569.25	**Average**	**36.82%**	**63.18%**		
	Main course										
223	Salmon	£10.25	£8.54	£2.20	£33.00	£128.13	15	25.76%	74.24%		yes
224	Pork fillet	£15.00	£12.50	£3.20	£89.60	£350.00	28	25.60%	74.40%		yes
225	Braised beef	£13.50	£11.25	£3.80	£45.60	£135.00	12	33.78%	66.22%		no
226	Chicken	£9.25	£7.71	£1.90	£104.50	£423.96	55	24.65%	75.35%	STAR	yes
227	Turkey escalopes	£11.50	£9.58	£2.25	£4.50	£19.17	2	23.48%	76.52%		yes
228	Aubergine	£7.95	£6.63	£1.50	£48.00	£212.00	32	22.64%	77.36%		yes
				Total	£325.20	£1,268.25	**Average**	**25.98%**	**74.02%**		
	Dessert										
330	Trifle	£4.95	£4.13	£1.25	£35.00	£115.50	28	30.30%	69.70%	STAR?	yes
331	Fruit salad	£5.80	£4.83	£1.90	£114.00	£290.00	60	39.31%	60.69%		yes
332	Brandy snaps	£6.00	£5.00	£2.20	£44.00	£100.00	20	44.00%	56.00%		no
333	Choc. mousse	£3.60	£3.00	£1.90	£3.80	£6.00	2	63.33%	36.67%		no
334	Bavarois	£4.50	£3.75	£1.70	£20.40	£45.00	12	45.33%	54.67%		no
335	Gateau	£4.90	£4.08	£1.20	£13.20	£44.92	11	29.39%	70.61%		yes
				Total	£230.40	£601.42	**Average**	**41.94%**	**58.06%**		
			Overall cost % average		**35%**			Overall GP% average	**65%**		
			Grand total of cost totals		**£804.00**			Grand total of selling totals	**£2,438.92**		

(Please note the following	Increase ALL starters by		(inc. VAT)	Ex. VAT:
increment boxes will only change	Increase ALL mains dishes by		(inc. VAT)	Ex. VAT:
the selling price ex. VAT)	Increase ALL desserts by		(inc. VAT)	Ex. VAT:

Menu engineering spreadsheet

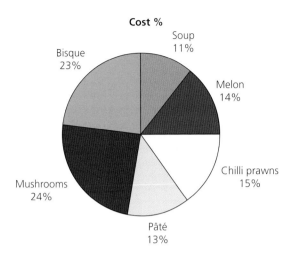

Menu engineering pie chart

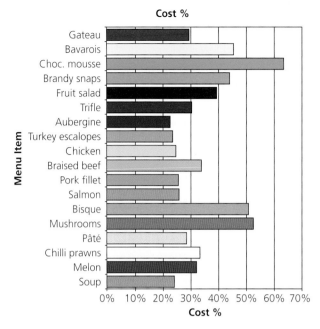

Menu engineering bar chart

2.4 Management of kitchen resources

Chefs working in modern kitchens are responsible for large budgets and expensive resources. These resources do not just include staff but raw materials such as food, equipment, energy, gas and electricity usage. It is important to control waste and unnecessary use of energy.

The effective and efficient management of resources requires knowledge and experience. In addition, it is necessary to keep up to date. This may require attending courses on management, the use of IT packages, hygiene, legislation, etc.

How the control of resources is administered will depend partly on the systems of the organisation but also on the way the person in control operates. Apart from knowledge and experience, respect from those for whom one is responsible is earned, not given, by the way staff are managed in the workplace. Having earned the respect and cooperation of staff, a system of controls and checks needs to be operated that is smooth running and not disruptive. Training and delegation may be required to ensure effective control and, periodically, it is essential to evaluate the system to see that the recording and monitoring are being effective.

The purpose of control is to make certain that:
- supplies of what is required are available
- supplies are of the right quality and quantity
- they are available on time
- there is a minimum of wastage
- there is no over-stocking
- there is no theft
- legal requirements are met.

Control in every catering organisation is crucial – in every establishment. The role of supervisors and managers, from food and beverage managers and their assistants, to executive chefs, sous chefs and chef de parties, is to organise themselves, other people, their time and their physical resources.

Physical resources often include financial control, depending on the establishment. An essential factor of good organisation is effective control of all the above mentioned elements. Successful control and management applies to all aspects of catering:
- purchasing of food
- security
- storage of food
- waste management , procedures are in place to prevent excess waste
- preparation of food, minimum amount of waste in the preparation in order to achieve the correct yield
- energy use, gas and electricity. When purchasing new equipment always assess the energy efficiency of the equipment. Make sure that equipment is not left on unnecessarily. Carry out regular audits to assess whether equipment is being used effectively and correctly
- water. Conserve were possible water. Make sure water is not wasted, taps are fully turned off when not in use
- production of food, no over production. Correct cooking methods used in order to achieve the exact yield
- presentation of food, correct size portions
- equipment, maximum use of equipment and safety of equipment. Making sure all equipment is not misused

- hygiene and food safety, all legislation is followed
- maintenance, all equipment is well maintained so as to prevent breakdown and loss of production
- health and safety, all health and safety procedures and systems are in place
- legal aspects all compliance legislation is covered.

Staffing

The management of staff is an important aspect of the management of kitchen resources. Staff scheduling must be carefully worked so that the correct number of staff are on duty at any one time in order to cope with the volume of the business. Too few staff can put unnecessary pressure on the kitchen and affect the efficient running of the kitchen and result in the guests having to wait too long for their orders thus damaging the overall guest experience. It is important to manage staff effectively through good communication, clear job descriptions, building good working relationships, good guidance and supervision at all times.

Understand financial management within professional kitchens

3.1 Purchasing considerations

Purchasing is a highly significant process that contributes towards the achievement of financial targets and outcomes. The accuracy and reliability of information available to the purchaser can make a vast difference to the overall financial performance of a business.

Analysing historical data can help predict future requirements. For example, analysing the sales history of various dishes can help to predict the likelihood of future sales. Certain menu items will be more popular than others, providing the purchaser with a reliable indication as to their likely requirements.

Other considerations include purchasing ingredients at the most appropriate times. Buying commodities and ingredients when they are at their peak (high in demand and price, for example, strawberries during Wimbledon tennis fortnight) or out of season, not only makes meeting financial targets more difficult, but it is also likely that the products themselves, particularly fresh and highly perishable products, are not of their highest natural quality .

Selecting suppliers

The selection of suppliers is an important part of the purchasing process. First, consider how a supplier will be able to meet the needs of your operation. Consider:

- price
- delivery
- quality/standards.

Information on suppliers can be obtained from other purchasers and visits to suppliers' establishments are to be encouraged. When interviewing prospective suppliers, question how reliable they will be under competition and assess their stability under varying market conditions.

Quantity and quality

Determining the quantity and quality of items to be purchased is important and is based on operational needs. The buyer must be informed of the products needed by their team of chefs or other members of the production team. The chef and his or her team must establish the quality, and they should be required to inspect the goods on arrival. The buyer then checks out the market, seeking the best quality and best price. Delivery arrangements and other factors will be discussed and agreed at this point.

When considering the quantity needed, the following factors should be considered:

- the number of people to be served in a given period
- sales history
- portion sizes (determined by yield-testing standard portion controls).

The buyer needs to know about production to be able to decide how many portions a given size or quantity may yield. They must also understand the various yields. Shrinkage during cooking may vary if not controlled carefully, causing problems in portion control and yield.

The chef must inform the buyer of quantities. The buyer must also be aware of different packaging sizes, such as jars, bottles and cans, and the yield from each package. Another consideration is that larger, cheaper sizes may not represent the best buy for an infrequently used product.

Grades, styles, appearance, composition, varieties and quality factors must be indicated, such as:

- colour
- bruising
- texture
- irregular shape
- size
- maturity
- absence of defects.

Quality standards should be established by the chef and management team when the menu is planned. Menus and recipes may be developed using standardised recipes that relate directly to the buying procedure and standard purchasing specifications. Other considerations include storage upon delivery, ensuring there are sufficient high-quality and hygienic conditions available.

It is also important to consider the handling of allergenic products, such as nuts. To prevent contamination, products of this type are handled and stored separately from other goods. In smaller establishments, the chef or catering manager often speaks directly to the supplier to stipulate quality and size and weight requirements, as well as pricing.

The product itself needs careful consideration. From a purchasing perspective, this is essential to ensure reliability and consistency of supply. Further considerations include the customer's perception and expectations of the business. The type of product purchased is inherently connected with the nature of the business and the market segment in which it is positioned.

Customers expect good value for money and will have an expectation of the product and service delivery. Food-based establishments position themselves in particular market segments (luxury, leisure, travel), offering products and services to meet the needs of their customers. They have to make decisions as to the most suitable products they will need to source to match their offer with their target customers' expectations. With this in mind, purchasing planning will require consideration of:

- quality
- customer expectations
- production output
- quantities required and yield
- time constraints
- labour requirements
- equipment requirements
- costs.

Purchasing decisions need to account for the right product depending on all of the points above. For example, in a large, mid-market restaurant with a small kitchen, would it be better to make 50 litres of fresh stock for soups and sauces every day, or would it be more appropriate to buy a high-quality convenience stock requiring much less labour, storage and energy to produce? Similarly, would it be better for a four-bedroom guest house to buy croissants for breakfast from a reliable bakery or to spend many hours producing fresh croissants? Alternatively, would a customer spending a lot of money in a fine dining establishment expect to be served part-baked rolls or frozen vegetables?

Such decisions have to be considered carefully and operations planned to ensure that the organisation is meeting the needs of its customers, but in a way that is manageable, operationally viable and, most importantly, financially sound. This will help to protect the viability of the business and its employees, owners and investors.

Other purchasing considerations

You will also need to consider:

- price
- storage requirements
- contract/conditions of supply.

3.2 Purchasing cycle

The purchasing cycle is essential to business performance, and a robust purchasing policy is the initial control point of the catering business.

Once a menu is planned, a number of activities must occur to bring it to reality. One of the most important stages is to purchase and receive the materials needed to produce the menu items. Skilful purchasing and detailed checks upon receiving goods will contribute significantly towards the production of high-quality dishes on the menu.

There are six important steps in the purchasing cycle:

1 Understand the market.
2 Design robust purchase procedures.
3 Determine purchasing needs.
4 Receive and check the goods.
5 Establish and use specifications.
6 Evaluate the purchasing task.

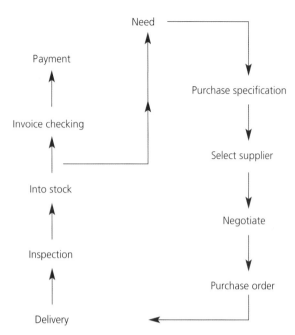

The purchasing cycle (Source: Drummond, 1998, reproduced by permission of Hodder Education)

Understanding the market

A market is a place in which ownership of a commodity changes from one person to another through an agreed transaction. This could occur, for example, in a retail or wholesale outlet, over the phone or via the internet.

Since markets vary considerably, a food and beverage buyer must know the characteristics of each market. The buyer must also have knowledge of the items to be purchased, such as:

- where they are grown
- seasons of production
- approximate costs
- conditions of supply and demand
- laws and regulations governing the market and the products
- marketing agents and their services
- processing
- storage requirements
- commodity and product, class and grade.

Principles of purchasing

A menu dictates an operation's needs. Based on this, the buyer searches for a market that can reliably supply the company. Once the right market has been located, the various products available that meet the buyer's needs are investigated. These products must also provide the quality expected by the establishment. Factors that affect production needs include:

- type and image of the establishment
- style of operation and system of service
- occasion for which the item is needed
- amount of storage available (dry, refrigerated or frozen)
- finance available and supply policies of the organisation
- availability, seasonality, price trends and supply
- skills, experience and number of staff
- equipment and resources available.

The skill level of employees, catering assistants and chefs must be taken into account during the production process. The condition of the product is significant, as is control during processing, in order to maximise the likelihood of the product resulting in the item or dish required. The storage life of the product is another factor that needs to be considered as part of this process.

Types of need

There are three types of need where product purchasing is concerned.

1 Perishable – fresh fruit and vegetables, dairy products, meat and fish; prices and suppliers may vary; informal needs of buying are frequently used; perishables should be purchased to meet short-term menu needs. Carrying a large stock of perishable items carries the risk of wastage, so careful assessment of supply is required to ensure all items are processed within use-by dates.

2 Staple – supplies that are canned, bottled, dehydrated or frozen; formal or informal purchasing may be used; because items are staple and can be stored easily, negotiation strategies are frequently used to take advantage of purchasing using 'quantity' as negotiating leverage. Staple commodities tend to have a long shelf life and best-before date, and can therefore be purchased to meet long-term needs based on storage capacities.

3 Daily use needs – daily use or contract items are delivered frequently on par-stock basis; stocks are kept up to the desired level and supply is often automated; supplies may arrive daily, several times a week, weekly or less often; most items are perishable, therefore, supplies must not be excessive, only sufficient to get through to the next delivery.

The buyer

The buyer is the key person who makes decisions regarding quality, quantity, price and what will satisfy customers. They must also consider the contribution to making profit. The buyer's decisions are often reflected in the success or failure of the operation. A buyer must not only be knowledgeable about the products, but have the necessary skills to deal with sales people, suppliers and other market agents. A chef and buyer must be prepared for difficult and sometimes aggressive negotiations. In larger organisations, groups or chain restaurants, purchasing decisions are often made by procurement departments, centrally managed and negotiated, and in such cases the chefs are told who to purchase from.

A buyer must also be able to make good use of market conditions. For example, if there is an abundance of fresh salmon at low cost, can the organisation make use of extra salmon purchases? Is there sufficient freezer space? Can the chef create a demand on the menu?

Another key consideration is the scale of purchasing, or 'purchasing power'. A large organisation spending large amounts of money with a particular supplier is in a stronger position to negotiate terms and conditions than a smaller organisation placing low-value orders. Furthermore, large organisations and operators often negotiate fixed market prices to prevent future fluctuations (particularly rises) in price. This puts them in a better position to predict financial forecasts and control costs. However, suppliers will also try to build in contingencies to ensure that they are not affected by unforeseen fluctuations: they will try to raise their fixed price above the lowest current market price to protect them in the case of a future rise in price. If a supplier guarantees low market prices in the long term without a contingency, this could leave them in a position where they are supplying goods at a loss due to the contractual conditions of a long-term fixed-price contract.

Buying methods

Purchasing procedures depend on the type of market and the kind of operation. They are either formal or informal. Both have advantages and disadvantages.

Informal methods are suitable for casual buying, where the amount involved is not large and speed and simplicity are desirable. Formal contracts are best for large contracts for commodities purchased over a long period of time; prices do not vary much during a year once the basic price has been established, although some price flexibility may

be built in. Prices and supply tend to fluctuate more with informal methods.

Informal buying usually involves oral negotiations, talking directly to sales people, face to face or on the phone. Informal methods vary according to market conditions.

Often referred to as competitive buying, formal buying involves giving suppliers written specifications and quantity needs. Negotiations are normally documented.

Liaising with suppliers

There are many important factors in a successful working relationship with food suppliers. Both parties have responsibilities to ensure proper food safety and quality. Food-borne illness incidents, regardless of their cause, have an impact on the reputation of caterers, and a supplier's business could be at risk due to a food scare or poisoning outbreak. A good working relationship and knowledge of each other's responsibilities is a major factor in avoiding such incidents.

Suppliers must be aware of what is expected. They must also make the caterer aware of any food safety limitations associated with the product. These are usually stated on all labelling. Such statements may simply instruct to 'keep refrigerated' or 'keep frozen', while others may include graphics or a chart of the projected shelf life at different storage temperatures. Meeting these specifications is an important factor in a successful catering operation.

Commercial documents

Delivery notes are sent with goods supplied as a means of checking that everything ordered has been delivered. The delivery note should also be checked against the duplicate or matching order sheet.

Invoices are bills sent to clients, setting out the cost of goods supplied or services rendered. An invoice should be sent on the day the goods are despatched or the services rendered, or as soon as possible afterwards. At least one copy of each invoice is made and used to inform accounts, stock records and so on.

```
INVOICE

Phone: 0208 574 1133                     No. 03957
Fax: 0208 574 1123              Vegetable Suppliers Ltd.,
Email: greend@veg.sup.ac.uk              D. Green
Website: www.greend.com          5 Warwick Road,
Messrs. L. Moriarty & Co.,               Southall,
597 High Street,                         Middlesex
Ealing,
London, W5                      Terms: 5% one month

Your order No. 67 dated 3rd September, 20...          £

Sept 26th    10 kg Potatoes bag 6.00                6.00
             5 kg Sprouts, net 8.00                 8.00
                                                   14.00
```

```
STATEMENT

Phone: 0208 574 1133            Vegetable Suppliers Ltd.,
Fax: 0208 574 1123                       D. Green
Email: greend@veg.sup.ac.uk      5 Warwick Road,
Website: www.greend.com                  Southall,
Messrs. L. Moriarty & Co.,               Middlesex
597 High Street,
Ealing,                         Terms: 5% one month
London, W5

20...                                               £
Sept 10th    Goods                                 45.90
     17th    Goods                                 32.41
     20th    Goods                                 41.30
     26th    Goods                                 16.15
                                                  135.76
     28th    Returns credited                       4.80
                                                  130.96
```

Example of an invoice and statement

Invoices contain the following information:
- name, address, telephone numbers (as a printed heading), fax numbers of the firm supplying the goods or services
- name and address of the firm to whom the goods or services have been supplied
- the word 'invoice'
- date on which the goods or services were supplied
- particulars of the goods or services supplied, together with the prices
- a note concerning the terms of settlement, such as 'Terms: 5% one month', which means that if the person receiving the invoice settles his or her account within one month, a deduction of five per cent discount will be applied for timely payment.

Credit notes are advice to clients, setting out allowances made for goods returned or adjustments made through errors of overcharging on invoices. They should also be issued when chargeable containers such as crates, boxes or sacks are returned. Credit notes have exactly the same format as invoices except that the words 'credit note' appear in place of the word 'invoice'. To make them more easily distinguishable they are usually printed in red, whereas invoices are printed in black. A credit note should be sent as soon as it is known that a client is entitled to the credit of a sum with which the company has previously been charged by invoice.

Statements are summaries of all invoices and credit notes sent to clients during the previous accounting period (usually one month). They also show any sums owing or paid from previous accounting periods and the total amount due. A statement is usually a copy of a client's ledger account and does not contain more information than is necessary to check invoices and credit notes.

Other documents that may be used during the purchasing cycle include:
- requisition
- purchase specification (see Section 2.2)
- quotation
- purchase order.

3.3 Income statements

In order to run a successful business you will need to understand financial data that is relevant to the different departments of a food operating business, including:
- sales
- cost of sales
- gross profit
- labour costs
- apportioned costs such as administration, marketing, rent/mortgage, insurances, energy costs and banking charges.

VAT (value added tax)

VAT is currently charged at the rate of 20 per cent, which has to be added to the value of any taxable sale.

VAT-inclusive price

The VAT-inclusive price is the price with VAT added. It may be calculated as follows:

Net VAT sale value £90.00

$$\frac{£90}{100} \times 120 = £108 \text{ or, more simply,}$$

$$£90 \times £1.20 = £108$$

Another way to calculate this is to take the selling price before tax and multiply it by 1 plus the rate of tax divided by 100. For example:

Selling price (SP) = £7.00

VAT = 20%

$$\frac{VAT}{100} = \frac{20}{100} = 0.2$$

$$1 + 0.2 = 1.2$$

$$SP \times 1.2 = £7.00 \times 1.2 = £8.40$$

If VAT changes, carry out the calculation in the same way. For example, if VAT is 17.5 per cent:

Selling price (SP) = £7.00

VAT = 17.5%

$$\frac{VAT}{100} = \frac{17.5}{100} = 0.175$$

1 + 0.175 5 1.175

SP × 1.175 = £7.00 × 1.175 5 £8.225

VAT-exclusive price

If a price is quoted including VAT, you might want to work out what it was before the tax was added; this is the VAT-exclusive price. In order to calculate the VAT exclusive element of a VAT-inclusive price of £108, it is not correct to take 20 per cent of the value of the inclusive selling price of £108 as this would also take 20 per cent from the VAT amount, giving an inaccurate answer. This would result in a net price of £86.40. Instead, the correct calculation is:

$$\frac{£108}{120}(100 + 20) \times 100 = £90.00 \text{ or, simply, } \frac{£108}{£1.20}$$
$$= £90$$

Another way to calculate this is to take the selling price including tax and multiply it by 1 plus the rate of tax divided by 100. For example:

Selling price including tax (SP) = £8.40

VAT = 20%

$$\frac{VAT}{100} = \frac{20}{100} = 0.2$$

1 + 0.2 = 1.2

$$\frac{SP}{1.2} = \frac{£8.40}{1.2} = £7.00$$

WORKED EXAMPLE

If a dish costs £2.80 to produce, what should its selling price be to achieve 60 per cent profit on sales?

£2.00 cost = £5.00 selling price (exclusive of VAT)

£0.80 cost = £2.00 selling price

Selling price = £7.00

Or, more commonly, $\frac{2.80}{40}$ (food cost as a percentage of sales) × 100 = £7.00

Add VAT (£7.00 3 .20) = £1.40 (£7.00 + £1.40 = £8.40) or £7.00 × 1.2 = £8.40

Either method results in the correct calculation and the final selling price of £8.40.

Profit and loss account (P&L)

The profit and loss (P&L) account measures the gains or losses from normal operations over a period of time. It measures total income and deducts total cost to establish the financial performance of your operations.

Remember:

Profits = Revenues – Expenses

Terms used in the profit and loss account

- Income – this is the total value of invoiced products or services supplied less VAT and trade discounts over a one year period. The words 'turnover' and 'sales' mean the same.
- Cost of sales – this is the direct cost of goods or services sold. All other goods left at the end of the accounting period are entered as 'stock' in the balance sheet (see Section 3.4).
- Gross profit – this figure is the total amount of profit on all sales after deducting the direct cost of making the goods or supplying the service. Expenses, tax and interest are yet to be deducted.
- Distribution – this covers postal and vehicle distribution, wages for sales and marketing staff and anything that is clearly involved with sales promotion.
- Expenses – all costs are listed here. Again, as with all costs in the P&L, figures exclude vat.
- Depreciation – the reduction in value of a fixed asset.

Small businesses only need a simple profit and loss statement such as the example laid out below.

WORKED EXAMPLE

The One Stop Sandwich Bar

Profit and loss account for the year ended 31 March 2016

	£	£
Sales		135,720
Less Cost of Sales		
Materials (food etc.)	6,500	
Direct costs	1,500	
		(8,000)
Gross profit		127,720
Less: Administrative expenses		
Wages and Salaries	8,000	
Rent	6,000	
Bad debts written off	300	
Marketing and advertising	512	
Accountancy fee	500	
Legal and professional fee	120	
Depreciation	280	
Travel expenses	475	
Hotel expenses	550	
Subsistence	100	
Electricity & Gas	3545	
Charitable donations	150	
Entertaining	235	
Postage & stationery	102	
Telephone	175	
Insurance	480	
Profit/loss on foreign currency	75	
Research and development cost written off	5,000	
		(26.599)
Net profit		101,121

Profit and loss account

3.4 Balance sheets

A balance sheet is a snapshot of a business' financial condition at a specific moment in time, usually at the close of an accounting period. A balance sheet comprises assets, liabilities and owners' or stockholders' equity. A balance sheet helps you quickly get a handle on the financial strength and capabilities of the business.

Assets and liabilities are divided into short- and long-term obligations, including cash accounts such as checking, money market or government securities.

At any given time, assets must equal liabilities plus owners' equity. An asset is anything the business owns that has monetary value. Liabilities are the claims of creditors against the assets of the business.

Cash account

The following are essential for a simple cash account:
● All entries must be dated.
● All monies received must be clearly named and entered on the left-hand, or debit side of the book/record.
● All monies paid out must also be clearly shown and entered on the right-hand, or credit side of the book/record at the end of a given period – either a day, week or month or at the bottom of each page.
● The book must be balanced, that is, both sides are totalled; the difference between the two is known as the balance. If, for example, the debit side (money received) is greater than the credit side (money paid out), then a credit or right-hand-side balance is shown, so that the two totals are then equal; a credit balance then means cash in hand.
● A debit balance cannot occur because it is impossible to pay out more than is received.

An example is given. The general rule is:
● debit = monies coming in
● credit = monies going out.

DR.	First week			Date	Payment	CR.
Date	Receipts	£		Date	Payment	£
Oct 3 4 5	to lunches " teas " tax rebate	400 100 60 560		Oct 1 2 6	by repairs " grocer " butcher " balance c/fwd	80 100 120 260 560
DR.	Second week			Date	Receipts	CR.
Date	Receipts	£		Date	Receipts	£
Oct 9 11	to balance b/fwd " sale of pastries " goods	260 100 200 560		Oct 8 10 11 12	by fishmonger " fuel " tax " balance c/fwd	50 50 40 420 560
DR.	Third week			Date	Receipts	CR.
Date	Receipts	£		Date	Receipts	£
Oct 15 17 24 26 29	To balance b/fwd " teas " pastries " goods " goods " goods	420 120 110 60 80 60 850		Oct 19 21	By butcher " grocer " balance c/fwd	80 60 710 850

Cash account

Terms used in the balance sheet

- Assets – these are subdivided into current and long-term assets to reflect the ease of liquidating each asset. Cash is considered the most liquid of all assets. Long-term assets, such as real estate or kitchen equipment, are less likely to sell overnight or have the capability of being quickly converted into a current asset such as cash.
- Current assets – an assets that can be easily converted into cash within one calendar year. Examples of current assets would be checking or money market accounts, accounts receivable and notes receivable that are due within one year's time. Current assets include:
  - cash – money available immediately, such as in checking accounts, is the most liquid of all short-term assets
  - accounts receivables – this is money owed to the business for purchases made by customers, suppliers and other vendors
  - notes receivables – those that are due within one year are current assets. Notes that cannot be collected on within one year should be considered long-term assets.

- Fixed assets – these include land, buildings, machinery and vehicles that are used in connection with the business. Types of fixed asset include:
  - land – unlike other fixed assets, land is not depreciated because it is considered an asset that never wears out
  - buildings – these are depreciated over time
  - office equipment – includes equipment such as copiers, fax machines, printers and computers used in the business
  - machinery – represents machines and equipment used in the business to produce its products. Examples of machinery include heavy-duty kitchen equipment
  - vehicles – any vehicles used in the business
  - total fixed assets – the total value of all fixed assets in the business, less any accumulated depreciation.
- Total assets – this figure represents the total value of both the short-term and long-term assets of your business.
- Liabilities and owners' equity – this includes all debts and obligations owed by the business to outside creditors, vendors or banks that are payable within one year, plus

the owners' equity. Often, this side of the balance sheet is simply referred to as 'liabilities'. Liabilities include:

- accounts payable – this includes all short-term obligations owed by the business to creditors, suppliers and other vendors. Accounts payable can include supplies and materials acquired on credit
- notes payable – this represents money owed on a short-term collection cycle of one year or less. It may include bank notes, mortgage obligations or vehicle payments
- accrued payroll and withholding – this includes any earned wages or withholdings that are owed to or for employees but have not yet been paid
- total current liabilities – this is the sum total of all current liabilities owed to creditors that must be paid within a one-year time frame
- long-term liabilities – these are any debts or obligations owed by the business that are due more than one year out from the current date.

- mortgage note payable – this is the balance of a mortgage that extends out beyond the current year. For example, a business may have paid off three years of a fifteen-year mortgage note, of which the remaining eleven years, not counting the current year, are considered long-term
- owners' equity – sometimes this is referred to as stockholders' equity. Owners' equity is made up of the initial investment in the business as well as any retained earnings that are reinvested in the business
- common stock – this is stock issued as part of the initial or later-stage investment in the business
- retained earnings – these are earnings reinvested in the business after the deduction of any distributions to shareholders, such as dividend payments.
- Total liabilities and owners' equity – this comprises all debts and monies that are owed to outside creditors, vendors or banks and the remaining monies that are owed to shareholders, including retained earnings reinvested in the business.

Balance sheet layout

Financial information	Sub total	Totals 1 Assets Current assets
	Cash (Current asset)	
	Accounts receivables (Current asset)	
	Notes receivables (Current asset)	
Fixed assets		
	Land (Fixed asset)	
	Buildings (Fixed asset)	
	Office equipment (Fixed asset)	
	Machinery (Fixed asset)	
	Vehicles (Fixed asset)	
	Total fixed assets	
	Total assets (current + fixed)	
2. Liabilities and owners' equity		
	Accounts payable (Total liability)	
	Notes payable (Total liability)	
	Accrued payroll and withholding	
	(Total liability)	
Total current liabilities (Total liability)		
Long-term liabilities (Total liability)		
Mortgage payable (Total liability)		
	Owners' equity (Total liability)	
	Common stock (Total liability)	
Retained earnings (Total liability)		
3. Total liabilities and owners' equity		
4. Total (please note that assets (current and fixed) must equal total liabilities and equity		

TEST YOURSELF

1 Why is it necessary for a chef to understand the importance of dish costing?

2 State the difference in the following:
 a) Variable costs
 b) Fixed costs
 c) Overheads

3 What is the difference between gross profit and net profit?

4 How can food waste be reduced in a commercial kitchen?

5 What is a cyclical menu? List the advantages of operating cyclical menus.

6 Name **four** items of portion control equipment.

7 In menu engineering, what is meant by the following:
 a) Stars
 b) Dogs
 c) Puzzles
 d) Plow horses

8 List **three** benefits of menu engineering.

9 Name **six** important steps in the purchasing cycle.

10 Name **three** types of purchasing need.

11 What is the purpose of a standard purchasing specification?

12 What is the purpose of a balance sheet?

13 What information does a profit and loss account contain?

14 What is meant by the breakeven analysis?

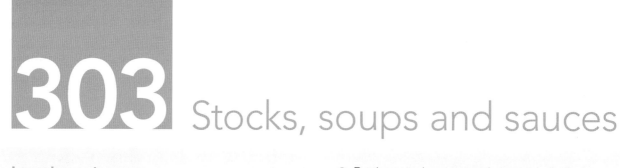

303 Stocks, soups and sauces

Learning outcomes

In this unit you will be able to:

1 Prepare stocks, soups and sauces, including:
  – knowing different types of stocks, soups and sauces
  – knowing the difference between mother sauces and their derivatives
  – knowing the quality points of commodities and understand how they affect the preparation and cooking of stocks, soups and sauces
  – knowing how and being able to use preparation techniques for stocks, soups and sauces, using preparation techniques appropriate to dish requirements
  – understanding the correct storage procedures to use throughout production and on completion of the final products.

2 Produce stocks soups and sauces, including:
  – knowing how and being able to use techniques to produce stocks, soups and sauces to meet dish requirements
  – knowing the degree of cooking time and temperature needed to produce stocks, soups and sauces to meet dish requirements
  – knowing and being able to use commodities to enhance or finish soups and sauces
  – knowing the combinations of sauces and dishes
  – understanding the correct storage procedures to use throughout production and on completion of the final products
  – being able to measure and evaluate against quality standards throughout preparation and cooking
  – being able to evaluate products against dish requirements and production standards and recognise any faults.

Recipes included in this chapter

No.	Name
	Stocks
	White stocks
1	White stock
2	White vegetable stock
3	Fish stock
4	Crab stock
	Brown stocks
5	Brown stock
6	Brown vegetable stock
7	Reduced veal stock for sauce
	Soups
	Roux-based soups
8	Pumpkin velouté
9	Carrot and butterbean soup
10	Chicken soup
11	Mushroom soup
12	Asparagus soup
13	Cream of tomato soup
14	Cream of spinach and celery soup
	Potage soups
15	Minestrone
16	Paysanne soup

No.	Name
	Fish/shellfish soups
17	Prawn bisque
18	New England clam chowder
19	Leek and potato soup (vichyssoise)
20	Chilled tomato and cucumber soup (gazpacho)
	Purée soups
21	Potato and watercress soup
22	Red lentil soup
23	Roasted red pepper and tomato soup
24	Vegetable purée soup
	Broths
25	Vegetable and barley soup
26	Scotch broth
27	Clear (consommé)
	Miscellaneous
28	Mulligatawny
	Sauces
	Jus and gravies
29	Lamb jus
30	Beef jus
31	Chicken jus
32	Red wine jus
33	Roast gravy (*jus rôti*)

34	Thickened gravy (*jus lié*)
	Béchamel derivatives
35	Béchamel sauce (white sauce)
36	Mornay sauce
37	Soubise sauce
38	Parsley sauce
	Velouté derivatives
39	Velouté (chicken, veal, mutton)
40	Aurore sauce
41	Ivory sauce
42	Mushroom sauce
43	Suprême sauce
44	Fish velouté
	Jus lié/brown sauce derivatives
45	Chasseur sauce
46	Piquant sauce (*sauce piquante*)
47	Robert sauce (*sauce Robert*)
48	Madeira sauce (*sauce Madère*)
49	Brown onion (*sauce Lyonnaise*)
50	Italian sauce (*sauce Italienne*)
51	Pepper sauce (*sauce poivrade*)
	Purées
52	Apple sauce
53	Cranberry and orange dressing
54	Green or herb sauce
	Reductions
55	Reduction of stock (glaze)
56	Reduction of wine, stock and cream
	Other sauces
57	Hollandaise sauce
58	Béarnaise sauce
59	Sauce diable
60	Tomato sauce (*sauce tomate*)
61	Sweet and sour sauce
	Cold sauces and dressings
62	Mayonnaise
63	Tartare sauce (*sauce tartare*)

64	Remoulade sauce (*sauce remoulade*)
65	Andalusian sauce (*sauce Andalouse*)
66	Thousand Island dressing
67	Tomato vinaigrette
68	Mint sauce
69	Yoghurt and cucumber raita
70	Balsamic vinegar and olive oil dressing
71	Horseradish sauce (*sauce raifort*)
72	Tomato and cucumber salsa
73	Salsa verde
74	Avocado and coriander salsa
75	Pear chutney
76	Tomato chutney
77	Pesto sauce
78	Tapenade
79	Mixed pickle
80	Gribiche sauce
	Butter sauces
81	Butter sauce (*beurre blanc*)
82	Melted butter (*beurre fondu*)
83	Compound butters
	Flavoured oils
84	Herb oil
85	Lemon oil
86	Mint oil
87	Vanilla oil
88	Basil oil
89	Walnut oil
90	Parsley oil
91	Roasted pepper oil
	Savoury foams
92	Pimms E'spuma
93	Pea E'spuma
94	Parsnip and vanilla E'spuma

Prepare stocks

Stock

Stock is the basis of all meat sauces, soups and purées. It is really just the juice of meat, extracted by long and gentle simmering, or the infusion/transfer of flavour from an ingredient such as bones, vegetables or shellfish.

When making stock, remember that the object is to draw the goodness out of the materials and into the liquor, giving it the right flavour and nutrients for the end product, whether it is a soup, sauce or a reduction. The type and quality of ingredients you use are important.

Stock can be used for a variety of recipes and preparations in the kitchen. It can be used for soups, sauces, cooking vegetables, braising and stewing both vegetables and meat and fish recipes.

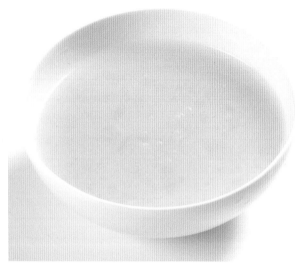

Stock

1.1 Types of stock

White stock

A basic preparation where vegetables and bones are blanched but not browned to give a clear **white stock**.

Brown stock

A basic preparation where vegetables and bones are browned, usually in the oven. Once browned, they are covered with water, boiled and simmered to give a rich brown colour.

Vegetable stock

Stock produced by simmering vegetables and herbs, used primarily for vegetarian dishes.

Nage

A nage is a well-seasoned stock (i.e. mushroom nage is a well-seasoned mushroom stock) and is usually used for cooking seafood and to enhance flavours.

Contemporary fish stock

This is made from a variety of fish and shellfish bones and is used in a variety of recipes.

Convenience stocks

Many establishments use prepared stocks to save time and labour, but also to reduce the food safety risks associated with the preparation, production and storage of stocks. These stocks are purchased as a chilled stock, frozen or in a paste or a dry powder. Many of the fresh chilled products are of a very high quality.

KEY TERMS

White stock: a stock produced by blanching but not browning vegetables and bones, to give a clear stock.

Blanch: plunging food into boiling water for a brief time before plunging into cold water to halt the cooking process. The purpose of blanching is to soften and/or partly cook the food without colouring it.

Brown stock: stock produced by browning vegetables and bones before covering in water, boiling and simmering.

Vegetable stock: stock produced by simmering vegetables and herbs.

1.2 Quality points

Stocks are the foundation of flavour of many recipes in the professional kitchen, so care and attention must be given to their production.

- Always use fresh ingredients. Never use mouldy, rotting vegetables as these will impair the flavour.
- For white stock it is advisable to **blanch** the bones, i.e. cover with cold water, bring to the boil then refresh under running cold water. This removes the surface impurities which are sometimes referred to as scum.
- For good quality **brown stock**, brown the bones well in the oven as this will give the stock a good quality brown colour.
- Well-produced stocks should not have an overpowering flavour of one ingredient.
- Stocks should be clear and not cloudy; this can be achieved by always simmering gently.
- Once the stock has simmered for the correct amount of time, strain, chill rapidly if not for immediate use, then store in a refrigerator. When required, remove from the refrigerator and bring to the boil before use.
- Remember stock is an ideal medium for the growth of bacteria.

1.3 Preparing stock

- Choose high quality ingredients, bones, carcasses and vegetables.

- Where possible, chop the bones as this will assist the cooking process.
- Fat should also be skimmed off; if it is not, the stock will taste greasy.
- When making chicken stock, the bones should first be soaked to remove the blood that is in the cavity.

1.4 and 2.3 Storage of stock

- Ideally, stock should be made fresh daily and discarded at the end of the day. Stock can be stored but it is not always advisable.
- Chilled stock should be stored in a refrigerator at a temperature below 5 °C.
- If it is to be deep-frozen it should be labelled and dated and stored below a temperature of between −20 °C and −18 °C. Check the humidity as refrigerators can vary and less airflow means higher humidity. Stocks are best stored in a refrigerator with high humidity. Always date

label if storing stock and use the correct stock rotation, using the oldest first.
- Always store stock at the bottom of the refrigerator away from raw food. Ideally stock should be stored in a separate refrigerator. When taken from storage it must be boiled for at least two minutes before being used.
- Stock must not be reheated more than once.
- Cover with suitable lids or clingfilm.

HEALTH AND SAFETY

Stocks can easily become contaminated and a risk to health, so be sure to use good ingredients and hygienic methods of preparation.

For information on maintaining a safe and secure working environment, a professional and hygienic appearance, clean food production areas, equipment and utensils, and food hygiene, refer to Unit 301.

Produce stocks

2.1 Producing stock

- Scum should be removed from the top of the stock as it cooks. If it is not removed it will boil into the stock and spoil the colour and flavour.
- Stock should always simmer gently; if it is allowed to boil quickly, it will evaporate and go cloudy or milky.
- Salt should not be added to stock. Stock is very often reduced and evaporated to give a stronger flavour; if salt is added it would be concentrated and the result would be far too salty.
- If stock is going to be kept, strain it and cool quickly, then place it in the refrigerator.

Reducing stocks: glazes

A **glaze** is a reduced stock. To achieve a glaze the stock is evaporated on the stove until it is of a consistency similar to treacle or a thick jelly. This thicker consistency is caused by collagen – which is found in the bones, tendons and connective tissue of meat – breaking down into gelatine as the stock reduces. This results in a more concentrated flavour. However, care must be taken not to over-reduce the stock otherwise it will become too concentrated and may burn. Glazes are used to enhance the flavour of soups and sauces. They are also a method of preservation.

Cooking times

Cooking times differ for different types of stocks, due to the type of bones used and how much flavour concentration is required. For example, beef bones require a great deal longer than chicken bones – generally 4–6 hours; for chicken bones only 1–2 hours may be needed for the flavour to be extracted. Fish bones are traditionally cooked for 20 minutes; however, the stock may be infused with the bones at a lower temperature than boiling point for longer.

KEY TERM

Glaze: a reduced stock with a concentrated flavour. Often used to enhance the flavour of soups and sauces.

Chilling stock

If stock is not going to be used immediately, it must be chilled. The best way to chill stock is to strain it and then blast chill it to 5 °C within 90 minutes. Alternatively, if no blast chiller is available, strain the stock into a clean container and place the container in ice water, stirring continuously until 5 °C is reached. Store immediately in a refrigerator.

2.2 Commodities to enhance stocks

For ways of enhancing stocks, see the recipes section at the end of this unit.

2.3 Storage of stock

See Section 1.4 above for information on storage of stock.

2.4 Evaluation

For evaluation of stocks, soups and sauces, see the Evaluation section on page 75.

Prepare soups

1.1 Types of soup

Table 3.1 provides some examples of soups their preparation and presentation.

Soups can be classified into the following categories.

Roux-based soups

A roux-based soup is a soup which is thickened with a traditional roux (fat and flour). Examples of roux-based soups include:

- Velouté – a velvety French sauce made with white stock and thickened with a **liaison** (see Section 2.1, 'Thickening' on page 73).
- Cream soup – a roux-based soup finished with cream before it is served.

Soup

Table 3.1 Soups, their preparation and presentation

Soup classification	Base	Passed or unpassed	Finish	Example
Clear	Stock	Strained	Usually garnished	Consommé
Broth	Stock; cut vegetables	Unpassed	Chopped parsley or chopped fresh herbs	Scotch broth; minestrone
Purée	Stock; fresh vegetables; pulses	Passed	Croutons	Lentil soup; potato soup
Velouté	Blond roux; vegetables; stock	Passed	Liaison of yolk and cream	Velouté of chicken
Cream	Stock and vegetables; vegetable purée; béchamel; velouté	Passed	Cream, milk or yoghurt	Cream of vegetable; cream of fresh pea; cream of tomato
Bisque	Shellfish stock	Passed	Cream	Lobster soup
Potage	Stock with vegetables	Unpassed; may be passed	Milk (optional)	Bonne Femme (leek and potato)
Chilled	Stock; potato and leek	Passed	Cream	Vichyssoise
Miscellaneous soups that do not fall into any other category	Stock; curry powder	Passed	Cream yoghurt or fromage frais	Mulligatawny

Potage soups

A French term referring to a thick soup. Examples of potage soups include:

- Minestrone – an Italian vegetable-based soup
- Potage Saint-Germain – green pea soup
- Potage bonne femme – an **unpassed** potage and leek soup
- Potage parmentier – potato soup (**passed**)
- Potage Crécy – carrot soup (passed).

Fish/shellfish soups

Examples of fish/shellfish soups include:

- Bisque – a very rich soup with a creamy consistency; usually made of lobster or shellfish (crab, shrimp, etc.)
- Bouillabaisse – a Mediterranean fish soup/stew, made of multiple types of seafood, olive oil, water and seasonings including garlic, onions, tomato and parsley
- Chowder: a hearty North American soup, usually with a seafood base.

Chilled soups

Examples of chilled soups include:

- Gazpacho – a Spanish tomato–vegetable soup served ice cold
- Vichyssoise – a simple, flavourful puréed potato and leek soup, thickened with the potato itself and served cold.

> **PROFESSIONAL TIP**
> No fat is added to vichyssoise when it is served cold, because if it were, the dish would leave a fatty residue on the palate and offer a less-than-clear mouth feel.

> **KEY TERMS**
> **Liaison:** a mixture of egg yolks and cream that is used to thicken a sauce. It also refers to a thickening agent.
>
> **Passed:** a thin soup such as a consommé, served by passing through a fine muslin cloth to remove the solid particles.
>
> **Unpassed:** a thin soup such as a broth which is served along with all the ingredients used.

Purée soups

A soup with a vegetable base that has been puréed in a food mill or blender. It is often altered after milling with the addition of broth, cream, butter, sour cream or coconut milk. These soups are named after the main puréed vegetable. They are soups which are passed and may be garnished. Examples include:

- Purée Dubarty (cauliflower) – garnished with small florets of cauliflower
- Purée Saint-Germain (also known as Potage Saint-Germain) – green pea
- Purée of vegetables – mixed vegetables, such as potatoes, leeks, carrots and celery
- Lentil soup.

Broths

A soup in which bones, meat, fish, grains or vegetables have been simmered in water or stock. Examples of broths include:

- Consommé – a completely clear double or triple broth (broth added to another broth) with a meat, rather than bone base; consommé is painstakingly strained to make it clear.
- Dashi – the Japanese equivalent of consommé; made of giant seaweed or *konbu*, dried bonito and water.

> **KEY TERMS**
> **Roux:** a soup thickened with a traditional roux of fat and flour.
>
> **Potage:** a usually thick soup.
>
> **Broth:** a soup consisting of a stock or water in which meat or fish and/or vegetables have been simmered.
>
> **Purée:** a soup with a vegetable base that has been puréed.
>
> **Consommé:** a completely clear broth.

> **Healthy Eating Tips**
> Look for clear soups containing vegetables, beans and lean protein such as chicken, fish or lean beef. Italian minestrone, bouillabaisse and gazpacho are excellent choices; cream-based soups can often be adapted to fit a healthier menu.

1.2 Quality points

- Soups should never be too thick and should easily flow from the spoon.
- Only the best quality stock of the appropriate flavour should be used to enhance the soup. Care should be taken to preserve the flavour of the main ingredients – they should not be overpowered by an over-strong stock.
- Always use fresh, carefully selected, quality ingredients.

- Follow the recipe carefully to achieve the right balance of flavours and consistency. Use the correct quantities and ingredients.
- Portion size is usually approximately 250 millilitres, but can be less if followed by a number of courses.
- Always check the flavour and consistency before service.
- Peel, prepare and weigh ingredients as part of the *mise en place*. Make sure you have the correct ratio of ingredients for the required number of portions.

1.3 Preparing soup

Most soup-making begins by preparing a stock (see page 62), made by slowly simmering vegetables with seasonings and then straining the liquid.

The underlying flavours of these foundation ingredients are enhanced by adding herbs and seasonings. In many cases, this flavour base begins by preparing a mixture of

flavouring elements cooked in a little fat or oil. Because of their aromatic properties and flavours, most soups begin this phase with a combination of onion, garlic, leeks and carrots (this is called a **mirepoix**) or **aromats**. The result is the foundation of all soups. To this foundation meat, fish, vegetables, fruits, seasonings, fats like butter or cream and vegetables or dried pulses are added in countless variations to create the wealth of soups available.

Healthy Eating Tips
Soup offers a very good, healthy option. For healthier soup, butter can be replaced with vegetable oil, reducing the cholesterol intake. Cream can be replaced with fromage frais or yoghurt. Broths and vegetable soups are healthy options as they are both light and filling.

KEY TERMS

Mirepoix: approximately 1-cm diced carrot, onion and celery cooked in fat or oil and used as a flavour base for soups.

Aromats: herbs such as parsley, chervil and basil used as a flavour base; may also include vegetables such as onions and celery.

Weighing and measuring

It is important to weigh every ingredient to determine the exact flavour and consistency of the soup, but more importantly to determine the exact nutritional content per portion. Measuring the ingredients accurately will also result in the correct yield and less wastage.

Cuts of vegetables

Soups which require small cuts of vegetables (such as paysanne for minestrone and gros brunoise for Scotch broth) are either cut by hand or by using a machine which cuts vegetables to the required size. (See Chapter 4 for more information on cuts of vegetables.)

PROFESSIONAL TIP
Always use strong, thick-bottomed pans to cook soup to avoid scalding and browning.

1.4 and 2.3 Storage of soup

If soup is being stored it must be blast chilled, stored in a refrigerator at 3–5 °C, covered and date labelled. When reheated it must be boiled or heated to at least 72 °C for two minutes. Stock rotation should be used; always use the oldest first. Soup is best positioned at the bottom of the refrigerator and kept away from any raw food. Check the humidity of the refrigerator, which should be high. If possible soup should be stored in a separate refrigerator. Take care not to over produce and store for too long.

Produce soups

2.1 Producing soup

Step by step: passing and straining soups

Conical strainers are used to pass the soup to obtain a fine consistency. Conical strainers come in a range of sizes from fine (chinois) to medium.

1 Slowly simmer vegetables and seasonings with stock to develop flavours.

2 Purée using a stick blender, liquidiser or food processor.

3 To ensure a smooth consistency, pass the soup through the conical strainer. Use a ladle to carefully pump the soup through.

> **PROFESSIONAL TIP**
> Preparing soups using stick blenders, liquidisers and food processors produces excellent purée soups with a smooth consistency.

Presenting soup

Soups are usually presented individually, garnished and finished with fresh herbs, cream, fromage frais, yoghurt and so on. Sauces may be presented in a sauce boat or as a cordon on the dish or they may mask the product. Whichever type of presentation is used, it must be clean.

Serving soup

- Serve hot soups at 63 °C or above to control the growth of pathogenic organisms. Discard soup if not sold after two hours.
- Cold soups should be served at 5 °C or below. Chill the soup in a blast chiller below 8°C within 90 minutes to slow down bacterial growth.
- Depending on the function and purpose, soups are usually served in portions of 250 millilitres maximum. Sometimes, for example, if speciality soup is served as an *amuse bouche*, the portion size is reduced to approximately 50 millilitres.

2.2 Commodities to enhance soup

Examples of garnishes that can be served with soup include:
- croutons – slices of white or brown bread cut into 1-centimetre dice, carefully shallow fried in vegetable oil or clarified butter.
- sippets – corners of bread cut from a flat tin loaf, finely sliced and toasted in the oven. Both croutons and sippets may be flavoured by rubbing chopped garlic into the bread.
- toasted flutes (*croutes de flute*) – very small thin baguette slices, toasted or sprinkled with olive oil or melted butter and baked in the oven until crisp.

Accompaniments that can be served with soup include:
- brunoise, julienne or paysanne of vegetables
- concassé
- fine noodles
- rice
- chopped herbs
- finely diced cooked chicken.

Most soups in the recipe section are ungarnished so choose from the above suggestions to finish your soup or devise a garnish of your own.

2.3 Storage of soup

See Section 1.4 on page 66 for storage of soup.

2.4 Evaluation

For evaluation of stocks, soups and sauces, see the Evaluation section on page 73.

Prepare sauces

A sauce is a liquid that has been thickened. Sauces add flavour and moisture to dishes; they also contribute to the overall texture of the dish. They provide a contrast to, or compliment the dish with which they are being served. Sauce can also be used to enhance the nutritional value of a dish.

Basic Sauces were traditionally known as Mother sauces as many derivatives are derived from these basic sauces. This term is not used today in modern professional kitchens.

1.1 Types of sauces

Béchamel sauces

Béchamel is a basic white sauce made using butter, flour and milk. Many sauces are derived from a basic béchamel sauce by the use of additional ingredients or garnishes. Table 3.2 provides some examples of béchamel sauces and dishes with which they are often served.

Table 3.2 Béchamel sauce derivatives

Sauce	Additions to basic béchamel	Served with
Egg	Hard-boiled eggs, diced	Poached or boiled fish
Cheese (mornay)	Grated cheese, egg yolk and cream	Fish or vegetables
Cream	Cream, milk, natural yoghurt or fromage frais	Poached fish and boiled vegetables
Onion	Chopped or diced, cooked onions; cream	Roast lamb or mutton
Soubise	As for onion sauce, but passed through a strainer	Roast lamb or mutton
Parsley	Chopped parsley	Poached or boiled fish and vegetables
Mustard	English or continental mustard	Grilled herring
Anchovy	Anchovy essence	Poached, boiled or fried fish

Velouté sauces

A basic velouté sauce is a blond roux, to which a white stock and chicken, veal, fish and so on are added. Sauces derived from velouté and dishes with which they are often served are shown in Table 3.3.

Video: Blond roux and velouté

http://bit.ly/2pCPbZT

Table 3.3 Velouté sauce derivatives

Sauce	Additions to basic velouté	Served with
Caper	Capers and single cream	Boiled or poached leg of mutton
Aurore	Mushroom trimmings, cream, egg yolk, lemon juice, tomato purée	Poached or boiled chicken, poached eggs
Mushroom	Sliced button mushrooms, egg yolk and cream	Poached or boiled chicken, sweetbreads
Ivory	Meat glaze and cream	Poached or boiled chicken
Suprême	Cream and lemon juice	

Jus lié and brown sauces (demi-glace)

Today, **demi-glace** is often replaced with **jus lié**, or a good reduced stock. A jus lié is a lightly thickened brown stock – traditionally veal – which is flavour enhanced with tomato and mushroom trimmings and thickened with arrowroot. In some cases a ham bone is added; this is optional. A good brown stock is reduced in some recipes with red wine and lightly thickened with arrowroot, similar to jus lié. A glaze is a reduced stock (usually brown). There are many variations of brown sauce. Some of these are shown in Table 3.4.

Table 3.4 Brown sauce derivatives

Sauce	Additions to basic brown sauce	Served with
Chasseur	Mushrooms, tomato concassé, white wine, tarragon	Sauté of chicken or shallow-fried steaks
Devilled	Cayenne pepper, white wine, mignonette pepper	Grilled chicken, sautéed kidneys
Bordelaise	Red wine, shallots, bone marrow	Shallow-fried steaks, grilled steaks
Charcutiere	White wine, gherkins	Grilled pork chops
Robert	White wine, mustard	Veal steaks, chops
Piquant	Gherkins, capers and herbs	Grilled chicken, shallow-fried steaks
Sherry	Sherry	Veal escalopes, veal chops, chicken suprêmes (shallow fried)
Madeira	Madeira	Veal and pork escalopes, fried calves' liver
Brown onion	Sliced brown onions, white wine	Burgers, sausages (shallow fried and grilled)
Lyonnaise	Sliced brown onions, white wine	Burgers, sausages (shallow fried and grilled)
Italian	Mushrooms, tomatoes and ham	Sautéed calves' and lambs' liver
Pepper	Crushed peppercorns	Grilled and shallow-fried steaks

Purée-based sauces

Purée-based sauces can be fruit based (for example, apple sauce or cranberry sauce), vegetable based (for example, tomato sauce) or herb based. A herb-based sauce is one where the herbs are puréed with the base and constitute the main flavour; an example is green sauce (a mayonnaise with herbs). Green sauce is served with cold poached salmon.

Gravies

Examples of gravies include:

- beef, lamb or roast chicken jus
- red wine
- roast gravy – jus de rôti
- thickened gravy – jus lié.

Other sauces and dressings

Other sauces include:

- horseradish
- sweet and sour
- bread
- salsas (for example, salsa verde, yoghurt and cucumber salsa, tomato and cucumber salsa, avocado and coriander salsa)
- chutneys (for example, pear chutney)
- pickles (for example, mixed pickle)
- dressings (for example, pesto, tapenade, cranberry and orange dressing for duck).

Bread sauce, which is often served as an accompaniment to roast turkey and roast chicken

Emulsions

Many sauces are **emulsions**. Foods which are called emulsions include cream, butter, mayonnaise and margarine. The oil and water in the food products are held together with an emulsifying agent which stops them from separating. Mayonnaise and hollandaise are examples of emulsions; the emulsifying agent is lecithin found in the egg yolk.

KEY TERMS

Béchamel: white sauce.

Jus lié: thickened gravy.

Demi-glace: brown refined sauce.

Emulsion: a mixture of two or more liquids that are normally immiscible (non-mixable or un-blendable).

Butter sauces

There are a number of butter sauces and these are shown in Table 3.5.

Table 3.5 Butter sauces

Sauce	Description	Served with
Clarified butter	Butter that has been melted and skimmed; the butter is carefully poured off to leave the milky residue behind, giving a clear fat that can reach higher temperatures than normal butter without burning.	Steamed vegetables or poached or grilled fish
Beurre noisette	This translates to 'nut butter'; its flavour comes from the caramelisation of the milk in the butter solids. It is achieved by placing diced hard butter into a moderately hot pan and bringing to a foam. The milk cooking in the fat creates a popping/cracking sound, which stops when it is caramelised.	Poached or steamed vegetables and fish; popularly used with shallow fried fish
Beurre noir (black butter)	*Beurre noisette* taken a little further (almost to burn the sediment), with vinegar added.	Traditionally skate but can be used for a variety of dishes
Sauce	Description	Served with
Beurre fondu/emulsion	An emulsion between fat and liquid; melted butter emulsified with any nage will give a slightly thicker sauce.	Used to coat vegetables or fish
Compound butter sauces	Made by mixing the flavouring ingredients into softened butter, which can then be shaped into a roll 2 cm in diameter, placed in wet greaseproof paper or foil, hardened in a refrigerator and cut into 0.5-cm slices when required. Examples include parsley butter, herb butter, chive butter, garlic butter, anchovy butter, shrimp butter, mustard and liver pâté.	Grilled and some fried fish; grilled meats

1.2 Quality points

A sauce should:
- be smooth
- look glossy
- have a definite flavour
- be of a light texture and free flowing
- use a thickening medium in moderation.

1.3 Preparing sauces

For details of preparing sauces, see Section 2.1 Producing sauces and the recipes section at the end of the chapter.

1.4 and 2.3 Storage of sauces

- Cool rapidly and store in a refrigerator at a temperature below 5 °C.

- If sauces are to be deep-frozen, any flour must be replaced with a modified starch to prevent curdling on thawing and reheating.
- If the sauce is to be deep-frozen, label and date and store between –20 °C and –18 °C.
- When taken from storage, boil the sauce for at least two minutes before using. Do not reheat the sauce more than once.

> **HEALTH AND SAFETY**
> Never store a stock, sauce, gravy or soup above eye level as someone could accidentally spill the contents over themselves if they don't see it.

Produce sauces

2.1 Producing sauces

Thickening

There are a number of methods of thickening sauces.

Beurre manié

A paste made from equal quantities of soft butter and flour which is mainly used for fish sauces. It is added to a simmering liquid, which should be whisked continuously to prevent it becoming lumpy.

Egg yolks and cream (liaison)

Using egg yolks for thickening is commonly known as a liaison, and is traditionally used to thicken a classic velouté. Egg yolks and cream are mixed together and added to the sauce off the boil. It is essential to keep stirring the sauce once you have added the eggs, otherwise they will curdle. Once the sauce is thickened it must be removed and served immediately. Do not allow the liquid to boil or simmer. Egg yolks are used in mayonnaise, hollandaise and custard sauces, although in a different way for each sauce.

Roux

A roux is a combination of fat and flour, which are cooked together. Liquid is then added to the cooked mixture to make the sauce. There are three degrees to which a roux may be cooked:

- White – used for white (béchamel) sauces and soups. Cook equal quantities of margarine or butter and flour together for a few minutes, without colouring, until the mixture is a sandy texture. Polyunsaturated vegetable margarine or vegetable oil can be used as an alternative; using oil gives a slack roux but means the liquid can be incorporated easily.
- Blond – used for veloutés, tomato sauce and soups. Cook equal quantities of margarine, butter or vegetable oil and flour for longer than a white roux, without colouring, until it is a sandy texture.
- Brown – traditionally used for brown (espagnole) sauce and soups. It is slightly browned in the roux-making process by cooking the fat and flour mixture for a bit longer than in the other methods.
- **Continental** – a very easy and straightforward thickening agent that can be frozen and used as a quick thickener during service or at the last minute. Mix equal quantities of flour and vegetable oil to a paste and place in the oven at 140°C. Cook the mixture, mixing it in on itself continually until a biscuit texture is achieved. Remove and allow to cool to room temperature. Form it into a sausage shape using a double layer of cling film. Chill, then freeze. To use, remove it from the freezer and shave a little off the end of the log. Whisk it into the boiling sauce (as the flour is already cooked it is not necessary to add it slowly to prevent lumping as this will not occur). Once the desired thickness has been achieved, pass the sauce (strain it) and serve.

Step by step: thickening using a roux

1 Add equal quantities of margarine or butter and flour together to the pan.

2 Cook for a few minutes, without colouring, until the mixture is a sandy texture. This is a white roux that can be used to thicken béchamel sauces.

3 Cooking the mixture for longer will result in a blond roux, which can be used for veloutés, tomato sauce and soups.

4 For a brown roux, brown the mixture in the roux-making process. This can be used for brown sauces and soups.

> **HEALTH AND SAFETY**
> Never add boiling liquid to a hot roux, as you may be scalded by the steam that is produced and the sauce may become lumpy.

> **PROFESSIONAL TIP**
> Do not allow a roux sauce to stand over a moderate heat for any length of time as it may become thin due to a chemical change (dextrinisation) in the flour.

Reduction

Some sauces are produced by reducing stock, wine or cream, using a single ingredient or combination. No starches are added.

Emulsification

Emulsified sauces mix together ingredients that do not easily mix or stay mixed. The most common sauces are mayonnaise (oil and egg yolks) and hollandaise (butter and egg yolks). An emulsion is a suspension of the two liquids which do not mix naturally.

Sweating

In some recipes, vegetables such as chopped onions and garlic, herbs and spices are sweated in oil or butter to extract flavour before the liquid is added to produce the sauce.

Deglaze or deglace

Deglazing involves adding wine, stock or water to the pan in which the bones or food was cooked to release the sediments or flavours in the pan.

Skimming

Skimming means to remove the impurities and scum from a cooking liquid.

Clarification

This involves turning cloudy liquid clear, as in the making of consommé

Liaison

A liaison is a thickening agent used to thicken liquids.

Cornflour, arrowroot or starch

These are used for thickening gravy and sauces. They are diluted with water, stock or milk and then stirred into the boiling liquid, which is allowed to re-boil for a few minutes and is then strained. For large-scale cooking and economy, flour may be used.

Sauce flour

Sauce flour is a specially milled flour that does not need to have any fat added to it to prevent it from going lumpy. Sauces may be thickened using this flour. It is useful when making low-fat sauces.

Blood

Traditionally used in recipes such as jugged hare, but is used rarely today.

Cooking liquor

Liquor from certain dishes and/or stock can be reduced to give a light sauce.

Rice

Rice is used to thicken some shellfish bisques.

Butter (*monter au beurre*) and olive oil

Monter au beurre means 'mounting' the sauce with small pieces of butter or oil to thicken and give the sauce a shine.

Checking and finishing sauces

When you have finished preparing a sauce, always check:
- consistency – the sauce should be light and free flowing. It should lightly coat the back of a metal spoon
- flavour – it should be seasoned but not over seasoned.

> **PROFESSIONAL TIP**
> It is important to develop your palate to recognise correct tastes and flavours.

2.2 Use commodities to enhance sauces

Combinations

Sauces accompany dishes and there are many classical combinations. Some examples are:
- bread-crumbed fish served with tartare sauce
- roast beef with horseradish sauce
- roast pork and apple sauce
- battered fish with tomato sauce
- roast lamb with mint sauce.

This provides opportunities for the chef to vary these combinations, for example:

- adding calvados to apple sauce
- adding chilli to tomato sauce
- substituting mint sauce for a redcurrant and port wine sauce.

2.3 Storage

See Section 1.4 on page 66 for storage of sauces.

2.4 Evaluation

For evaluation of stocks, soups and sauces, see the Evaluation section below.

2.5 Evaluation of stocks, soups and sauces

- Portion size – this is important for soups and sauces. The portion size must be adequate and sufficient but not too large as this could affect the gross profit margin.
- Colour – soups and sauces must be attractive to the eye. Colour denotes quality and freshness. Stocks should not be cloudy.

- Taste – the taste must denote the description of the soup or sauce, be well seasoned and full of flavour. Stocks should not be seasoned as they often are reduced in the preparation of dishes.
- Texture – soups and sauces must not be too thick, sticky or starchy.
- Consistency – soups and sauces should flow easily and be easy to serve. Sauces should coat the back of a soup and mask the products with ease. Reduced stocks should not be reduced so much that they become too strong and sticky.
- Temperature – this is crucial. Hot soups must be at least 72 °C, cold soups at 5 °C. Sauces must also be served at the same temperatures. Stocks must be chilled once produced and stored at 3 °C. Before use they must be boiled at 100 °C for approximately three minutes.
- Aroma – this is crucially important. Stocks, soups and sauces must smell fresh with no unusual or recognisable smells detected.
- Presentation – all food must be well presented on the plate or serving dish. The serving plates and dishes must be clean with no smears. Any garnish must be well placed and appropriate. Where possible, take photos of all dishes and evaluate how these can be improved.

TEST YOURSELF

1 List the ingredients and describe the cooking principles for brown beef stock.

2 State **four** quality points you should look for in a white vegetable stock.

3 Give **two** examples of each of the following:
   a) purée soups   c) chilled soups
   b) potage soups   d) broths.

4 Name **two** alternatives to cream when finishing a soup.

5 At what temperature should hot soup be served?

6 Describe how you would cook lobster bisque.

7 Name **two** derivatives of béchamel.

8 Give an example of a dish with which chasseur sauce is traditionally served.

9 Describe how you should store stocks and sauces.

PRACTICAL COOKERY for the Level 3 Advanced Technical Diploma in Professional Cookery

1 White stock

Energy	Calories	Fat	Saturated fat	Carbohydrates	Sugar	Protein	Fibre	Sodium
105 kJ	25 kcal	0.3 g	0.1 g	4.9 g	4 g	1 g	2.4 g	0.022 g

	4.5 litres	10 litres
Raw, meaty bones	1 kg	2.5 kg
Water	5 litres	10.5 litres
Onion, carrot, celery, leek	400 g	1.5 kg
Bouquet garni	½	1½
Peppercorns	8	16

Mise en place

1 Chop the bones into small pieces, and remove any fat or marrow.

Cooking

1 Place the bones in a large stock pot, cover with cold water and bring to the boil.

2 Wash off the bones under cold water, then clean the pot.

3 Return the bones to the cleaned pot, add the water and reboil.

4 Skim as and when required, wipe round inside the pot and simmer gently.

5 After 2 hours, add the washed, peeled whole vegetables, bouquet garni and peppercorns.

6 Simmer for 6–8 hours. Skim, strain and, if to be kept, cool quickly and refrigerate.

2 White vegetable stock

Energy	Calories	Fat	Saturated fat	Carbohydrates	Sugar	Protein	Fibre	Sodium
11 kJ	3 kcal	0.0 g	0.0 g	0.4 g	0.4 g	0.0 g	0.2 g	2.9 g

	4 portions	10 portions
Onions	100 g	250 g
Carrots	100 g	250 g
Celery	100 g	250 g
Leeks	100 g	250 g
Water	1.5 litres	3.75 litres

Mise en place

1 Roughly chop all the vegetables.

Cooking

1 Place all the ingredients into a saucepan, add the water, bring to the boil.

2 Allow to simmer for approximately 1 hour.

3 Skim if necessary. Strain and use.

> **VARIATION**
> ● **White fungi stock:** add 200–400 g white mushrooms, stalks and trimmings (all well washed) to the recipe.

3 Fish stock

Energy	Calories	Fat	Saturated fat	Carbohydrates	Sugar	Protein	Fibre	Sodium
1523 kJ	371 kcal	25.0 g	3.7 g	3.2 g	2.1 g	1.0 g	1.6 g	15.0 g

	2 litres
Fish bones, no heads, gills or roe (turbot, sole or brill bones are best)	5 kg
Olive oil	100 ml
Onions, finely chopped	3
Leeks, finely chopped	3
Celery sticks, finely chopped	3
Fennel bulb, finely chopped	1
Dry white wine	350 ml
Parsley stalks	10
Thyme	3 sprigs
White peppercorns	15
Lemons, finely sliced	2

Mise en place

1 Wash off the bones in cold water for 1 hour.

Cooking

1 Heat the olive oil in a pan that will hold all the ingredients and still have a 1-cm gap at the top for skimming. Add all the vegetables and sweat without colour for 3 minutes.

2 Add the fish bones and sweat for a further 3 minutes.

3 Add the white wine and enough water to cover. Bring to a simmer, skim off the impurities and add the herbs, peppercorns and lemon. Turn off the heat.

4 Infuse for 25 minutes, then pass into another pan and reduce by half. The stock is now ready for use.

4 Crab stock

Energy	Calories	Fat	Saturated fat	Carbohydrates	Sugar	Protein	Fibre	Sodium
9472 kJ	2302 kcal	157.0 g	23.0 g	30.0 g	24.0 g	5.8 g	14.4 g	165.0 g

	2 litres
Crab shells, smashed	2 kg
Prawns, with shells still on	1.5 kg
Corn oil	50 ml
Brandy (optional)	200 ml
Pernod (optional)	100 ml
Carrots, peeled and chopped for mirepoix	250 g
Leeks, peeled and chopped for mirepoix	250 g
Celery, chopped for mirepoix	150 g
Garlic cloves, smashed	2
Shallot, peeled	180 g
Tomato paste	150 ml
Fish stock	2.5 litres
Small sprig of thyme	
Bay leaf	1

Cooking

1 Roast the shells in the oil and deglaze with the brandy and Pernod (or, if not using these, some of the stock).

2 In a separate pan, roast the vegetables, then add the tomato paste, stock and herbs, add the roasted shells and the prawns, and simmer for 20 minutes.

3 Turn off the heat and allow to infuse for 30 minutes. Pass, and reduce by half. The stock is now ready for use.

5 Brown stock

Energy	Calories	Fat	Saturated fat	Carbohydrates	Sugar	Protein	Fibre	Sodium
13 kJ	3 kcal	0.3 g	0.1 g	0.0 g	0.0 g	0.0 g	0.0 g	0.4 g

	4.5 litres	10 litres
Raw, meaty bones	1 kg	2.5 kg
Water	5 litres	10.5 litres
Onion, carrot, celery, leek	400 g	1.5 kg
Bouquet garni	½	1½
Peppercorns	8	16

Mise en place

1 Chop the beef bones and brown well on all sides either by placing in a roasting tin in the oven, or carefully browning in a little fat in a frying pan.

2 Drain off any fat.

Cooking

1 Place the bones in a stock pot.

2 Brown any sediment that may be in the bottom of the tray, deglaze (swill out) with 0.5 litre of boiling water, simmer for a few minutes and add to the bones.

3 Add the cold water, bring to the boil and skim. Simmer for 2 hours.

4 Wash, peel and roughly cut the vegetables, fry in a little fat until brown, strain and add to the bones.

5 Add the bouquet garni and peppercorns.

6 Simmer for 6–8 hours. Skim and strain.

> **PROFESSIONAL TIP**
> A few squashed tomatoes and washed mushroom trimmings can also be added to brown stocks to improve flavour, as can a calf's foot and/or a knuckle of bacon. If bacon is used, dishes made with the stock will not be suitable for some religious diets.

Video: Making brown stock

http://bit.ly/2oP1TRb

6 Brown vegetable stock

Energy	Calories	Fat	Saturated fat	Carbohydrates	Sugar	Protein	Fibre	Sodium
205 kJ	49 kcal	3.9 g	0.5 g	1.6 g	1.5 g	0.1 g	0.8 g	170.0 g

	4 litres approximately
Onions	300 g
Carrots	300 g
Celery	300 g
Leeks	300 g
Sunflower oil	180 ml
Tomatoes	150 g
Mushroom trimmings	150 g
Peppercorns	18
Water	4 litres
Yeast extract	15 g

Cooking

1 Fry the mirepoix in the oil until golden brown.

2 Drain and place in a suitable saucepan. Add all the other ingredients except the yeast extract and water.

3 Cover with the water and bring to the boil.

4 Add the yeast extract and simmer gently for approximately 1 hour. Then skim if necessary and use.

> **VARIATION**
> ● **Brown fungi stock:** add 200–400 grams open or field mushrooms, stalks and trimmings (all well washed) to the recipe.

Mise en place

1 Cut the vegetables into mirepoix.

7 Reduced veal stock for sauce

Energy	Calories	Fat	Saturated fat	Carbohydrates	Sugar	Protein	Fibre	Sodium
72 kJ	17 kcal	0.1 g	0.0 g	3.3 g	3.3 g	0.0 g	1.3 g	7.9 g

No flour is used in the thickening process and consequently a lighter textured sauce is produced. Care needs to be taken when reducing this type of sauce so that the end product is not too strong or bitter.

	4 litres
Veal bones, chopped	4 kg
Water	4 litres
Calves' feet, split lengthways (optional)	2
Carrots, roughly chopped	400 g
Onions, roughly chopped	200 g
Celery, roughly chopped	100 g
Tomatoes, blanched, skinned, quartered	1 kg
Mushrooms, chopped	200 g
Large bouquet garni	1
Unpeeled cloves of garlic (optional)	4

Cooking

1 Brown the bones and calves' feet in a roasting tray in the oven.

2 Place the browned bones in a stock pot, cover with cold water and bring to simmering point.

3 Using the same roasting tray and the fat from the bones, brown off the carrots, onions and celery.

4 Drain off the fat, add the vegetables to the stock and deglaze the tray.

5 Add the remainder of the ingredients; simmer gently for 4–5 hours. Skim frequently.

6 Strain the stock into a clean pan and reduce until a light consistency is achieved.

8 Pumpkin velouté

Energy	Calories	Fat	Saturated fat	Carbohydrates	Sugar	Protein	Fibre
2883 kJ	689 kcal	71.4 g	43.9 g	6.7 g	1.5 g	5.3 g	1.5 g

Parmesan, grated	30 g	70 g
Truffle oil	1 tbsp	2 tbsp
Salt, pepper		
Chicken stock	400–600 ml	1–1.5 litres

Cooking

1 Sweat the shallots in the butter, without colour, until cooked and soft.

2 Add the garlic, pumpkin, Parmesan and truffle oil. Correct the seasoning and cook for 5 minutes.

3 Add the chicken stock, bring to the boil, simmer for 5 minutes.

4 Blitz, pass, correct the seasoning, then blast chill if to be stored.

	4 portions	10 portions
Shallots, sliced	1	3
Butter	320 g	800 g
Clove of garlic, sliced (optional)	½	1
Large squash or pumpkin (300 g), flesh diced	1	2

9 Carrot and butterbean soup

Energy	Calories	Fat	Saturated fat	Carbohydrates	Sugar	Protein	Fibre
891 kJ	213 kcal	5.9 g	0.7 g	33.4 g	19.3 g	8.5 g	8.1 g

Cooking

1 Cook the onion and garlic in the oil for a few minutes, without colour, then add the carrots and stir well.

2 Add the carrot juice and vegetable stock. Bring to the boil, turn down to a simmer and cook for about 15 minutes until the carrot is cooked through.

3 Add the beans and cook for a further 5 minutes or so until they are heated through.

4 Liquidise in a food processor until smooth; check seasoning.

	4 portions	10 portions
Onions, peeled and chopped	1	2
Cloves of garlic, chopped	2	5
Sunflower or other vegetable oil	15 ml	35 ml
Large to medium carrots, brunoise	6	15
Carrot juice	500 ml	1.25 litres
Vegetable stock	500 ml	1.25 litres
Butter beans, cooked	400 g	1 kg
Seasoning		

PROFESSIONAL TIP

With any soup recipe, it is important to simmer the soup gently. Do not let it boil vigorously, because too much water will evaporate. If the soup has boiled, add more stock or water to make up for this.

10 Chicken soup (*crème de volaille* or *crème reine*)

Energy	Calories	Fat	Saturated fat	Carbohydrates	Sugar	Protein	Fibre *
836 kJ	199 kcal	13.6 g	6.2 g	14.0 g	4.2 g	5.9 g	1.0 g

* Using hard margarine

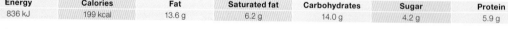

	4 portions	10 portions
Onion, leek and celery, sliced	100 g	250 g
Butter or oil	50 g	125 g
Flour	50 g	125 g
Bouquet garni	1	2
Salt, pepper		
Milk or cream	250 ml or 125 ml	625 ml or 300 ml
Cooked dice of chicken (garnish)	25 g	60 g

Cooking

1 Gently cook the sliced onions, leek and celery in a thick-bottomed pan, in the butter or oil, without colouring.

2 Mix in the flour; cook over a gentle heat to a sandy texture without colouring.

3 Cool slightly; gradually mix in the hot stock. Stir to the boil.

4 Add the bouquet garni and season.

5 Simmer for 30–45 minutes; skim when necessary. Remove the bouquet garni.

6 Liquidise or pass firmly through a fine strainer.

7 Return to a clean pan, reboil and finish with milk or cream; correct the seasoning.

8 Add the chicken garnish and serve.

VARIATION
- Natural yoghurt, skimmed milk or non-dairy cream may be used in place of dairy cream.
- Add cooked small pasta or sliced mushrooms.

Healthy Eating Tips
- Use soft margarine or sunflower/vegetable oil in place of the butter.
- Use the minimum amount of salt.
- The least fatty option is to use a combination of semi-skimmed milk and yoghurt or fromage frais, not cream.

11 Mushroom soup (*crème de champignons*)

Energy	Calories	Fat	Saturated fat	Carbohydrates	Sugar	Protein	Fibre *
712 kJ	170 kcal	11.8 g	5.2 g	12.6 g	3.0 g	3.8 g	1.6 g

* Using hard margarine

Cooking

1 Gently cook the sliced onions, leek and celery in the butter or oil in a thick-bottomed pan, without colouring.

2 Mix in the flour and cook over a gentle heat to a sandy texture without colouring.

3 Remove from the heat and cool slightly.

4 Gradually mix in the hot stock. Stir to the boil.

5 Add the well-washed, chopped mushrooms, the bouquet garni and season.

6 Simmer for 30–45 minutes. Skim when needed.

7 Remove the bouquet garni. Pass through a sieve or liquidise.

8 Pass through a medium strainer. Return to a clean saucepan.

9 Reboil, correct the seasoning and consistency; add the milk or cream.

	4 portions	10 portions
Onion, leek and celery, sliced	100 g	250 g
Butter or oil	50 g	125 g
Flour	50 g	125 g
White stock (preferably chicken)	1 litre	2.5 litres
White mushrooms, washed and chopped	200 g	500 g
Bouquet garni	1	2
Salt, pepper		
Milk (or cream)	125 ml (or 60 ml)	300 ml (or 150 ml)

12 Asparagus soup (*crème d'asperges*)

Energy	Calories	Fat	Saturated fat	Carbohydrates	Sugar	Protein	Fibre
919 kJ	223 kcal	13.1 g	8.0 g	18.9 g	8.8 g	8.4 g	2.5 g

	4 portions	10 portions
Onion, sliced	50 g	125 g
Celery, sliced	50 g	125 g
Butter or oil	50 g	125 g
Flour	50 g	125 g
White stock (preferably chicken)	1 litre	2.5 litres
Asparagus stalk trimmings, or Tin of asparagus	200 g / 150 g	500 g / 325 g
Bouquet garni	1	2
Salt, pepper		
Milk or cream	250 ml or 125 ml	625 ml or 300 ml

Cooking

1 Gently sweat the onions and celery, without colouring, in the butter or oil.

2 Remove from the heat, mix in the flour, return to a low heat and cook out, without colouring, for a few minutes. Cool.

3 Gradually add the hot stock. Stir to the boil.

4 Add the well-washed asparagus trimmings, or the tin of asparagus, and bouquet garni. Season.

5 Simmer for 30–40 minutes, then remove bouquet garni.

6 Liquidise and pass through a strainer.

7 Return to a clean pan, reboil, correct seasoning and consistency. (Milk with a little cornflour can be added to adjust the consistency.)

8 Add the milk or cream and serve.

Healthy Eating Tips
● Use an unsaturated oil (sunflower/vegetable) to lightly oil the pan. Drain off any excess after the frying is complete and skim the fat from the finished dish.

13 Cream of tomato soup (*crème de tomates fraiche*)

Energy	Calories	Fat	Saturated fat	Carbohydrates	Sugar	Protein	Fibre
1159 kJ	277 kcal	16.2 g	4.8 g	28.5 g	13.6 g	6.3 g	6.6 g

	4 portions	10 portions
Butter or oil	50 g	125 g
Bacon trimmings, optional	25 g	60 g
Onion, diced	100 g	250 g
Carrot, diced	100 g	250 g
Flour	50 g	125 g
Fresh, fully ripe tomatoes	1 kg	2.5 kg
Stock	1 litre	2.5 litres
Bouquet garni	1	2
Salt, pepper		
Croutons – sliced stale bread	1	3
Butter	50 g	125 g

Cooking

1 Melt the butter or heat the oil in a thick-bottomed pan.

2 Add the bacon, onion and carrot (mirepoix) and brown lightly.

3 Mix in the flour and cook to a sandy texture.

4 Gradually add the hot stock.

5 Stir to the boil.

6 Remove the eyes from the tomatoes, wash them well, and squeeze them into the soup after it has come to the boil.

7 If colour is lacking, add a little tomato purée soon after the soup comes to the boil.

8 Add the bouquet garni and season lightly.

9 Simmer for approximately 1 hour. Skim when required.

10 Remove the bouquet garni and mirepoix.

11 Liquidise or pass firmly through a sieve, then through a conical strainer.

12 Return to a clean pan, correct the seasoning and consistency. Bring to the boil.

PROFESSIONAL TIP

Flour may be omitted from the recipe if a thinner soup is required.

A slightly sweet/sharp flavour can be added to the soup by preparing what is known as a gastric (*gastrique*). In a thick-bottomed pan, reduce 100 ml of malt vinegar and 35 g caster sugar until it is a light caramel colour. Mix this into the completed soup.

Some tomato purée can be stronger than others, so you may have to add a little more or less when making this soup.

Healthy Eating Tips
● Toast the croutons rather than frying them.
● Use the minimum amount of salt – there is plenty in the bacon.

VARIATION
● Without fresh tomatoes: substitute 150 g of tomato purée for the fresh tomatoes (375 g for 10 portions). When reboiling the soup (step 12), add 500 ml of milk or 125 ml of cream, yoghurt or fromage frais.
● Try adding the juice and lightly grated peel of 1–2 oranges or cooked rice or a chopped fresh herb, such as chives.

ACTIVITY

1 Prepare, cook and taste the recipe for tomato soup with and without a gastric. Discuss and assess the two versions.

2 Name and prepare a variation of your own.

3 Review the basic recipe for tomato soup and adjust to meet dietary requirements for a customer who requires a low fat vegetarian version.

14 Cream of spinach and celery soup

Energy	Calories	Fat	Saturated fat	Carbohydrates	Sugar	Protein	Fibre
886 kJ	214 kcal	12.7 g	4.7 g	15.1 g	11.3 g	10.1 g	6.1 g

Serve fried or toasted croutons separately.

	4 portions	10 portions
Shallots, peeled and chopped (small mirepoix)	2	5
Leeks, washed and chopped (small mirepoix)	1	2
Cloves of garlic, peeled and chopped	5	7
Corn or other vegetable oil	2 tbsp	5 tbsp
Celery sticks, washed and chopped (small mirepoix)	4	10
Flour	15 g	35 g
Fresh spinach, well washed	500 g	1.25 kg
Soya milk	600 ml	1.5 litres
Vegetable stock (see recipe 3)	600 ml	1.5 litres
Salt, to taste		

Cooking

1 Cook the shallots, leeks and garlic in the oil for a few minutes, without colour.

2 Add the celery and cook for another few minutes until starting to soften.

3 Add the flour and mix well, then throw in the spinach and mix around. Add the soya milk and vegetable stock slowly, ensuring there are no lumps.

4 Stir continuously, bring to a simmer, then switch off and remove from heat. Cover and leave for a few minutes.

5 Blend until smooth in a food processor. Check seasoning and serve.

15 Minestrone

Energy	Calories	Fat	Saturated fat	Carbohydrates	Sugar	Protein	Fibre
1115 kJ	22.9 kcal	22.9 g	5.8 g	11.9 g	4.2 g	3.8 g	4.1 g

* Using sunflower oil

	4 portions	10 portions
Mixed vegetables (onion, leek, celery, carrot, turnip, cabbage), peeled	300 g	750 g
Butter or oil	50 g	125 g
White stock or water	0.5 litres	1.5 litres
Bouquet garni	1	2
Salt, pepper		
Peas	25 g	60 g
French beans	25 g	60 g
Spaghetti	25 g	60 g
Potatoes, peeled	50 g	125 g
Tomato purée	1 tsp	3 tsp
Tomatoes, skinned, deseeded, diced	100 g	250 g
Fat bacon (optional)	50 g	125 g
Parsley (optional)		
Clove of garlic (optional)	1	2½

Mise en place

1 Cut the peeled and washed mixed vegetables into paysanne.

Cooking

1 Cook slowly without colour in the oil or butter in the pan with a lid on.

2 Add the stock, bouquet garni and seasoning; simmer for approximately 20 minutes.

3 Add the peas and the beans cut into diamonds and simmer for 10 minutes.

4 Add the spaghetti in 2-cm lengths, the potatoes cut into paysanne, the tomato purée and the tomatoes, and simmer gently until all the vegetables are cooked.

5 Meanwhile, finely chop the fat bacon, parsley and garlic, and form into a paste.

6 Mould the paste into pellets the size of a pea and drop into the boiling soup.

7 Remove the bouquet garni, correct the seasoning.

8 Serve grated Parmesan cheese and thin toasted flute (French loaf) slices separately.

Vegetables chopped into paysanne. Fry or sweat the vegetables.

Simmer the ingredients in the stock. Add the pellets of paste to the soup.

Healthy Eating Tips
- Use an unsaturated oil (sunflower or vegetable) to lightly oil the pan. Drain off any excess after the frying is complete and skim the fat from the finished dish.
- Season with the minimum amount of salt as the bacon and cheese are high in salt.

16 Paysanne soup (*potage*)

Energy	Calories	Fat	Saturated fat	Carbohydrates	Sugar	Protein	Fibre
772 kJ	186 kcal	14 g	7.8 g	10.8 g	5.7 g	4.8 g	3.4 g

	4 portions	10 portions
Chicken or vegetable stock	1 litre	2.5 litres
Diced streaky bacon (optional)	50 g	125 g
Celery sticks	50 g	125 g
Onion	100 g	250 g
Carrots	100 g	250 g
Turnips	50 g	125 g
Leeks	50 g	125 g
Cabbage (optional)	50 g	125 g
Potatoes	100 g	250 g
Butter, margarine or oil	50 g	125 g
Chopped parsley, chopped basil		

Mise en place

1 Cut the celery, onion, carrots, leeks, cabbage and potatoes into paysanne.

Cooking

1 In a suitable saucepan, add the butter, margarine or oil. Fry the streaky bacon until lightly cooked.
2 Add the onion, leeks and celery and sweat for 2–3 minutes.
3 Add the rest of the vegetables.
4 Add the stock. Bring to the boil. Simmer until all the vegetables are cooked.
5 The soup may be finished with 125 ml boiled milk (300 ml for 10 portions).
6 Sprinkle on the chopped parsley and basil.

SERVING SUGGESTION
Diced tomato concassé may also be added: 50 g for 4 portions, 125 g for 10.

17 Prawn bisque

Energy	Calories	Fat	Saturated fat	Carbohydrates	Sugar	Protein	Fibre
1578 kJ	381 kcal	31.2 g	13.0 g	8.2 g	4.0 g	12.9 g	0.8 g

	4 portions	10 portions
Oil	50 ml	125 ml
Butter	30 g	75 g
Unshelled prawns	250 g	625 g
Flour	20 g	50 g
Tomato purée	1 tbsp	2 tbsp
Shellfish nage (4 portions)	1 litre	2.5 litres
Fish stock	150 ml	375 ml
Whipping cream	120 ml	300 ml
Dry sherry	75 ml	180 ml
Paprika, pinch		
Seasoning		
Chives, chopped (optional)		

Cooking

1 Heat the oil and the butter. Add the prawns and cook for 3–4 minutes on a moderately high heat.

2 Sprinkle in the flour and cook for a further 2–3 minutes.

3 Add the tomato purée and cook for a further 2 minutes.

4 Meanwhile, bring the nage up to a simmer and, once the tomato purée has been cooked in, slowly add to the prawn mix, being mindful that you have formed a roux; stir in the fish stock to prevent lumping.

5 Once all the stock has been added, bring to the boil and simmer for 3–4 minutes.

6 Pass through a fine sieve, return the shells to the pan and pound to extract more flavour and more colour.

7 Pour over the fish stock, bring to the boil, then pass this back onto the already passed soup.

8 Bring to the boil, add the cream and sherry, correct the seasoning and served with chopped chives.

---

18 New England clam chowder

Energy	Calories	Fat	Saturated fat	Carbohydrates	Sugar	Protein	Fibre
1109 kJ	269 kcal	14.9 g	7.7 g	24.5 g	2.5 g	14.9 g	1.8 g

* Using bacon or salt pork

	4 portions	10 portions
Salt pork, cut into 0.5-cm dice	50 g	125 g
Onion, finely chopped	50 g	125 g
Cold water	300 ml	750 ml
Potatoes, cut into 0.5-cm dice	500 g	1.25 kg
Fresh trimmed clams, or tinned clams, and their juices	200 g	500 g
Cream	180 ml	450 ml
Thyme, crushed or chopped	1 g	2 g
Salt, white pepper		
Butter, softened	20 g	50 g
Paprika		

Cooking

1 Dry-fry the pork in a thick-bottomed saucepan for about 3 minutes, stirring constantly until a thin film of fat covers the bottom of the pan.

2 Stir in the chopped onion and cook gently until a light golden brown.

3 Add the water and potatoes, bring to the boil and simmer gently until the potatoes are cooked but not mushy.

4 Add the chopped clams and their juice, the cream and thyme, and heat until almost boiling. Season with salt and pepper.

5 Correct the seasoning, stir in the softened butter and serve, dusting each soup bowl with a little paprika.

> **SERVING SUGGESTION**
> The traditional accompaniment is salted cracker biscuits. An obvious variation would be to use scallops in place of clams.

19 Leek and potato soup (vichyssoise)

Energy	Calories	Fat	Saturated fat	Carbohydrates	Sugar	Protein	Fibre
1653 kJ	397 kcal	26.4 g	16.3 g	36.9 g	4.5 g	5.3 g	4.0 g

Cooking

1 Sweat the sliced onion and leek without colour in the butter. Cook until very tender.

2 Add the potatoes and bring quickly to the boil with the water or vegetable stock.

3 Liquidise in food processor and allow to cool.

4 Check seasoning. Note that seasoning needs to reflect the serving temperature.

5 Serve cold with whipped cream and chopped chives. Or, alternatively, finish soup with 125 mls single cream for 4 portions, 300 mls for 10 portions, or substitute the cream for créme fraiche or fromage frais or natural yogurt.

	4 portions	10 portions
Onions, finely sliced	160 g	400 g
Leeks, finely sliced	175 g	430 g
Butter	125 g	300 g
Potatoes, diced small	750 g	1.8 kg
Water or vegetable stock	1.8 litres	4.5 litres
Salt	15 g	30 g
Pepper to taste		
Garnish		
Whipped cream		
Chives, chopped		

20 Chilled tomato and cucumber soup (gazpacho)

Energy	Calories	Fat	Saturated fat	Carbohydrates	Sugar	Protein	Fibre
382 kJ	91 kcals	6.71 g	1 g	6.2 g	AQ ?	1.83 g	1.56 g

Preparing the dish

1 Mix all the ingredients together.
2 Season and add crushed chopped garlic to taste.
3 Stand in a cool place for an hour.
4 Correct the consistency with iced water and serve chilled.

VARIATION

- Instead of serving with all the ingredients finely chopped, the soup can be liquidised and garnished with chopped tomato, cucumber and pepper.
- The soup may also be finished with chopped herbs.
- A tray of garnishes may accompany the soup, e.g. chopped red and green pepper, chopped onion, tomato, cucumber and croutons.

This soup has many regional variations. It is served chilled and has a predominant flavour of cucumber, tomato and garlic.

	4 portions	10 portions
Tomato juice	500 ml	1.25 litres
Tomatoes, skinned, de-seeded and diced	100 g	250 g
Cucumber, peeled and diced	100 g	250 g
Green pepper, diced	50 g	125 g
Onion, chopped	50 g	125 g
Mayonnaise	1 tbsp	2–3 tbsp
Vinegar	1 tbsp	2–3 tbsp
Seasoning		
Clove garlic	1	2–3

21 Potato and watercress soup (*purée cressonnière*)

Energy	Calories	Fat	Saturated fat	Carbohydrates	Sugar	Protein	Fibre
583 kJ	139 kcal	5.5 g	3.3 g	21 g	3.8 g	2.7 g	3.4 g

Mise en place

1 Pick off 12 neat leaves of watercress and plunge into a small pan of boiling water for 1–2 seconds. Refresh under cold water immediately; these leaves are to garnish the finished soup.

Cooking

1 Melt the butter or heat the oil in a thick-bottomed pan.

2 Add the peeled and washed sliced onion and leek, cook for a few minutes without colour with the lid on.

3 Add the stock, the peeled, washed and sliced potatoes, the rest of the watercress, including the stalks, and the bouquet garni. Season.

4 Simmer for approximately 30 minutes. Remove the bouquet garni, skim off all fat.

5 Liquidise or pass the soup firmly through a sieve then pass through a medium conical strainer.

6 Return to a clean pan, reboil, correct the seasoning and consistency, and serve.

7 Garnish with watercress.

	4 portions	10 portions
Butter or oil	25 g	60 g
Onion, peeled and sliced	50 g	125 g
White of leek, sliced	50 g	125 g
White stock or water	1 litre	2.5 litres
Potatoes, peeled and sliced	400 g	1.5 kg
Watercress	small bunch	small bunch
Bouquet garni	1	2
Salt, pepper		
Parsley, chopped		
Croutons		
Stale bread	1 slice	3 slices
Butter, margarine or oil	50 g	125 g

22 Red lentil soup

Energy	Calories	Fat	Saturated fat	Carbohydrates	Sugar	Protein	Fibre
1807 kJ	432 kcal	71.4 g	43.9 g	6.7 g	1.5 g	5.3 g	1.5 g

	4 portions	10 portions
Ham hock	320 g	800 g
Onion, peeled	25 g	60 g
Whole carrot, peeled	½	1
Baby shallots	1	3
Leeks	50 g	125 g
Celery	50 g	125 g
Oil	40 ml	100 ml
Red lentils	200 g	500 g
Cooking liquid from the hock	1 litre	2.5 litres
Milk, cream or crème fraiche	120 ml	300 ml

Cooking

1 Place the ham hock, onion and carrot in a pan and cover with about 3 litres of water.

2 Bring to the boil and then turn down to a slow simmer. (When the hock is cooked, the centre bone will slide out in one smooth motion.)

3 Slice the shallots, leek and celery into 1-cm dice.

4 Heat a pan with the oil, add the vegetables and cook until they are slightly coloured; add the lentils and cover them with the ham stock.

5 Bring to the boil, then turn the heat down to a very slow simmer.

6 Cook until all the lentils have broken down.

7 Allow to cool for 10 minutes and then purée until smooth.

8 Correct the consistency as necessary, and finish with boiled milk, cream or crème fraiche.

> **VARIATION**
>
> **Meat-free version**: omit the ham hock and use a vegetable stock or water instead of the cooking liquid.

23 Roasted red pepper and tomato soup

Energy	Calories	Fat	Saturated fat	Carbohydrates	Sugar	Protein	Fibre
983 kJ	235 kcal	16.8 g	7.1 g	18.3 g	16.4 g	3.6 g	4.5 g

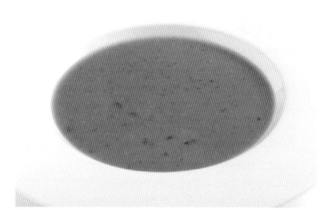

	4 portions	10 portions
Red peppers	4	10
Plum tomatoes	400 g	1.25 kg
Butter or oil	50 g	125 g
Onion, chopped	100 g	250 g
Carrot, chopped	100 g	250 g
Stock	500 ml	1.5 litres
Crème fraiche	2 tbsp	5 tbsp
Basil	25 g	75 g
Croutons		
Slice stale bread	1	3
Butter	50 g	125 g

Mise en place

1 Core and deseed the peppers and halve the tomatoes.

Cooking

1 Lightly sprinkle with oil and place on a tray into a hot oven or under a grill until the pepper skins are blackened.

2 Allow the peppers to cool in a plastic bag.

3 Remove the skins and slice the flesh.

4 Place the butter or oil in a pan, add the onions and carrots and fry gently for 5 minutes.

5 Add the stock, peppers and tomatoes and bring to the boil.

6 Simmer for 30 minutes, correct the seasoning and blend in a food processor until smooth.

7 Add crème fraiche and torn basil leaves and serve with croutons. In place of créme fraiche you may use single cream.

24 Vegetable purée soup

Energy	Calories	Fat	Saturated fat	Carbohydrates	Sugar	Protein	Fibre	*
601 kJ	143 kcal	10.3 g	4.4 g	11.4 g	1.8 g	1.9 g	1.9 g	

* Using hard margarine

	4 portions	10 portions
Onions, leek and celery, sliced	100 g	250 g
Other suitable vegetables**, sliced	200 g	500 g
Butter or oil	50 g	125 g
Flour	50 g	125 g
White stock or water	1 litre	2.5 litres
Bouquet garni	1	2
Salt, pepper		

** Suitable vegetables include Jerusalem artichokes, cauliflower, celery, leeks, onions, parsnips, turnips and fennel.

Cooking

1 Gently cook all the sliced vegetables in the fat under a lid, without colour.

2 Mix in the flour and cook slowly for a few minutes without colour. Cool slightly.

3 Gradually mix in the hot stock. Stir to the boil.

4 Add the bouquet garni and season.

5 Simmer for approximately 45 minutes; skim when necessary.

6 Remove the bouquet garni; liquidise or pass firmly through a sieve and then through a medium strainer.

7 Return to a clean pan, reboil, and correct the seasoning and consistency.

VARIATION
- Add a little spice, sufficient to give a subtle background flavour, e.g. garam masala with parsnip soup.
- Just before serving add a little freshly chopped herb(s), e.g. parsley, chervil, tarragon or coriander.

Healthy Eating Tips
- Try using more vegetables to thicken the soup in place of the flour.

Video: Making a purée soup

http://bit.ly/2oPu2rC

25 Vegetable and barley soup

Energy	Calories	Fat	Saturated fat	Carbohydrates	Sugar	Protein	Fibre
905 kJ	217 kcal	13.3 g	1.9 g	21.7 g	7.9 g	4.1 g	2.5 g

	4 portions	10 portions
Corn or other vegetable oil	50 ml	125 ml
Onions, finely diced	1	2
Leeks, cut into rounds	1	2
Carrots, peeled and roughly chopped	2	3
Celery sticks, cut into ½ cm dice	2	3
Cloves of garlic, crushed	2	4
Large potatoes, peeled and cut into 0.5-cm dice	3	7
Pearl barley (cooked)	150 g	375 g
Vegetable stock	1.5 litres	3.75 litres
Head of Swiss chard, washed and shredded (including stalks)	1	2
Seasoning		

Cooking

1 Heat the oil in a pan large enough to hold all the ingredients.

2 Place the onions, leeks, carrots and celery in the oil and cook until slightly golden.

3 Add the garlic and cook for a further 2 minutes. Add the potato and cook for a further 2 minutes.

4 Add the barley and vegetable stock. Bring to the boil, then simmer for 15 minutes, until the potatoes are just soft.

5 Stir in the Swiss chard and cook for a further 2 minutes.

6 Check the seasoning, correct if necessary then serve.

26 Scotch broth

Energy	Calories	Fat	Saturated fat	Carbohydrates	Sugar	Protein	Fibre
655 kJ	157 kcal	7.8 g	3.1 g	4.9 g	2.7 g	16.6 g	1.6 g

	4 portions	10 portions
Lean beef	200 g	500 g
Beef stock	1 litre	2.5 litres
Barley	25 g	60 g
Vegetables (carrot, turnip, leek, celery, onion), cut into paysanne	200 g	500 g
Bouquet garni	1	2
Salt, pepper		
Chopped parsley		

Cooking

1 Place the beef, free from fat, in a saucepan and cover with cold water.

2 Bring to the boil, then immediately wash off under running water.

3 Clean the pan, replace the meat, cover with cold water, bring to the boil and skim.

4 Add the washed barley, simmer for 1 hour.

5 Add the vegetables, bouquet garni and seasoning.

6 Skim when necessary; simmer for approximately 30 minutes, until tender.

7 Remove the meat, allow to cool and cut from the bone, remove all fat and cut the meat into neat dice the same size as the vegetables; return to the broth.

8 Correct the seasoning, skim off all the fat, add the chopped parsley and serve.

Healthy Eating Tips
- Remove all fat from the meat.
- Use only a small amount of salt.
- There are lots of healthy vegetables in this dish and the addition of a large bread roll will increase the starchy carbohydrate.

27 Clear soup (consommé)

Energy	Calories	Fat	Saturated fat	Carbohydrates	Sugar	Protein	Fibre
126 kJ	30 kcal	0.0 g	0.0 g	1.8 g	0.0 g	5.6 g	0.0 g

	4 portions	10 portions
Chopped or minced beef (shin)	200 g	500 g
Salt, to taste		
Egg whites	1–2	3–5
Cold white or brown beef stock	1 litre	2.5 litres
Mixed vegetables (onion, carrot, celery, leek)	100 g	250 g
Bouquet garni	1	2
Peppercorns	3–4	8–10

Mix one quarter of the stock with the beef, salt and egg whitel.

Once the remaining stock and the vegetables have been added, bring slowly to the boil.

After the soup has simmered for at least 1½ hours, strain it carefully.

Mise en place

1 Thoroughly mix the beef, salt, egg white and a quarter of the cold stock in a thick-bottomed pan. Peel, wash and finely chop the vegetables.

Cooking

1 Add to vegetables to the beef with the remainder of the stock, the bouquet garni and the peppercorns.

2 Place over a gentle heat and bring slowly to the boil, stirring occasionally.

3 Allow to boil rapidly for 5–10 seconds. Give a final stir.

4 Lower the heat so that the consommé is simmering very gently.

5 Cook for 1½–2 hours without stirring.

6 Strain carefully through a double muslin.

7 Remove all fat, using both sides of 8-cm square pieces of kitchen paper.

8 Correct the seasoning and colour, which should be a delicate amber.

9 Degrease again, if necessary. Bring to the boil and serve.

PROFESSIONAL TIP

A consommé should be crystal clear. The clarification process is caused by the albumen of the egg white and meat coagulating, rising to the top of the liquid and carrying other solid ingredients. The remaining liquid beneath the coagulated surface should be gently simmering. Cloudiness is due to some or all of the following:
- poor quality stock
- greasy stock
- unstrained stock
- imperfect coagulation of the clearing agent
- whisking after boiling point is reached, whereby the impurities mix with the liquid
- not allowing the soup to settle before straining
- lack of cleanliness of the pan or cloth
- any trace of grease or starch.

Healthy Eating Tips
- This soup is fat free!
- Keep the salt to a minimum and serve as a low-calorie starter for anyone wishing to reduce the fat in their diet.

VARIATION
- Consommés are varied in many ways by altering the flavour of the stock (chicken, chicken and beef, game, etc.), also by the addition of numerous garnishes (julienne or brunoise of vegetables, shredded savoury pancakes or pea-sized profiteroles) added at the last moment before serving, or small pasta.
- Cold, lightly jellied consommés, served in cups, with or without garnish (diced tomato), may be served in hot weather.

28 Mulligatawny

Energy	Calories	Fat	Saturated fat	Carbohydrates	Sugar	Protein	Fibre
952 kJ	227 kcals	17.1 g	5.0 g	16.3 g	4.1 g	3.3 g	2.6 g

* Using sunflower oil

	4 portions	10 portions
Chopped onions	100 g	250 g
Clove of garlic, chopped	1	3
Butter or oil	50 g	125 g
Flour, white or wholemeal	50 g	125 g
Curry powder or garam masala	10 g	25 g
Tomato puree	10 g	25 g
Brown stock, meat or vegetable	1 litre	2.5 litres
Chopped apple	25 g	60 g
Finely chopped fresh ginger	6 g	15 g
Chopped chutney	10 g	25 g
Desiccated or freshly chopped coconut	25 g	60 g
Salt to taste		
Cooked rice, white or wholegrain	10 g	25 g
Cream, yoghurt or fromage frais to finish	60 mls	180 mls

Cooking

1. Lightly brown the onion and the garlic in the fat or oil.
2. Mix in the flour and the curry powder, cook out for a few minutes, browning slightly.
3. Mix in the tomato puree. Cool slightly.
4. Gradually add the brown stock, stirring continuously, and bring to the boil.
5. Add the remainder of the ingredients and season with salt.
6. Simmer for 30–45 minutes.
7. Liquidise and pass through a strainer, or blitz.
8. Return to a clean pan, reboil.
9. Correct the seasoning and consistency.
10. Add the cooked rice, serve in individual soup cups with a thread of cream, yoghurt or fromage frais.

SERVING SUGGESTION

A small naan bread or popadums may be served with the soup.

The soup may also be garnished with a dice of chicken sautéed in a little oil and curry powder.

29 Lamb jus

Energy	Kcal	Fat	Saturated fat	Carbohydrates	Sugar	Protein	Fibre	Sodium
960 kJ	227 kcal	0.6 g	0.1 g	27.1 g	26.6 g	1.0 g	10.3 g	102.0 g

Cooking

1 Pre-heat the oven to 175 °C. Place the herbs, garlic and wine in a large, deep container. Place all the bones on to a roasting rack on top of the container of herbs and wine and roast in the oven for 50–60 minutes. When the bones are completely roasted and have taken on a dark golden-brown appearance, remove from oven.

2 Place all the ingredients in a large pot and cover with cold water. Put the pot onto the heat and bring to the simmer; immediately skim all fat that rises to the surface.

3 Turn the heat off and allow the bones and vegetables to sink. Once this has happened, turn the heat back on and bring to just under a simmer, making as little movement as possible to create more of an infusion than a stock.

4 Skim continuously. Leave to infuse for 12 hours, then pass through a fine sieve, place in the blast chiller until cold and then in the refrigerator overnight. Next day, reduce down rapidly, until you have about 2 litres remaining.

	Makes 2 litres
Thyme	bunch
Bay leaves, fresh	4
Garlic	2 bulbs
Red wine	1 litre
Lamb bones	2 kg
Veal bones	1 kg
White onions, peeled	6
Large carrots, peeled	8
Celery sticks	7
Leeks, chopped	4
Tomato purée	6 tbsp

30 Beef jus

Energy	Calories	Fat	Saturated fat	Carbohydrates	Sugar	Protein	Fibre	Sodium
4523 kJ	1088 kcal	61.0 g	31.6 g	40.0 g	22.2 g	4.3 g	2.2 g	2,639.0 g

	Makes 1 litre
Mushrooms, finely sliced	750 g
Butter	100 g
Shallots, finely sliced	350 g
Beef trim diced	350 g
Sherry vinegar	100 ml
Red wine	700 ml
Chicken stock	500 ml
Beef stock	1 litre

Cooking

1 Caramelise the mushrooms in foaming butter, strain, then put aside in pan.

2 Caramelise the shallots in foaming butter, strain, then put aside in pan.

3 In another pan, caramelise the beef trim until golden brown.

4 Place the mushrooms, shallots and beef trim in one of the pans. Deglaze the other two pans with the vinegar, then add to the pan with the beef, shallots and mushrooms in it.

5 In a separate pan, reduce the wine by half and add to the main pan.

6 Add the stock, then reduce to sauce consistency.

7 Pass through a sieve, then chill and store until needed.

31 Chicken jus

Energy	Calories	Fat	Saturated fat	Carbohydrates	Sugar	Protein	Fibre	Sodium
4284 kJ	1034 kcal	88.0 g	18.8 g	22.5 g	22.1 g	1.1 g	8.3 g	1,220.0 g

	Makes 1 litre
Chicken stock	600 ml
Lamb jus	600 ml
Chicken wings, chopped small	300 g
Vegetable oil	60 ml
Shallots, sliced	100 g
Butter	50 g
Tomatoes, chopped	200 g
White wine vinegar	40 ml
Red wine vinegar	75 ml
Tarragon (chopped)	1 tsp
Chervil (chopped)	1 tsp

Cooking

1 Put the jus and stock in a pan and reduce to 1 litre.

2 Roast the chicken wings in oil until slightly golden.

3 Add the shallots and butter, and cook until lightly browned (do not allow the butter to burn).

4 Strain off the butter and return the bones to the pan; deglaze with the vinegar and add tomatoes.

5 Ensure the bottom of the pan is clean. Add the reduce stock/jus and simmer for 15 minutes.

6 Pass through a sieve, then reduce to sauce consistency.

7 Remove from the heat and infuse with the aromats for 5 minutes.

8 Pass through a chinois and then muslin cloth.

32 Red wine jus

Energy	Calories	Fat	Saturated fat	Carbohydrates	Sugar	Protein	Fibre	Sodium
1530 kJ	368 kcal	23.6 g	13.8 g	10.2 g	9.9 g	1.1 g	3.6 g	771.0 g

* Using lamb jus

	Makes 1 litre
Shallots, sliced	150 g
Butter	50 g
Garlic, halved	10 g
Red wine vinegar	100 ml
Red wine	250 ml
Chicken stock	350 ml
Lamb or beef jus	250 ml
Bay leaves	2
Sprig of thyme	1

Cooking

1 Caramelise the shallots in foaming butter until golden, adding the garlic at the end.

2 Strain through a colander and then put back into the pan and deglaze with the vinegar.

3 Reduce the red wine by half along with the stock and jus, at the same time as colouring the shallots.

4 When everything is done, combine and simmer for 20 minutes.

5 Pass through a sieve and reduce to sauce consistency.

6 Infuse the aromats for 5 minutes.

7 Pass through muslin cloth and store until needed.

33 Roast gravy (*jus rôti*)

Energy	Calories	Fat	Saturated fat	Carbohydrates	Sugar	Protein	Fibre	Sodium
309 kJ	74 kcal	3.9 g	1.5 g	3.3 g	2.7 g	6.2 g	1.3 g	0.1 g

	Serves 4	Serves 10
Raw veal bones or beef and veal trimmings	200 g	500 g
Stock or water	500 ml	1.25 litres
Onions, chopped	50 g	125 g
Celery, chopped	25 g	60 g
Carrots, chopped	50 g	125 g

Cooking

1 Chop the bones and brown in the oven, or brown in a little oil on top of the stove in a frying pan. Drain off all the fat.

2 Place the bones in a saucepan with the stock or water.

3 Bring to the boil, skim and allow to simmer.

4 Add the lightly browned vegetables, which may be fried in a little fat in a frying pan or added to the bones when partly browned.

5 Simmer for 1½–2 hours.

6 Remove the joint from the roasting tin when cooked.

7 Return the tray to a low heat to allow the sediment to settle.

8 Carefully strain off the fat, leaving the sediment in the tin.

9 Return the joint to the stove and brown carefully; deglaze with the brown stock.

10 Allow to simmer for a few minutes.

11 Correct the colour and seasoning. Strain and skim off all fat.

PROFESSIONAL TIP

For preference, use beef bones for roast beef gravy and the appropriate bones for lamb, veal, mutton and pork.

Healthy Eating Tips
- Use an unsaturated oil (sunflower or vegetable). Lightly oil the pan.
- Season with the minimum amount of salt.

34 Thickened gravy (*jus lié*)

Energy	Calories	Fat	Saturated fat	Carbohydrates	Sugar	Protein	Fibre	Sodium
247 kJ	59 kcal	2.6 g	1.0 g	4.7 g	1.4 g	4.1 g	0.7 g	0.1 g

Cooking

1 Start with roast gravy (recipe 33) or reduced veal stock (recipe 7). Add a little tomato purée, a few mushroom trimmings and a pinch of thyme and simmer for 10–15 minutes.

2 Stir some arrowroot diluted in cold water into the simmering gravy.

3 Reboil, simmer for 5–10 minutes and pass through a strainer.

VARIATION
- Add a little rosemary, thyme or lavender.

35 Béchamel sauce (white sauce)

Energy	Calories	Fat	Saturated fat	Carbohydrates	Sugar	Protein	Fibre	Sodium
6510 kJ	1556 kcal	101.0 g	63.0 g	119.0 g	46.0 g	0.0 g	4.0 g	1,162.0 g

* Using salted butter and semi-skimmed milk

	1 litre	4.5 litres
Butter or oil	100 g	400 g
Plain white flour	100 g	400 g
Milk, warmed	1 litre	4.5 litres
Onion, studded with cloves	1	2–3

Cooking

1 Melt the butter or heat the oil in a thick-bottomed pan.

2 Mix in the flour with a heat-proof plastic spoon.

3 Cook for a few minutes, stirring frequently. As you are making a white roux, do not allow the mixture to colour.

4 Remove the pan from the heat and allow the roux to cool.

5 Return the pan to the stove and, over a low heat, gradually mix the milk into the roux.

6 Add the studded onion.

7 Allow the mixture to simmer gently for 30 minutes, stirring frequently to make sure the sauce does not burn on the bottom.

8 Remove the onion and pass the sauce through a conical strainer.

PROFESSIONAL TIP
To prevent a skin from forming, brush the surface of the sauce with melted butter. When ready to use, stir this into the sauce. Alternatively, cover the sauce with greaseproof paper or cling film.

36 Mornay sauce

Energy	Calories	Fat	Saturated fat	Carbohydrates	Sugar	Protein	Fibre	Sodium
2685 kJ	639 kcal	31.9 g	17.9 g	52.0 g	22.7 g	0.0 g	1.6 g	569.0 g

* Using semi-skimmed milk and regular wheat flour

Mise en place

1 The milk may be first infused with a studded onion clouté, carrot and a bouquet garni. Allow to cool.

Cooking

1 Place the milk in a suitable saucepan, gradually whisk in the sauce flour. Bring slowly to the boil until the sauce has thickened.

2 Mix in the cheese and egg yolk when the sauce is boiling.

3 Remove from the heat. Strain if necessary.

4 Do not allow the sauce to reboil at any time.

	500 ml
Milk	500 ml
Grated cheese	50 g
Egg yolk	1
Sauce flour	40 g

PROFESSIONAL TIP
Add a little cornflour prior to heating to stabilise the sauce.

37 Soubise sauce

Energy	Calories	Fat	Saturated fat	Carbohydrates	Sugar	Protein	Fibre	Sodium
1748 kJ	411 kcal	9.2 g	5.6 g	58.0 g	28.7 g	0.0 g	4.4 g	2,183.0 g

* Using semi-skimmed milk and regular wheat flour

Cooking

1 Cook the onions without colouring them, either by boiling or sweating in butter.

2 The milk may be first infused with a studded onion clouté, carrot and a bouquet garni. Allow to cool.

3 Place the milk in a suitable saucepan, gradually whisk in the sauce flour. Bring slowly to the boil until the sauce has thickened.

4 Add the onions, season, simmer for approximately 5–10 minutes.

5 Blitz well. Pass through a strainer.

	500 ml
Onion, chopped or diced	100 g
Milk	500 ml
Sauce flour	40 g
Seasoning	

38 Parsley sauce

Energy	Calories	Fat	Saturated fat	Carbohydrates	Sugar	Protein	Fibre	Sodium
1609 kJ	379 kcal	9.2 g	5.6 g	52.0 g	22.8 g	0.0 g	2.6 g	2,183.0 g

* Using semi-skimmed milk and regular wheat flour

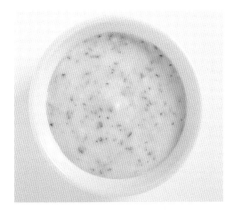

Mise en place

1 The milk may be first infused with a studded onion clouté, carrot and a bouquet garni. Allow to cool.

Cooking

1 Place the milk in a suitable saucepan, gradually whisk in the sauce flour. Bring slowly to the boil until the sauce has thickened.

2 Season, simmer for approximately 5–10 minutes. Use as required.

	500 ml
Milk	500 ml
Sauce flour	40 g
Parsley, chopped	1 tbsp
Seasoning	

39 Velouté (chicken, veal, mutton)*

Energy	Calories	Fat	Saturated fat	Carbohydrates	Sugar	Protein	Fibre	*
4594 kJ	1094 kcal	82.6 g	35.4 g	79.0 g	1.6 g	13.3 g	3.6 g	

* Using hard margarine, for 1 litre. Using sunflower oil instead, this recipe provides, for 1 litre: 5304 kJ/1263 kcal Energy; 101.5 g Fat; 13.3 g Sat Fat; 78.9 g Carb; 1.5 g Sugar; 13.2 g Protein; 3.6 g Fibre

3 Cook out to a sandy texture over gentle heat without colouring.

4 Allow the roux to cool.

5 Gradually add the boiling stock.

6 Stir until smooth and boiling.

7 Allow to simmer for approximately 1 hour.

8 Pass it through a fine conical strainer.

NOTE

This is a basic white sauce made from white stock and a blond roux.

A velouté sauce for chicken or veal dishes is usually finished with cream and, in some cases, egg yolks. The finished sauce should be of a light consistency, barely coating the back of a spoon.

Alternatively, these sauces can be made with a béchamel sauce made from a white roux.

Healthy Eating Tips

● Make sure all the fat has been skimmed from the stock before adding it to the roux.

	4 portions	10 portions
Butter or oil	100 g	400 g
Flour	100 g	400 g
Stock (chicken, veal, mutton) as required	1 litre	4.5 litres

Cooking

1 Melt the butter or heat the oil in a thick-bottomed pan.

2 Add the flour and mix in.

40 Aurore sauce

Energy	Calories	Fat	Saturated fat	Carbohydrates	Sugar	Protein	Fibre	Sodium
478 kJ	115 kcal	8.6 g	4.7 g	5.8 g	2.0 g	0.1 g	0.9 g	376.0 g

	4 portions	10 portions
Chicken or veal velouté	250 ml	625 ml
Mushroom trimmings, washed and dried	50 g	125 g
Fresh cream, double or single	50 ml	125 ml
Egg yolks	1	3
Lemon juice	½ lemon	1½ lemon
Tomato puree	1 tbsp	1½ tbsp
Seasoning		

Cooking

1 Place the velouté in a suitable saucepan and bring to the boil.
2 Add the mushroom trimmings. Allow to simmer for 10 minutes.
3 Add the tomato purée; mix well.
4 Strain through a fine chinois.
5 In a basin mix the egg yolks and cream.
6 Add a little of the hot sauce to this mix; whisk well.
7 Return to the main sauce and stir. Do NOT allow to re-boil.
8 Finish with lemon juice.

SERVING SUGGESTION
Serve with poached chicken or poached eggs. If serving with poached or steamed fish use fish velouté.

41 Ivory sauce

Energy	Calories	Fat	Saturated fat	Carbohydrates	Sugar	Protein	Fibre	Sodium
351 kJ	85 kcal	6.8 g	4.3 g	4.9 g	0.8 g	1.3 g	0.6 g	0.0 g

	4 portions	10 portions
Chicken or veal velouté	250 ml	625 ml
Mushroom trimmings, washed and dried	50 g	125 g
Cream, single or double	50 ml	125 ml
Lemon juice	½ lemon	1½ lemons
Chicken or beef meat glaze	1 tbsp	1½ tbsp
Seasoning		

Cooking

1 Place the velouté into a suitable saucepan and bring to the boil.
2 Add the mushroom trimmings; simmer for 10 minutes.
3 Pass the sauce through a fine chinois.
4 Add the meat glaze; stir well to achieve an ivory colour.
5 Add the lemon juice and finish with cream. Correct the seasoning.

SERVING SUGGESTION
Serve with grilled veal or pork chops, grilled or poached chicken.

42 Mushroom sauce

Energy	Calories	Fat	Saturated fat	Carbohydrates	Sugar	Protein	Fibre	Sodium
6923 kJ	1664 kcal	126.0 g	72.0 g	85.0 g	4.3 g	0.3 g	4.9 g	2,349.0 g

* Using chicken stock, salted butter and single cream

	1 litre	4.5 litres
Butter or oil	100 g	400 g
Flour	110 g	400 g
Stock (chicken, veal, fish, mutton) as required	1 litre	4.5 litres
Mushroom trimmings	50 g	225 g
White button mushrooms, well-washed, sliced, sweated	200 g	900 g
Cream	120 ml	540 ml
Egg yolk	2	9
Lemon, juice of	½	1

Cooking

1 Melt the butter or heat the oil in a thick-bottomed pan.
2 Add the flour and mix in.
3 Cook out to a sandy texture over gentle heat without colouring.
4 Allow the roux to cool.
5 Gradually add the boiling stock.
6 Stir until smooth and boiling.
7 Add the mushrooms and trimmings. Allow to simmer for approximately 1 hour.
8 Pass it through a fine conical strainer.
9 Simmer for 10 minutes, then add the egg yolk, cream and lemon juice.

43 Suprême sauce

Energy	Calories	Fat	Saturated fat	Carbohydrates	Sugar	Protein	Fibre	Sodium
1339 kJ	322 kcal	25.2 g	15.3 g	21.1 g	1.8 g	4.1 g	1.6 g	0.2 g

	4 portions	10 portions
Margarine, butter or oil	100 g	400 g
Flour	100 g	400 g
Stock (chicken, veal, fish or mutton)	1 litre	4.5 litres
Mushroom trimmings	25 g	60 g
Egg yolk	1	2
Cream	60 ml	150 ml
Lemon juice	2–3 drops	5–6 drops

2 Add the flour and mix in. Cook to a sandy texture over a gentle heat, without colouring.
3 Allow the roux to cool, then gradually add the boiling stock. Stir until smooth and boiling.
4 Add the mushroom trimmings. Simmer for approximately 1 hour.
5 Pass through a final conical strainer.
6 Finish with a liaison of the egg yolk, cream and lemon juice. Serve immediately.

SERVING SUGGESTION
This sauce can be served hot with boiled chicken, vol-au-vents, etc. The traditional stock base is a good chicken stock.

PROFESSIONAL TIP
Once the liaison has been added, do not re-boil this sauce

Cooking

1 Melt the fat or oil in a thick-bottomed pan.

44 Fish velouté

Energy	Calories	Fat	Saturated fat	Carbohydrates	Sugar	Protein	Fibre
4805 kJ	1144 kcal	90.4 g	39.0 g	77.8 g	1.6 g	9.5 g	3.6 g

	1 litre	2.5 litres
Butter	100 g	250 g
Flour	100 g	250 g
Fish stock	1 litre	2.5 litres

Cooking

1 Prepare a blond roux using the butter and flour.

2 Gradually add the stock, stirring continuously until boiling point is reached.

3 Simmer for approximately 1 hour.

4 Pass through a fine conical strainer.

NOTE
This will give a thick sauce that can be thinned down with the cooking liquor from the fish for which the sauce is intended.

45 Chasseur sauce (*sauce chasseur*)

Energy	Calories	Fat	Saturated fat	Carbohydrates	Sugar	Protein	Fibre
227 kJ	55 kcal	5.3 g	2.5 g	1.4 g	1.2 g	0.5 g	0.5 g

	4 portions	10 portions
Butter or oil	25 g	60 g
Shallots, chopped	10 g	25 g
Garlic clove, chopped (optional)	1	1
Button mushrooms, sliced	50 g	125 g
White wine, dry	60 ml	150 ml
Tomatoes, skinned, deseeded, diced	100 g	250 g
Demi-glace, jus-lié or reduced stock	250 ml	625 ml
Parsley and tarragon, chopped		

3 Add the garlic and the mushrooms, cover and cook gently for 2–3 minutes.

4 Strain off the fat.

5 Add the wine and reduce by half. Add the tomatoes.

6 Add the demi-glace; simmer for 5–10 minutes.

7 Correct the seasoning. Add the tarragon and parsley.

SERVING SUGGESTION
May be served with fried steaks, chops, chicken, etc.

Cooking

1 Melt the butter or heat the oil in a small sauteuse.

2 Add the shallots and cook gently for 2–3 minutes without colour.

46 Piquant sauce (*sauce piquante*)

Energy	Calories	Fat	Saturated fat	Carbohydrates	Sugar	Protein	Fibre	Sodium
151 kJ	36 kcal	1.3 g	0.5 g	3.0 g	1.3 g	2.4 g	0.7 g	0.0 g

	4 portions	10 portions
Vinegar	60 ml	150 ml
Shallots, chopped	50 g	125 g
Demi-glace, jus lié or reduced stock	250 ml	625 ml
Gherkins, chopped	25 g	60 g
Capers, chopped	10 g	25 g
Chervil, tarragon and parsley, chopped	½ tbsp	1½ tbsp

Cooking

1 Place the vinegar and shallots in a small sauteuse and reduce by half.

2 Add the demi-glace; simmer for 15–20 minutes.

3 Add the rest of the ingredients.

4 Skim and correct the seasoning.

SERVING SUGGESTION
May be served with made up dishes, sausages and grilled meats.

47 Robert sauce (*sauce Robert*)

Energy	Calories	Fat	Saturated fat	Carbohydrates	Sugar	Protein	Fibre	Sodium
401 kJ	96 kcal	5.7 g	3.1 g	8.3 g	6.5 g	2.5 g	0.4 g	0.2 g

	4 portions	10 portions
Oil or butter	20 g	50 g
Onions, finely chipped	10 g	25 g
Vinegar	60 ml	150 g
Demi-glace, jus lié or reduced stock	250 ml	625 ml
English or continental mustard	1 level tbsp	2½ level tbsp
Caster sugar	1 level tbsp	2½ level tbsp

Cooking

1 Melt the fat or oil in a small sauteuse.

2 Add the onions. Cook gently without colour.

3 Add the vinegar and reduce completely.

4 Add the demi-glace jus lié reduced stock; simmer for 5–10 minutes.

5 Remove from the heat and add the mustard, diluted with a little water and the sugar; do not boil.

6 Skim and correct the seasoning.

SERVING SUGGESTION
May be served with fried sausages and burgers or grilled pork chops.

48 Madeira sauce (*sauce Madère*)

Energy	Calories	Fat	Saturated fat	Carbohydrates	Sugar	Protein	Fibre	Sodium
365 kJ	88 kcal	6.4 g	3.7 g	3.5 g	1.4 g	2.1 g	0.3 g	0.1 g

	4 portions	10 portions
Demi-glace, jus lié or reduced stock	250 ml	625 ml
Madeira wine	2 tbsp	5 tbsp
Butter	25 g	60 g

Cooking

1 Boil the demi-glace in a small sauteuse.

2 Add the Madeira and re-boil.

3 Correct the seasoning.

4 Pass through a fine conical strainer.

5 Gradually mix in the butter.

> **SERVING SUGGESTION**
> May be served with braised ox tongue, ham or veal steak.

> **VARIATION**
> Dry sherry or port wine may be substituted for Madeira and the sauce renamed accordingly.

49 Brown onion sauce (*sauce Lyonnaise*)

Energy	Calories	Fat	Saturated fat	Carbohydrates	Sugar	Protein	Fibre	Sodium
374 kJ	90 kcal	6.8 g	3.9 g	3.5 g	1.8 g	3.7 g	1.0 g	0.1 g

	4 portions	10 portions
Margarine, oil or butter	25 g	60 g
Onions, sliced	100 g	250 g
Vinegar	2 tbsp	5 tbsp
Demi-glace, jus lié or reduced stock	250 ml	625 ml

Cooking

1 Melt the fat or oil in a sauteuse.

2 Add the onions; cover with a lid and cook gently until tender and golden in colour.

3 Remove the lid and colour lightly.

4 Add the vinegar and completely reduce.

5 Add the demi-glace; simmer for 5–10 minutes.

6 Skim and correct the seasoning.

> **SERVING SUGGESTION**
> May be served with burgers, fried liver or sausages.

50 Italian sauce (*sauce Italienne*)

Energy	Calories	Fat	Saturated fat	Carbohydrates	Sugar	Protein	Fibre	Sodium
359 kJ	86 kcal	6.5 g	3.7 g	4.4 g	2.2 g	2.4 g	0.8 g	0.1 g

	4 portions	10 portions
Margarine, oil or butter	25 g	60 g
Shallots, chopped	10 g	25 g
Mushrooms, chopped	50 g	125 g
Demi-glaze, jus-lié or reduced stock	250 ml	625 ml
Lean ham, chopped	25 g	60 g
Tomatoes, skinned, de-seeded	100 g	250 g
Diced parsley, chervil and tarragon, chopped		

3 Add the mushrooms and cook gently for a further 2–3 minutes.

4 Add the demi-glace, ham and tomatoes. Simmer for 5–10 minutes.

5 Correct the seasoning; add the chopped herbs.

Healthy Eating Tips
- Trim as much fat as possible from the ham.
- The ham is salty, so do not add more salt; flavour will come from the herbs.

Cooking

1 Melt the fat or oil in a small sauteuse.

2 To make a duxelle, add the shallots and cook gently for 2–3 minutes.

51 Pepper sauce (*sauce poivrade*)

Energy	Calories	Fat	Saturated fat	Carbohydrates	Sugar	Protein	Fibre	Sodium
195 kJ	47 kcal	4.6 g	1.6 g	0.5 g	0.5 g	0.0 g	0.5 g	46.0 g

* Using margarine

Cooking

1 Melt the fat or oil in a small sauteuse.

2 Add the vegetables and herbs (mirepoix) and allow to brown.

3 Add the wine, vinegar and pepper. Reduce by half.

4 Add the demi-glace and simmer for 20–30 minutes.

5 Correct the seasoning.

6 Pass through a fine conical strainer.

SERVING SUGGESTION
Usually served with joints or cuts of venison.

	4 portions	10 portions
Margarine, oil or butter	25 g	60 g
Onions	50 g	125 g
Carrots	50 g	125 g
Celery	50 g	125 g
Bay leaf	1	1
Sprig of thyme		
Vinegar	2 tbsp	5 tbsp
Mignonette pepper	5 g	12 g
Demi-glace, jus lié or reduced stock	250 ml	625 ml

52 Apple sauce

Energy	Calories	Fat	Saturated fat	Carbohydrates	Sugar	Protein	Fibre
276 kJ	65 kcal	2.6 g	1.6 g	11.0 g	11.0 g	0.2 g	1.1 g

	8 portions
Cooking apples	400 g
Sugar	50 g
Butter or margarine	25 g

Cooking

1 Peel, core and wash the apples.

2 Place with other ingredients in a covered pan and cook to a purée.

3 Pass through a sieve or liquidise.

53 Cranberry and orange dressing

Energy	Calories	Fat	Saturated fat	Carbohydrates	Sugar	Protein	Fibre
398 kJ	93 kcal	0.2 g	0.0 g	22.7 g	22.7 g	1.3 g	4.2 g

Cooking

1 Place the cranberries in a suitable saucepan with the rest of the ingredients.

2 Bring to the boil and simmer gently for approximately 1 hour, stirring from time to time.

3 Remove from the heat and leave to cool. Use as required.

PROFESSIONAL TIP
The dressing may also be liquidised if a smooth texture is required.

	4 portions	10 portions
Cranberries	400 g	1 kg
Granulated sugar	50 g	125 g
Red wine	125 ml	250 ml
Red wine vinegar	2 tbsp	5 tbsp
Orange zest and juice	2	4

54 Green or herb sauce

Energy	Calories	Fat	Saturated fat	Carbohydrates	Sugar	Protein	Fibre	Sodium
1082 kJ	263 kcal	28.0 g	4.3 g	0.9 g	0.6 g	0.8 g	0.2 g	0.2 g

Cooking

1 Pick, wash, blanch and refresh the green leaves.

2 Squeeze dry.

3 Pass through a very fine sieve.

4 Mix with the mayonnaise.

> **SERVING SUGGESTION**
> May be served with cold salmon or trout.

	4 portions	10 portions
Spinach, tarragon, chervil, chives, watercress	50 g	125 g
Mayonnaise	250 g	625 g

55 Reduction of stock (glaze)

A glaze is a stock, fond or nage that has been reduced: that is, much of the water content is removed by gently simmering. The solid content, and all the flavour, stays in the glaze.

Any kind of stock can be used, but it is important to be careful if using meat stock. Meat stock contains collagen; if the stock is cooked at boiling temperature, there will be a lot of collagen in the glaze. This means the sauce will become thick more quickly than non-meat glazes. It will then be impossible to reduce it any more without burning it.

Glazes have a strong flavour and contain a lot of salt, so only use small amounts.

56 Reduction of wine, stock and cream

Energy	Calories	Fat	Saturated fat	Carbohydrates	Sugar	Protein	Fibre	Sodium
2360 kJ	573 kcal	50.0 g	31.5 g	4.5 g	4.5 g	0.7 g	0.3 g	40.0 g

* Using dry white wine and whipping cream

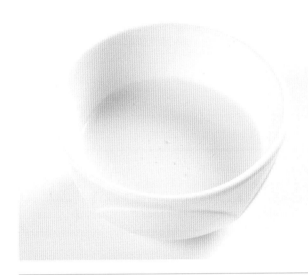

	Approximately 250 ml
White stock	500 ml
White wine	125 ml
Double or whipping cream	125 ml

Cooking

1 Place the stock and white wine in a suitable saucepan.

2 Reduce by at least two-thirds to a slightly syrup consistency and finish with cream.

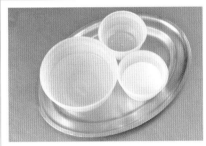
Measure out all the ingredients.

Reduce the wine and stock.

Add the cream.

57 Hollandaise sauce

Energy	Calories	Fat	Saturated fat	Carbohydrates	Sugar	Protein	Fibre	Sodium
11562 kJ	2809 kcal	301.0 g	179.0 g	2.1 g	2.0 g	0.1 g	1.7 g	2,432.0 g

* Using salted butter

	500 g
Peppercorns, crushed	12
White wine vinegar	3 tbsp
Egg yolks	6
Melted butter	325 g
Salt and cayenne pepper	

Cooking

1 Place the peppercorns and vinegar in a small pan and reduce to one third.

2 Add 1 tablespoon of cold water and allow to cool. Add the egg yolks.

3 Put on a bain marie and whisk continuously to a sabayon consistency.

4 Remove from the heat and gradually whisk in the melted butter.

5 Add seasoning and pass through muslin or a fine chinois.

6 Store in an appropriate container at room temperature.

HEALTH AND SAFETY

Egg-based sauces should not be kept warm for more than 2 hours. After this time, they should be thrown away, but are best made fresh to order.

VARIATION

- **Mousseline sauce**: hollandaise base with lightly whipped cream.
- **Maltaise sauce**: hollandaise base with lightly grated zest and juice of one blood orange.

FAULTS

If you add oil or butter when making hollandaise or mayonnaise, the sauce may curdle because the lecithin has had insufficient time to coat the droplets. This can be rectified by adding the broken sauce to more egg yolks.

58 Béarnaise sauce

Energy	Calories	Fat	Saturated fat	Carbohydrates	Sugar	Protein	Fibre	Sodium
11523 kJ	2799 kcal	301.0 g	179.0 g	3.2 g	2.8 g	0.1 g	0.8 g	2,433.0 g

* Using salted butter

	500 g
Shallots, chopped	50 g
Tarragon	10 g
Peppercorns, crushed	12
White wine vinegar	3 tbsp
Egg yolks	6
Melted butter	325 g
Salt and cayenne pepper, chervil and tarragon to finish, chopped	

Cooking

1 Place the shallots, tarragon, peppercorns and vinegar in a small pan and reduce to one-third.

2 Add 1 tablespoon of cold water and allow to cool.

3 Add the egg yolks.

4 Put on a bain-marie and whisk continuously to a sabayon consistency.

5 Remove from the heat and gradually whisk in the melted butter. Add seasoning.

6 Pass through muslin or a fine chinois.

7 To finish, add the chopped chervil and tarragon.

8 Store in an appropriate container at room temperature.

> **VARIATION**
> - **Choron sauce**: 200-g tomato concassé, well dried. Do not add the chopped tarragon and chervil to finish.
> - **Foyot or valois sauce**: 25-g warm meat glaze.
> - **Paloise sauce**: made as for béarnaise sauce, using chopped mint stalks in place of the tarragon in the reduction. To finish, add chopped mint instead of the chervil and tarragon.

59 Sauce diable

Energy	Calories	Fat	Saturated fat	Carbohydrates	Sugar	Protein	Fibre	Sodium
1306 kJ	317 kcal	21.7 g	13.2 g	2.6 g	1.9 g	0.9 g	0.8 g	192.0 g

* Using salted butter and brown stock

Cooking

1 In a suitable pan, add the butter and gently sweat the shallots without colour.

2 Add the white wine; reduce by half.

3 Add the brown stock or jus lié; bring to the boil and simmer for two minutes.

> **SERVING SUGGESTION**
> Serve with grilled chicken and steaks.

> **PROFESSIONAL TIP**
> The sauce should have a background flavour of white wine, with a sharp after taste from the cayenne pepper. The sauce may also be improved with the addition of 15-g of soft butter before serving.

	250 g
Butter	25 g
Shallots finely chopped	25 g
Dry white wine	150 ml
Reduced brown stock or jus lié	250 ml
Cayenne pepper	2 g

60 Tomato sauce (*sauce tomate*)

Energy	Calories	Fat	Saturated fat	Carbohydrates	Sugar	Protein	Fibre	Sodium
234 kJ	56 kcal	2.4 g	0.8 g	6.7 g	4.5 g	0.0 g	2.0 g	296.0 g

* Using margarine

	4 portions	10 portions
Margarine, butter or oil	10 g	25 g
Onions (for mirepoix)	50 g	125 g
Carrots (for mirepoix)	50 g	125 g
Celery (for mirepoix)	25 g	60 g
Bay leaf (for mirepoix)	1	3
Bacon scraps (optional)	10 g	25 g
Flour	10 g	25 g
Tomato purée	50 g	125 g
Stock	375 ml	1 litre
Clove of garlic	½	2
Salt and pepper		

Cooking

1 Melt the fat or oil in a small sauteuse.

2 Add the vegetables, herbs (mirepoix) and bacon scraps and brown slightly.

3 Mix in the flour and cook to a sandy texture. Allow to colour slightly.

4 Mix in the tomato purée; allow to cool.

5 Gradually add the boiling stock; stir to the boil.

6 Add the garlic and season then simmer for one hour.

7 Correct the seasoning and cool.

8 Pass through a fine conical strainer.

> **VARIATION**
> - The amount of tomato purée used may need to vary according to its strength.
> - The sauce can also be made without using flour, by adding 400 g of fresh, fully ripe tomatoes or an equivalent tin of tomatoes for four portions.

61 Sweet and sour sauce

Energy	Calories	Fat	Saturated fat	Carbohydrates	Sugar	Protein	Fibre	Sodium
791 kJ	185 kcal	0.1 g	0.0 g	46.0 g	46.0 g	46.0 g	0.4 g	583.0 g

	4 portions	10 portions
White vinegar	375 ml	1 litre
Brown sugar	150 g	375 g
Tomato ketchup	125 ml	300 ml
Worcester sauce	1 tbsp	2½ tbsp
Seasoning		

Cooking

1 Boil the vinegar and sugar in a suitable pan.

2 Add the tomato ketchup, Worcester sauce and seasoning.

3 Simmer for a few minutes then use as required. This sauce may also be lightly thickened with cornflour or another thickening agent.

62 Mayonnaise

Energy	Calories	Fat	Saturated fat	Carbohydrates	Sugar	Protein	Fibre	Sodium
1153 kJ	280 kcal	27.6 g	4.7 g	0.1 g	0.1 g	0.1 g	0.0 g	144.0 g

* Using vinegar and corn oil

	8 portions
Egg yolks, very fresh or pasteurised	3
Vinegar or lemon juice	2 tsp
Small pinch of salt	
English or continental mustard	½ tsp
A mild-flavoured oil such as corn oil or the lightest olive oil	250 ml
Water, boiling	1 tsp

> ⚠️ **HEALTH AND SAFETY**
> Because of the risk of salmonella food poisoning, it is strongly recommended that pasteurised egg yolks are used.

1 Place yolks, vinegar, salt and mustard in the bowl of a food mixer.

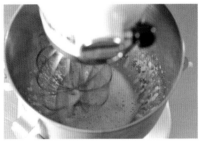

2 Whisk until thoroughly mixed.

3 Continue to whisk vigorously and start to add the oil – this needs to be done slowly.

4 Keep whisking until all the oil has been added.

5 Whisk in the boiling water.

6 Taste and correct seasoning if necessary.

VARIATION

Add:
- fresh chopped herbs
- garlic juice – peel a clove garlic and press it using a garlic press
- thick tomato juice.

PROFESSIONAL TIP

If the mayonnaise becomes too thick while you are making it, whisk in a little water or vinegar.

FAULTS

Mayonnaise may separate, turn or curdle for several reasons:
- the oil has been added too quickly
- the oil is too cold
- not enough whisking
- the egg yolks were stale and weak.

To reconstitute (bring it back together):
- Take a clean basin, pour 1 teaspoon of boiling water and gradually but vigorously whisk in the curdled sauce a little at a time.
- Alternatively, in a clean basin, whisk a fresh egg yolk with half a teaspoon of cold water then gradually whisk in the curdled sauce.

63 Tartare sauce (*sauce tartare*)

Energy	Calories	Fat	Saturated fat	Carbohydrates	Sugar	Protein	Fibre	Sodium
911 kJ	221 kcal	24.0 g	3.6 g	0.8 g	0.6 g	0.6 g	0.1 g	0.2 g

Preparing the dish

1 Combine all the ingredients.

SERVING SUGGESTION

This sauce is usually served with deep-fried fish.

PROFESSIONAL TIP

Finely chop the gherkins and capers and use a blender to give the desired texture and consistency.

Healthy Eating Tips

Proportionally reduce the fat by adding some low-fat yoghurt instead of some of the mayonnaise.

	8 portions
Mayonnaise	250 ml
Capers, chopped	25 g
Gherkins, chopped	50 g
Sprig of parsley, chopped	

64 Remoulade sauce (*sauce remoulade*)

Energy	Calories	Fat	Saturated fat	Carbohydrates	Sugar	Protein	Fibre	Sodium
907 kJ	220 kcal	24.0 g	3.6 g	0.8 g	0.6 g	0.7 g	0.1 g	0.3 g

Preparing the dish

1 Prepare as for tartare sauce (recipe 63), adding 1 teaspoon of anchovy essence and mixing thoroughly. Makes 8 portions.

SERVING SUGGESTION

This sauce may be served with fried fish. It can also be mixed with a fine julienne of celeriac to make an accompaniment to cold meats, terrines, etc.

65 Andalusian sauce (*sauce Andalouse*)

Energy	Calories	Fat	Saturated fat	Carbohydrates	Sugar	Protein	Fibre	Sodium
6299 kJ	1428 kcal	160.0 g	24.9 g	14.3 g	13.8 g	13.3 g	0.7 g	1,286.0 g

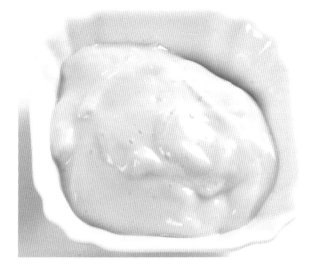

Preparing the dish

1 Take 250 ml of mayonnaise and add 2 tbsp of tomato juice or ketchup and 1 tbsp of red pepper cut into julienne. Makes 250 ml.

SERVING SUGGESTION

May be served with cold salads.

PROFESSIONAL TIP

For Andalusian sauce, Thousand Island dressing, green sauce and other similar recipes, use a blender to achieve the desired texture and flavour.

66 Thousand Island dressing

Energy	Calories	Fat	Saturated fat	Carbohydrates	Sugar	Protein	Fibre	Sodium
928 kJ	225 kcal	24.0 g	3.6 g	2.6 g	2.3 g	0.5 g	0.1 g	0.3 g

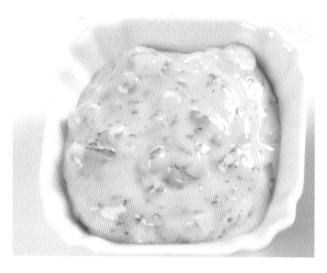

Finely chopped clove of garlic	1
Tabasco sauce	2 drops
Worcester sauce	½ tsp
Gherkins finely chopped	2

Preparing the dish

1 Mix all ingredients together.

2 Use as a dressing for fish cocktails and salads.

> **NOTE**
> Thousand Island dressing is an example of a dressing which has become a standard condiment.

	4–6 portions
Mayonnaise	250 mls
Tomato ketchup	30 mls
White wine vinegar	1 tbsp
Caster sugar	1 tsp
Finely chopped onion	15 g

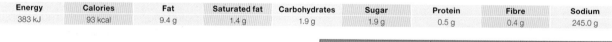

67 Tomato vinaigrette

Energy	Calories	Fat	Saturated fat	Carbohydrates	Sugar	Protein	Fibre	Sodium
383 kJ	93 kcal	9.4 g	1.4 g	1.9 g	1.9 g	0.5 g	0.4 g	245.0 g

	4 portions	10 portions
Tomatoes	200 g	500 g
Caster sugar	½ tsp	1¼ tbsp
White wine vinegar	1 tbsp	2 ½ tsp
Extra virgin olive oil	3 tbsp	8 tbsp
Seasoning		

Preparing the dish

1 Blanch and deseed the tomatoes; purée in a food processor.

2 Add the sugar, vinegar, olive oil and seasoning; whisk well to emulsify.

3 The vinaigrette should be smooth and well emulsified.

68 Mint sauce

Energy	Calories	Fat	Saturated fat	Carbohydrates	Sugar	Protein	Fibre	Sodium
26 kJ	6 kcal	0.0 g	0.0 g	0.7 g	0.6 g	0.1 g	0.0 g	0.0 g

	8 portions
Mint	2–3 tbsp
Caster sugar	1 tsp
Vinegar	125 ml

Preparing the dish

1 Chop the washed, picked mint and mix with the sugar.

2 Place in a china basin and add the vinegar.

3 If the vinegar is too sharp, dilute it with a little water.

SERVING SUGGESTION
Serve with roast lamb.

PROFESSIONAL TIP
A less acid sauce can be produced by dissolving the sugar in 125 ml boiling water and, when cold, adding the chopped mint and 1–2 tablespoon vinegar to taste.

69 Yoghurt and cucumber raita

Energy	Calories	Fat	Saturated fat	Carbohydrates	Sugar	Protein	Fibre	Sodium
366 kJ	87 kcal	1.4 g	0.8 g	11.9 g	11.7 g	7.2 g	1.3 g	0.1 g

	4 portions
Cucumbers	2
Salt	
Spring onions, chopped	2 tbsp
Yoghurt	500 ml
Cumin seeds	1 ½ tsp
Lemon juice	1
Fresh coriander or mint, chopped	

2 Sprinkle the dice with salt and leave for 15 minutes, then drain away the liquid and rinse the cucumbers quickly in cold water. Drain well.

3 Combine the onion, yoghurt and lemon juice; taste to see if more salt is required.

4 Roast the cumin seeds in a dry pan, shaking the pan or stirring constantly until brown.

5 Bruise or crush the seeds and sprinkle over the yoghurt mixture.

SERVING SUGGESTION
Serve chilled, garnished with mint and coriander.

NOTE
This is a dish from Punjab, northern India. Serve as an accompaniment to curry.

Preparing the dish

1 Peel the cucumbers, halve them lengthways and remove the seeds. Cut into small dice.

PROFESSIONAL TIP
Sprinkling the cucumber with salt removes the juices, which are hard to digest. Remember to wash off the salt before use.

70 Balsamic vinegar and olive oil dressing

Energy	Calories	Fat	Saturated fat	Carbohydrates	Sugar	Protein	Fibre	Sodium
1180 kJ	287 kcal	30.9 g	4.4 g	1.6 g	1.5 g	1.5 g	0.0 g	124.0 g

Preparing the dish

1 Whisk all ingredients together and correct the seasoning.

PROFESSIONAL TIP
The amount of balsamic vinegar needed will depend on its quality, age and so on. Add more or less as required.

This dressing works well because it is not an emulsion. The oil and vinegar provide a stark contrast and can be stirred just before serving.

	8 portions
Water	62 ml
Olive oil	250 ml
Balsamic vinegar	62 ml
Sherry vinegar	2 tbsp
Caster sugar	½ tsp
Seasoning	

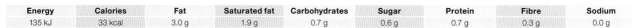

71 Horseradish sauce (*sauce raifort*)

Energy	Calories	Fat	Saturated fat	Carbohydrates	Sugar	Protein	Fibre	Sodium
135 kJ	33 kcal	3.0 g	1.9 g	0.7 g	0.6 g	0.7 g	0.3 g	0.0 g

Preparing the dish

1 Wash, peel and rewash the horseradish. Grate finely.

2 Mix all the ingredients together.

SERVING SUGGESTION
Serve with roast beef, smoked trout, eel or halibut.

PROFESSIONAL TIP
It is essential to blend the ingredients without over-mixing them, in order to get a good flavour.

	8 portions
Horseradish	25 g
Vinegar or lemon juice	1 tbsp
Salt, pepper	
Cream or crème fraiche, lightly whipped	125 ml

72 Tomato and cucumber salsa

Energy	Calories	Fat	Saturated fat	Carbohydrates	Sugar	Protein	Fibre
202 kJ	49 kcal	4.3 g	0.7 g	2.0 g	2.0 g	0.6 g	0.7 g

Preparing the dish

1 In a large bowl, mix all the ingredients together.

2 Correct seasoning and serve.

Healthy Eating Tips
Rely on the herbs for flavour, with the minimum amount of salt.

VARIATION
This recipe may be varied by using any chopped salad ingredients and fresh herbs (e.g. tarragon, chervil). Do not be afraid to experiment.

Extra vegetables can be added and the salsa used liberally with grilled fish or chicken. Rice could be served with this, or the salsa used to fill a tortilla.

	8 portions
Ripe tomatoes, skinned, deseeded, chopped	400 g
Cucumber, peeled, chopped	½
Spring onions, chopped	6
Fresh basil, chopped	1 tbsp
Fresh parsley, chopped	1 tbsp
Olive oil	3 tbsp
Lemon or lime (juice of)	1
Salt and pepper	

73 Salsa verde

Energy	Calories	Fat	Saturated fat	Carbohydrates	Sugar	Protein	Fibre	
281 kJ	69 kcal	7.5 g	1.1 g	0.2 g	0.1 g	0.1 g	0.0 g	*

* Per tablespoon

Preparing the dish

1 In a large bowl, mix all the ingredients together and check the seasoning.

SERVING SUGGESTION
Serve with grilled fish.

	8 portions
Mint, coarsely chopped	1 tbsp
Parsley, coarsely chopped	3 tbsp
Capers, coarsely chopped	3
Garlic clove (optional)	1
Dijon mustard	1 tsp
Lemon, juice of	½
Extra virgin olive oil	120 ml
Salt (optional)	

74 Avocado and coriander salsa

Energy	Calories	Fat	Saturated fat	Carbohydrates	Sugar	Protein	Fibre
361 kJ	87 kcal	8.6 g	1.4 g	1.8 g	1.4 g	0.8 g	1.0 g

Preparing the dish

1 In a large bowl, mix all the ingredients together, check seasoning and serve.

SERVING SUGGESTION
Serve with cold dishes such as salads or terrines.

	8 portions
Ripe avocado, peeled and diced	1
Ripe tomatoes, peeled, deseeded and diced	3
Shallot, peeled and cut into rings	1
Fresh coriander, chopped	1 tsp
Pine kernels, toasted	10 g
Cucumber, diced	25 g
Lemon or lime (juice)	1
Virgin olive oil	3 tbsp
Salt and pepper to taste	

Healthy Eating Tips
- Although avocado is rich in fat, it is unsaturated fat and therefore healthier.
- Try using the salsa to fill a tortilla, and add grilled fish or chicken to make a healthy meal.

75 Pear chutney

Energy	Calories	Fat	Saturated fat	Carbohydrates	Sugar	Protein	Fibre	Sodium
24870 kJ	5824 kcal	7.5 g	1.8 g	1,446.0 g	1,438.0 g	920.0 g	82.0 g	275.0 g

Cooking

1 Make a thick syrup with the white wine vinegar and the sugar.

2 Add the ginger, onion, nutmeg, saffron, cinnamon, sultanas and concassé, and reduce to a thick syrup.

3 Add the diced pears and reduce again to a sticky consistency.

4 Cool the chutney and then store in a Kilner jar.

SERVING SUGGESTION
Serve with cheese, cold meats, pâtés or terrines.

	Makes approximately 5 kg
White wine vinegar	900 g
Demerara sugar	900 g
Ginger, brunoise	125 g
Onion, diced	375 g
Nutmeg	5 g
Saffron	0.25 g
Cinnamon	5 g
Golden sultanas	375 g
Tomato concassé	750 g
Pears, diced	2 kg

76 Tomato chutney

Energy	Calories	Fat	Saturated fat	Carbohydrates	Sugar	Protein	Fibre	Sodium
7232 kJ	1702 kcal	4.0 g	0.2 g	379.0 g	376.0 g	299.0 g	23.3 g	4,157.0 g

Cooking

1 Peel and coarsely chop the tomatoes, then combine with the remaining ingredients in a large heavy-duty saucepan.

2 Stir over heat without boiling until the sugar dissolves. Simmer uncovered, stirring occasionally until the mixture thickens (about 1½ hours).

3 Place in hot, sterilised jars. Seal while hot.

SERVING SUGGESTION
Serve with cheese, cold meats or terrines.

	1 litre
Tomatoes, peeled	1.5 kg
Onions, finely chopped	450 g
Brown sugar	300 g
Malt vinegar	375 ml
Mustard powder	1 ½ tsp
Cayenne pepper	½ tsp
Coarse salt	2 tsp
Mild curry powder	1 tbsp

77 Pesto sauce

Energy	Calories	Fat	Saturated fat	Carbohydrates	Sugar	Protein	Fibre	Sodium
4218 kJ	1020 kcal	95.0 g	24.9 g	7.1 g	1.2 g	0.0 g	0.7 g	1,339.0 g

Preparing the dish

1 Place all ingredients into a food processor and mix to a rough-textured sauce.

2 Transfer to a bowl and leave for at least 1 hour to enable the flavours to develop.

SERVING SUGGESTION
Pesto is traditionally served with large flat pasta called trenetta.

Pesto is also used as a cordon in various fish and meat-plated dishes, e.g. grilled fish, medallions of veal.

VARIATION
● Use flat leaf parsley in place of basil and walnuts in place of pine nuts.

	250 ml
Fresh basil leaves	100 ml
Pine nuts (lightly toasted)	1 tbsp
Garlic (picked and crumbled)	2 cloves
Parmesan cheese (grated)	40 g
Pecorino cheese (grated)	40 g
Olive oil	5 tbsp
Salt and pepper	

78 Tapenade

Energy	Calories	Fat	Saturated fat	Carbohydrates	Sugar	Protein	Fibre
773 kJ	188 kcal	19.8 g	2.9 g	0.3 g	0.1 g	2 g	2.7 g

Preparing the dish

1 Mix all the ingredients together, adding the olive oil to make a paste.

2 For a smoother texture, place garlic, lemon juice, capers and anchovies into a food processor and process until a smooth texture. Add the olives and parsley, and sufficient oil to form a smooth paste.

3 Season, if required.

4 Garnish with a sprinkle of roast cumin and chopped red chilli.

Serve chilled.

	4 portions	10 portions
Capers	45 g	110g
Lemon juice	1 tbsp	2 tbsp
Anchovy fillets, chopped	6	15
Black olives, copped	250 g	625g
Parsley, chopped	1 tsp	2 tsp
Salt and pepper	Pinch	Pinch
Virgin olive oil	4 tbsp	10 tbsp
Garlic cloves crushed/chopped	1	3
Roast cumin	5 g	12 g
Chopped red chilli	½	1

NOTE
Tapenade is a Provençale dish consisting of pureed or finely chopped black olives, capers, anchovies and olive oil. It may also contain garlic, herbs, tuna, lemon juice or brandy. Its name comes from the Provençale word for capers: tapeno. It is popular in the South of France, where it is generally eaten as an hors d'oeuvre, spread on toast.

79 Mixed pickle

Energy	Calories	Fat	Saturated fat	Carbohydrates	Sugar	Protein	Fibre	Sodium
2126 kJ	512 kcal	6.0 g	0.9 g	41.0 g	39.7 g	5.9 g	22.1 g	6,006.0 g

1 small cauliflower, 1 cucumber, 3 green tomatoes, 1 onion, 1 small marrow or 3 courgettes	
For the spiced vinegar	
Vinegar	1 litre
Blade mace	5 g
Allspice	5 g
Cloves	5 g
Stick cinnamon	5 g
Peppercorns	6
Root ginger (for hot pickle)	5 g

Cooking

1 To make the spiced vinegar, tie the spices in muslin, place them in a covered pan with the vinegar and heat slowly to boiling point.

2 Remove from the heat and stand for 2 hours, then remove the bag.

3 Prepare the vegetables, with the exception of the marrow, and soak them in brine for 24 hours.

4 Peel the marrow, remove the seeds and cut into small squares, sprinkle and salt, and let it stand for 12 hours.

5 Drain the vegetables, pack them into jars, and cover with cold spiced vinegar.

6 Cover the jars and allow the pickle to mature for at least a month before use.

80 Gribiche sauce

Energy	Calories	Fat	Saturated fat	Carbohydrates	Sugar	Protein	Fibre	Sodium
10264 kJ	2494 kcal	269.0 g	19.8 g	1.8 g	1.3 g	0.3 g	1.0 g	2,281.0 g

	500 g
Sieved hard boiled eggs	2
Vegetable oil	325 ml
French mustard	1 tsp
White wine vinegar	100 ml
Chopped capers	30 g
Chopped gherkins	30 g
Chopped parsley	½ tsp
Chopped chives	½ tsp
Chopped basil	½ tsp
Seasoning	

2 Whisk together with a balloon whisk.

3 Slowly add the oil, whisking continuously.

4 When the emulsion is formed, add the remainder of the ingredients and season.

> **NOTE**
> This is an emulsified sauce. The mustard is a natural emulsifying agent.

> **SERVING SUGGESTION**
> Serve with steamed fish, fried fish, cold meats, chicken and gammon.

Preparing the dish

1 Place the sieved egg yolks in a basin and add the mustard and vinegar.

81 Butter sauce (*beurre blanc*)

Energy	Calories	Fat	Saturated fat	Carbohydrates	Sugar	Protein	Fibre	Sodium
1568 kJ	381 kcal	41.1 g	26.0 g	0.9 g	0.9 g	0.6 g	0.2 g	0.4 g

	4 portions	10 portions
Water	125 ml	300 ml
Wine vinegar	125 ml	300 ml
Shallot, finely chopped	50 g	125 g
Unsalted butter	200 g	500 g
Lemon juice	1 tsp	2½ tsp
Salt and pepper		

Cooking

1 Reduce the water, vinegar and shallots in a thick-bottomed pan to approximately 2 tablespoons.

2 Allow to cool slightly.

3 Gradually whisk in the butter in small amounts, whisking continually until the mixture becomes creamy.

4 Whisk in the lemon juice, season lightly and keep warm in a bain marie.

VARIATION

The sauce may be strained if desired. It can be varied by adding, for example, freshly shredded sorrel or spinach, or blanched fine julienne of lemon or lime. It is suitable for serving with fish dishes.

82 Melted butter (*beurre fondu*)

Energy	Calories	Fat	Saturated fat	Carbohydrates	Sugar	Protein	Fibre	Sodium
388 kJ	94 kcal	10.3 g	6.5 g	0.1 g	0.1 g	0.1 g	0.0 g	

	4 portions	10 portions
Butter	200 g	500 g
Water or white wine	2 tbsp	5 tbsp

Cooking

Method 1:

1 Boil the butter and water together gently until combined, then pass through a fine strainer.

Method 2:

1 Melt the butter in the water and carefully strain off the fat, leaving the water and sediment in the pan.

SERVING SUGGESTION

Usually served with poached fish and certain vegetables, for example, blue trout, salmon; asparagus, sea kale.

83 Compound butter

Compound butters are made by mixing the flavouring ingredients into softened butter, which can then be shaped into a roll two centimetres in diameter, placed in wet greaseproof paper or foil, hardened in a refrigerator and cut into slices when required.

Herb butter

Parsley butter: chopped parsley and lemon juice.
Herb butter: mixed herbs (chives, tarragon, fennel, dill) and lemon juice.

Chive butter: chopped chives and lemon juice.
Garlic butter: garlic juice and chopped parsley or herbs.
Anchovy butter: few drops anchovy essence.
Shrimp butter: finely chopped or pounded shrimps.
Garlic: mashed to a paste.
Mustard: continental-type mustard.
Liver pâté: mashed to a paste with the butter.

SERVING SUGGESTION
Compound butters are served with grilled and some fried fish, and with grilled meats.

84 Herb oil

Energy	Calories	Fat	Saturated fat	Carbohydrates	Sugar	Protein	Fibre	Sodium
8482 kJ	2063 kcal	226.0 g	32.5 g	3.0 g	2.2 g	0.0 g	5.4 g	149.0 g

Cooking

1 Blanch all the herbs and spinach for 1½ minutes.

2 Drain well, place with the oil in a liquidiser and blitz for 2½ minutes. Pass and decant when rested.

SERVING SUGGESTION
Uses include salads, salmon mi cuit and other fish dishes.

	200 ml
Picked flat leaf parsley	25 g
Chives	10 g
Picked basil leaves	10 g
Picked spinach	100 g
Corn oil	250 ml

85 Lemon oil

Energy	Calories	Fat	Saturated fat	Carbohydrates	Sugar	Protein	Fibre	Sodium
9340 kJ	2272 kcal	250.0 g	28.6 g	1.2 g	0.0 g	0.0 g	2.5 g	1.7 g

	250 ml
Lemons, rind (with no pith – the whitish layer between skin and fruit)	3
Lemon grass stick, cut lengthways and chopped into 2-cm strips	1
Grapeseed oil	250 ml
Olive oil	2 tbsp

Preparing the dish

1 Place all the ingredients into a food processor and pulse the mix until the lemon peel and grass are approximately 3 mm thick.

2 Allow to stand for 2 days. Decant and store in the fridge until ready for use (or freeze for longer if you wish).

86 Mint oil

Energy	Calories	Fat	Saturated fat	Carbohydrates	Sugar	Protein	Fibre	Sodium
5171 kJ	1257 kcal	136.0 g	8.9 g	5.3 g	0.0 g	0.0 g	0.0 g	15.0 g

* Using vegetable oil

	150 ml
Mint	100 g
Oil	150 ml
Salt	3 tbsp

3 Place in a blender and slowly add the oil.

4 Allow to settle overnight and decant into bottles.

SERVING SUGGESTION
Uses include lamb dishes, salads and fish dishes.

Cooking

1 Blanch the mint for 30 seconds.

2 Refresh and squeeze the water out.

87 Vanilla oil

Energy	Calories	Fat	Saturated fat	Carbohydrates	Sugar	Protein	Fibre	Sodium
7149 kJ	1737 kcal	180.0 g	11.9 g	1.1 g	1.1 g	1.1 g	0.0 g	2.0 g

	200 ml
Vegetable oil	200 ml
Vanilla pods, whole	5
Vanilla pods, used	2
Vanilla extract	50 ml

Preparing the dish

1 Warm the oil to around 60 °C; add the vanilla in its various forms and infuse, scraping all the seeds into the oil.

2 Store in a plastic or glass bottle.

SERVING SUGGESTION
Uses include salads, salmon mi cuit and other fish dishes.

88 Basil oil

Energy	Calories	Fat	Saturated fat	Carbohydrates	Sugar	Protein	Fibre	Sodium
7434 kJ	1808 kcal	200.0 g	13.2 g	1.3 g	0.0 g	0.0 g	0.0 g	804.0 g

* Using vegetable oil

	200 ml
Fresh basil	25 g
Oil	200 ml
Salt, to taste	
Mill pepper	

Cooking

1 Blanch and refresh the basil; purée with the oil.

2 Allow to settle overnight and decant.

3 Store in bottles with a sprig of blanched basil.

> **VARIATION**
> Basil extract can be used in place of fresh basil; 50 g of grated Parmesan or Gorgonzola cheese may also be added to the basil oil.

89 Walnut oil

Energy	Calories	Fat	Saturated fat	Carbohydrates	Sugar	Protein	Fibre	Sodium
20057 kJ	4873 kcal	523.0 g	84.0 g	3.0 g	2.5 g	0.0 g	4.4 g	500.0 g

	500 ml
Olive or walnut oil	500 ml
Walnuts, finely crushed	75 g
Parmesan cheese	75 g
Salt, to taste	50 ml
Mill pepper	

Preparing the dish

1 Mix all the ingredients together and bottle until required.

90 Parsley oil

Energy	Calories	Fat	Saturated fat	Carbohydrates	Sugar	Protein	Fibre	Sodium
8473 kJ	2061 kcal	226.0 g	32.5 g	2.8 g	2.5 g	0.0 g	5.7 g	95.0 g

	200 ml
Picked flat leaf parsley	75 g
Picked spinach	50 g
Corn oil	250 ml

Cooking

1 Blanch the parsley and spinach for 1½ minutes, drain well and place with the oil in a liquidiser.

2 Liquidise for 2½ minutes, place in the fridge and allow the sediment to settle overnight.

3 The next day, decant when rested and use.

> **SERVING SUGGESTION**
> Uses include fish and meat dishes.

91 Roasted pepper oil

Energy	Calories	Fat	Saturated fat	Carbohydrates	Sugar	Protein	Fibre	Sodium
16830 kJ	4093 kcal	454.0 g	65.0 g	1.1 g	1.0 g	0.0 g	0.5 g	984.0 g

Infusing the ingredients

	500ml
Virgin olive oil	500 ml
Mirepoix (carrot, onion, celery)	200 g
Red pepper (cleaned and roasted)	2
Yellow pepper (cleaned and roasted)	2
Green pepper (cleaned and roasted)	2
Bay leaves	2
Black peppercorns	8
Garlic cloves	2
Sea salt	½ tsp
Olive oil	1 tsp
White wine vinegar	1 tsp

Cooking

1 Heat 1 teaspoon of olive oil in a pan.

2 Sweat the mirepoix and one of each of the peppers until golden brown.

3 Add 1 bay leaf, the black peppercorns, one clove of garlic and the sea salt, and cook for a further 3–4 minutes.

4 Add half the olive oil and bring to simmering point.

5 Take off the heat and allow to cool.

6 When completely cool, pass through a chinois into a clean Kilner jar.

7 Add the white wine vinegar and the remaining olive oil, roasted garlic, roasted peppers and remaining bay leaf.

8 For full flavour, leave for at least one month before use. To use, strain the oil, discarding the ingredients.

Recipe supplied by Mark McCann

92 Pimms espuma

Energy	Calories	Fat	Saturated fat	Carbohydrates	Sugar	Protein	Fibre	Sodium
2824 kJ	669 kcal	0.2 g	0.0 g	119.0 g	119.0 g	79.0 g	0.0 g	75.0 g

Apple juice	150 g
Stock syrup	130 g
Tonic water	250 g
Pimms	140 g
Leaves gelatine, soaked	3

Preparing the dish

1 Soak the gelatine in ice cold water.
2 Warm 100 g tonic water, add the gelatine and stir until dissolved.
3 Mix in the rest of the ingredients.
4 Place into an iSi gourmet whip and charge with 2 x N$_2$O charges.
5 Chill in fridge for 6 hours.
6 Shake the container every 2 hours to ensure an even set.

NOTE
An iSi gourmet whip uses gas and pressure to create foams, light sauces and soups or infusions. It is a cylindrical canister with a handle and assorted tips for dispensing. If using a 0.5-litre whipper, use only one charge and if using a 1-litre whipper then use two charges. Recipes 92, 93 and 94 are for the 1-litre whipper.

93 Pea espuma

Energy	Calories	Fat	Saturated fat	Carbohydrates	Sugar	Protein	Fibre	Sodium
3826 kJ	921 kcal	70.0 g	42.0 g	42.0 g	25.6 g	0.0 g	19.1 g	934.0 g

* Using salted butter

Frozen peas	400 g
Chilled chicken stock	250 g
Butter, soft	10 g
Double cream	100 g
Full fat milk	100 g
Seasoning	

Cooking

1 Blanch the peas for 3 minutes in salted boiling water, then strain.
2 Blitz with the chicken stock until smooth, add the cream and butter, and blitz for another minute.
3 Pass through a fine sieve.
4 Heat to 60 °C in a heavy based saucepan and season.
5 Place into an iSi gourmet whip and charge with 2 x N$_2$O charges.

94 Parsnip and vanilla espuma

Energy	Calories	Fat	Saturated fat	Carbohydrates	Sugar	Protein	Fibre	Sodium
6161 kJ	1492 kcal	144.0 g	89.0 g	33.3 g	19.8 g	0.1 g	11.3 g	1,130.0 g

* Using salted butter

Cooking

1 Peel, core and thinly slice parsnips.

2 Place the butter and vanilla in a heavy-based saucepan and cook the parsnips slowly until very soft (approximately 40 minutes).

3 Remove vanilla and blitz into a purée in a food processor.

4 Add the warm chicken stock, milk and cream.

5 Remove from food processor and pass through a sieve into a heavy based saucepan.

6 Heat to 60 °C. Add sherry vinegar and season to taste.

7 Place into an iSi gourmet whip and charge with 2 x N_2O charges.

Parsnips	300 g
Vanilla pod	½
Butter	100 g
Chicken stock (warm)	200 g
Full fat milk	100 g
Double cream	100 g
Sherry vinegar	3 g
Seasoning	

304 Fruit and vegetables

Learning outcomes

In this unit you will be able to:

1 Prepare fruits and vegetables, including:
   - identifying common classifications and types of fruits and vegetables and their typical portion sizes or average weights
   - knowing the quality points of different fruits and vegetables
   - knowing the seasons for fruit and vegetables
   - knowing which fruits and vegetables are classed as allergens
   - understanding and being able to apply the correct preparation methods, using the correct tools and equipment to meet the dish specifications
   - knowing which preservation methods could be used to increase shelf life of fruit and vegetables, and understanding the benefits and drawbacks for using preservation methods
   - understanding how coatings and stuffings are used to enhance the end product to meet the dish requirements
   - understanding the correct storage procedures to use throughout production and on completion of the final products.

2 Produce fruit and vegetable dishes, including:
   - understanding and being able to use techniques to produce fruit and vegetable dishes to meet the recipe specifications
   - being able to use techniques to produce liquids, sauces and accompaniments to finish fruit and vegetable dishes
   - being able to measure and evaluate against quality standards throughout preparation and cooking
   - being able to evaluate products against dish requirements and production standards and recognise any faults.

Recipes included in this chapter

No.	Name
	Fruit
1	Grapefruit
2	Avocado pear
3	Fresh fruit salad
4	Tropical fruit plate
5	Caribbean fruit curry
	Vegetables
	Root vegetables
6	Buttered celeriac, turnips or swedes
7	Goats' cheese and beetroot tarts, with salad of watercress
8	Golden beetroot with Parmesan
9	Parsnips
10	Salsify
11	Vichy carrots
	Potatoes
12	Baked jacket potatoes
13	Chateau potatoes
14	Croquette potatoes
15	Delmonico potatoes
16	Duchess potatoes
17	Fondant potatoes
18	Fried or chipped potatoes
19	Hash brown potatoes
20	Macaire potatoes
21	Mashed potatoes
22	Parmentier potatoes
23	Potatoes cooked in milk with cheese
24	Potatoes with bacon and onions
25	Roast potatoes
26	Sauté potatoes
27	Sauté potatoes with onions
28	Swiss potato cakes (rösti)
	Other tubers
29	Puree of Jerusalem artichokes
	Bulbs
30	Braised onions
31	Caramelised button onions
32	Onion bhajias
33	Poached fennel
34	Roast garlic
	Leaves and brassicas
35	Braised red cabbage
36	Pickled red cabbage

37	Spinach purée
38	Stir-fried cabbage with mushrooms and beansprouts
Pods and seeds	
39	Broad beans
40	Corn on the cob
41	Haricot bean salad
42	Mangetout
43	Ladies' fingers (okra) in cream sauce
44	Peas French-style
Vegetable fruits	
45	Courgette and potato cakes with mint and feta cheese
46	Deep-fried courgettes
47	Shallow-fried courgettes
48	Fettuccini of courgette with chopped basil and balsamic vinegar
49	Roast squash or pumpkin with cinnamon and ginger
50	Spiced aubergine purée
51	Stuffed aubergine
52	Stuffed tomatoes
53	Tomato concassé
Stems and shoots	
54	Asparagus points or tips
55	Asparagus wrapped in puff pastry with Gruyere
56	Globe artichokes
57	Shallow-fried chicory
Fungi and mushrooms	
58	Sauté of wild mushrooms
59	Greek-style mushrooms
Flowers	
60	Broccoli

61	Cauliflower au gratin
62	Cauliflower polonaise
Vegetable protein	
63	Crispy deep-fried tofu
64	Oriental stir-fry Quorn
Sea vegetables	
65	Deep-fried seaweed
66	Japanese sea vegetable and noodle salad
67	Sea kale
Mixed vegetable dishes	
68	Alu-chole (vegetarian curry)
69	Chinese vegetables with noodles
70	Stuffed peppers
71	Coleslaw
72	Mixed vegetables
73	Ratatouille
74	Tempura
75	Vegetarian strudel
76	Crisp polenta and roasted Mediterranean vegetables
Salads	
77	Niçoise salad
78	Potato salad
79	Vegetable salad/Russian salad
80	Waldorf salad
81	Caesar salad
Stuffings	
82	Apple and prune stuffing

Prepare vegetables

Introduction

Fresh vegetables and fruit are important foods, both from an economic and nutritional point of view. On average, each person consumes 125–150 kilograms fruit and vegetables per year.

Buying fruit and vegetables is difficult because they are highly perishable and lose quality quickly if not properly stored and handled. Availability of fruit and vegetables also varies as a result of seasonal fluctuations and supply and demand. Automation in harvesting and packaging speeds up the handling process, which along with use of preservation methods, helps to retain quality.

Fruit and vegetables are an excellent source of antioxidants, which can lower the risk of heart disease, cancer and diabetes-related damage, and even slow down the body's natural ageing process.

Video: vegetable quality and buying points

http://bit.ly/1DCYHek

1.1 Classifications of vegetables

Vegetables can include tubers (potatoes), leaves and brassicas (lettuce), stems (asparagus), roots (carrots), flowers (broccoli), bulbs (garlic), pods and seeds (peas and beans) and botanical fruits such as cucumbers, squash, pumpkins and capsicums.

Other vegetable classifications include vegetable fruits, fungi and mushrooms, vegetable protein, nuts and seaweed or sea vegetables.

'Vegetable' is a culinary term that generally refers to the edible part of a plant. All parts of herbaceous plants (plants that have leaves and stems, and which die down at the end of the growing season) eaten as food by humans are normally considered to be vegetables. Although mushrooms actually belong to the biological family of fungi, they are also commonly considered to be vegetables.

In general, vegetables are thought of as being savoury, rather than sweet, although there are many exceptions (for example, parsnips, sweet potatoes, pumpkin). Some vegetables are scientifically classed as fruits (for example, tomatoes and avocados), but are commonly used as vegetables because they are not sweet.

Vegetables are eaten in a variety of ways – as part of main meals (served with poultry, meat or fish), as ingredients and as snacks.

Composition

Most vegetables contain at least 80 per cent water (the remainder is made up of carbohydrate, protein and fat). The water content varies depending on the type of vegetable: squashes contain a high percentage of water; potatoes contain a great deal of starch. Corn, carrots, parsnips and onions contain invert sugars.

Because of the high water content, most fruits and vegetables provide only minerals, certain vitamins and some roughage. However, these additions – particularly of vitamins such as ascorbic acid – can be of immense value in our diet.

A typical plant cell is like a balloon, not blown up with air but with water. The internal pressure in the cell is called **turgor pressure** and can be as high as nine times that of atmospheric pressure. When plant tissues are in a state of maximum water content they are said to be **turgid** or in a complete state of turgor. In a normal cell in full turgor the pressure is equalised by the elasticity of the cell wall.

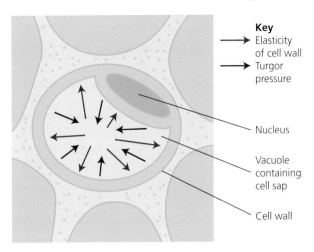

Key
→ Elasticity of cell wall
→ Turgor pressure

Nucleus

Vacuole containing cell sap

Cell wall

Equalised pressures within a plant cell

Once the supply of water is reduced or cut off to the cell, water is gradually lost from the plant by **transpiration** and the turgor pressure cannot be maintained. The elasticity of the cell wall now exceeds the outward turgor pressure and the cell starts to collapse. This is repeated throughout the plant, causing it to wilt. The rate of wilting varies: harvested leafy vegetables, such as lettuce, for example, only last short periods before becoming limp.

As plant cells become older, a complex substance called lignin is deposited in the cell wall, making it tougher. Water evaporates and sugars become concentrated – old, raw carrots, for example, are sweeter than young ones. Sugars change as soon as the vegetable is separated from the plant. (A good example is corn, which is often rushed straight from the stalk to the pot in order to preserve its taste.)

The cell wall of fruit and vegetable cells is made of carbohydrates, mainly cellulose. Cells are held together by **pectins** and **hemicellulose**; these change as the vegetable ripens, so that the cells part easily, giving a softer texture. Starch is the second most common carbohydrate after cellulose and is the main food material in most vegetables.

KEY TERMS

Pectin: a complex polysaccharide present in primary cells. Pectin is formed in the middle of plants and it helps bind cells together.

Cellulose: an insoluble substance; the main constituent of plant cell walls and of vegetable fibres such as cotton. The plant uses this for support in stems, leaves, husks of seeds and bark.

Hemicellulose: a polysaccharide similar to cellulose but consists of many different sugar building blocks in shorter chains than cellulose.

Transpiration: gradual loss of water from a plant cell.

Turgor pressure: the internal pressure inside a plant cell.

Turgid: plant cells that are in a state of maximum water content.

Factors affecting the composition of vegetables

- Origin – where vegetables are grown affects their composition and nutritive value. The type of soil (acid, alkali, rocky, sandy or clay) is also a factor.
- Season – most fruit and vegetables are seasonal and dependent on climate. The climate and the weather affect the size of fruits and vegetables and the yield per acre. Adverse weather conditions can reduce yields dramatically. Good weather conditions can increase yields and produce high quality fruits and vegetables.

1.2 Quality points

When selecting vegetables, the following quality points should be considered.

- Aroma – check the freshness through smell, which should be fresh, pungent and without any mouldy or musty smells.
- Freshness – vegetables should be firm to touch, with no bruising or withered leaves or parts.
- Type – check the fruit and vegetables are of the right type, degree of ripeness and correct specification for the chosen dish.
- Size – the vegetables should be of the correct size and grade against the specification.
- Damage – the vegetables should not be damaged in anyway. If they are, send them back to the supplier and obtain a credit note.
- Colour – make sure the vegetables are the correct colour. They should be at their best, ripe and ready to eat.
- Packaging – check the packaging is not damaged. If the packaging is damaged, discard the vegetables, return to the supplier and obtain a credit note.
- Temperature – fresh vegetables must be at ambient temperature. Frozen goods should be at –18 °C to –20 °C and chilled goods at –5 °C. These temperatures must be checked on delivery. Any produce out of the correct temperature range must be discarded and sent back to the supplier with a request for a credit note.
- Texture – all vegetables must be of the correct texture, ripe and ready for use.

Nutritional content

The nutritional content of different types of vegetables varies considerably. With the exception of pulses, vegetables do not contain much protein or fat. They contain water soluble vitamins such as vitamin B and vitamin C, fat soluble vitamins including vitamin A and vitamin D, as well as carbohydrates and minerals. Root vegetables contain starch or sugar for energy, a small but valuable amount of protein, some mineral salts and vitamins. They are also useful sources of **cellulose** (fibre) and water. Green vegetables are rich in mineral salts and vitamins, particularly vitamin C and carotene. The greener the leaf, the larger the quantity of vitamins it contains. The chief mineral salts are calcium and iron.

Fresh vegetables are an important part of our diet, so it is essential to pay attention to quality, purchasing, storage and efficient preparation and cooking if their nutritional content is to be conserved.

The EU has a quality grading system for vegetables:
- **Extra class:** produce of the highest quality
- **Class 1:** produce of good quality
- **Class 2:** produce of reasonably good quality
- **Class 3:** produce of low market quality.

Purchasing and selection

The purchasing of vegetables is affected by:
- the perishable nature of the products
- varying availability owing to seasonal fluctuations and supply and demand
- the effects of preservation, for example, freezing, drying and canning.

Fresh vegetables are living organisms and will lose quality quickly if not stored and handled properly. Automation in harvesting and packaging speeds the handling process and helps retain quality.

Roots

A root is the part of the plant growing down into the soil to support the plant and take up water and nutrients. Quality points for root vegetables will vary depending on type, but in general root vegetables must be:
- clean and free from soil
- firm and not soft or spongy
- sound and free from blemishes
- of an even size and shape.

Table 4.1 Root vegetables

Vegetable	Description	Uses	Quality points	Cooking methods
Beetroot	Two main types: round and long.	Soups, salads and as a vegetable. The tops of the beetroot can be eaten like spring greens.	Good colour without blemishes (when raw).	Steam, boil, roast, purée.
Carrot	Grown in numerous varieties and sizes.	Soups, sauces, stocks, stews, salads and as a vegetable.	Firm, without blemishes or any discolouration.	Boil, steam, roast.
Celeriac	Large, light brown celery-flavoured root.	Soups, salads and as a vegetable.	Firm, small to medium bulbs without any discoloration or blemishes.	Boil, steam, roast.
Horseradish	Long, light brown, narrow root.	Grated and used for horseradish sauce. Becoming more widely used in culinary work, in creams, foams and chutneys.	Fresh, firm roots without blemishes and discolouration.	Sauces, steaming.
Mooli	Long, white, thick member of radish family.	Soups, salads or as a vegetable.	Firm roots without any discolouration or blemishes.	Boil, steam, braise.
Parsnip	Long, white root, tapering to a point. It has a unique nut-like flavour.	Soups added to casseroles and as a vegetable (roasted, puréed, etc.).	Roots firm to touch, without any blemishes; creamy white in colour.	Boil, steam, roast.
Radish	Small summer variety; round or oval.	Served with dips, in salads or as a vegetable in white or cheese sauce.	Good red colour, firm to touch.	Salads, stir-fry.
Salsify and scorzonera (black salsify)	Also called oyster plant because of similarity of taste; long, narrow, white, or black-skinned root (scorzonera), slightly astringent in flavour.	Soups, salads and as a vegetable.	Firm to touch; good brownish colour.	Braise, boil.
Swede	Large root with yellow flesh.	As a vegetable mashed or parboiled and roasted; may be added to stews.	Unblemished without discoloration.	Boil, steam, roast.
Turnip	Two main varieties: long and round.	Soups, stews and as a vegetable.	Firm without blemishes.	Boil, steam, roast.

Tubers

A tuber is an underground stem, used for storing food, which is also able to produce new plants.

Table 4.2 Tubers

Vegetable	Description	Uses	Quality points	Cooking methods
Artichokes, Jerusalem	Potato-like tuber with a bittersweet flavour.	Soups, salads and as a vegetable.	Firm, without discolouration.	Roast, boil, steam.
Potatoes	Several named varieties of potato are grown and these will be available according to the season. The various varieties fall into four categories: floury, firm, waxy or salad. The different varieties have differing characteristics (see below).	Varies depending on variety.	Firm, smooth, not excessively wrinkled, withered or cracked. Do not buy those with a lot of sprouts or green areas.	Some varieties are more suitable for certain methods of cooking than others (see below).
Sweet potatoes	Long tubers with purple or sand coloured skins and orange or cream flesh; flavour is sweet and aromatic.	As a vegetable (fried, puréed, creamed, candied) or made into a sweet pudding.	Firm with no discolouration.	Boil, steam, roast, purée..
Yams	Similar to sweet potatoes; usually cylindrical, often knobbly in shape; popular in Caribbean cookery.	As for sweet potatoes.	Large and firm.	Boil, steam, roast, purée.

Potato varieties

All potatoes should be sold by name (for example, King Edward, Désirée, Maris Piper); this is important as the caterer needs to know which varieties are best suited for specific cooking purposes.

Cooked potatoes may have different textures, depending on whether they are waxy or floury varieties. This is due to changes that happen to the potato cells during cooking. Waxy potatoes are translucent and may feel moist and pasty. Floury potatoes are brighter, look grainier and feel drier. These differences affect the performance of the potato when cooked in different ways (for example, boiling versus roasting).

Floury potatoes are especially popular in the UK. They are suitable for baking, mashing and chipping as they have a soft, dry texture when cooked. They are not suitable for boiling, however, because they tend to disintegrate. Popular varieties of floury potato include King Edward and Maris Piper.

Waxy potatoes are more solid than floury potatoes and hold their shape when boiled, but do not mash well. They are particularly suitable for baked and layered potato dishes such as boulangère potatoes. Popular varieties include Cara and Charlotte.

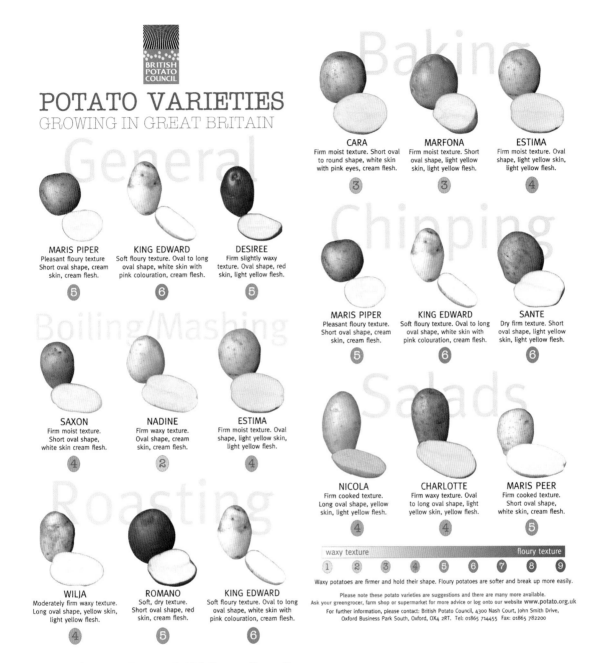

Varieties of potato (Source: British Potato Council)

Ready prepared potatoes

Potatoes can be bought in many convenience forms: peeled, turned, cut into various shapes for frying, scooped into balls (Parisienne) or an olive shape. Chips are available fresh, frozen, chilled or vacuum-packed. Frozen potatoes are available as croquettes, hash browns, sauté and roast. Mashed potato powder is also available. The advantages of using ready prepared potatoes are that they are labour saving and reduce wastage. However, they cost more than fresh potatoes and may not taste as good.

PROFESSIONAL TIP

Potato yields:
- 0.5 kg of old potatoes will yield approximately 3 portions.
- 0.5 kg of new potatoes will yield approximately 4 portions.
- 1.5 kg of old potatoes will yield approximately 10 portions.
- 1.25 kg of new potatoes will yield approximately 10 portions.

Table 4.3 Potato varieties and recommended cooking methods

Potato variety	Cooking methods
Cara	Boil, bake, chip, wedge
Charlotte	Boil, salad use
Desiree	Boil, roast, bake, chip, mash, wedge
Golden	Wonder Boil, roast, crisps
King Edward	Boil, bake, roast, mash, chip
Maris Piper	Boil, roast, bake, chip
Pink Fir Apple	Boiled, salad use
Premiere	Boil
Record	Crisps
Romano	Boil, bake, roast, mash
Saxon	Boil, bake, chip
Wilja	Boil, bake, chip, mash

Bulbs

A bulb is an organ found in some plants which consists of an underground axis with many thick overlapping leaves.

Table 4.4 Bulbs

Vegetable	Description	Uses	Quality points	Cooking methods
Fennel	The bulb is the swollen leaf base and has a pronounced flavour.	Raw in salads and cooked.	Firm bulbs without any discolouration.	Boil, steam, braise, stir fry, roast, and grill.
Garlic	An onion-like bulb with a papery skin, inside of which are small individually wrapped cloves. Garlic has a pungent, distinctive flavour and should be used sparingly.	Used extensively in many forms of cookery; may be roasted and served as a garnish.	Firm to touch, no discolouration.	Roast, sauté.
Leeks	Summer leeks have long white stems, bright green leaves and a milder flavour than winter leeks, which have a stockier stem and a stronger flavour.	Stocks, soups, sauces, stews, hors d'oeuvre and as a vegetable.	Fresh, no signs of woodiness (lignin).	Boil, steam, braise, stir fry.
Onions	There are numerous varieties with different coloured skins and varying strengths.	After salt, the onion is probably the most frequently used flavouring in cookery; it can be used in almost every type of food except sweet dishes.	Firm to touch.	Boil, steam, roast, purée, grill.
Shallot	Has a similar but more refined flavour than the onion and is therefore more often used in a wide range of culinary work.	May be roasted, used with grated vegetables and stir fries.	Firm to the touch.	Sauté, braise.
Spring onions	Small and slim, like miniature leeks. Ramp looks like a spring onion but is stronger.	Soups, salads and Chinese and Japanese cookery, with grilled and roasted vegetables.	Tops must be a bright green colour, bulbs firm to touch.	Stir fry.

Leaves

A leaf is part of the plant that lives off water and makes food by photosynthesis.

Table 4.5 Leaves

Vegetable	Description	Uses	Quality points
Chicory	A lettuce with coarse, crisp leaves and a sharp, bitter taste in the outer leaves, inner leave are milder.	Braise or roast.	A good white yellowish colour.
Chinese leaves	Long, white, densely packed leaves with a mild flavour, resembling celery.	Makes a good substitute for lettuce; boiled, braised or stir fried as a vegetable.	Purchase fresh, firm leaves without blemishes.
Cress	There are 15 varieties of cress with different flavours.	Suitable for a large range of foods.	Leaves must be firm and bright in colour.
Lettuce	Many varieties including cabbage, cos, little gem, iceberg, oak leaf, Webb's.	Chiefly for salads, or used as a wrapping for other foods, e.g. fish fillets.	Leaves must be bright in colour and firm.
Nettles	Once cooked, the sting disappears.	Soups.	Should be picked young.
Sorrel	A type of cress with larger leaves and a peppery taste.	Salads, garnishes.	Leaves must be firm, not withered or bruised.
Spinach	Bright green, sour leaves, which can be overpowering if used on their own.	Salad and soups.	Best when tender and young. Leaves should be bright green, not withered or bruised.
Swiss chard	Tender, dark green leaves with a mild musky flavour.	Soups, garnishing egg and fish dishes, as a vegetable and raw in salads. May also be used in stir-fries.	Leaves must be bright green, not withered or bruised.
Vine leaves	All leaves from grape vines can be eaten when young.	To wrap food such as meat or fish before further cooking.	Must be a good, smooth, dark green colour.
Watercress	Long stems with round, dark, tender green leaves and a pungent peppery flavour.	Soups, salads, and for garnishing roasts and grills of meat and poultry.	Leaves must be bright green colour, not withered or bruised.

Flowers

A flower is the reproductive part of the plant. It is often brightly coloured to attract insects.

Table 4.6 Flowers

Vegetable	Description	Uses	Quality points	Cooking method
Broccoli	Various types including calabrese, white, green, purple sprouting, delicate vegetable with a gentle flavour.	Soups, salads, stir-fry dishes and cooked and served in many ways as a vegetable.	Flower must be bright in colour, green without any discolouration.	Steam or boil.
Cauliflower	Heads of creamy white florets with a distinctive flavour.	Soup, and cooked and served in various ways as a vegetable.	Flowers must be firm without any discolouration.	Various. May also be puréed.

Brassicas

Brassicas are plants that are part of the mustard family.

Table 4.7 Brassicas

Vegetable	Description	Uses	Quality points	Cooking method
Brussels sprouts	Small green buds growing on thick stems.	Can be used for soup but are mainly used as a vegetable; can be cooked and served in a variety of ways.	Small sprouts are better, with a fresh clean flavour that is not overpowering. Bright green colour, no discoloration, blemishes or bruising.	Boil, steam, shred or shallow fry.
Cabbage	Three main types including green, white and red, many varieties of green cabbage available at different seasons of the year; early green cabbage is deep green and loosely formed; later in the season they firm up, with a solid heart; Savoy is considered the best of the winter green cabbage.	Green and red as a vegetable; white cabbage is used for coleslaw.	Good colour, no discolouration or withering.	Boiled, braised or stir fried.
Kale and curly kale	Thick green leaves; the curly variety is the most popular.	Stir-fries.	Leaves must be firm, without any discoloration or withering.	Boiled, steamed or braised.
Pak choi	Chinese cabbage with many varieties.	Stir-fries.	Leaves must be crisp and firm, good colour without any withering.	Boiled, steamed or braised.

Pods and seeds

A pod is a fruit or seed case that usually splits along two seams to release its seeds when mature. A seed is the reproductive part of flowering plants.

Table 4.8 Pods and seeds

Vegetable	Description	Uses	Quality points	Cooking method
Broad beans	Pale green, oval-shaped beans contained in a thick fleshy pod.	Young broad beans can be removed from the pods, cooked in their shells and served as a vegetable in various ways; old broad beans will toughen when removed from the pods and will have to be shelled before being served. Used as a vegetable, garnish, stir-fries and stews.	Must be firm to touch, good colour.	Boil, steam.
Butter or lima beans	Butter beans are white, large, and flattish and oval shaped; lima beans are smaller.	As a vegetable or salad, stew or casserole ingredient.	Good colour, firm.	Boil, steam.
Mangetout	Also called snow peas or sugar peas; flat pea pod with immature seeds that, after topping, tailing and stringing, may be eaten in their entirety.	As a vegetable, in salads and for stir fry dishes.	Must be bright green in colour, firm without any discoloration.	Boil, steam, shallow fry, stir-fry.
Okra	Curved and pointed seed pods with a flavour similar to aubergines.	Cooked as a vegetable or in Creole-type stews.	Must be green in colour and firm.	Boil, steam, shallow fry or braise.
Peas	Garden peas are normal size, petit pois are a dwarf variety, marrowfat peas are dried.	Popular as a vegetable. Peas are also used for soups, salads, stews and stir-fry dishes.	Must be a good green colour.	Usually boiled.
Runner beans	Popular vegetable that must be used when young.	As a vegetable, used in salads and stir-fries.	Bright green colour and a pliable velvety feel; if coarse, wilted or older beans are used they will be stringy and tough. Must be bright green colour and firm.	Boil, steam, also used in stir-fries.
Sweetcorn	Also known as maize or Sudan corn; available 'on the cob', fresh or frozen or in kernels, canned or frozen.	A versatile commodity and used as a first course, in soups, salads, casseroles and as a vegetable.	Bright yellow in colour, no discoloration.	Boil, steam, also used in stir-fries.

Stems and shoots

A stem is the part of the plant usually visible above the ground which bears the buds, leaves, flowers and fruit, however, the stem can also be underground. A shoot is a young stem.

Table 4.9 Stems and shoots

Vegetable	Description	Uses	Quality points	Cooking method
Asparagus	Three main types: white, with creamy white stems and a mild flavour; French, with violet or bluish tips and a stronger more astringent flavour; green, with what is considered a delicious aromatic flavour.	Used on every course of the menu, except the sweet course.	Firm, crisp with no discoloration.	Boil, steam or stir-fry.
Bean sprouts	Slender young sprouts of the germinating soya or mung bean.	As a vegetable accompaniment, in stir-fry dishes and salads.	Crisp.	Stir-fry, salads.
Cardoon	Longish plant with root and fleshy ribbed stalks, white to light green in colour.	Soups, stocks, sauces, cooked as a vegetable and raw in salads and dips.	Firm to touch without discoloration.	Boil, steam or braise.
Celery	Long-stemmed bundles of fleshy, ribbed stalks, white to light green in colour.	Soups, stocks, sauces, cooked as a vegetable and raw in salads and dips.	Crisp and firm.	Boil, steam or braise.
Chicory	Also known as Belgian endive; conical heads of crisp white, faintly bitter leaves.	Cooked as a vegetable and raw in salads and dips.	No discoloration or withered leaves.	Steam or braise.
Globe artichokes	Resemble fat pine cones with overlapping fleshy, green, inedible leaves, all connected to an edible fleshy base or bottom.	Used as a first course, hot or cold, stuffed, as a vegetable and in casseroles. The fronds may be used in casseroles, stews and as a garnish.	Good bright green colour; leaves must be firm and crisp.	Boil, baked, fry, or in casseroles.
Kohlrabi	Stem that swells to turnip shape above the ground, those about the size of a large egg are best for cookery purposes (other than soup or purées).	May be cooked as a vegetable, stuffed and baked and added to stews and casseroles.	Firm to touch with no discoloration.	Boil, steam, braise.
Palm hearts	The buds of cabbage palm trees. Firm ivory colour.	Salads, pasta dishes, stews, casseroles.	Good colour, tightly closed.	Boil, steam, braise, salads.
Samphire	The two types are marsh samphire, which grows in estuaries and salt marshes, and white rock samphire (sometimes called sea fennel), which grows on rocky shores. Marsh samphire is also known as glass wort and sometimes sea asparagus.	As a garnish and in stir-fries.	Good colour, firm to touch.	Boil or steam.
Water chestnuts	Common name for a number of aquatic herbs and their nut-like fruit; best known type is the Chinese water chestnut, sometimes known as the Chinese sedge.	Popular in Chinese and Thai dishes.	Firm, without discoloration.	Boil, steam and stir-fry.

Vegetable fruits

These are fruits of plants that are used as a vegetable.

Table 4.10 Vegetable fruits

Vegetable	Description	Uses	Quality points	Cooking method
Aubergine	Firm, elongated, varying in size with smooth shiny skin, ranging in colour from purple red to purple black; inner flesh is white with tiny soft seeds, almost without flavour, it requires other seasonings, e.g. garlic, lemon juice, herbs, to enhance its taste. Varieties include baby, Japanese, white, striped, Thai.	Popular sliced in layered dishes such as moussaka. Also in pastes and dips.	No discoloration or blemishes.	Sliced and fried or baked, steamed or stuffed.
Avocado	Fruit that is mainly used as a vegetable because of its bland, mild, nutty flavour. Two main types are the summer variety, which is green when unripe and purple black when ripe, with golden yellow flesh, and the winter variety, which is more pear-shaped with smooth green skin and pale green to yellow flesh.	Eaten as first courses and used in soups, salads, dips and as garnishes to other dishes, hot and cold.	Purchase when slightly unripe, without bruises.	Salads.
Courgettes	Baby marrow, light to dark green in colour, with a delicate flavour, becoming stronger when cooked with other ingredients, e.g. herbs, garlic, spices.	As a vegetable or ingredient in other dishes. Grated in salads.	Smooth shiny skins, firm.	Boil, steam, fry, bake, stuff and stir-fry.
Cucumber	A long, smooth-skinned fruiting vegetable, ridged and dark green in colour.	Salads, soups, sandwiches, garnishes and as a vegetable. May also be, puréed with fromage frais, yoghurt or cream and used as a dip.	Green, firm and crisp.	Braise, purée.
Marrow	Long, oval-shaped edible gourds with ridged green skins and a bland flavour.	As a vegetable, used in stir-fries.	No discoloration or bruising.	As for courgettes.
Pepper	Available in three colours. Green peppers are unripened and they turn yellow to orange and then red (they must remain on the plant to do this).	Used raw and cooked in salads, vegetable dishes, stuffed and baked, casseroles and stir-fry dishes.	Bright colour, firm and crisp.	Sauté, braise, stir-fry.
Pumpkin	Vary in size and can weigh up to 50 kg; associated with Halloween as a decoration.	Soups or pumpkin pie.	Firm to touch, no discoloration.	Steam, boil, braise.
Squash	Many varieties, e.g. acorn, butternut, summer crookneck, delicate, hubbond, kuboche, onion.	Soups, salads, vegetable dishes.	Flesh must be firm and glowing; all types must have a good colour without bruising or discoloration.	Boil, bake, steam or purée.
Tomatoes	Along with onions, probably the most frequently used 'vegetable' in cookery; several varieties, including cherry, yellow, globe, large ridged (beef) and plum. Vine-ripened tomatoes are allowed to ripen on the vine before being picked. Commercial tomatoes are picked when green, so that they are easier to handle. They don't bruise or break; they will turn red en route to the suppliers.	Soups, sauces stews, salads, sandwiches and as a vegetable. Sundried for use when fresh tomatoes lack flavour.	Firm, even colour, no bruising or discoloration.	Grill, roast, fry, salads.

Mushrooms and fungi

All mushrooms, both wild and cultivated, have a great many uses in cookery, in soups, stocks, salads, vegetables, savouries and garnishes. Wild mushrooms are also available in dried form. All mushrooms must be firm to touch, with no presence of slime or discolouration or unpleasant smell. They should have a good colour.

- Mushrooms – field mushrooms are found in meadows from late summer to autumn; they have a creamy white cap and stalk and a strong earthy flavour. Used in a variety of dishes; can be grilled or shallow fried.
- Cultivated mushrooms – available in three types: button (small, succulent, weak in flavour), cap and open or flat mushrooms. Used in a variety of dishes; can be grilled or shallow fried.
- Ceps – wild mushrooms with short, stout stalks with slightly raised veins and tubes underneath the cap in which the brown spores are produced. Grill or shallow fry.
- Chanterelles or girolles – wild, funnel-shaped, yellow-capped mushrooms with a slightly ribbed stalk that runs up under the edge of the cap. Grill or shallow fry.
- Horns of plenty – trumpet-shaped, shaggy, almost black wild mushrooms. Grill or shallow fry.
- Morels – delicate, wild mushrooms varying in colour from pale beige to dark brown–black with a flavour that suggests meat. Grill, shallow fry, often used as a garnish.
- Oyster – creamy gills and firm flesh; delicate with shorter storage life than regular mushrooms. Grill or shallow fry.
- Shiitake – solid texture with a strong, slightly meaty flavour. Grill or shallow fry.
- Truffles – black (French) and white (Italian) are rare, expensive but highly esteemed for the unique flavour they can give to so many dishes. Black truffles from France are sold fresh, canned or bottled; white truffles from Italy are never cooked, but grated or finely sliced over certain foods, such as pasta and risotto.

Clockwise from top left: girolles, chestnut mushrooms, hedgehog mushrooms, large cup (Portobello) mushroom, button mushrooms, brown chanterelles and shiitake mushrooms

Vegetable protein

- Soya products – these are high in protein and produced from leguminous seeds. They contain important essential amino acids that cannot be made in the body. The soya bean is an important world product from which many products are made. Soya flour is used to boost the protein in a variety of dishes.
- Tofu – this is produced from soya beans. The beans are boiled, puréed and pressed through a sieve to produce soya milk. The milk is then processed in a similar way to soft cheese. Always keep tofu in a refrigerator and follow the instructions on the packet. Tofu is also known as bean curd. Silken tofu is a Japanese-style tofu with a softer consistency than regular tofu. It is used in soups, sauces and desserts, and is seen as a healthier option to cream.
- Tempeh – this is a fermented soya bean product that is cooked and pressed. It is similar to tofu, but has a firmer texture and is purchased chilled.
- Seitan – this is produced from wheat; it has a high gluten content and is used as a vegetarian alternative.
- Textured vegetable protein (TVP) – this is a meat substitute manufactured from protein derived from wheat, oats, cottonseed, soya beans and other sources. The main source of TVP is the soya bean, because of its high protein content. TVP is used mainly as a meat extender, to make meat go further. The TVP content can vary from 10 to 60 per cent replacement of fresh meat. Some caterers on very tight budgets use it, but it is mainly used in food manufacturing. By partially replacing the meat in certain dishes – such as casseroles, stews, pies, pasties, sausage rolls, hamburgers, meat loaf and pâté – it is possible to reduce costs, meet nutritional targets and serve food that looks acceptable.

Vegetable proteins

Mycoprotein

Mycoprotein is a meat substitute produced from a plant that is a distant relative of the mushroom. It contains protein and fibre, and is the result of a fermentation process similar to the way yoghurt is made. It may be used as an alternative to chicken or beef or in vegetarian dishes.

Quorn is a mycoprotein. It is a low-fat food that can be used in a variety of dishes (for example, oriental stir-fry). Quorn does not shrink during preparation and cooking. Quorn mince or pieces can be substituted for chicken or minced meats. Its mild savoury flavour means that it complements the herbs and spices in a recipe and it is able to absorb flavour. Frozen Quorn may be cooked straight from the freezer, or may be defrosted overnight in the refrigerator. Once thawed, it must be stored in the refrigerator and used within 24 hours.

Seaweed or sea vegetables

Seaweed is commonly used as a food in Japan, but less so in Europe and America. Some examples of edible seaweed are described below.

- Nori or purple laver – a purplish-black seaweed often seen wrapped round sushi rice in sushi. Nori grows as a very thin, flat reddish-black seaweed and is found in most temperate zones around the world.
- Honori or green laver – occurs naturally in the bays and gulfs of southern areas of Japan.
- Kombu or huidai – the Japanese name for the dried seaweed that is derived from a mixture of *Laminaria* species.
- Wakame or quandai-cai – this is a brown, stringy seaweed, deep green colour that occurs on rocky shores and bays in the temperate zones of Japan, the Republic of Korea and China. Wakame has a high dietary fibre content, higher than nori or kombu. It is used in miso soup.
- Hiziki (*Hizikia fusiforme is*) – a brown seaweed collected from the wild in Japan.
- Mozuku – a brown seaweed that is harvested from natural populations in the more tropical climate of the southern islands of Japan.
- Dulse – a red algae with leathery fronds (leaves). It is found in Eastern Canada and is especially abundant around Grand Manan Island. In Nova Scotia and Maine, dried dulse is often served as a salty cocktail snack.

- Irish moss or carrageenan – used in Irish foods and as a setting agent.
- Sea moss (ogo) – collected and sold as a salad vegetable in Hawaii.
- Arame – a seaweed that comes in small, flaky strips. It is used in soups, salads and stir-fries. It goes well with sushi and tofu.

Nuts

There are many varieties of nuts available, including:

- almonds
- Brazil nuts
- cashew nuts
- coconuts
- hazelnuts
- macadamia nuts
- peanuts
- pecans
- pine nuts
- pistachios
- sweet chestnuts
- walnuts.

Nuts in the shell have longer storage potential than shelled nuts. Broken pieces are more perishable than halves or whole kernels. A rancid nut has a flat, metallic taste and a lingering aftertaste. Because of their high fat content, nuts can easily absorb odours from external sources and therefore should not be stored around other foods that have strong odours, unless they are stored in tightly sealed containers.

Seasons for commonly used vegetables

With the advancement and development of transportation and refrigeration, vegetables and fruit can be purchased from all over the world. In many cases, seasons no longer exist; for example, we can enjoy strawberries all year round.

However, knowing where fruit and vegetables come from (**provenance**) has become increasingly important, because today's consumer is concerned about the environment, carbon omissions and restricting the amount of food travelling around the world using aircraft cargo. Many consumers now prefer to purchase from local suppliers and therefore knowledge of when fruits and vegetables are in season is important. Table 4.11 shows the UK seasons for commonly used vegetables.

> **KEY TERM**
>
> **Provenance:** where food comes from, for example, where it is grown, reared, produced or finished.

Table 4.11 UK seasons for commonly used vegetables

	Spring (March, April, May)	Summer (June, July, August)	Autumn (September, October, November)	Winter (December, January, February)
Asparagus	■			
Aubergine		■		
Beetroot		■		
Broad beans		■		
Brussels sprouts				■
Cabbage				■
Carrots		■	■	
Cauliflower	■			
Celeriac			■	■
Chicory				■
Courgettes		■		
Cucumber		■		
Field mushrooms			■	
Fennel		■		
Fresh peas		■		
Garlic		■		
Green beans		■		
Jersey Royal new potatoes	■			
Jerusalem artichoke				■
Kale			■	■
Leeks			■	
Lettuce and salad leaves		■	■	
Marrow			■	
New potatoes		■		
Parsnips				■
Potatoes			■	
Pumpkin			■	
Purple sprouting broccoli	■			
Radishes		■		
Red cabbage				■
Rocket		■		
Runner beans		■	■	
Salad onions		■		
Savoy cabbage	■			
Sorrel	■	■		
Spring greens	■			
Spring onion	■			
Squashes			■	
Swede				■
Sweetcorn			■	
Tomatoes		■		
Turnips				■
Watercress	■	■		

1.3 Preparing and preserving vegetables

Preparation methods

- Washing – vegetables should be washed to remove dirt and foreign bodies if they are not going to be peeled.
- Sorting – vegetables are sorted into grade quality and size, sometimes country of origin and producer.
- Peeling – removing the skin of a vegetable when it is raw or cooked.
- Trimming – trimming vegetables into shape and size for even cooking and presentation.
- Cutting – dicing, slicing and shredding vegetables. Chopping and grating are also ways of cutting vegetables.
- Crushing – used for vegetables such as garlic; also covers the crushing of nuts.
- Scoring – this is when vegetables are scored prior to cooking to assist with the penetration of heat, for example potatoes before baking.
- Deseeding – removal of the seeds from vegetables
- Coating – coating vegetables in flour, batter or breadcrumbs before cooking.

- Stuffing – vegetables can be stuffed with a filling before or after cooking, for example stuffed peppers, stuffed tomatoes.
- Blanching – vegetables are blanched to remove the skin or to cook for a short time to deactivate the enzymes prior to freezing.
- Soaking – generally refers to the soaking of dried vegetables and pulses prior to cooking.
- Concassé – generally refers to a dice of tomato flesh, but may also be applied to other vegetables, such as peppers and mushrooms.
- Marinating – vegetables are immersed in liquid such as wine, vinegar, fruit juice, herbs and spices to add flavour prior to cooking.
- Tying – vegetables are tied with string or twine to form a shape or to keep their shape during the preparation and cooking process.
- Portioning – vegetables are cut into portion sizes during preparation or after cooking. Portioning is important for cost control and to prevent waste.

Cuts of vegetables

The size to which vegetables are cut may vary according to their use. However, the shape does not change unless overcooked.

Step by step: julienne (strips)

1 Cut the vegetables into 2-cm lengths (short julienne).

2 Cut the lengths into thin slices.

3 Cut the slices into thin strips.

4 Double the length gives a long julienne, used for garnishing (e.g. salads, meats, fish and poultry dishes).

Step by step: brunoise (small dice)

1 Cut the vegetables into convenient-sized lengths.

2 Cut the lengths into 2-mm slices.

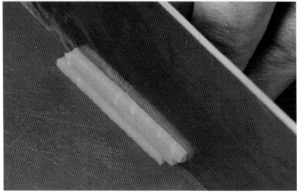

3 Cut the slices into 2-mm strips.

4 Cut the strips into 2-mm squares.

Step by step: jardinière (batons)

1 Cut the vegetables into 1.5-cm lengths.

2 Cut the lengths into 3-mm slices.

3 Cut the slices into batons (3 × 3 × 18 mm).

Step by step: macédoine (0.5-cm dice)

1 Cut the vegetables into convenient lengths.

2 Cut the lengths into 0.5-cm slices.

3 Cut the slices into 0.5-cm strips.

4 Cut the strips into 0.5-cm squares.

Paysanne

There are at least four accepted methods of cutting paysanne. In order to cut economically, the shape of the vegetables should dictate which method to choose. All are cut thinly. Options include:

- 1-cm sided triangles
- 1-cm sided squares
- 1-cm diameter rounds
- 1-cm diameter rough-sided rounds.

Paysanne vegetables

Concassé

This means roughly chopped (e.g. skinned and deseeded tomatoes are roughly chopped for many food preparations).

> **PROFESSIONAL TIP**
> It is important to cut the vegetables into even-sized pieces to enable them to cook evenly.

Tomato concassé

Preservation methods

- Canning – some vegetables are preserved in tins, for example, artichokes, asparagus, carrots, celery, beans, peas (fine, garden, processed), tomatoes (whole, purée), mushrooms and truffles. Note that canning involves a process of sterilisation, and this causes the vegetables to become overcooked.
- Dehydration – onions, carrots, potatoes and cabbage are shredded and quickly dried until they contain only five per cent water. Drying and rehydration affects the quality of the vegetables due to the processing technology.
- Drying – the seeds of legumes (peas and beans) have their moisture content reduced to 10 per cent.
- Pickling – onions and red cabbage are examples of vegetables preserved in spiced vinegar. Pickling changes the taste of the vegetables.
- Salting – French and runner beans may be sliced and preserved in dry salt, which affects the flavour.
- Freezing – many vegetables, such as peas, beans, sprouts, spinach and cauliflower, are deep frozen. Frozen vegetables have to be cooked carefully to avoid overcooking.
- Blanching – a technique which is used to partly cook and preserve the colour of green vegetables by plunging them in boiling water and then running them under cold water. It is also used to remove skin from tomatoes. Blanching is also used to cook potato chips without colour; they can then be stored and, as they are required, are plunged into hot oil to make crisp.
- Vacuum-packing – a technique which involves placing perishable food in a plastic film package, removing the air from inside and sealing the package. This is a good method for preserving the quality of vegetables.
- Chilling – vegetables should be chilled at approximately 3°C–5°C. This process preserves vegetables for on average three to five days, with a maximum of eight days depending on the type of vegetable.
- Bottling – a method of preservation which uses a vinegar (usually acetic acid) and pasteurisation. For sauerkraut the preserving medium is lactic acid.

1.4 Coatings and stuffings

There are various ways of cooking vegetables with a coating. The most common is in a light batter (tempura) and deep fried. Some vegetables, such as cauliflower, may be dipped in milk and flour and deep fried. The flour used can be white, wholemeal or gramflour. Other vegetables may be coated in flour, egg and breadcrumbs and shallow or deep fried. Examples of coatings include:

- tempura batter – used with a range of vegetables, peppers, mushrooms, celery, French beans, cauliflower, onions, etc.
- breadcrumbs (pané à l'Anglais) – used with mushrooms, potatoes, butternut squash
- milk and flour (pané à la Française) – used with broccoli and cauliflower.

Mushroom duxelle, which is used to stuff tomatoes, may be used to coat vegetables such as braised celery and sprinkled with breadcrumbs, toasted and served.

Breadcrumbs with additions (chopped nuts, herbs and spices) may be used to coat vegetables before they are served, or brushed with melted butter or vegetable oil and toasted and served. For more examples of coatings and stuffings used with vegetables, see the recipe section at the end of the chapter.

> **PROFESSIONAL TIP**
> Freshly chopped herbs may be added to the breadcrumbs or batter (for example, chopped parsley, chervil, basil). Ground spices (for example, cinnamon, nutmeg, cardamom, chopped nuts, almonds, pine nuts or Brazil nuts) may also be added to the breadcrumbs.

1.5 Storage

Vegetables do not stay fresh for very long and this causes particular problems. Fresh vegetables are living organisms and will lose quality quickly if they are not properly stored and handled. The fresher the vegetables, the better their flavour, so ideally they should not be stored at all. However, as storage is often necessary it should be for the shortest time possible.

Many root and non-root vegetables that grow underground can be stored over winter in a root cellar or other similarly cool, dark and dry place, to prevent the growth of mould, greening and sprouting. It is important to understand the properties and weaknesses of the particular roots to be stored (i.e. what will help them to last longer and what will damage them quickly). Stored correctly, these vegetables can last through to early spring and be almost as nutritious as when they were fresh.

Potatoes can be stored for several months without affecting their quality, but they should be stored at a constant temperature of 3°C. If this is not possible, buying fresh potatoes regularly is best practice. There are three essential rules to bear in mind when storing potatoes: dry, dark and cool. Light should be avoided as this will cause sprouting and eventually the greening effect that contains mild toxins.

Note that when storing fresh cooked and blanched vegetables they must be date labelled and covered with cling film or placed in a sealed container. Nearly all fresh fruit and vegetables stored at approximately 3 °C require a relative humidity of 95 per cent. Relative humidity is the amount of water vapour present in the air expressed as a percentage of vapour needed for saturation at the same temperature.

During storage, leafy vegetables lose moisture and vitamin C. They should be stored for as short a time as possible in a container, such as a plastic bag or a sealed plastic container in a cool place,

- Store all vegetables in a cool, dry, well-ventilated room at an even temperature of 4–8 °C to help minimise spoilage.
- Check vegetables daily and discard any that are unsound.
- Remove root vegetables from their sacks and store in bins or racks.
- Store green vegetables on well-ventilated racks.
- Store salad vegetables in a cool place and leave in their containers.
- Store frozen vegetables at –18 °C or below.
- Check for use-by dates, damaged packaging and any signs of freezer burn.

Produce vegetable dishes

2.1 Cooking methods

Only approximate times for cooking vegetables are given in the recipes in this chapter - quality, age, freshness and size all affect the length of cooking time required. Young, freshly picked vegetables will need to be cooked for a shorter time than vegetables that have been allowed to grow older and that may have been stored after picking.

Most vegetables are cooked **al dente**, which means the vegetables are cooked until they are firm, crisp and with a bite. This retains some of the vital vitamins such as vitamin C.

Vegetables that contain high levels of starch (e.g. potatoes) need to be cooked through completely, to soften the starch.

Boiling and blanching

As a general rule, all root vegetables (except new potatoes) are put in cold salted water, which is then heated until boiling. Vegetables that grow above the ground are put directly into boiling salted water, so they are cooked for the minimum amount of time. Cooking them as quickly as possible helps them to keep their flavour, food value and colour.

Delicate vegetables – particularly green vegetables – must be blanched in salted boiling water and then refreshed in ice-cold water to stop the cooking process. The main reason for this is because, between the temperatures of 66 °C and 77 °C, chlorophyll (the pigment in green plants) is unstable. To retain the colour of the vegetables it is important to get through this temperature zone as quickly as possible.

Steaming and stir-frying

All vegetables cooked by boiling may also be cooked by steaming. The vegetables are prepared in exactly the same way as for boiling, then placed into steamer trays, lightly seasoned with salt and steamed under pressure. As with boiling, they should be cooked for the minimum period of time in order to conserve maximum food value and retain colour. High-speed steam cookers are ideal for this purpose and, because they cook so quickly, can be used for batch cooking (cooking in small quantities throughout the service) so that large quantities do not have to be cooked prior to service, then refreshed and reheated.

Many vegetables can be cooked from raw by stir-frying, which is a quick and nutritious method of cooking.

> **HEALTH AND SAFETY**
> Make sure all vegetables are dry before placing into hot oil. Always have a spider and basket at hand when deep frying in order to remove the food quickly and safety if there is a problem. Wear protective clothing (a chef's jacket with long sleeves) and make sure every chef is confident of the procedures and systems for frying and knows the fire drill.

Roasting

The vegetables are washed, peeled and cut to regular size, placed into a roasting tray with a little vegetable oil and seasonings (for example salt, pepper, chopped herbs, crushed garlic; spices such as all spice, cinnamon and nutmeg) may also be added. The oven should be at approximately 200 °C.

Roast potatoes are cut into chunks and parboiled slowly until just cooked on the inside before draining, tossing in a liberal amount of olive oil or other vegetable oil and seasoning and roasting. They are roasted in a hot oven (220 °C) for 15 minutes, then the temperature is turned down to 180 °C and the potatoes are basted every five minutes to ensure a crisp coat all over. Once cooked, serve immediately, as keeping them warm for too long will cause the inside to steam, making the outside soft and leathery.

Grilling or barbequing

This is a healthy way of cooking vegetables as little or no fat is used. Some vegetables are better parboiled before grilling (for example, potatoes, carrots, fennel, yams and butternut squash). The vegetables may also be sprinkled with fresh chopped herbs or spices. Vegetables may be grilled under the salamander, on grill plates or on a barbeque.

> **PROFESSIONAL TIP**
> When grilling, roasting or stir-frying vegetables, the oil or fat chosen will add to the flavour and give a distinctive background taste to the vegetables. Fats and oils carry important flavour components.

Stewing

Stewing is a traditional method of cooking vegetables. The vegetables are cut into even pieces, gently sweated on the stove in a little oil, then a stock or tomato juice is added and they are allowed to simmer on the stove or in an oven. The stewed vegetables may be flavoured with chopped herbs and/or spices. A classical stewed vegetable dish is ratatouille (this consists of aubergines, peppers, courgettes, onions and tomatoes in their juices).

Braising

Vegetables most suitable for braising include potatoes, onions, cabbage, fennel and celery. Braising requires most vegetables to be blanched (with the exception of potatoes) and placed in a shallow pan with stock (this should come two-thirds of the way up the pan, so as not to cover the vegetables).

Cooking liquids

Appropriate cooking liquids for vegetables include:
- boiling water
- boiling stock – vegetable, meat or fish
- wine – white or red
- some fruit juices – for example, elderflower, orange, cranberry, pomegranate
- mineral water – Vichy is the classical mineral water used to cook carrots
- vegetable oil – used for shallow and deep frying

> **TAKE IT FURTHER**
> In order to keep up to date it is advisable to regularly look at different restaurant menus to see how fruit and vegetables are being used on the menu, for example, to see what new fruit and vegetables are being introduced. Employers and guest chefs should be invited into colleges and training establishments to show how they are being innovative and creative with fruits and vegetables.

2.2 Presenting finished dishes

It is important to finish vegetable dishes at the point of service to retain their temperature and for green vegetables to retain their colour. There are a variety of finishes for vegetable dishes based on individual recipes. Some vegetables are simply brushed with butter, (although this is not recommended for heathy eating dishes and olive oil is a good substitute). Vegetable dishes may be finished with pine nuts, sesame seeds, chopped mixed herbs, cream or fromage frais, a suitable sauce such as cheese sauce, hollandaise, red wine jus, toasted almonds, toasted breadcrumbs or toasted brioche crumbs with mixed herbs or spices.

Holding for service

The professional way to hold vegetables for service is to cook them in advance, normally al dente, chill them quickly either in ice water or cold running water, or more appropriately in a blast chiller. Hold them chilled until required for service, when they are reheated depending on the requirements of type of vegetable and the recipe. Reheat either in a chauffent, a saucepan of boiling water, or quickly in a little oil or butter. The recommended methods of reheating is in a combination oven, with an injection of steam, or quickly in a microwave.

When holding vegetables for service it is important not to leave vegetables hot for too long in deep containers as they will lose some of their nutritional value and colour.

Storing vegetables after cooking

It is important to store vegetables not for immediate use in a refrigerator at 5 °C.

> **HEALTH AND SAFETY**
> To prevent bacteria from raw vegetables passing on to cooked vegetables, store them in separate areas. Thaw out frozen vegetables correctly and *never* refreeze them once they have thawed out.

2.3 Evaluation

See page 158 for Evaluation of fruit and vegetables.

Prepare fruit

1.1 Classifications of fruit

Fruits are usually classified into the following categories:
- Soft fruits – for example, strawberries, raspberries, blackcurrants, blueberries, cranberries, blackberries and gooseberries.
- Hard fruits – for example, apples, pears and coconuts.
- Stone fruits – for example, apricots, plums, peaches, cherries and avocados.
- Citrus fruits – examples include lemons, limes, oranges, grapefruit, kumquats and mandarins.
- Tropical fruits – for example, bananas, breadfruit, passion fruit, mangoes, custard apples, cape gooseberries, dates, dragon fruit, lychee, pineapples, melons and paw paw (papaya).

1.2 Quality points

Fresh fruit should be:
- whole and fresh looking (for maximum flavour the fruit must be ripe but not overripe)
- firm, according to type and variety
- clean and free from traces of pesticides and fungicides
- free from external moisture
- free from any unpleasant foreign smell or taste
- free from pests or disease
- sufficiently mature; it must be capable of being handled and travelling without being damaged
- free from any defects characteristic of the variety in shape, size and colour
- free of bruising and any other damage due to weather conditions.

Soft fruits deteriorate quickly, especially if they are not sound. Take care to see that they are not damaged or overripe when purchased. Soft fruits should look fresh; there should be no signs of wilting, shrinking or mould. The colour of certain soft fruits is an indication of their ripeness (for example, strawberries or dessert gooseberries).

> **PROFESSIONAL TIP**
> Care must be taken when buying melons to ensure they are not over- or underripe. This can be assessed by carefully pressing the top or bottom of the fruit and smelling the outside skin for sweetness. There should be a slight degree of softness to cantaloupe and charentais melons. The stalk should be attached, otherwise the melon deteriorates quickly.

Seasons for commonly used fruits

Table 4.12 Seasons for commonly used fruits

	Spring (March, April and May)	Summer (June, July and August)	Autumn (September, October and November)	Winter (December, January and February)
Apples				
Blackberries				
Blueberries				
Currants (white, red and black)				
Damsons				
Elderberries				
Pears				
Plums				
Quince				
Raspberries				
Rhubarb				
Sloes				
Strawberries				

1.3 Preparing fruits

- Washing – fruit should be washed to remove dirt and foreign bodies if they are not going to be peeled.
- Sorting – fruit are sorted into grade quality and size and sometimes country of origin and producer.
- Coring – removal of the core of the fruit, for example, apples and pears. This is normally done with a corer.
- Segmenting – portioning fruit, such as orange or grapefruit, into segments using a knife.
- Peeling – removing the skin of a fruit when it is raw or cooked.
- Trimming – trimming fruit into shape and size for even cooking and presentation.
- Deseeding – removal of the seeds from fruit
- Coating – coating fruit in flour, batter or breadcrumbs before cooking.
- Stuffing – fruit can be stuffed with a filling before or after cooking, for example stuffed apples.
- Blanching – fruit which are blanched to remove the skin or to cook for a short time to deactivate the enzymes prior to freezing.
- Marinating – fruit are immersed in liquid such as wine, fruit juice, herbs and spices to add flavour prior to cooking.
- Tying – fruit are tied with string or twine to form a shape or to keep their shape during the preparation and cooking process.
- Portioning – fruit are cut into portion sizes during preparation or after cooking. Portioning is important for cost control and to prevent waste.

Step by step: segmenting (grapefruit)

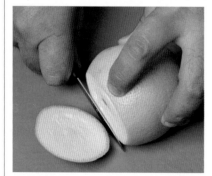

1 Using a sharp knife, use a straight cut to remove the top.

2 Repeat to remove the other end of the grapefruit.

3 Cut around the flesh, keeping very close to the skin to remove both the pith and the skin.

Step by step: segmenting (grapefruit)

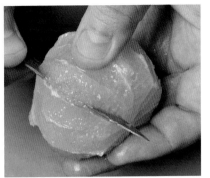

4 Remove all of the pith and skin to reveal the pink fruit.

5 Cut down one side of the vein in the fruit and then down the other side of the vein to remove the segment.

6 Repeat the process to remove all segments. Some segments will be larger than others.

1.4 Coatings and stuffings

An example of a fruit stuffing is apple and prune stuffing (page 227). Fruit stuffings can also be used in pastry work, for example, in Eccles cakes or mince pies. Coatings are also used with pastry, for example apple fritters (page 477). You will find more examples in the recipes section at the end of this chapter and in Unit 303 Stocks, soups and sauces and Unit 309 Desserts and puddings.

1.5 Storage

- Hard fruits, such as apples, should be left in boxes and kept in a cool store.
- Soft fruits, such as raspberries and strawberries, should be left in their punnets or baskets in a cold room.
- Stone fruits, such as apricots and plums, are best placed in trays so that any damaged fruit can be seen and discarded.
- Peaches and citrus fruits are left in their delivery trays or boxes.
- Bananas should not be stored in too cold a place because their skins will turn black.

Produce fruit dishes

2.1 Cooking methods

Poaching or stewing

Traditionally fruit is poached in a light syrup of sugar and water. The density of the syrup (the amount of sugar to water) will depend on the ripeness of the fruit – the riper the fruit, the denser the syrup needs to be to hold the fruit in shape.

Traditionally this method is known as stewing fruit due to the gentle slow method of cooking. The finished product is known as a compote of fruit, which is traditionally served at breakfast as a mixture of poached fruit such as prunes, pears, apples and so on. Poached fruit is also served as a dessert, usually as a garnish or accompaniment.

Apart from water and sugar, fruit juice (for example, orange, cranberry, apple and mango) is often used to poach fruit.

Various wines are also used (white, red, sparkling, rosé and Marsala).

Baking

Fruit may be baked in the oven (for example, baked pineapple, apple and pear). The fruit is usually sprinkled with sugar (white or brown) and spices such as cinnamon or mixed spice. Often thin slices of apple, pineapple or pear are baked in the oven until crisp and used as a garnish for desserts.

Grilling or barbequeing

Fruit such as pineapple may be marinated in wine or spirits, such as rum, and grilled on an oiled barbeque or under the salamander.

2.2 Presenting finished dishes

For details of how to present fruit dishes, see the recipe section at the end of this unit.

2.3 Evaluation of fruit and vegetables

The finished fruit or vegetable dish should be evaluated against the recipe specification and production standards and should also take into account any faults with regard to:

- portion size
- colour
- taste – the dish should not be over seasoned; the real flavour of the fruit and vegetables should be clearly identifiable
- texture
- consistency – this refers to the recipe being consistent every time it is produced. Any sauces accompanying the dish or incorporated in the dish should also be of the correct consistency
- seasoning
- degree of cooking
- presentation – the dish should be pleasing to the eye, colourful, attractive and neatly presented
- temperature
- aroma.

TEST YOURSELF

1 Name **six** varieties of potato suitable for roasting and boiling.

2 State **two** quality points to look for when buying leeks.

3 Describe the following cuts of vegetables:
   a) macedoine
   b) jardinière
   c) julienne
   d) paysanne

4 Describe the preparation and cooking of braised red cabbage.

5 How should fresh green vegetables be stored?

6 What is meant by 'blanching' of vegetables?

7 Explain what is meant by 'textured vegetable protein'.

8 Which fruit or vegetable classification does each of the following belong to?
   a) carrot
   b) sweet potato
   c) cauliflower
   d) peach
   e) grapefruit

9 Describe how fresh strawberries should be stored to help retain their quality.

10 State **three** advantages of canning as a preservation method for vegetables.

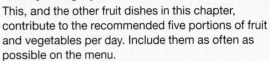

1 Grapefruit (*pamplemousse*)

Energy	Calories	Fat	Saturated fat	Carbohydrates	Sugar	Protein	Fibre
101 kJ	24 kcal	0.1 g	0 g	5.4 g	5.4 g	0.6 g	1.4 g

Preparing the dish

The fruit should be peeled with a sharp knife in order to remove all the white pith and yellow skin. Cut into segments and remove all the pips. The segments and the juice should then be dressed in a cocktail glass or grapefruit coupe and chilled. A cherry (or other fruit, for example raspberry of strawberry) may be added. Allow ½ to 1 grapefruit per head. The common practice of sprinkling with caster sugar is incorrect, as some customers prefer their grapefruit without sugar.

PROFESSIONAL TIP

Make sure all the pith is removed from the grapefruit: it tastes very bitter.

Healthy Eating Tip

This, and the other fruit dishes in this chapter, contribute to the recommended five portions of fruit and vegetables per day. Include them as often as possible on the menu.

VARIATION

- **Grapefruit**: serve hot, sprinkled with rum and demerara sugar.
- **Grapefruit and orange cocktail**: allow half an orange and half a grapefruit per head.
- **Orange cocktail**: use oranges in place of grapefruit.
- **Florida cocktail**: a mixture of grapefruit, orange and pineapple segments.

HEALTH AND SAFETY

Make sure you use the correct coloured chopping board.

SERVING SUGGESTION

To improve presentation try using different styles of cocktail glasses. Alternatively, serve on attractive plates.

2 Avocado pear (l'avocat)

Energy	Calories	Fat	Saturated fat	Carbohydrates	Sugar	Protein	Fibre
568 kJ	138 kcal	14.1 g	3 g	1.4 g	0.4 g	1.4 g	3.3 g

The pears must be ripe (test by pressing gently – the pear should give slightly). Allow half an avocado per portion.

1 Cut the avocado in half.

2 Remove the stone.

3 Peel the skin before slicing the avocado (if required).

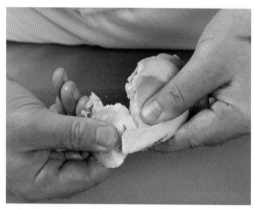

4 Slice the avocado.

5 Serve garnished with lettuce or other salad leaves accompanied by vinaigrette or variations on vinaigrette.

VARIATION

Avocado pears are sometimes filled with shrimps or crabmeat bound with a shellfish cocktail sauce or other similar fillings. They may be served hot or cold using a variety of fillings and sauces.

Avocados may also be halved lengthwise, the stone removed, the skin peeled and the flesh sliced and fanned on a plate. Garnish with a simple or composed salad.

Healthy Eating Tip

Serve with plenty of salad vegetables and bread or toast (optional butter or spread).

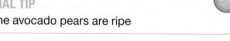
PROFESSIONAL TIP

Make sure the avocado pears are ripe

HEALTH AND SAFETY

Make sure your knives are sharp and use the correct coloured chopping board

3 Fresh fruit salad

Energy	Calories	Fat	Saturated fat	Carb	Sugar	Protein	Fibre
493 kJ	117 kcal	0.0 g	0.0 g	30.3 g	29.5 g	0.9 g	3.0 g

5 Stone the cherries, leave whole.

6 Cut the grapes in half, peel if required, and remove the pips.

7 Mix carefully and place in a glass bowl in the refrigerator to chill.

8 Just before serving, peel and slice the banana and mix in.

	4 portions	**10 portions**
Orange	1	2–3
Dessert apple	1	2–3
Dessert pear	1	2–3
Cherries	50 g	125 g
Grapes	50 g	125 g
Banana	1	2–3
Stock syrup		
Caster sugar	50 g	125 g
Water	125 ml	375 ml
Lemon, juice of	½	1

> **VARIATION**
> - All the following fruits may be used: dessert apples, pears, pineapple, oranges, grapes, melon, strawberries, peaches, raspberries, apricots, bananas, cherries, Kiwi fruit, plums, mangoes (see recipe 4), pawpaws and lychees. Allow about 150 g of unprepared fruit per portion.
> - Kirsch, Cointreau or Grand Marnier may be added to the syrup. All fruit must be ripe.
> - A fruit juice (e.g. apple, orange, grape or passion fruit) can be used instead of stock syrup.

> **HEALTH AND SAFETY**
> Make sure knives are sharp and correct coloured chopping boards are used.

Preparing the dish

1 For the syrup, boil the sugar with the water and place in a bowl.

2 Allow to cool, add the lemon juice.

3 Peel and cut the orange into segments.

4 Quarter the apple and pear, remove the core, peel and cut each quarter into two or three slices, place in the bowl and mix with the orange.

> **Healthy Eating Tip**
> Use fruit juice, orange or apple, in place of sugar and water syrup.

> **FAULTS**
> Make sure that the fruit is cut evenly.

> **ON A BUDGET**
> Stick to the basic recipe and avoid using the more exotic fruits such as paw paw, passion fruit, mango etc.

4 Tropical fruit plate

An assortment of fully ripe fruits, for example pineapple, papaya, mango (see photos), peeled, deseeded, cut into pieces and neatly dressed on a plate. An optional accompaniment could be yoghurt, vanilla ice cream, créme fraiche, fresh or clotted cream.

Healthy Eating Tips
This colourful dessert helps to meet the recommended target of five portions of fruit and vegetables

SERVING SUGGESTION
Serve on a banana leaf.

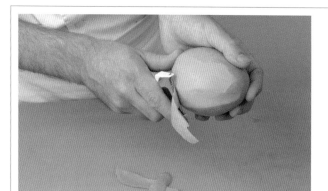

1 Peeling a mango.

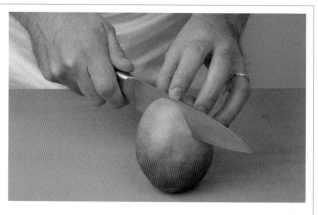

2 To dice a mango, first slice through it, keeping lateral to the stone.

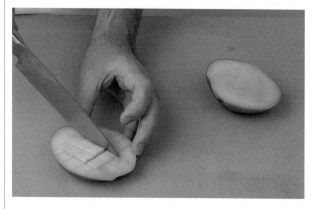

3 Next, score the flesh.

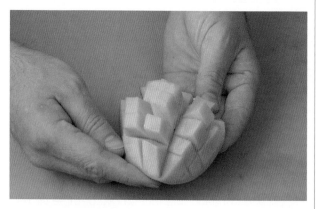

4 Bend out the cubes.

5 Caribbean fruit curry

Energy	Calories	Fat	Saturated fat	Carb	Sugar	Protein	Fibre *
2856 kJ	682 kcal	39.1 g	14.9 g	78.7 g	68.3 g	8.7 g	13.2 g

* Using single cream

Mise en place

1 Peel and cut vegetables and gather all the necessary equipment together.

2 Skin and cut the pineapple in half, remove the tough centre. Cut into 1-cm chunks. Peel the apples and pears, remove the cores, cut into 1-cm pieces. Peel and slice the mangoes. Skin and cut the bananas into 1-cm pieces. Cut the guavas and pawpaws in half, remove the seeds, peel, dice into 1-cm pieces.

3 Marinate the fruit in the lime juice.

Cooking

1 Fry the onion in the sunflower margarine and oil until lightly brown, add the curry powder, sweat together, add the wholemeal flour and cook for 2 minutes.

2 Add the ginger, coconut, tomato concassé, tomato puree and sultanas.

3 Gradually add sufficient boiling fruit juice to make a light sauce.

4 Add yeast extract, stir well. Simmer for 10 minutes.

5 Add the sultanas, fruit juice and cashew nuts, stir carefully, allow to heat through.

6 Finish with cream, smetana for fromage frais.

	4 portions	10 portions
Pineapple	1 small	1 large
Dessert apples	2	5
Small dessert pears	2	5
Mangoes	2	5
Bananas	2	5
Guava	1	2–3
Pawpaw	1	2–3
Rind and juice of lime, grated	1	2–3
Onion, chopped	50 g	125 g
Sunflower margarine or butter	25 g	60 g
Sunflower or vegetable oil	60 ml	150 ml
Madras curry powder	50 g	125 g
Wholemeal flour	25 g	60 g
Ginger, freshly grated	10 g	25 g
Desiccated coconut	50 g	125 g
Tomato, skinned, deseeded and diced	100 g	250 g
Tomato puree	25 g	60 g
Yeast extract	5 g	12 g
Sultanas	50 g	125 g
Fruit juice	0.5 litre	1.25 litres
Cashew nuts	50 g	125 g
Single cream or smetana	60 ml	150 ml

SERVING SUGGESTION

Serve in a suitable dish; separately serve poppadoms, wholegrain pilaff rice and a green salad.

Healthy Eating Tips
- Lightly oil a well-seasoned pan with the sunflower oil to fry the onion.
- No added salt is needed.
- Try finishing the dish with low-fat yoghurt in place of the cream.
- Serve with plenty of starchy carbohydrate and salad.

FAULTS

Make sure that fruits and vegetables are not unevenly cut and overcooked

6 Buttered celeriac, turnips or swedes

Energy	Calories	Fat	Saturated fat	Carbohydrates	Sugar	Protein	Fibre
253 kJ	60 kcal	5.4 g	3.3 g	2.5 g	2.5 g	0.7 g	1.9 g

	4 portions	10 portions
Celeriac, turnips or swedes	400 g	1 kg
Salt, sugar		
Butter	25 g	60 g
Parsley, chopped		

Cooking

1 Place in a pan with a little salt, a pinch of sugar and the butter. Barely cover with water.

2 Cover with a buttered paper and allow to boil steadily in order to evaporate all the water.

3 When the water has completely evaporated, check that the vegetables are cooked; if not, add a little more water and continue cooking. Do not overcook.

4 Toss the vegetables over a fierce heat for 1–2 minutes to glaze.

5 Drain well, and serve sprinkled with chopped parsley. Serve in individual dishes.

PROFESSIONAL TIP
Do not overcook the vegetables. For best results make sure they are slightly undercooked

VARIATION
● In place of butter, drizzle with olive oil or rapeseed oil

Mise en place

1 Peel and wash the vegetables.

2 Cut into neat pieces or turn barrel shaped.

7 Goats' cheese and beetroot tarts, with salad of watercress

Energy	Calories	Fat	Saturated fat	Carbohydrates	Sugar	Protein	Fibre
2377 kJ	568 kcal	37.7 g	9.0 g	43.6 g	7.5 g	18.6 g	1.8 g

	4 portions	10 portions
Puff pastry	400 g	1 kg
Shallots	150 g	375 g
Beetroot, cooked	200 g	500 g
Goats' cheese	200 g	500 g
Watercress, bunch	1	2

Mise en place

1 Make the pastry.

2 Cook the beetroot.

Cooking

1 Roll the puff pastry to a thickness of 3 mm.

2 Chill the rolled puff pastry for 10 minutes.

3 Finely slice the shallots and sweat down without colour.

4 Cut the puff pastry into four discs approximately 150 mm in diameter.

5 Chill the pastry discs for 10 minutes.

6 Dice the cooked beetroot into pieces 10 × 10 mm.

7 To make the tarts, place the shallots on the pastry discs.

8 Cook at 180 °C for 12 minutes.

9 Once cooked, remove from the oven and top with the diced beetroot and crumbled goats' cheese.

10 To finish the dish, place the tarts on plates and finish with picked watercress and vinaigrette.

VARIATION
- Use different types of cheese such as cheddar or stilton. Sprinkle chopped mint on the beetroot

FAULTS
Take care not to overbake the tarts.

ON A BUDGET
Use cheaper types of cheese.

SERVING SUGGESTION
Alternatively, serve with a horseradish mayonnaise.

---

8 Golden beetroot with Parmesan

Energy	Calories	Fat	Saturated fat	Carbohydrates	Sugar	Protein	Fibre
370 kJ	88 kcal	2.8 g	2.4 g	7.7 g	7.1 g	6.2 g	2.5 g

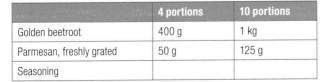

Mise en place

1 Peel the golden beetroot and cut into 5-mm slices.

Cooking

1 Either steam or plain boil the beetroot until tender.

2 Drain well and place in a suitable serving dish. Sprinkle with Parmesan and grill under the salamander or in the oven.

Healthy Eating Tip
Reduce the amount of Parmesan by 50 per cent and replace with 50 per cent toasted brown breadcrumbs mixed together

FAULTS
Overcooking the beetroot will make it mushy.

SERVING SUGGESTION
Serve on individual plates or as a garnish.

	4 portions	10 portions
Golden beetroot	400 g	1 kg
Parmesan, freshly grated	50 g	125 g
Seasoning		

9 Parsnips (*panais*)

Energy	Calories	Fat	Saturated fat	Carbohydrates	Sugar	Protein	Fibre
235 kJ	56 kcal	0.0 g	0.0 g	13.5 g	2.7 g	1.3 g	2.5 g

2 Cut into neat pieces and cook in lightly salted water until tender, or steam.

3 Drain and serve with melted butter or in a cream sauce.

> **VARIATION**
> - Parsnips may be roasted in the oven in a little fat (as shown here) or in with a joint, and can be cooked and prepared as a purée.

> **PROFESSIONAL TIP**
> For great roast parsnips, blanch them for 2 minutes, drain, then roast in hot olive oil.

> **Healthy Eating Tip**
> Omit the butter or cream, leave plain or drizzle on olive oil or rape seed oil.

Mise en place

1 Wash the parsnips well. Peel and re-wash well.

Cooking

1 Cut into quarters lengthwise, remove the centre root if tough.

> **FAULTS**
> Parsnips are overcooked, burnt or undercooked.

> **SERVING SUGGESTION**
> Serve in individual dishes or as a garnish.

10 Salsify (*salsifi*)

Energy	Calories	Fat	Saturated fat	Carbohydrates	Sugar	Protein	Fibre
76 kJ	18 kcal	0.0 g	0.0 g	2.8 g	2.8 g	1.9 g	0.0 g

Half a kilogram will yield 2–3 portions.

Mise en place

1 Wash, peel and rewash the salsify.
2 Cut into 5-cm lengths.

Cooking

1 Cook in a blanc (see below). Do not overcook.

> **SERVING SUGGESTION**
> Salsify may be served brushed with melted butter, or coated in mornay sauce (see page 100), sprinkled with grated cheese and browned under a salamander. It may also be passed through batter and deep-fried.

To make the blanc:

	4 portions	10 portions
Flour	10 g	20 g
Cold water	0.5 litres	1 litre
Salt, to taste		
Lemon, juice of	½	1

Mise en place

1 Peel the salsify.
2 Make the batter.

Cooking

1 Mix the flour and water together.
2 Add the salt and lemon juice. Pass through a strainer.
3 Place in a pan, bring to the boil, stirring continuously.

> **Healthy Eating Tip**
> Omit the cheese or use a reduced fat cheese.

> **VARIATION**
> ● Instead of batter, pass through milk and flour.

> **FAULTS**
> Salsify can be overcooked, the batter may be soggy or the frying oil too cool.

11 Vichy carrots (*carottes Vichy*)

Energy	Calories	Fat	Saturated fat	Carbohydrates	Sugar	Protein	Fibre *
338 kJ	82 kcal	5.4 g	3.4 g	8.0 g	7.5 g	0.7 g	2.4 g

* Using water only

Cooking

1 Place in a pan with a little salt, add sugar and butter. Barely cover with Vichy water.

2 Cover with a buttered paper and allow to boil steadily in order to evaporate all the water.

3 When the water has completely evaporated, check that the carrots are cooked; if not, add a little more water and continue cooking. Do not overcook.

4 Toss the carrots over a fierce heat for 1–2 minutes in order to give them a glaze.

5 Serve sprinkled with chopped parsley.

> **NOTE**
> Vichy water is water from the French town of Vichy. This dish is characterised by the glaze produced by reducing the butter and sugar.

	4 portions	10 portions
Carrots	400 g	1 kg
Salt, pepper		
Sugar	1 tsp	$2\frac{1}{2}$ tsp
Butter	25 g	60 g
Vichy water (optional)		
Parsley, chopped		

Healthy Eating Tips
- Use the minimum amount of salt.
- Omit the butter and replace with olive oil.
- Reduce the amount of sugar by half.

Mise en place

1 Peel and wash the carrots (which should not be larger than 2 cm in diameter).

2 Cut into 2-mm-thin slices on the mandolin.

HEALTH AND SAFETY
If using a mandolin, make sure a safety guard is used. Use the correct colour chopping board.

FAULTS
Carrots are unevenly sliced and overcooked.

12 Baked jacket potatoes (*pommes au four*)

Energy	Calories	Fat	Saturated fat	Carbohydrates	Sugar	Protein	Fibre *
401 kJ	94 kcal	0.2 g	0.0 g	21.8 g	0.9 g	2.7 g	1.6 g

* Using medium potato, 180 g analysis given per potato.

Mise en place

1 Select good-sized potatoes; allow 1 potato per portion.

2 Scrub well and make a 2-mm-deep incision round the potato.

Cooking

1 Place the potato, on a small mound of ground (sea) salt to help keep the base dry, on a tray in a hot oven at 230–250 °C for about 1 hour. Turn the potatoes over after 30 minutes.

2 Test by holding the potato in a cloth and squeezing gently; if cooked it should feel soft.

PROFESSIONAL TIP

Lightly scoring the potatoes will help to make sure that they cook evenly.

If the potatoes are being cooked in a microwave, you must prick the skins first.

Healthy Eating Tips
- Potatoes can be baked without sea salt.
- Fillings based on vegetables with little or no cheese or meat make a healthy snack meal.

VARIATION
- Split and filled with any of the following: grated cheese, minced beef or chicken, baked beans, chilli con carne, cream cheese and chives, mushrooms, bacon, ratatouille, prawns in mayonnaise, coleslaw.
- The cooked potatoes can also be cut in halves lengthwise, the potato spooned out from the skins, seasoned, mashed with butter, returned to the skins, sprinkled with grated cheese and reheated in an oven or under the grill.

FAULTS

Ensure the potatoes are not overbaked.

13 Château potatoes (*pommes château*)

Energy	Calories	Fat	Saturated fat	Carbohydrates	Sugar	Protein	Fibre
839 kJ	200 kcal	2.9 g	0.7 g	34.4 g	1.2 g	4.2 g	3.5 g

Mise en place

1 Select small, even-sized potatoes and wash them.
2 Turn the potatoes into barrel-shaped pieces about the same size as fondant potatoes (recipe 17).

Cooking

1 Place in a saucepan of boiling water for 2–3 minutes, then refresh immediately. Drain in a colander.
2 Finish as for roast potatoes. Use a non-stick tray, lightly oiled greaseproof paper or non-stick mat for the roasting.

HEALTH AND SAFETY
Use correct colour chopping boards. Do not overheat the oil.

Healthy Eating Tip
Use a low calorie oil spray. Toss the potatoes well in the spray. This reduces the amount of oil used.

14 Croquette potatoes (*pommes croquettes*)

Energy	Calories	Fat	Saturated fat	Carbohydrates	Sugar	Protein	Fibre	*
1699 kJ	405 kcal	25.4 g	6.6 g	40.8 g	1.1 g	6.0 g	2.2 g	

* Using hard margarine and frying in peanut oil.

Cooking

1 Use a duchess mixture (see recipe 16) moulded into cylinder shapes 5 × 2 cm.
2 Pass through flour, eggwash and breadcrumbs.
3 Reshape with a palette knife and deep-fry in hot deep oil (185 °C) in a frying basket.
4 When the potatoes are a golden colour, drain well and serve.

Healthy Eating Tips
● Add the minimum amount of salt.
● Use peanut or sunflower oil to fry the croquettes, and drain on kitchen paper.

VARIATION
● Croquette potatoes may be stuffed with cooked spinach. The purée may be flavoured with mixed chopped herbs or horseradish.
● Try mixing chopped nuts or oats into the breadcrumbs.

FAULTS
Insufficient coating with the egg and breadcrumbs.

15 Delmonico potatoes (*pommes Delmonico*)

Energy	Calories	Fat	Saturated fat	Carbohydrates	Sugar	Protein	Fibre	*
900 kJ	214 kcal	6.3 g	3.7 g	37.5 g	2.7 g	4.4 g	2.0 g	

* Using old potatoes and whole milk

Mise en place

1 Wash, peel and rewash the potatoes.

2 Cut into 6-mm dice.

Cooking

1 Barely cover with milk, season lightly with salt and pepper and allow to cook for 30–40 minutes.

2 Place in an earthenware dish, sprinkle with breadcrumbs and melted butter, brown in the oven or under the salamander and serve.

Healthy Eating Tip
Use skimmed milk in place of full fat milk and olive oil in place of butter.

VARIATION
Add chopped crushed garlic and/or mixed chopped fresh herbs.

16 Duchess potatoes (*pommes duchesse*)

Energy	Calories	Fat	Saturated fat	Carbohydrates	Sugar	Protein	Fibre *
819 kJ	195 kcal	8.2 g	3.3 g	28.6 g	0.6 g	3.5 g	1.5 g

* Using old potatoes, whole milk, hard margarine

	4 portions	10 portions
Floury potatoes	600 g	1.5 kg
Egg yolks	1	3
Butter	25 g	60 g
Salt, pepper		

Mise en place

1 Wash, peel and rewash the potatoes. Cut to an even size.

Cooking

1 Cook potatoes in lightly salted water.

2 Drain off the water, cover and return to a low heat to dry out the potatoes.

3 Pass through a medium sieve or a special potato masher or mouli.

4 Place the potatoes in a clean pan.

5 Add the egg yolks and stir in vigorously with a kitchen spoon.

6 Mix in the butter. Correct the seasoning.

7 Place in a piping bag with a large star tube and pipe out into neat spirals, about 2 cm in diameter and 5 cm tall, on to a lightly greased baking sheet.

8 Place in a hot oven at 230 °C for 2–3 minutes in order to firm the edges slightly.

9 Remove from the oven and brush with eggwash.

10 Brown lightly in a hot oven or under the salamander.

1 Add egg yolks to the mashed potato.

2 Pipe the potato into neat spirals, ready for baking.

NOTES

At step 3, it is important to return the drained potatoes to the heat, so that they are as dry as possible.

VARIATION

Add chopped fresh mixed herbs to the potato puree.

FAULTS

If the mashed potato is too wet the shapes will collapse in the oven.

Video: Making duchess potatoes

http://bit.ly/2pD2jyn

17 Fondant potatoes (*pommes fondantes*)

Energy	Calories	Fat	Saturated fat	Carbohydrates	Sugar	Protein	Fibre	*
956 kJ	228 kcal	7.0 g	2.1 g	39.6 g	0.9 g	4.1 g	1.5 g	

* Using old potatoes and hard margarine for 1 portion (125 g raw potato).

Video: Making fondant potatoes

http://bit.ly/2oMGgjE

1 Turn the potatoes into barrel shapes.

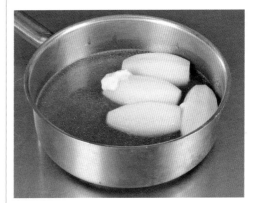

2 Place in a pan, half covered with stock, to cook.

Barrel shape (left) and cake shape (right)

Mise en place

1 Select small or even-sized medium potatoes.

2 Wash, peel and rewash.

3 Turn into 8-sided barrel shapes, allowing 2–3 per portion, about 5 cm long, end diameter 1.5 cm, centre diameter 2.5 cm.

Cooking

1 Brush with melted butter or oil.

2 Place in a pan suitable for the oven.

3 Half cover with white stock, season lightly with salt and pepper.

4 Cook in a hot oven at 230–250 °C, brushing the potatoes frequently with melted butter or oil.

5 When cooked, the stock should be completely absorbed by the potatoes.

6 Brush with melted butter or oil and serve.

PROFESSIONAL TIP

To give the potatoes a good glaze, use a high-quality stock and baste during cooking.

VARIATION

Fondant potatoes can be lightly sprinkled with:
- thyme, rosemary or oregano (or this can be added to the stock)
- grated cheese (Gruyère and Parmesan or Cheddar)
- chicken stock in place of white stock.

Healthy Eating Tip
- Use a little unsaturated oil to brush over the potatoes before and after cooking.
- No added salt is needed; rely on the stock for flavour.

FAULTS

Insufficiently glazed or poorly and unevenly shaped potatoes.

18 Fried or chipped potatoes (*pommes frites*)

Energy	Calories	Fat	Saturated fat	Carbohydrates	Sugar	Protein	Fibre	*
1541 kJ	367 kcal	15.8 g	2.8 g	54.1 g	0.0 g	5.5 g	1.5 g	

* Using old potatoes and peanut oil.

Mise en place

1 Prepare and wash the potatoes.

2 Cut into slices 1 cm thick and 5 cm long.

3 Cut the slices into strips 5 × 1 × 1 cm.

4 Wash well and dry in a cloth.

Cooking

1 Cook without colour in a frying basket in moderately hot fat (165 °C).

2 Drain and place on kitchen paper on trays until required.

3 When required, place in a frying pan and cook in hot fat (185 °C) until crisp and golden.

4 Drain well, season lightly with salt and serve.

SERVING SUGGESTION

For crisp chips fry at the very last minute. Serve immediately as individual portions.

FAULTS

Potatoes should be evenly cut. If the oil is too cool for frying the chips may turn out soggy.

NOTES

Because chips are so popular, the following advice from the Potato Marketing Board is useful.

● Cook chips in small quantities; this will allow the oil to regain its temperature more quickly; chips will then cook faster and absorb less fat.

● Do not let the temperature of the oil exceed 199 °C as this will accelerate the fat breakdown.

● Use oils high in polyunsaturates for a healthier chip.

● Ideally use a separate fryer for chips and ensure that it has the capacity to raise the fat temperature rapidly to the correct degree when frying chilled or frozen chips.

● Although the majority of chipped potatoes are purchased frozen, the Potato Marketing Board recommends the following potatoes for those who prefer to make their own chips: Maris Piper, Cara, Désirée. King Edward and Santé are also good choices.

Healthy Eating Tips

● Chipped potatoes may be blanched twice, first at 140 °C, followed by a re-blanch at 160 °C until lightly coloured.

● Blanch in a steamer until just cooked, drain and dry well – final temperature 165 °C.

19 Hash brown potatoes

Energy	Calories	Fat	Saturated fat	Carbohydrates	Sugar	Protein	Fibre
954 kJ	228 kcal	11.3 g	5.3 g	25.8 g	0.9 g	7.1 g	2.0 g

	4 portions	10 portions
Potatoes	600 g	2 kg
Butter	25 g	60 g
Lardons of bacon	100 g	250 g
Seasoning		

Mise en place

1 Wash, peel and rewash the potatoes.
2 Coarsely grate the potatoes, rewash quickly and then drain well.

Cooking

1 Melt the fat in a suitable frying pan. Add the lardons of bacon, fry until crisp and brown, remove from the pan and drain.
2 Pour the fat back into the frying pan, add the grated potato and season.
3 Press down well, allow 2 cm thickness, and cook over a heat for 10–15 minutes or in a moderate oven at 190°C, until a brown crust forms on the bottom.
4 Turn out into a suitable serving dish and sprinkle with the lardons of bacon and chopped parsley.

Healthy Eating Tip
- Dry-fry the lardons in a well-seasoned pan.
- Brush a little oil over the potatoes and cook in a hot oven. Use the minimum amount of salt.
- Replace the butter with olive oil or rapeseed oil.

VARIATION
Add a little chopped chilli to the potatoes, garlic and mixed chopped herbs.

FAULTS
Potatoes may be too brown and overcooked.

20 Macaire potatoes (potato cakes) (*pommes Macaire*)

Energy	Calories	Fat	Saturated fat	Carbohydrates	Sugar	Protein	Fibre	*
4392 kJ	1047 kcal	65.7 g	14.7 g	109.8 g	2.7 g	11.4 g	10.8 g	

* Using hard margarine and sunflower oil

Half a kilogram will yield 2–3 portions.

Mise en place

1 Prepare and cook as for baked jacket potatoes (recipe 12).
2 Cut in halves, remove the centre with a spoon, and place in a basin.
3 Add 25 g butter per 0.5 kg, a little salt and milled pepper.

Cooking

1 Mash and mix as lightly as possible with a fork.
2 Using a little flour, mould into a roll, then divide into pieces, allowing one or two per portion.

3 Mould into 2-cm round cakes. Quadrille with the back of a palette knife. Flour lightly.
4 Shallow-fry on both sides in very hot oil and serve.

PROFESSIONAL TIP

Make sure the potato mixture is firm enough to be shaped, and fry the cakes in very hot oil, or they will lose their shape.

VARIATION

Additions to potato cakes can include:
- chopped parsley, fresh herbs or chives, or duxelle
- cooked chopped onion
- grated cheese.

Healthy Eating Tips

Use less salt. Replace butter with olive or rapeseed oil.

FAULTS

Mash potato may be insufficiently dry. Cakes may be poorly and unevenly shaped. Frying oil may be insufficiently hot for frying, causing the cakes to break up and be poorly coloured.

SERVING SUGGESTION

Serve garnished with micro herbs on individual dishes.

1 Place the potatoes on a baking tray with salt.

2 Once baked, cut in half and scoop out the centre.

3 Mould the potato into cakes.

21 Mashed potatoes (*pommes purées*)

Energy	Calories	Fat	Saturated fat	Carbohydrates	Sugar	Protein	Fibre	*
763 kJ	182 kcal	7.1 g	4.4 g	29.0 g	1.1 g	2.4 g	1.5 g	

* Using old potatoes, butter and whole milk.

	4 portions	10 portions
Floury potatoes	0.5 kg	1.25 kg
Butter	25 g	60 g
Milk, warm	30 ml	80 ml

Mise en place

1 Wash, peel and rewash the potatoes. Cut to an even size.

Cooking

1 Cook in lightly salted water, or steam.
2 Drain off the water, cover and return to a low heat to dry out the potatoes.
3 Pass through a medium sieve or a ricer.
4 Return the potatoes to a clean pan.
5 Add the butter and mix in with a wooden spoon.
6 Gradually add warm milk, stirring continuously until a smooth creamy consistency is reached.
7 Correct the seasoning and serve.

PROFESSIONAL TIP

Drain the potatoes as soon as they are cooked. If they are left standing in water, they will become too wet and spoil the texture of the dish.

Healthy Eating Tips

● Add a minimum amount of salt.
● Add a little olive oil in place of the butter, and use semi-skimmed milk.

VARIATION

Try:
● dressing in a serving dish and surrounding with a cordon of fresh cream
● placing in a serving dish, sprinkling with grated cheese and melted butter, and browning under a salamander
● adding 50 g diced cooked lean ham, 25 g diced red pepper and chopped parsley
● adding lightly sweated chopped spring onions
● using a good-quality olive oil in place of butter
● adding a little garlic juice (use a garlic press)
● adding a little fresh chopped rosemary or chives
● mixing with equal quantities of parsnip
● adding a little freshly grated horseradish or horseradish cream.

FAULTS

Do not allow the potatoes to overcook and become too watery as when mashed they will have a gluey texture.

1 Pass the cooked potato through a sieve.

2 Add milk to the potatoes and combine.

22 Parmentier potatoes (*pommes parmentier*)

Energy	Calories	Fat	Saturated fat	Carbohydrates	Sugar	Protein	Fibre
1819 kJ	433 kcal	33.5 g	6.3 g	32.8 g	0.7 g	2.3 g	1.7 g

Half a kilogram will yield 2–3 portions

Mise en place

1 Select medium to large potatoes.
2 Wash, peel and rewash.
3 Trim on three sides and cut into 1-cm slices.
4 Cut the slices into 1-cm strips.
5 Cut the strips into 1-cm dice.
6 Wash well and dry in a cloth.

Cooking

7 Cook in hot shallow oil in a frying pan until golden brown.
8 Drain, season lightly and serve sprinkled with chopped parsley.

Healthy Eating Tip
Fry in the minimum amount of olive oil.

VARIATION
In place of parsley use mixed chopped herbs. Add a little chopped onion and chopped chilli during the end of the frying process.

23 Potatoes cooked in milk with cheese

Energy	Calories	Fat	Saturated fat	Carbohydrates	Sugar	Protein	Fibre
747 kJ	178 kcal	5.4 g	3.4 g	5.4 g	4.0 g	8.0 g	1.8 g

Cooking

1 Place in an ovenproof dish and just cover with milk.

2 Season, sprinkle with grated cheese and cook in a moderate oven (190 °C) until the potatoes are cooked and golden brown.

Healthy Eating Tip
Use skimmed milk in place of full fat milk.

VARIATION
Add a little chopped onion and chopped herbs to the milk.

SERVING SUGGESTION
Where possible, bake in individual dishes.

	4 portions	10 portions
Potatoes	500 g	1.25 kg
Milk	250 ml	600 ml
Salt and pepper		
Grated cheese	50 g	125 g

Mise en place

1 Slice the peeled potatoes to ½ cm thick.

24 Potatoes with bacon and onions (*pommes au lard*)

Energy	Calories	Fat	Saturated fat	Carbohydrates	Sugar	Protein	Fibre
836 kJ	199 kcal	10.1 g	3.8 g	22.2 g	1.8 g	6.4 g	2.5 g

* Using old potatoes

	4 portions	10 portions
Potatoes, peeled	400 g	1.25 kg
Streaky bacon (lardons)	100 g	250 g
Button onions	100 g	250 g
White stock	250 ml	600 ml
Salt, pepper		
Parsley, chopped		

Healthy Eating Tip
- Dry-fry the bacon in a well-seasoned pan and drain off any excess fat.
- Add little or no salt.

VARIATION
Add a little chopped chilli and some mixed chopped herbs to the stock.

Mise en place

1 Cut the potatoes into 1-cm dice.
2 Cut the bacon into 0.5-cm lardons.

Cooking

1 Lightly fry the bacon in a little fat together with the onions and brown lightly.
2 Add the potatoes, half cover with stock, season lightly with salt and pepper. Cover with a lid and cook steadily in the oven at 230–250 °C for approximately 30 minutes.
3 Correct the seasoning, serve in a vegetable dish, sprinkled with chopped parsley.

FAULTS
Take care not to over fry the bacon, make sure potatoes are evenly cut and not overcooked to avoid the potatoes breaking.

SERVING SUGGESTION
Sprinkle with chopped parsley or chopped herbs

25 Roast potatoes (*pommes rôties*)

Energy	Calories	Fat	Saturated fat	Carbohydrates	Sugar	Protein	Fibre
956 kJ	228 kcal	7.0 g	1.1 g	39.6 g	0.9 g	4.1 g	1.5 g

* Using old potatoes and peanut oil for 1 portion (125 g raw potato).

Mise en place

1 Wash, peel and rewash the potatoes.
2 Cut into even-sized pieces (allow 3–4 pieces per portion).

Cooking

1 Heat a good measure of oil or dripping in a roasting tray.
2 Add the well-dried potatoes and lightly brown on all sides.
3 Season lightly with salt and cook for about 1 hour in a hot oven at 230–250 °C.
4 Turn the potatoes over after 30 minutes.
5 Cook to a golden brown. Drain and serve.

> **PROFESSIONAL TIP**
> Roast potatoes can be parboiled for 10 minutes, refreshed and well dried before roasting. This will cut down on the cooking time and can also give a crisper potato.

> **Healthy Eating Tips**
> ● Brush the potatoes with peanut or sunflower oil, with only a little in the roasting tray.
> ● Drain off all the fat when cooked.

26 Sauté potatoes (*pommes sautées*)

Energy	Calories	Fat	Saturated fat	Carbohydrates	Sugar	Protein	Fibre	*
1249 kJ	297 kcal	11.4 g	1.3 g	46.8 g	0.4 g	4.9 g	1.7 g	

* Using old potatoes and sunflower oil

Mise en place

1 Select medium even-sized potatoes. Scrub well.

Cooking

1 Plain boil or cook in the steamer. Cool slightly and peel.

2 Cut into 3-mm slices.

3 Toss in hot shallow oil in a frying pan until lightly coloured; season lightly with salt.

4 Serve sprinkled with chopped parsley.

NOTES

Maris piper or Cara potatoes are good varieties for this dish.

Healthy Eating Tips

● Use a little hot sunflower oil to fry the potatoes.
● Add little or no salt; the customer can add more if required.

FAULTS

If potatoes are over boiled or steamed, they will break up in the frying process. If the oil is too hot, the potatoes will burn.

SERVING SUGGESTION

Serve well drained of fat or oil in individual dishes.

Cut the potatoes into 3-mm slices.

Toss the slices in hot oil to cook (sauté).

27 Sauté potatoes with onions (*pommes lyonnaise*)

Energy	Calories	Fat	Saturated fat	Carbohydrates	Sugar	Protein	Fibre
1098 kJ	261 kcal	9.4 g	1.1 g	41.5 g	6.4 g	4.5 g	5.3 g

Allow 0.25 kg onion to 0.5 kg potatoes.

Mise en place

1 Boil the potatoes and peel and shred onions.

Cooking

1 Shallow-fry the onions slowly in 25–50 g oil, turning frequently, until tender and nicely browned; season lightly with salt.

2 Prepare sauté potatoes as for recipe 26.

3 Combine the two and toss together.

4 Serve as for sauté potatoes.

Finely slice the onions.

Sauté the potatoes and onions together.

28 Swiss potato cakes (rösti)

Energy	Calories	Fat	Saturated fat	Carbohydrates	Sugar	Protein	Fibre	*
700 kJ	168 kcal	10.5 g	6.5 g	17.3 g	0.7 g	2.2 g	1.3 g	

* Using butter

2 Cool, then shred into large flakes on a grater.

3 Heat the oil or butter in a frying pan.

4 Add the potatoes, and season lightly with salt and pepper.

5 Press the potato together and cook on both sides until brown and crisp.

The potato can be made in a 4-portion cake or in individual rounds.

Healthy Eating Tips
- Lightly oil a well-seasoned pan with sunflower oil to fry the rösti.
- Use the minimum amount of salt.

VARIATION
- The potato cakes may also be made from raw potatoes.
- Add sweated chopped onion.
- Add sweated lardons of bacon.
- Use 2 parts of grated potato to 1 part grated apple.

	4 portions	10 portions
Potatoes, unpeeled	400 g	1 kg
Oil or butter	50 g	125 g
Salt, pepper		

Cooking

1 Parboil in salted water (or steam) for approximately 5 minutes.

Shred the parboiled potatoes on a grater.

Press into shape in the frying pan.

Turn out carefully after cooking.

29 Purée of Jerusalem artichokes (*topinambours en purée*)

Energy	Calories	Fat	Saturated fat	Carbohydrates	Sugar	Protein	Fibre
502 kJ	108 kcal	5.3 g	3.3 g	16.0 g	2.5 g	2.5 g	7.0 g

Mise en place

1 Wash, peel and rewash the artichokes.

2 Cut in pieces if necessary. Barely cover with water; add a little salt.

Cooking

1 Simmer gently until tender. Drain well.

2 Pass through a sieve or mouli or liquidise.

3 Return to the pan, reheat and mix in the butter; correct the seasoning and serve.

> **VARIATION**
> 125 ml (300 ml for 10 portions) cream or natural yoghurt may be mixed in before serving.

> **FAULTS**
> Make sure artichokes are sufficiently drained. If they are insufficiently drained the purée will be too wet.

	4 portions	10 portions
Jerusalem artichokes	600 g	1.5 kg
Salt, pepper		
Butter	25 g	60 g

30 Braised onions (*oignons braisés*)

Energy	Calories	Fat	Saturated fat	Carbohydrates	Sugar	Protein	Fibre
245 kJ	58 kcal	0.4 g	0.1 g	10.9 g	10.4 g	3.4 g	2.8 g

Mise en place

1 Select even medium-sized onions; allow ½ kg per 2–3 portions.

Cooking

1 Peel, wash and cook in lightly salted boiling water for 30 minutes, or steam.

2 Drain and place in a pan or casserole suitable for use in the oven.

3 Add bouquet garni, half cover with stock; put on the lid and braise gently in the oven at 180–200 °C until tender.

4 Drain well and dress neatly in a vegetable dish.

5 Reduce the cooking liquor with an equal amount of jus lié, reduced stock or demi-glace. Correct the seasoning and consistency and pass. Mask the onions and sprinkle with chopped parsley.

> **SERVING SUGGESTION**
> Serve individually or as a garnish. In place of parsley you can use micro herbs.

> **FAULTS**
> If the onions are overcooked , the jus-lie will be over reduced giving too strong a flavour

31 Caramelised button onions

Energy	Calories	Fat	Saturated fat	Carbohydrates	Sugar	Protein	Fibre
713 kJ	171 kcal	10.8 g	6.5 g	18.4 g	16.7 g	1.2 g	1.2 g

3 Barely cover with water or brown stock. Add the sugar. Cook the button onions until they are tender and the liquid has reduced with the sugar to a light caramel glaze. Carefully coat the onions with the glaze.

> **PROFESSIONAL TIP**
> The important thing is to reduce the stock and sugar until they form a light caramel syrup.

> **VARIATION**
> ● Use honey in place of sugar.

> **Healthy Eating Tip**
> The sugar can be omitted.

> **FAULTS**
> Onions can be too coloured and overcooked in the stock or water.

	4 portions
Butter or vegetable oil	50 g
Button onions	250 g
Water or brown stock	
Sugar	50 g

Cooking

1 Place the butter or oil in a shallow pan.

2 Fry the button onions quickly to a light golden brown colour.

32 Onion bhajias

Energy	Calories	Fat	Saturated fat	Carbohydrates	Sugar	Protein	Fibre
630 kJ	152 kcal	12.1 g	1.3 g	8.5 g	1.7 g	2.9 g	2.4 g

	4 portions	10 portions
Bessan or gram flour	45 g	112 g
Hot curry powder	1 tsp	2½ tsp
Salt		
Water	75 ml	187 ml
Onion, finely shredded	100 g	250 g

Mise en place

1 Mix together the flour, curry powder and salt.

2 Blend in the water carefully to form a smooth, thick batter.

3 Stir in the onion, stir well.

Cooking

1 Drop a tablespoon of the mixture into deep oil at 200 °C. Fry for 5–10 minutes until golden brown.

2 Drain well and serve as a snack with mango chutney as a dip.

PROFESSIONAL TIP
- The oil must be at the correct temperature before the bhajias are fried.

Healthy Eating Tips
- Use the minimum amount of salt.
- Make sure the oil is hot so that less is absorbed into the surface. Drain on kitchen paper.

HEALTH AND SAFETY
Make sure frying oil is at the correct temperature. Take care when placing the bhajias in the deep fat fryer.

VARIATION
- Add some finely chopped chilli to the mixture.

FAULTS
If the oil is too hot this will burn the bhajias. If the oil is too cold they will become soggy.

SERVING SUGGESTION
Serve as individual portions with a yoghurt, cucumber and mint dip or a mango chutney.

33 Poached fennel

Energy	Calories	Fat	Saturated fat	Carbohydrates	Sugar	Protein	Fibre
47 kJ	11 kcal	0.2 g	0 g	1.5 g	1.4 g	0.9 g	3.1 g

Mise en place

1 Trim the fennel heads and remove any blemishes.

Cooking

1 Gently place in a suitable pan of boiling stock or water.

2 Remove to the side of the stove and gently poach until tender.

> **NOTE**
> Poached fennel may be served as an accompanying vegetable or used as a garnish.

34 Roast garlic

Energy	Calories	Fat	Saturated fat	Carbohydrates	Sugar	Protein	Fibre
255 kJ	61 kcal	3.2 g	0.4 g	5.7 g	0.6 g	2.8 g	1.9 g

Mise en place

1 Peel the garlic.

2 Divide it into natural segments and place in a suitable roasting tray.

Cooking

1 Sprinkle with olive oil and roast in the oven until golden brown and tender.

> **SERVING SUGGESTION**
> Usually served as a garnish.

35 Braised red cabbage (*choux à la flamande*)

Energy	Calories	Fat	Saturated fat	Carbohydrates	Sugar	Protein	Fibre
754 kJ	180 kcal	15.2 g	8.4 g	7.8 g	7.7 g	3.4 g	3.2 g

	4 portions	10 portions
Red cabbage	300 g	1 kg
Salt, pepper		
Butter	50 g	125 g
Cooking apples	100 g	250 g
Caster sugar	10 g	25 g
Vinegar or red wine	125 ml	300 ml
Bacon trimmings (optional)	50 g	125 g

Mise en place

1 Quarter, trim and shred the cabbage. Wash well and drain.

Cooking

1 Season lightly with salt and pepper.

2 Place in a well-buttered casserole or pan suitable for placing in the oven (not aluminium or iron, because these metals will cause a chemical reaction that will discolour the cabbage).

3 Add the peeled and cored apples. Cut into 1-cm dice and sugar.

4 Add the vinegar and bacon (if using), cover with a buttered paper and lid.

5 Cook in a moderate oven at 150–200 °C for 1½ hours.

6 Remove the bacon (if used) and serve.

> **VARIATION**
> Other flavourings include 50 g sultanas, grated zest of one orange, pinch of ground cinnamon.

Strain off most of the liquid after braising.

> **Healthy Eating Tip**
> The fat and salt content will be reduced by omitting the bacon.

> **FAULTS**
> The bottom of the braising pan burns due to the liquid evaporating too fast.

> **SERVING SUGGESTION**
> Serve in individual dishes. The cabbage may be garnished with segments of orange.

36 Pickled red cabbage

Energy	Calories	Fat	Saturated fat	Carbohydrates	Sugar	Protein	Fibre
1246 kJ	301 kcal	2.1 g	0.01 g	28 g	27.2 g	10.5 g	20.3 g

* Using 700 ml vinegar per recipe. Discounted spices.

Mise en place

1 Remove the outer leaves of the cabbage and shred the rest finely.

Preparing the dish

1 Place in a deep bowl, sprinkle each layer with dry salt and leave for 24 hours.

2 Rinse and drain, cover with spiced vinegar (see recipe for mixed pickle, page 124) and leave for a further 24 hours, mixing occasionally. Pack and cover.

> **VARIATION**
> ● Try using different types of vinegar, sherry, red wine, etc.

37 Spinach purée (*epinards en purée*)

Energy	Calories	Fat	Saturated fat	Carbohydrates	Sugar	Protein	Fibre	
713 kJ	171 kcal	10.8 g	6.5 g	18.4 g	16.7 g	1.2 g	1.2 g	*

* Using 25 g butter per 0.5 kg

Half a kilogram will yield two portions.

Mise en place

1 Remove the stems and discard them.

2 Wash the leaves very carefully in plenty of water, several times if necessary.

Cooking

1 Wilt for 2–3 minutes, taking care not to overcook.

2 Place on a tray and allow to cool.

3 Pass through a sieve or mouli, or use a food processor.

4 Reheat in 25–50 g butter, mix with a kitchen spoon, correct the seasoning and serve.

> **VARIATION**
> ● Creamed spinach purée can be made by mixing in 30 ml cream and 60 ml béchamel sauce or natural yoghurt before serving. Serve with a border of cream.
> ● An addition would be 1-cm triangle-shaped croutons fried in butter.
> ● Spinach may also be served with toasted pine kernels or finely chopped garlic.

> **FAULTS**
> Spinach cooked in too much water will be too wet and overcooked.

38 Stir-fried cabbage with mushrooms and beansprouts

Energy	Calories	Fat	Saturated fat	Carbohydrates	Sugar	Protein	Fibre
413 kJ	100 kcal	6.9 g	0.8 g	4.9 g	3.9 g	4.8 g	5.8 g

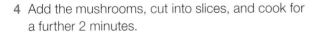

	4 portions	10 portions
Sunflower or vegetable oil	2 tbsp	5 tbsp
Spring cabbage or pak choi, shredded	400 g	1 kg
Soy sauce	2 tbsp	5 tbsp
Mushrooms	200 g	500 g
Beansprouts	100 g	250 g
Freshly ground pepper		

Cooking

1 Heat the oil in a suitable pan (e.g. a wok).

2 Add the cabbage and stir for 2 minutes.

3 Add the soy sauce, stir well. Cook for a further minute.

4 Add the mushrooms, cut into slices, and cook for a further 2 minutes.

5 Stir in the beansprouts and cook for 1–2 minutes.

6 Stir well. Season with freshly ground pepper and serve.

PROFESSIONAL TIP

The cabbage must be shredded evenly and the mushrooms cut evenly, so that they will cook at the same rate.

Healthy Eating Tips

● Reduce the oil by half when cooking the cabbage.

● No added salt is needed as soy sauce is used.

HEALTH AND SAFETY

Keep arms covered when quickly stir-frying.

VARIATION

● Add chopped spring onions to the dish.

FAULTS

Stir-frying is quick cooking. Take care not to overcook.

SERVING SUGGESTION

Serve in attractive deep plates.

39 Broad beans (*fèves*)

Energy	Calories	Fat	Saturated fat	Carbohydrates	Sugar	Protein	Fibre *
344 kJ	81 kcal	0.6 g	0.1 g	5.0 g	0.4 g	7.9 g	6.5 g

* 100 g portion

Half a kilogram will yield about two portions.

Remove the inner shells of broad beans.

Cooking

1 Shell the beans and cook in boiling salted water for 10–15 minutes until tender. Do not overcook.

2 If the inner shells are tough, remove before serving.

PROFESSIONAL TIP

The modern technique is to take the beans out of their pod and outer skin before serving – this reveals the bright green, tender beans.

VARIATION

Try:
- brushing with butter
- brushing with butter then sprinkling with chopped parsley
- binding with ½-litre cream sauce or fresh cream.

40 Corn on the cob (*maïs*)

Energy	Calories	Fat	Saturated fat	Carbohydrates	Sugar	Protein	Fibre
646 kJ	154 kcal	2.9 g	0.5 g	28.5 g	2.1 g	5.1 g	5.9 g

Mise en place

1 Allow one cob per portion.
2 Trim the stem.

Cooking

1 Cook in lightly salted boiling water for 10–20 minutes or until the corn is tender. Do not overcook.

2 Remove the outer leaves and fibres.

3 Serve with a sauceboat of melted butter.

NOTE

Creamed sweetcorn can be made by removing the corn from the cooked cobs, draining well and binding lightly with cream (fresh or non-dairy), béchamel sauce or yoghurt.

Healthy Eating Tip

In place of melted butter, serve warm olive oil.

SERVING SUGGESTION

Serve on individual dishes, usually with corn skewers.

41 Haricot bean salad (*salade de haricots blancs*)

Energy	Calories	Fat	Saturated fat	Carbohydrates	Sugar	Protein	Fibre
278 kJ	66 kcal	2.1 g	0.4 g	9.0 g	0.7 g	3.3 g	3.1 g

Preparing the dish

1 Combine all the ingredients.

This recipe can be used for any type of dried bean.

Healthy Eating Tips

- Lightly dress with vinaigrette.
- Add salt sparingly.

VARIATION

- Use chopped spring onions in place of chives.
- Add some chopped chilli and 50 g tomato concassé.

SERVING SUGGESTION

Serve in individual dishes or in a salad bowl.

	4 portions	10 portions
Haricot beans, cooked	200 g	500 g
Vinaigrette	1 tbsp	2½ tbsp
Parsley, chopped		
Onion, chopped and blanched if necessary	¼	½
Chives (optional)	15 g	40 g
Salt, pepper		

42 Mangetout

Energy	Calories	Fat	Saturated fat	Carbohydrates	Sugar	Protein	Fibre
94 kJ	22 kcal	0.1 g	0 g	2.8 g	2.4 g	3.1 g	0 g

Cooking

1 Cook in boiling salted water for 2–3 minutes, until slightly crisp.

2 Serve whole, brushed with butter.

Video: Boiling vegetables

http://bit.ly/2oMnZTQ

Healthy Eating Tip
Omit the butter or drizzle with olive oil.

Half a kilogram of mangetout will yield 4–6 portions.

Mise en place

1 Top and tail, wash and drain.

FAULTS
Mangetout take only seconds to cook and are very easily overcooked.

43 Ladies' fingers (okra) in cream sauce (*okra à la crème*)

Energy	Calories	Fat	Saturated fat	Carbohydrates	Sugar	Protein	Fibre	*
928 kJ	221 kcal	20.2 g	9.8 g	5.7 g	5.7 g	4.4 g	3.2 g	

* Using hard margarine

Cooking

1 Blanch in lightly salted boiling water, or steam; drain.

2 Sweat in the butter for 5–10 minutes, or until tender.

3 Carefully add the cream sauce.

4 Bring to the boil, correct the seasoning and serve in a suitable dish.

SERVING SUGGESTION
Okra may also be served brushed with butter or sprinkled with chopped parsley.

	4 portions	10 portions
Ladies' fingers (okra)	400 g	1.25 kg
Butter or margarine	50 g	125 g
Cream sauce	250 ml	625 ml

PROFESSIONAL TIP
Okra can become glutinous (slimy) when it is cooked in a sauce. To avoid this, wash the okra and let it dry before cooking.

Mise en place

1 Top and tail the ladies' fingers (okra).

Healthy Eating Tips
● Use a little unsaturated oil to sweat the okra.
● Try using half cream sauce and half yoghurt, adding very little salt.

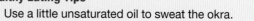

44 Peas French-style (*petit pois à la française*)

Energy	Calories	Fat	Saturated fat	Carbohydrates	Sugar	Protein	Fibre
515 kJ	123 kcal	5.6 g	3.4 g	12.9 g	5.8 g	5.9 g	5.7 g

2 Peel and wash the onions, shred the lettuce and add to the peas with half the butter, a little salt and the sugar.

Cooking

1 Barely cover with water. Cover with a lid and cook steadily, preferably in the oven, until tender.

2 Correct the seasoning.

3 Mix the remaining butter with the flour, making a *beurre manié*, and place it into the boiling peas in small pieces until thoroughly mixed; serve.

	4 portions	**10 portions**
Peas (in the pod)	1 kg	2.5 kg
Spring or button onions	12	40
Lettuce, small	1	2–3
Butter	25 g	60 g
Salt		
Caster sugar	½ tsp	1 tsp
Flour	5 g	12 g

NOTE

When using frozen peas, allow the onions to almost cook before adding the peas.

VARIATION

Using frozen peas will reduce the cooking time. Add fried lardons of bacon.

Mise en place

1 Shell and wash the peas and place in a sauteuse.

1 Shell the peas and discard the pods.

2 Place the prepared ingredients into the sauteuse.

3 Add a *beurre manié* at the end of the cooking time.

45 Courgette and potato cakes with mint and feta cheese

Energy	Calories	Fat	Saturated fat	Carbohydrates	Sugar	Protein	Fibre
930 kJ	223 kcal	14.2 g	7.5 g	15.2 g	2.3 g	9.4 g	2.1 g

	6 portions
Courgettes	3 large
Potatoes	350 g
Fresh mint, chopped	2 tbsp
Spring onions, finely chopped	2
Feta cheese	200 g
Eggs	1
Plain flour	25 g
Butter	25 g
Olive oil	1 tbsp
Salt, pepper	

Mise en place

1 Lightly scrape the courgettes to remove the outside skin.

2 Purée in a food processor. Remove, sprinkle with salt to remove the excess moisture, leave for 1 hour.

3 Rinse under cold water, squeeze out all excess moisture, dry on a clean cloth.

Cooking

1 Steam or parboil the potatoes for 8–10 minutes. Cool and peel.

2 Carefully grate the potatoes, place in a bowl, then season.

3 Add the courgettes, mint, spring onion, chopped feta cheese and the beaten egg. Mix well.

4 Divide the mixture into six and shape into cakes approximately 1-cm thick.

5 Dust with flour.

6 Brush the cakes with melted butter and oil, place on a baking sheet, cook in an oven at 200 °C for 15 minutes; turn over and continue to cook for a further 15 minutes.

7 Serve on suitable plates garnished with fresh blanched mint leaves and green sauce (see page 119).

PROFESSIONAL TIP

Make sure that all excess moisture is removed from the courgettes at step 1. If they are too moist, the mixture will be difficult to handle.

Healthy Eating Tips
- Make sure the puréed courgettes are rinsed well to remove the added salt.
- Use a little sunflower oil to brush the cakes before cooking.

VARIATION
- Use other types of cheese, such as cheddar or stilton.

SERVING SUGGESTION

In place of mint leaves use basil leaves or coriander leaves.

46 Deep-fried courgettes (*courgettes frites*)

Energy	Calories	Fat	Saturated fat	Carbohydrates	Sugar	Protein	Fibre *
481 kJ	111 kcal	11.4 g	1.4 g	1.8 g	1.7 g	1.8 g	0.9 g

* Using vegetable oil

Cooking

1 Pass through flour, or milk and flour, or batter.

2 Deep-fry in hot fat at 185 °C. Drain well and serve.

> **PROFESSIONAL TIP**
> Make sure the oil is very hot before adding the courgette. Fry it quickly and drain it before serving.

> **FAULTS**
> Courgettes will be soggy if the cooking oil is at too low a temperature.

Mise en place

1 Wash. Top and tail, and cut into round slices 3–6-cm thick.

47 Shallow-fried courgettes (*courgettes sautées*)

Energy	Calories	Fat	Saturated fat	Carbohydrates	Sugar	Protein	Fibre *
456 kJ	111 kcal	10.7 g	6.6 g	1.9 g	1.8 g	1.9 g	0.9 g

* Using butter

Mise en place

1 Wash. Top and tail, and cut into round slices 3–6-cm thick.

Cooking

1 Gently fry in hot oil or butter for 2 or 3 minutes, drain and serve.

48 Fettuccini of courgette with chopped basil and balsamic vinegar

Energy	Calories	Fat	Saturated fat	Carbohydrates	Sugar	Protein	Fibre
745 kJ	181 kcal	17.9 g	2.6 g	1.9 g	1.8 g	1.8 g	1.2 g

Mise en place

1 Slice the courgettes finely lengthwise, using a mandolin (Japanese slicer).

Cooking

1 Heat the olive oil in a suitable pan. Sauté the courgette slices quickly without colour for 35 seconds.

2 Place on suitable plates. Drizzle with olive oil and balsamic vinegar, and top with shredded basil leaves.

SERVING SUGGESTION
This may be served as a vegetarian starter or as a garnish for fish and meat dishes.

HEALTH AND SAFETY
Make sure the Japanese slicer has a guard.

VARIATION
Coriander can be used in place of basil.

	4 portions	10 portions
Courgettes, large	2	5
Olive oil	50 ml	125 ml
Olive oil, to finish	20 ml	125 ml
Balsamic vinegar, to finish	20 ml	50 ml
Basil leaves, shredded	2	5

49 Roast butternut squash or pumpkin with cinnamon and ginger

Energy	Calories	Fat	Saturated fat	Carbohydrates	Sugar	Protein	Fibre
342 kJ	81 kcal	4.1 g	0.5 g	10.4 g	5.6 g	1.4 g	2.5 g

Cooking

1 Place the butter into a suitable roasting pan and heat gently. Add the olive oil, ginger and cinnamon.

2 Add the squash gently and stir until the squash is coated in the spice mixture. Season and add the sugar.

3 Place in an oven at 200 °C until tender and golden brown.

4 When cooked, sprinkle with lemon juice.

	4 portions	10 portions
Butternut squash or pumpkin	500 g	1.25 kg
Butter or margarine	25 g	60 g
Olive oil	2 tbsp	5 tbsp
Ground cinnamon	½ tsp	1 tsp
Ginger, peeled and freshly chopped	½ tsp	1 tsp
Caster sugar	10 g	25 g
Lemon, juice of	½	1

VARIATION

● Garlic may be added to the spice mixture, and mixed spice may be used in place of cinnamon.

Healthy Eating Tip

Omit the butter or margarine and just use olive or rapeseed oil.

SERVING SUGGESTION

Serve on individual plates or as a garnish. May be garnished with micro herbs.

Mise en place

1 Peel the squash, cut it in half and remove the seeds.

2 Cut the squash into 1.5-cm dice or into small wedges.

Peel and halve the squash, then remove the seeds.

Cut the squash into even pieces.

Stir the squash pieces into the warm butter and spices before roasting.

50 Spiced aubergine purée

Energy	Calories	Fat	Saturated fat	Carbohydrates	Sugar	Protein	Fibre
176 kJ	41 kcal	1 g	0.2 g	6.5 g	6 g	2.7 g	6 g

Mise en place

1 Dice the aubergine and mix with the salt and cumin in a suitable bowl.

2 Allow to stand for 30 minutes.

Cooking

1 Dry in a cloth and deep-fry for 10 minutes.

2 Purée with the rest of the ingredients and pass.

VARIATION

This purée can be added to rice or couscous, or used as a filling for stuffed vegetables. It can also be used as a garnish for meat dishes.

SERVING SUGGESTION

Use as a filling with other vegetables, e.g. tomatoes. Use as a garnish or serve in individual dishes.

	4–6 portions
Diced aubergine	1 kg
Salt	20 g
Cumin	5 g
Tomato purée	45 g
Water	200 ml
Rose harissa	15 g
Vegetable nage	200 ml

51 Stuffed aubergine (*aubergine farcie*)

Energy	Calories	Fat	Saturated fat	Carbohydrates	Sugar	Protein	Fibre
557 kJ	134 kcal	12.2 g	1.5 g	4.3 g	3.8 g	2.4 g	4.5 g

	4 portions	10 portions
Aubergines	2	5
Shallots, chopped	10 g	25 g
Oil or fat, to fry		
Mushrooms sliced or diced	100 g	250 g
Parsley, chopped		
Tomato concassé	100 g	250 g
Salt, pepper		
Demi-glace or jus-lié	125 ml	300 ml

Mise en place

1 Cut the aubergines in two lengthwise.

2 With the point of a small knife, make a cut round the halves approximately 0.5 cm from the edge, then make several cuts 0.5 cm deep in the centre.

Cooking

1 Deep-fry in hot fat at 185 °C for 2–3 minutes; drain well.

2 Scoop out the centre pulp and chop it finely.

3 Cook the shallots in a little oil or fat without colouring.

4 Add the well-washed mushrooms. Cook gently for a few minutes.

5 Mix in the pulp, parsley and tomato; season. Replace in the aubergine skins.

6 Sprinkle with breadcrumbs and melted butter. Brown under the salamander.

7 Serve with a cordon of demi-glace or jus-lié.

VARIATION

● Crushed chopped garlic and chopped chilli may be added to the mix.

FAULTS

If the skins of the aubergines are broken then the filling will not be held in. Make sure that there is sufficient filling in the skins.

52 Stuffed tomatoes (*tomates farcies*)

Energy	Calories	Fat	Saturated fat	Carbohydrates	Sugar	Protein	Fibre
430 kJ	102 kcal	5.9 g	3.5 g	10.6 g	5.7 g	2.5 g	2.2 g

	4 portions	10 portions
Tomatoes, medium-sized	8	20
Duxelle		
Shallots, chopped	10 g	25 g
Butter or oil	25 g	60 g
Mushrooms	150 g	375 g
Salt, pepper		
Clove of garlic, crushed (optional)	1	2–3
Breadcrumbs (white or wholemeal)	25 g	60 g
Parsley, chopped		

Mise en place

1 Wash the tomatoes, remove the eyes.
2 Remove the top quarter of each tomato with a sharp knife.
3 Carefully empty out the seeds without damaging the flesh.

4 Place on a greased baking tray.

Cooking

1 Cook the shallots in a little butter or oil without colour.
2 Add the washed chopped mushrooms; season with salt and pepper; add the garlic if using. Cook for 2–3 minutes.
3 Add a little of the strained tomato juice, the breadcrumbs and the parsley; mix to a piping consistency. Correct the seasoning. At this stage, several additions may be made (e.g. chopped ham, cooked rice).
4 Place the mixture in a piping bag with a large star tube and pipe into the tomato shells. Replace the tops.
5 Brush with oil, season lightly with salt and pepper.
6 Cook in a moderate oven at 180–200 °C for 4–5 minutes.
7 Serve garnished with fresh picked parsley or fresh basil or rosemary.

Healthy Eating Tips
- Use a small amount of an unsaturated oil to cook the shallots and brush over the stuffed tomatoes.
- Add the minimum amount of salt.
- Adding cooked rice to the stuffing will increase the amount of starchy carbohydrate.

VARIATION
Add chopped mixed herbs to the duxelle.

1 Cut out the eye of the tomato.

2 Slice off the top and remove the seeds from inside.

3 Pipe in the filling and then replace the top.

53 Tomato concassé (*tomate concassé*)

Energy	Calories	Fat	Saturated fat	Carbohydrates	Sugar	Protein	Fibre	Sodium
29.2 kJ	70 kcal	6.3 g	0.4 g	2.9 g	2.9 g	05 g	1 g	247 g

* Using onions, vegetable oil and half teaspoon salt. Based on 4 portions.

	4 portions	10 portions
Tomatoes	400 g	1.25 kg
Shallots or onions, chopped	25 g	60 g
Butter or oil	25 g	60 g
Salt, pepper		

SERVING SUGGESTION
Can be used as a garnish, a filling or in sauces.

Cooking

1 Plunge the tomatoes into boiling water for 5–10 seconds – the riper the tomatoes, the less time is required. Refresh immediately.
2 Remove the skins, cut in quarters and remove all the seeds.
3 Roughly chop the flesh of the tomatoes.
4 Meanwhile, cook the chopped onion or shallots without colour in the butter or oil.
5 Add the tomatoes and season lightly.
6 Simmer gently on the side of the stove until the moisture is evaporated.

NOTE
This is a cooked preparation that is usually included in the normal *mise en place* of a kitchen as it is used in a great number of dishes.

Uncooked tomato concassé is often used for mise en place.

1 Blanch the tomatoes and then peel them.

2 Cut each tomato into quarters.

3 Remove the seeds from each petal.

4 Roughly chop the tomatoes.

54 Asparagus points or tips (*pointes d'asperges*)

Energy	Calories	Fat	Saturated fat	Carbohydrates	Sugar	Protein	Fibre
124 kJ	29 kcal	0.9 g	0.1 g	1.6 g	1.6 g	3.5 g	1.6 g

Mise en place

1 Using the back of a small knife, carefully remove the tips of the leaves.

2 Scrape the stem, either with the blade of a small knife or a peeler.

3 Wash well. Tie into bundles of about 12 heads.

4 Cut off the excess stem.

Cooking

1 Cook in lightly salted boiling water for approximately 5–8 minutes.

2 Test if cooked by gently pressing the green part of the stem, which should be tender; do not overcook.

3 Lift carefully out of the water. Remove the string, drain well and serve.

NOTE

Young thin asparagus, 50 pieces to the bundle, is known as 'sprew' or 'sprue'. It is prepared in the same way as asparagus except that when it is very thin the removal of the leaf tips is dispensed with. It may be served as a vegetable, perhaps brushed with butter or olive oil.

SERVING SUGGESTION

Asparagus tips are also used in numerous garnishes for soups, egg dishes, fish, meat and poultry dishes, cold dishes, salad and so on.

PROFESSIONAL TIP

As the flavour of asparagus is mild and can be leached out very easily through the cooking medium, a method of cookery that ensures that no flavour is lost in the cooking process is microwaving.

1 To microwave, place a piece of cling film over a plate that will fit in the microwave and, more importantly, is microwave safe.

2 Spread the cling film with a little oil and salt, evenly place the asparagus on the plate in a single layer.

3 Cover the plate and asparagus with another piece of cling film, and microwave for 30-second stints until the asparagus is tender; serve immediately.

The benefit of this method is that it retains flavour and colour, and it can be cooked in minutes, as opposed to batch cooking, which will, invariably, cause the asparagus to lose flavour and colour the longer it is stored.

If larger-scale cooking is required, the more traditional method, boiling in lightly salted water, should be used: cooking, say, 100 portions of asparagus in the microwave should be avoided for obvious reasons!

FAULTS

Asparagus is easily over cooked. Keep it undercooked and crisp (al dente).

SERVING SUGGESTION

Serve on individual plates with hollandaise or vinaigrette sauce.

55 Asparagus wrapped in puff pastry with Gruyère

Energy	Calories	Fat	Saturated fat	Carbohydrates	Sugar	Protein	Fibre
2017 kJ	485 kcal	37.7 g	15.0 g	23.4 g	2.5 g	15.9 g	0.8 g

	4 portions	10 portions
Gruyère cheese	175 g	400 g
Parmesan, freshly grated	3 tbsp	7 tbsp
Crème fraiche	250 ml	625 ml
Puff pastry (page 518)	350 g	875 g
Eggwash or milk, for brushing		
Asparagus, freshly cooked	350 g	875 g
Salt, pepper		
Watercress, for garnish		

Mise en place

1 Cut the Gruyère cheese into 1-cm dice.

2 In a suitable bowl, mix the Parmesan cheese and crème fraiche; season.

Cooking

1 Roll out the puff pastry to approximately 0.25 cm thick and cut into squares approximately 18 x 18 cm.

2 Brush the edges with eggwash or milk.

3 Divide the crème fraiche, putting equal amounts onto the centre of each square. Lay the asparagus on top. Place the diced Gruyère cheese firmly between the asparagus.

4 Fold the opposite corners of each square to meet in the centre, like an envelope. Firmly pinch the seams together to seal them. Make a small hole in the centre of each one to allow the steam to escape. Place on a lightly greased baking sheet.

5 Allow to relax for 20 minutes in the refrigerator. Brush with eggwash or milk, sprinkle with Parmesan.

6 Bake in a hot oven at 200 °C for approximately 20–25 minutes until golden brown.

SERVING SUGGESTION
Serve garnished with watercress.

PROFESSIONAL TIP
Make sure the pastry parcels are well sealed so that the mixture does not escape during cooking.

Healthy Eating Tips
The puff pastry and cheese make this dish high in fat. Serve with plenty of starchy carbohydrate to dilute it.

VARIATION
● Other types of cheese can be used, e.g. Parmesan, Cheddar.

56 Globe artichokes (*artichauts en branche*)

Energy	Calories	Fat	Saturated fat	Carbohydrates	Sugar	Protein	Fibre	*
32 kJ	8 kcal	0.0 g	0.0 g	1.4 g	1.4 g	0.6 g	0.0 g	

* Not including sauce

Mise en place

1 Allow 1 artichoke per portion.
2 Cut off the stems close to the leaves.
3 Cut off about 2 cm across the tops of the leaves.
4 Trim the remainder of the leaves with scissors or a small knife.

Cooking

1 Place a slice of lemon at the bottom of each artichoke.
2 Secure with string.
3 Simmer in gently boiling, lightly salted water (to which a little ascorbic acid – one vitamin C tablet – may be added) until the bottom is tender (20–30 minutes).
4 Refresh under running water until cold.
5 Remove the centre of the artichoke carefully.
6 Scrape away all the furry inside (the choke) and leave clean.
7 Replace the centre, upside down.
8 Reheat by placing in a pan of boiling salted water for 3–4 minutes.
9 Drain and serve accompanied by a suitable sauce.

PROFESSIONAL TIP
Do not cook artichokes in an iron or aluminium pan because these metals cause a chemical reaction that will discolour them.

Healthy Eating Tips
Serve with a hollandaise sauce made with olive oil.

SERVING SUGGESTION
Serve hot with hollandaise sauce or cold with vinaigrette sauce.

57 Shallow-fried chicory (*endive meunière*)

Energy	Calories	Fat	Saturated fat	Carbohydrates	Sugar	Protein	Fibre	*
484 kJ	118 kcal	12.0 g	7.3 g	4.8 g	1.3 g	0.9 g	1.5 g	

* Using 37.5 g butter

Mise en place

1 Trim the stem, remove any discoloured leaves, wash.

Cooking

1 Cook as for braised chicory in a little water and lemon juice.
2 Drain, shallow-fry in a little butter and colour lightly on both sides.
3 Serve with 10 g per portion nut-brown butter, lemon juice and chopped parsley.

Healthy Eating Tip
Omit the nut brown butter, drizzle with olive oil.

	4 portions	10 portions
Fish or chicken stock	200 ml	500 ml
Chicory heads	8	20
Fresh lemon juice	3 tbsp	8 tbsp
Caster sugar	3 tbsp	8 tbsp
Sea salt, black pepper		
Butter	25 g	60 g

58 Sauté of wild mushrooms

Energy	Calories	Fat	Saturated fat	Carbohydrates	Sugar	Protein	Fibre
322 kJ	78 kcal	8.1 g	5.3 g	0.1 g	0.1 g	1.2 g	1 g

Cooking

1 Allow 50 g of mixed wild mushrooms per portion.
2 Shallow fry in butter or oil.

Healthy Eating Tip
Fry in olive or rapeseed oil.

SERVING SUGGESTION
Serve as a garnish or in individual dishes, sprinkle with chopped parsley or chopped mixed herbs or sprinkle with micro herbs.

59 Greek-style mushrooms (*champignons à la grecque*)

Energy	Calories	Fat	Saturated fat	Carbohydrates	Sugar	Protein	Fibre
587 kJ	142 kcal	15.2 g	2.2 g	0.4 g	0.3 g	1.0 g	0.6 g

	4 portions	10 portions
Water	250 ml	625 ml
Olive oil	60 ml	150 ml
Lemon, juice of	1	1 ½
bay leaf	½	1
Sprig of thyme		
Peppercorns	6	18
Coriander seeds	6	18
Salt		
Small white button mushrooms, cleaned	200 g	500 g

Mise en place

1 Combine all the ingredients except the mushrooms, to create a Greek-style cooking liquor.

Cooking

1 Cook the mushrooms gently in the cooking liquor for 3 to 4 minutes.

2 Serve cold with the unstrained liquor.

PROFESSIONAL TIP

Simmer the vegetables carefully so that they are correctly cooked and absorb the flavours.

VARIATION

Other vegetables such as artichokes and cauliflower can also be cooked in this style, using the same liquor:

- For artichokes, peel and trim 6 artichokes for 4 portions (or 15 for 10); cut the leaves short and remove the chokes. Blanch the artichokes in water with a little lemon juice for 10 minutes, refresh, then simmer in the Greek-style liquor for 15–20 minutes.
- For cauliflower, trim and wash one medium cauliflower for 4 portions (2 for 10); break it into small sprigs about the size of cherries. Blanch the sprigs for about 5 minutes, refresh, then simmer in the Greek-style liquor for 5–10 minutes. Keep the cauliflower slightly undercooked and crisp.

SERVING SUGGESTION

Serve in individual dishes with thyme and bay leaf.

60 Broccoli

Energy	Calories	Fat	Saturated fat	Carbohydrates	Sugar	Protein	Fibre
125 kJ	30 kcal	1 g	0.3 g	1.4 g	1.1 g	3.9 g	4.6 g

Cooking

1 Cook in lightly salted water, or steam.

NOTE

Because of their size, green and purple broccoli need less cooking time than cauliflower. Broccoli is usually broken down into florets and, as such, requires very little cooking: once brought to the boil, 1–2 minutes should be sufficient. This leaves the broccoli slightly crisp.

Tenderstem broccoli is increasingly popular. It has long, slim, evenly sized stems and small flower heads, so it cooks quickly. The distinctive flavour is similar to asparagus. Green and purple varieties are available, and, because of the uniform shape, they can be presented very attractively.

61 Cauliflower au gratin (*chou-fleur mornay*)

Energy	Calories	Fat	Saturated fat	Carbohydrates	Sugar	Protein	Fibre
632 kJ	150 kcal	10.4 g	3.9 g	8.6 g	3.8 g	6.3 g	2.0 g

Cooking

1 Reheat in a pan of hot salted water (chauffant), or reheat in butter in a suitable pan.

2 Place in vegetable dish or on a greased tray.

3 Coat with 250 ml of mornay sauce (see page 100).

4 Sprinkle with grated cheese.

5 Brown under the salamander and serve.

Healthy Eating Tip
No additional salt is needed as cheese is added.

PROFESSIONAL TIP
In order to achieve a good even glaze, add an egg yolk to the mornay sauce

Mise en place

1 Cut the cooked cauliflower into four or into florets.

62 Cauliflower polonaise (*chou-fleur polonaise*)

Energy	Calories	Fat	Saturated fat	Carbohydrates	Sugar	Protein	Fibre
575 kJ	139 kcal	11.9 g	6.9 g	4.1 g	1.9 g	4.0 g	1.7 g

Cooking

1 Cut the cooked cauliflower into four or into florets. Reheat in a chauffant or in butter in a suitable pan.

2 Heat 50 g butter, add 10 g fresh breadcrumbs in a frying pan and lightly brown. Pour over the cauliflower, sprinkle with sieved hardboiled egg and chopped parsley.

Healthy Eating Tip
Use olive oil instead of and use wholemeal breadcrumbs.

FAULTS
Take care not to over colour the breadcrumbs.

63 Crispy deep-fried tofu

Energy	Calories	Fat	Saturated fat	Carbohydrates	Sugar	Protein	Fibre	*
543 kJ	131 kcal	8.9 g	0.0 g	1.0 g	0.5 g	11.8 g	0.0 g	

* For a 50 g portion

Mise en place

1 Coat the tofu cubes with any of the following: flour, egg and breadcrumbs; milk and flour; cornstarch; arrowroot.

Cooking

1 Deep-fry the tofu at 180 °C, until golden brown. Drain.

2 Serve garnished with freshly grated ginger and julienne of herbs.

SERVING SUGGESTION
Serve with a tomato sauce flavoured with coriander.

Healthy Eating Tips
- Use an unsaturated oil to fry the tofu.
- Make sure the oil is hot so that less is absorbed.
- Alternatively, try dry-frying the tofu.

	4 portions	10 portions
Firm tofu, cut into cubes	200 g	500 g

64 Oriental stir-fry Quorn

Energy	Calories	Fat	Saturated fat	Carbohydrates	Sugar	Protein	Fibre
836 kJ	200 kcal	11.8 g	1.3 g	11.6 g	7.3 g	12.7 g	5.6 g

Mise en place

1 Prepare a marinade by mixing the soy sauce with the ginger, and season with black pepper.

2 Add the Quorn pieces, mix well and chill for 1 hour.

3 Strain the Quorn from the marinade.

Cooking

1 In a wok, add half the vegetable oil and stir-fry the Quorn quickly for approximately 4 minutes. Remove from the wok.

2 Add the remaining oil, and fry the spring onion and peppers for another 1–2 minutes.

3 Return the Quorn to the wok.

4 Add the strained marinade, sherry, stock and sugar. Bring to the boil.

5 Thicken lightly with the cornflour. Add the blanched almonds or cashews and stir gently to enable the ingredients to be covered with the sauce.

	4 portions	10 portions
Soy sauce	62 ml	156 ml
Ginger, freshly grated	12 g	30 g
Black pepper, to taste		
Quorn pieces, defrosted	200 g	500 g
Vegetable oil	1 tbsp	3 tbsp
Spring onions	8	20
Red pepper, halved, deseeded and finely sliced	1	3
Yellow pepper, halved, deseeded and finely sliced	1	3
Green pepper, halved, deseeded and finely sliced	1	3
Dry sherry	1 tbsp	3 tbsp
Vegetable stock	62 ml	156 ml
Sugar	¼ tsp	1 tsp
Cornflour	6 g	15 g
Blanched almonds or cashews	50 g	100 g

SERVING SUGGESTION
Serve with noodles or rice.

Healthy Eating Tips
- Use a little unsaturated oil to fry the Quorn, onions and peppers.
- No added salt is needed; there is plenty of flavour from the soy sauce, ginger and stock.

PRACTICAL COOKERY for the Level 3 Advanced Technical Diploma in Professional Cookery

65 Deep-fried seaweed

Energy	Calories	Fat	Saturated fat	Carbohydrates	Sugar	Protein	Fibre
686 kJ	166 kcal	15.9 g	0 g	4.2 g	0 g	1.9 g	0 g

Cooking

1 Quickly fry in hot, deep fat (approximately 175–190 °C).

2 Remove and drain on absorbent kitchen paper.

PROFESSIONAL TIP
The seaweed must be cooked quickly in hot oil, until it is crisp.

FAULTS
Make sure the oil is at the correct temperature otherwise the seaweed will be soggy.

Mise en place

1 Carefully pick over the seaweed. Wash and thoroughly drain and dry on a cloth.

66 Japanese sea vegetable and noodle salad

Energy	Calories	Fat	Saturated fat	Carbohydrates	Sugar	Protein	Fibre
1138 kJ	271 kcal	9.4 g	1.3 g	39 g	8.3 g	10.2 g	7.5 g

	4 portions	10 portions
Dried arame	125 g	320 g
Shredded green cabbage	500 g	1.25 kg
Coarsely chopped parsley	2 tbsp	5 tbsp
Julienne of carrots	125 g	320 g
Finely chopped celery	50 g	125 g
Chopped chives	50 g	125 g
Cooked udon noodles	400 g	1 kg
Sesame oil	2 tbsp	5 tbsp
Rice vinegar	3 tbsp	7 tbsp
Smoked tofu, diced in 1-cm dice	75 g	180 g

3 Add the cooked noodles, oil and vinegar; season and mix well.

4 Serve garnished with diced tofu.

Healthy Eating Tip
● Replace sesame oil with olive oil.

Preparing the dish

1 In a large bowl, cover the arame with water and soak for 10 minutes. Drain well.

2 Place the arame back in the bowl, add the cabbage, parsley, carrots, celery, chives and mix well.

212

67 Sea kale (*chou de mer*)

Energy	Calories	Fat	Saturated fat	Carbohydrates	Sugar	Protein	Fibre
33 kJ	8 kcal	0.0 g	0.0 g	0.6 g	0.6 g	1.4 g	0.0 g

Half a kilogram will yield about 3 portions.

Mise en place

1 Trim the roots and remove any discoloured leaves.

2 Wash well and tie into a neat bundle.

Cooking

1 Cook in boiling lightly salted water for 15–20 minutes. Do not overcook.

2 Drain well, serve accompanied with a suitable sauce (e.g. melted butter, hollandaise).

Healthy Eating Tip

- In place of melted butter or hollandaise, serve a warm vinaigrette made with balsamic vinegar and olive oil flavoured with crushed garlic and chopped mixed herbs.

68 Alu-chole (vegetarian curry)

Energy	Calories	Fat	Saturated fat	Carbohydrates	Sugar	Protein	Fibre	*
1214 kJ	290 kcal	17.5 g	1.6 g	26.6 g	5.3 g	10.6 g	5.5 g	

* Using lemon juice and vegetable oil

	4 portions	10 portions
Vegetable ghee or vegetable oil	3 tsp	7 ½ tsp
Small cinnamon sticks	4	10
Bay leaves	4	10
Cumin seeds	1 tsp	2½ tsp
Onion, finely chopped	100 g	250 g
Cloves garlic, finely chopped and crushed	2	5
Plum tomatoes, canned, chopped	400 g	1 kg
Hot curry paste	3 tsp	7½ tsp
Salt, to taste		
Potatoes in 1-cm dice	100 g	250 g
Water	125 ml	312 ml
Chickpeas, canned, drained	400 g	1 kg
Coriander leaves, chopped	50 g	125 g
Tamarind sauce or lemon juice	2 tbsp	3 tbsp

Cooking

1 Heat the ghee in a suitable pan.

2 Add the cinnamon, bay leaves and cumin seeds; fry for 1 minute.

3 Add the onion and garlic. Fry until golden brown.

4 Add the chopped tomatoes, curry paste and salt, and fry for a further 2–3 minutes.

5 Stir in the potatoes and water. Bring to the boil. Cover and simmer until the potatoes are cooked.

6 Add the chickpeas; allow to heat through.

7 Stir in the coriander leaves and tamarind sauce or lemon juice; serve.

NOTE

This is a dish from northern India. It can be served as a vegetarian dish with rice, or to accompany meat and chicken dishes.

PROFESSIONAL TIP

Fry the spices well to extract the maximum flavour from them.

Healthy Eating Tips
- Use a small amount of unsaturated oil to fry the spices and onion.
- Replace the ghee with olive oil

---

69 Chinese vegetables and noodles

Energy	Calories	Fat	Saturated fat	Carbohydrates	Sugar	Protein	Fibre	*
2332 kJ	554 kcal	21.6 g	1.8 g	80.6 g	5.7 g	14.3 g	1.9 g	

* Using canned bamboo shoots

Cooking

1 Cook the noodles in boiling salted water for about 5–6 minutes until al dente. Refresh and drain.

2 Heat the oil in a wok and stir-fry all the vegetables, except the beansprouts, for 1 minute. Then add the beansprouts and cook for a further minute.

3 Add the drained noodles, stirring well; allow to reheat through.

4 Correct the seasoning.

5 Serve in a suitable dish, garnished with the spring onions.

Healthy Eating Tips
- Keep added salt to a minimum.
- Use an unsaturated oil (olive or sunflower) and reduce the quantity used.

	4 portions	10 portions
Chinese noodles	400 g	1.250 kg
Oil	60 ml	150 ml
Celery	100 g	250 g
Carrot, cut in paysanne	100 g	250 g
Bamboo shoots	50 g	125 g
Mushrooms, finely sliced	75 g	180 g
Chinese cabbage, shredded	75 g	180 g
Beansprouts	100 g	250 g
Soy sauce	30 ml	75 ml
Garnish (spring onions, sliced lengthways and quickly stir-fried)		

70 Stuffed peppers (*piment farci*)

Energy	Calories	Fat	Saturated fat	Carbohydrates	Sugar	Protein	Fibre
1291 kJ	308 kcal	11.4 g	6.7 g	48.8 g	5.3 g	5.4 g	3.1 g

	4 portions	10 portions
Red peppers, medium-sized	4	10
Carrots, sliced	50 g	125 g
Onions, sliced	50 g	125 g
Bouquet garni	1	2
White stock	0.5 litres	1.25 litres
Salt, pepper		
Pilaff		
Rice (long grain)	200 g	500 g
Salt, pepper		
Onion, chopped	50 g	125 g
Butter	50 g	125 g

Cooking

1 Place the peppers on a tray in the oven or under the salamander for a few minutes, or deep-fry in hot oil at 180 °C, until the skin blisters.

2 Remove the skin; carefully cut off the top and empty out all the seeds.

3 Stuff with a well-seasoned pilaff of rice (ingredients as listed), which may be varied by the addition of mushrooms, tomatoes, ham, and so on.

4 Replace the top of the peppers.

5 Place the peppers on the sliced carrot and onion in a pan suitable for the oven; add the bouquet garni, stock and seasoning. Cover with a buttered paper and a lid.

6 Cook in a moderate oven at 180–200 °C for 1 hour or until tender.

7 Serve garnished with picked parsley.

Healthy Eating Tips

- This dish is low in fat if the peppers are placed in the oven or under the salamander, not deep-fried, and the butter or oil is kept to a minimum.
- Add little or no salt.
- If extra vegetables are added to the rice, and a vegetable stock used, this dish can be a useful vegetarian starter.

VARIATION

- In place of rice use couscous or other suitable grains such as quinoa. Try a little lemon grass added to the rice.

FAULTS

Do not overfill the peppers otherwise they will burst open.

SERVING SUGGESTION

Serve on individual plates. In place of parsley use micro herbs, fresh basil, coriander or lemongrass.

1 Briefly heat the peppers and then peel them.

2 Cut off the top and empty out the seeds.

3 Fill the pepper and then replace the stem.

71 Coleslaw

Energy	Calories	Fat	Saturated fat	Carbohydrates	Sugar	Protein	Fibre	*
2514 kJ	599 kcal	59.0 g	8.8 g	11.7 g	11.4 g	5.9 g	7.2 g	

* Using mayonnaise, for 4 portions

	4 portions	10 portions
White or Chinese cabbage	200 g	500 g
Carrot	50 g	125 g
Onion (optional)	25 g	60 g
Mayonnaise, natural yoghurt or fromage frais	125 ml	300 ml

Mise en place

1 Trim off the outside leaves of the cabbage.
2 Cut into quarters. Remove the centre stalk.

Preparing the dish

1 Wash the cabbage, shred finely and drain well.
2 Mix with a fine julienne of raw carrot and shredded raw onion. To lessen the harshness of raw onion, blanch and refresh.
3 Bind with mayonnaise, natural yoghurt or vinaigrette.

PROFESSIONAL TIP
Cut the cabbage into fine julienne to give the coleslaw a good, even texture.

Healthy Eating Tip
Replace some or all of the mayonnaise with natural yoghurt and/or fromage frais.

HEALTH AND SAFETY
If using a slicing machine make sure the guard is attached.

72 Mixed vegetables (*macédoine or jardinière de légumes*)

Energy	Calories	Fat	Saturated fat	Carbohydrates	Sugar	Protein	Fibre
58 kJ	14 kcal	0.1 g	0.0 g	2.5 g	1.7 g	1.0 g	2.1 g

	4 portions	10 portions
Carrots	100 g	250 g
Turnips	50 g	125 g
Salt		
French beans	50 g	125 g
Peas	50 g	125 g

Mise en place

1 Peel and wash the carrots and turnips; cut into 0.5-cm dice (*macédoine*) or batons (*jardinière*).

Cooking

1 Cook the carrots and turnips separately in lightly salted water, do not overcook. Refresh.

2 Top and tail the beans; cut into 0.5-cm dice, cook and refresh, do not overcook.

3 Cook the peas and refresh.

4 Mix the vegetables and, when required, reheat in hot salted water.

5 Drain well, serve brushed with melted butter.

Video: Cutting vegetables into *macédoine*

http://bit.ly/2ppuWOo

VARIATION
● Add sweetcorn for additional colour.

73 Ratatouille

Energy	Calories	Fat	Saturated fat	Carbohydrates	Sugar	Protein	Fibre
579 kJ	138 kcal	12.6 g	1.7 g	5.2 g	4.6 g	1.3 g	2.4 g

	4 portions	10 portions
Baby marrow (courgette)	200 g	500 g
Aubergines	200 g	500 g
Tomatoes	200 g	500 g
Oil	50 ml	125 ml
Onions, finely sliced	50 g	125 g
Clove of garlic, peeled and chopped	1	2
Red peppers, diced	50 g	125 g
Green peppers, diced	50 g	125 g
Salt, pepper		
Parsley, chopped	1 tsp	2–3 tsp

Mise en place

1 Trim off both ends of the marrow and aubergines.
2 Remove the skin using a peeler.

3 Cut into 3-mm slices.
4 Concassé the tomatoes (peel, remove seeds, roughly chop).

Cooking

1 Place the oil in a thick-bottomed pan and add the onions.
2 Cover with a lid and allow to cook gently for 5–7 minutes without colour.
3 Add the garlic, marrow and aubergine slices, and the peppers.
4 Season lightly with salt and mill pepper.
5 Allow to cook gently for 4–5 minutes, toss occasionally and keep covered.
6 Add the tomato and continue cooking for 20–30 minutes or until tender.
7 Mix in the parsley, correct the seasoning and serve.

NOTE
The vegetables need to be cut evenly so that they will cook evenly; it also improves the texture of the dish.

Healthy Eating Tips
● Use a little unsaturated oil to cook the onions.
● Use the minimum amount of salt.

1 Ingredients for ratatouille.

2 Add the tomato to the vegetables during cooking.

74 Tempura

Energy	Calories	Fat	Saturated fat	Carbohydrates	Sugar	Protein	Fibre
3397 kJ	815 kcal	55.9 g	7.4 g	67.4 g	5.3 g	14.8 g	5.0 g

	4 portions
Vegetable oil	500 ml
Courgettes, sliced	2
Sweet potato, scrubbed and sliced	1
Green pepper, seeds removed and cut into strips	1
Shiitake mushroom, stalks removed and halved if large	4
Onion, sliced as half moons	1
Parsley sprigs, to garnish	4
Batter	
(NB all ingredients must be stored in the fridge until just before mixing)	
Egg yolk	1
Ice-cold sparkling water	200 ml
Plain flour, sifted	100 g
Tentsuyu dipping sauce (optional)	
Dashi stock	200 ml
Mirin	3 tbsp
Soy sauce	3 tbsp
Ginger, grated	½ tsp

Mise en place

1 To prevent splattering during the frying, make sure to dry all deep-fry ingredients thoroughly first with a kitchen towel.

2 For the batter, beat the egg yolk lightly and mix with the ice-cold water.

3 Add half the flour to the egg and water mixture. Give the mixture a few strokes. Add the rest of the flour all at once. Stroke the mixture a few times with chopsticks or a fork until the ingredients are loosely combined. The batter should be very lumpy. If over-mixed, tempura will be oily and heavy.

Cooking

1 Heat the oil to 160 °C.

2 Dip the vegetables into plain flour and then into the batter, a few pieces at a time. Fry until just crisp and golden (about 1½ minutes).

3 Drain the cooked vegetables on a kitchen towel.

4 Serve immediately with a pinch of salt, garnished with parsley sprigs and lemon wedges, dry-roasted salt or with tentsuyu dipping sauce in a small bowl with grated ginger. This dish can also be served with an accompaniment of grated white radish.

5 To make the tentsuyu sauce (if required), combine the ingredients in a small saucepan; heat it through and leave to one side.

1 Use sparkling water to make tempura batter.

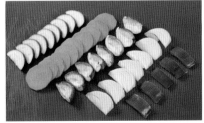

2 Cut a variety of vegetables into even pieces.

3 Dip each piece into the batter and then deep-fry it.

NOTE
Any vegetables with a firm texture may be used for tempura.

Healthy Eating Tips
Use sunflower or groundnut oil for frying.

75 Vegetarian strudel

Energy	Calories	Fat	Saturated fat	Carbohydrates	Sugar	Protein	Fibre
2117 kJ	504 kcal	27.6 g	4.0 g	54.1 g	10.5 g	14.3 g	9.7 g

	4 portions	10 portions
Strudel dough		
Strong flour	200 g	500 g
Pinch of salt		
Sunflower oil	25 g	60 g
Egg	1	2–3
Water at 37°C	83 ml	125 ml
Filling		
Large cabbage leaves	200 g	500 g
Sunflower oil	4 tbsp	10 tbsp
Onion, finely chopped	50 g	125 g
Cloves garlic, chopped	2	5
Courgettes	400 g	1 kg
Carrots	200 g	500 g
Turnips	100 g	250 g
Tomato, skinned, deseeded and diced	300 g	750 g
Tomato purée	25 g	60 g
Toasted sesame seeds	25 g	60 g
Wholemeal breadcrumbs	50 g	125 g
Fresh chopped basil	3 g	9 g
Seasoning		
Eggwash		

Cooking

1 To make the strudel dough, sieve the flour with the salt and make a well.

2 Add the oil, egg and water, and gradually incorporate the flour to make a smooth dough; knead well.

3 Place in a basin, cover with a damp cloth; allow to relax for 3 minutes.

4 Meanwhile, prepare the filling: take the large cabbage leaves, wash and discard the tough centre stalks, blanch in boiling salted water for 2 minutes, until limp. Refresh and drain well in a clean cloth.

5 Heat the oil in a sauté pan, gently fry the onion and garlic until soft.

6 Peel and chop the courgettes into ½-cm dice, blanch and refresh. Peel and dice the carrots and turnips, blanch and refresh.

7 Place the well-drained courgettes, carrots and turnips into a basin, add the tomato concassé, tomato puree, sesame seeds, breadcrumbs and chopped basil, and mix well. Season.

8 Roll out the strudel dough to a thin rectangle, place on a clean cloth and stretch until extremely thin.

9 Lay the drained cabbage leaves on the stretched strudel dough, leaving approximately a 1-cm gap from the edge.

10 Place the filling in the centre. Eggwash the edges.

11 Fold in the longer side edges to meet in the middle. Roll up.

12 Transfer to a lightly oiled baking sheet. Brush with the sunflower oil.

13 Bake for 40 minutes in a preheated oven at 180–200 °C.

14 When cooked, serve hot, sliced on individual plates with a cordon of tomato sauce made with vegetable stock.

PROFESSIONAL TIP

It is essential to roll out and stretch the strudel dough so that it is very thin, but without breaking it.

The water and other batter ingredients must be ice cold. The batter should be lumpy to give it texture. Do not over-mix it; this will make it oily and heavy.

Healthy Eating Tips
- Use the minimum amount of salt.
- Use a little unsaturated oil to cook the onion and garlic.

VARIATION

Other vegetables may also be used, for example, sweet potatoes, butternut squash, cauliflower, broccoli. These can also be made individually.

FAULTS

Make sure that the strudel dough has no cracks or holes so that the filling does not burst out in the baking process.

76 Crisp polenta and roasted Mediterranean vegetables

Energy	Calories	Fat	Saturated fat	Carbohydrates	Sugar	Protein	Fibre
3267 kJ	790 kcal	71.4 g	18.9 g	28.6 g	14.0 g	10.1 g	6.6 g

	4 portions	10 portions
Polenta		
Water	200 ml	500 ml
Butter	30 g	75 g
Polenta flour	65 g	160 g
Parmesan, grated	25 g	60 g
Egg yolks	1	2
Crème fraiche	110 g	275 g
Seasoning		
Roasted vegetables		
Red peppers	2	5
Yellow peppers	2	5
Courgettes	2	5
Red onions	2	5
Vegetable oil	200 ml	500 ml
Seasoning		
Clove of garlic	1	3
Thyme, sprigs	2	5

Cooking

Polenta:

1 Bring the water and the butter to the boil.

2 Season the water well and whisk in the polenta flour.

3 Continue to whisk until very thick.

4 Remove from the heat and add the Parmesan, egg yolk and crème fraiche.

5 Whisk until all incorporated; check the seasoning.

6 Set in a lined tray.

7 Once set, cut using a round cutter or cut into squares.

8 Reserve until required.

Roasted vegetables:

1 Roughly chop the vegetables into large chunks. Ensure the seeds are removed from the peppers.

2 Toss the cut vegetables in the oil and season well.

3 Place the vegetables in an oven with the aromats for 30 minutes at 180 °C.

4 Remove from the oven and drain. Reserve until required.

To serve:

1 To serve the dish, shallow-fry the polenta in a non-stick pan until golden on both sides.

2 Warm the roasted vegetables and place them in the middle of the plate. Place the polenta on top.

3 Serve with rocket salad and balsamic dressing.

PROFESSIONAL TIP

Line the tray with cling film and silicone paper before pouring in the polenta – this will stop it from sticking to the tray when it sets.

Healthy Eating Tip

● Use low fat crème fraiche for a lower fat version.

VARIATION

Vary the types of roasted vegetables.

FAULTS

Do not overcook the roasted vegetables as they will lose colour, flavour and texture.

77 Niçoise salad

Energy	Calories	Fat	Saturated fat	Carbohydrates	Sugar	Protein	Fibre	*
867 kJ	207 kcal	9.6 g	1.5 g	25.0 g	4.9 g	6.9 g	9.9 g	

* For 4 portions

	4 portions	10 portions
Tomatoes	100 g	250 g
French beans, cooked	200 g	500 g
Diced potatoes, cooked	100 g	250 g
Salt, pepper		
Vinaigrette	1 tbsp	2½ tbsp
Anchovy fillets	10 g	25 g
Capers	5 g	12 g
Stoned olives	10 g	25 g

Healthy Eating Tips
- Lightly dress with vinaigrette.
- The anchovies are high in salt, so no added salt is necessary.

VARIATION
Add sliced cooked chicken for a chicken Niçoise salad

Preparing the dish

1 Peel the tomatoes, deseed and cut into neat segments.
2 Dress the beans, tomatoes and potatoes neatly.
3 Season with salt and pepper. Add the vinaigrette.
4 Decorate with anchovies, capers and olives.

78 Potato salad (*salade de pommes de terre*)

Energy	Calories	Fat	Saturated fat	Carbohydrates	Sugar	Protein	Fibre *
2013 kJ	479 kcal	34.9 g	5.1 g	40.0 g	1.3 g	4.0 g	2.6 g

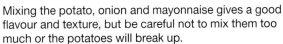

* Using mayonnaise, for 4 portions

NOTE
This is not usually served as a single hors d'oeuvre or main course.

PROFESSIONAL TIP
Mixing the potato, onion and mayonnaise gives a good flavour and texture, but be careful not to mix them too much or the potatoes will break up.

VARIATION
- Potato salad can also be made by dicing raw peeled or unpeeled potato, cooking them – preferably by steaming (to retain shape) – and mixing with vinaigrette while warm.
- Try adding two chopped hard-boiled eggs or 100 g of peeled dessert apple mixed with lemon juice, or a small bunch of picked watercress leaves.
- Potatoes may be cooked with mint and allowed to cool with the mint.
- Cooked small new potatoes can be tossed in vinaigrette with chopped fresh herbs (e.g. mint, parsley, chives).

	4 portions	10 portions
Potatoes, cooked	200 g	500 g
Vinaigrette	1 tbsp	2½ tbsp
Onion or chive (optional), chopped	10 g	25 g
Mayonnaise or natural yoghurt	125 ml	300 ml
Salt, pepper		
Parsley or mixed fresh herbs, chopped		

Preparing the dish

1 Cut the potatoes into ½–1-cm dice; sprinkle with vinaigrette.

2 Mix with the onion or chive, add the mayonnaise and correct the seasoning. (The onion may be blanched to reduce its harshness.)

3 Dress neatly and sprinkle with chopped parsley or herbs.

FAULTS
If potatoes are over cooked then when mixed with the mayonnaise they will turn to mash.

79 Vegetable salad/Russian salad (*salade de légumes/salade russe*)

Energy	Calories	Fat	Saturated fat	Carbohydrates	Sugar	Protein	Fibre
1566 kJ	373 kcal	35.0 g	5.2 g	10.1 g	8.2 g	5.0 g	11.9 g

* Using mayonnaise, for 4 portions

Cooking

1 Cook separately in salted water, refresh and drain well.

2 Top and tail the beans, and cut into 0.5-cm dice; cook, refresh and drain well.

3 Cook the peas, refresh and drain well.

4 Mix all the well-drained vegetables with vinaigrette and then mayonnaise.

5 Correct the seasoning. Dress neatly.

NOTE
Do not overcook the vegetables. Drain them well before adding the dressing – otherwise the salad will be too wet.

Healthy Eating Tips
- Try half mayonnaise and half natural yoghurt.
- Season with the minimum amount of salt.

VARIATION
Other vegetables may be used e.g. butternut squash, sweet potatoes, pumpkin.

	4 portions	10 portions
Carrots	100 g	250 g
Turnips	50 g	125 g
French beans	50 g	125 g
Peas	50 g	125 g
Vinaigrette	1 tbsp	2–3 tbsp
Mayonnaise or natural yoghurt	125 ml	300 ml
Salt, pepper		

Mise en place

1 Peel and wash the carrots and turnips, cut into 0.5-cm dice or batons.

80 Waldorf salad

Energy	Calories	Fat	Saturated fat	Carbohydrates	Sugar	Protein	Fibre
657 kJ	159 kcal	14 g	1.8 g	7 g	6.9 g	1.6 g	2.2 g

3 Dress on quarters or leaves of lettuce (may also be served in hollowed-out apples).

NOTE
When mixing in the mayonnaise, add just enough to give the right texture and flavour.

Healthy Eating Tip
Try using some yoghurt in place of the mayonnaise, which will proportionally reduce the fat.

VARIATION
● Add sultanas or grapes. Try decorating with pumpkin seeds.

Preparing the dish

1 Dice celery or celeriac and crisp russet apples.
2 Mix with shelled and peeled walnuts and bind with mayonnaise.

81 Caesar salad

Energy	Calories	Fat	Saturated fat	Carbohydrates	Sugar	Protein	Fibre *
1494 kJ	361 kcal	32.2 g	7.9 g	5.1 g	2.0 g	12.9 g	1.2 g

* Using toast for croutons

	4 portions	10 portions
Cos lettuce (medium size)	2	4
Croutons, 2-cm square	16	40
Eggs, fresh	2	4
Dressing		
Garlic, finely chopped	1 tsp	2 tsp
Anchovy fillets, mashed	4	8
Lemon juice	1	2
Virgin olive oil	6 tbsp	15 tbsp
White wine vinegar	1 tbsp	2 tbsp
Salt, black mill pepper		
To serve		
Parmesan, freshly grated	75 g	150 g

Mise en place

1 Separate the lettuce leaves, wash, dry thoroughly and refrigerate.

Cooking

1 Lightly grill or fry (in good fresh oil) the croutons on all sides.
2 Plunge the eggs into boiling water for 1 minute, remove and set aside.
3 Break the lettuce into serving-sized pieces and place into a salad bowl.
4 Mix the dressing, break the eggs, spoon out the contents, mix with a fork, add to the dressing and mix into the salad.
5 Mix in the cheese, scatter the croutons on top and serve.

PROFESSIONAL TIP
Because the eggs are only lightly cooked, they must be perfectly fresh, and the salad must be prepared and served immediately. In the interests of food safety, the eggs are sometimes hard boiled.

VARIATION
- Alternatively, the salad may be garnished with hard-boiled gull's eggs.
- Add sliced cooked breast of chicken for chicken Caesar salad.

Healthy Eating Tips
- No added salt is needed; anchovies and cheese are high in salt.
- Oven bake the croutons.
- Serve with fresh bread or rolls.

82 Apple and prune stuffing

Energy	Calories	Fat	Saturated fat	Carbohydrates	Sugar	Protein	Fibre	Sodium
1101 kJ	260 g	4.2 g	2.1 g	47 g	30.1 g	4.6 g	5.7 g	354 g

* Using unsalted butter and half teaspoon salt. Based on 4 portions.

	4 portions	10 portions
Finely chopped onion	100 g	250 g
Soaked prunes, well drained and chopped	225 g	560 g
Cooking apple	300 g	750 g
Butter or olive oil	15 g	40 g
Port	60 ml	150 ml
Fresh breadcrumbs	100 g	250 g
Sage, finely chopped	15 g	40 g
Seasoning		

Cooking

1 Melt the butter or oil in a suitable frying pan, add the onion, sweat without colour.

2 Add the prunes, the port and the peeled, cored and finely chopped apple. Cook for 5 minutes.

3 Stir in the sage and the breadcrumbs and season with salt and pepper.

4 Serve as a dressing for roast goose.

> **VARIATION**
> ● Use fresh plums or apricots in place of prunes.

305 Meat and offal

Learning outcomes

In this unit you will be able to:

1 Prepare meat and offal, including:
   - knowing the cuts and joints of meat from beef, lamb and pork
   - knowing the quality points for different cuts and joints of meat and understanding how they affect cooking methods
   - understanding how different factors affect the quality points
   - knowing how texture, structure, fat content and muscle development influences the choice of processes and preparation methods needed to meet dish requirements
   - being able to interpret a dish specification in order to determine the methodology, precise requirements, timings, presentation and balance of ingredients required
   - being able to use preparation techniques on cuts and joints of meat, including offal, which should be appropriate to dish requirements
   - understanding the correct storage procedures to use throughout production and on completion of final products

2 Produce meat and offal dishes, including:
   - understanding how texture, structure, fat content and muscle structure of meat and offal affect the cooking requirements
   - understanding the effect of cooking methods on fats, muscle tissues, connective tissues and nutritional value
   - understanding how to produce different types of sauces and dressings
   - being able to use cooking techniques on different cuts and joints of meat and offal
   - producing dishes that include sauces and/or dressings and using techniques including stewing, braising, roasting, pot roasting, pan frying, stir frying, deep frying and grilling
   - being able to measure and evaluate against quality standards throughout preparation and cooking
   - being able to evaluate products against dish requirements and production standards and recognise any faults.

Recipes included in this chapter

No.	Name
Beef	
1	Chateaubriand with Roquefort butter
2	Roast wing rib of beef
3	Yorkshire pudding
4	Sirloin steak with red wine sauce
5	Boiled silverside, carrots and dumplings
6	Boeuf bourguignonne
7	Beef olives
8	Beef stroganoff
9	Steak pudding
10	Steak pie
11	Carbonnade of beef
12	Hamburg or Vienna steak
13	Goulash
Lamb	
14	Best end or rack of lamb with breadcrumbs and parsley
15	Roast leg of lamb with mint, lemon and cumin

No.	Name
16	Pot roast shoulder of lamb
17	Sautéed forcemeat (gratin forcemeat)
18	Grilled lamb cutlets
19	Grilled loin or chump chops, or noisettes of lamb
20	Lamb kebabs
21	Irish stew
22	White lamb stew
23	Braised lamb shanks
24	Roast saddle of lamb with rosemary mash
25	Mixed grill
26	Brown lamb or mutton stew
27	Hotpot of lamb or mutton
28	Shepherd's pie (cottage pie)
29	Samosas
Veal	
30	Grilled veal cutlet
31	Braised shin of veal (osso buco)
32	Escalope of veal

33	Veal escalopes with Parma ham and mozzarella cheese
Pork and bacon	
34	Roast leg of pork
35	Sage and onion dressing for pork
36	Roast pork belly with shallots and champ potatoes
37	Spare ribs of pork in barbecue sauce
38	Pork loin chops with pesto and mozzarella
39	Pork escalope with Calvados sauce
40	Raised pork pie
41	Stir-fried pork fillet
42	Sweet and sour pork

43	Boiled bacon (hock, collar or gammon)
44	Griddled gammon with apricot salsa
Offal and mixed meats	
45	Potted meats
46	Braised oxtail
47	Braised lambs' hearts
48	Grilled lambs' kidneys
49	Kidney sauté
50	Fried lambs' liver and bacon
51	Braised veal sweetbreads (white)
52	Grilled veal sweetbreads

Additional recipes available online at www.hoddereducation.co.uk/practical-cookery-resources

Prepare meat and offal

Meat is the flesh of an animal considered to be edible; it is normally called butchers' meat by caterers and usually comes from cattle, sheep and pigs. Meat is a product of selective breeding and feeding techniques, which means that animals are produced to a high standard in terms of welfare, shape and yield, to produce tender flesh.

Present day demand is for a lean and tender meat; therefore, modern cattle tend to be well fleshed yet a compact animal, with specific named breeds regularly shown on menus.

Structure of meat

To cook meat properly it is important to understand its structure.

- Meat is made of fibres bound together by **connective tissue**.
- Connective tissue is made up of **elastin** (yellow) and **collagen** (white). Yellow tissue needs to be removed.
- Small fibres are present in tender cuts and young animals.
- Coarser fibres are present in tougher cuts and older animals.
- Fat helps to provide flavour, and moistens meat in roasting and grilling.
- Tenderness, flavour and moistness are increased if meat is hung after slaughter and before being used.

Meat varies considerably in its fat content. The fat is found around the outside of meat, in **marbling** and inside the meat fibres. The visible fat (saturated) should be trimmed off as much as possible before cooking.

> **TAKE IT FURTHER**
>
> Lots of useful information about meat can be obtained at www.qmscotland.co.uk.

> **KEY TERMS**
>
> **Connective tissue:** animal tissue that binds together the fibres of meat.
>
> **Elastin:** yellow protein found in connective tissue. This needs to be removed.
>
> **Collagen:** white protein found in connective tissue.
>
> **Marbling:** white flecks of fat within meat.

1.1 Quality points for different types of meat

Meat is a natural product and is therefore not a uniform product, varying in quality from carcass to carcass. Flavour, texture and appearance are determined by the type of animal and the way it has been fed or farmed.

Fat gives a characteristic flavour to meat and helps to keep it moist during roasting, but meat does not have to be fatty to be good quality and flavoursome.

The colour of meat is not a guide to quality. Consumers tend to choose light-coloured meat – bright red beef, for example – because they think it will be fresher than a dark red piece. Freshly butchered beef is bright red because the pigment (**myoglobin**) in the tissues has been chemically affected by the oxygen in the air, not because the meat itself is fresh. After several hours, the colour changes to dark red or brown as the pigment reacts further with oxygen in the air (is oxidised) to become **metamyoglobin**; darker meat can therefore still be fresh.

The colour of fat can vary from almost pure white in lamb, to bright yellow in beef. Colour depends on the feed, the breed and, to a certain extent, on the time of year.

KEY TERMS

Myoglobin: pigment in the tissues which gives meat its bright red colour.

Metamyoglobin: created when myoglobin is oxidised (reacts with oxygen in the air). This changes the colour of meat to dark red or brown.

The most useful guide to tenderness and quality is knowledge of the cuts of meat and their location on the carcass. The various cuts for each type of meat are described below, but a few principles can be followed.

- The leanest and most tender cuts – the '**prime**' cuts – come from the **hindquarters**.
- The parts of the animal that have had plenty of muscular exercise and where fibres have become hardened – the '**coarse**' cuts – come from the neck, legs and **forequarters**. These provide meat for braising and stewing, and many consider them to have more flavour, although they require slow cooking to make them tender.
- Farming techniques or the way the animal has been reared will also influence quality, texture and flavour of specific meats, for example marsh-fed lamb, or **free range** cattle, as well as the slaughter technique used or even how long it has been hung.
- The meat from young animals is generally more tender. Tenderness is a prime factor, so sometimes animals are injected with an **enzyme** such as **papin** before slaughter, which softens the fibres and muscles. This simply speeds up the natural process, as meat contains its own enzymes that gradually break down the protein

cell walls as the carcass ages, which is where the claim for aged meat comes from on some menus.

- It is for this tenderisation process that meat is hung for 10 to 20 days in controlled conditions (temperature and humidity) before it is sold. The longer meat is aged, the more expensive it becomes, as the cost of refrigeration is high and the meat itself shrinks because of evaporation and the trimming of the outside hardened edges.

KEY TERMS

Prime cut: the leanest and most tender cuts of meat; these come from the hindquarters.

Coarse cut: cuts from the neck, legs and forequarters; these are tougher cuts and therefore often cooked using slower methods such as braising and stewing.

Hindquarter: the back part of the side of meat.

Forequarter: the front section of the side of meat.

Enzyme: proteins that speed up the rate of a chemical reaction, causing food to ripen, deteriorate and spoil.

Papin: an enzyme that is sometimes injected into animals before slaughter to speed up the softening of fibres and muscles.

Free range: animals kept in natural conditions, with freedom of movement.

Joints and cuts of beef

Cuts of beef vary considerably, from very tender fillet steak to tough brisket or shin, and there is a greater variety of cuts in beef than for any other type of meat. While their names may vary, there are 14 primary cuts from a side of beef, each one composed of muscle, fat, bone and connective tissue. The least developed muscles, usually from the inner areas, can be roasted or grilled, while leaner and more sinewy meat is cut from the more highly developed external muscles. Exceptions are rib and loin cuts, which come from external but basically immobile muscles.

Knowing where the cuts come from (the hindquarter or the forequarter) helps to decide which method of cooking to use.

A whole side weighs approximately 180 kilograms and is divided into the wing ribs and the fore ribs.

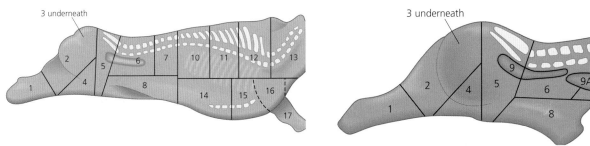

Side of beef

Hindquarter of beef

Table 5.1 Joints, uses and weights of the hindquarter (numbers in left-hand column refer to the picture of the side of beef)

Joint	Description	Uses	Preparation	Approximate weight
(1) Shin	Firm, lower leg muscle	Consommé, beef tea, stewing	Bone out, remove excess sinew; cut or chop as required	7 kg
(2) Topside	A lean, tender cut	Braising, stewing, second-class roasting	Roasting and braising: remove excess fat, cut into joints and tie with string Stewing: cut into dice or steaks	10 kg
(3) Silverside	A coarse joint.	Pickled in brine then boiled	Remove thigh bone; this joint is usually kept whole	14 kg
(4) Thick flank	A boneless, coarse joint.	Braising and stewing	As for topside	12 kg
(5) Rump	A good quality cut, though less tender than fillet or sirloin.	Grilling and frying as steaks, braised in the piece.	Bone out; cut off the first outside slice for pies and puddings. Cut into approximately 1.5 cm slices for steaks. The point steak – considered the most tender – is cut from the pointed end of the slice.	10 kg
(6) Sirloin	A boneless cut; more tender than rump, but not as tender as fillet.	Roasting, grilling and frying in steaks	Roasting whole on the bone: cut back the covering fat in one piece for approximately 10 cm. Trim off the sinew, replace the covering fat and tie with string if necessary. Ensure that the fillet has been removed. Roasting boned out: remove fillet and bone out; remove sinew as before. Remove the excess fat and sinew from the boned side. Roast open, or rolled and tied with string. Grilling and frying: as for roasting and cut into steaks as required (see below).	9 kg
(7) Wing ribs	Usually consists of the last three rib bones which, because of their curved shape, act as a natural trivet. A prime quality, first class roasting joint.	Roasting, grilling and frying in steaks.	Cut seven-eighths of the way through the spine or chine bone, remove the nerve, saw through the rib bones on the underside 5–10 cm from the end. Tie firmly with string. When the joint is cooked, remove the chine bone to make the meat easier to carve.	5 kg
(8) Thin flank	Also known as the skirt. A boneless, gristly cut.	Stewing, boiling, sausages	Trim off excessive fat and cut or roll as required.	10 kg
(9) Fillet	The leanest, most tender cut; from the centre of the sirloin.	Roasting, grilling and frying in steaks.	Roasting and pot roasting: remove head and tail of the fillet, leaving an even centre piece from which all the nerve and fat should be removed. This may be larded by using a larding needle to insert pieces of fatty bacon cut into long strips. Grilling and frying: Varies depending on weight and the number of steaks obtained from it. See below for more detail.	3 kg

Shin

Topside

Silverside

Sirloin steaks

- Minute steaks: cut into 1-centimetre slices, flatten with a cutlet bat dipped in water, making it as thin as possible, then trim.
- Sirloin steaks (entrecôte): cut into 1-centimetre slices and trim (approximate weight 150 grams).
- Double sirloin steaks: cut into 21-centimetre-thick slices and trim (approximate weight 250–300 grams).
- Porterhouse and T-bone steak: porterhouse steaks are cut including the bone from the rib end of the sirloin; T-bone steaks are cut from the rump end of the sirloin, including the bone and fillet.

Sirloin steaks

T-bone steaks

Video: Beef
http://bit.ly/1G36Ecv

Fillet

A typical breakdown of a 3-kilogram fillet would be:

- Chateaubriand: double fillet steak 3–10-centimetre thick, 2–4 portions. Average weight 300-gram–1-kilogram. Cut from the head of the fillet, trim off all the nerve and leave a little fat on the steak.
- Fillet steaks: approximately four steaks of 100–150 grams each, 1.5–2 centimetre thick. These are trimmed as for chateaubriand.
- Tournedos: approximately 6–8 at 100 grams each, 2–4 centimetres thick. Continue cutting down the fillet. Remove all the nerve and all the fat, and tie each tournedos with string.
- Tail of fillet: approximately 0.5 kilograms. Remove all fat and sinew, and slice or mince as required.

Fillet steaks

Forequarter of beef

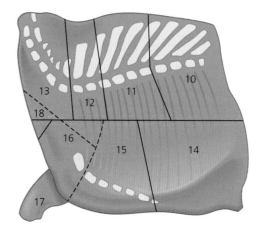

Forequarter of beef

232

Table 5.2 Joints, uses and weights of the forequarter (numbers in left-hand column refer to the picture of a forequarter of beef)

Joint	Description	Uses	Preparation	Approximate weight
(10) Fore rib	First class roasting joint	Roasting and braising	As for wing ribs.	8 kg
(11) Middle rib		Stewing and braising	As for wing ribs.	10 kg
(12) Chuck rib		Stewing and braising	Bone out, remove excess fat and sinew, and use as required	15 kg
(13) Sticking piece (neck area)	Lean and high in favour	Stewing and sausages	Bone out, remove excess fat and sinew, and use as required	9 kg
(14) Plate	Belly	Stewing and sausages	Bone out, remove excess fat and sinew, and use as required	10 kg
(15) Brisket	Quite a fatty joint; sold on or off the bone, or salted	Pickled in brine and boiled, pressed beef, slow roasting	Bone out, remove excess fat and sinew, and use as required	19 kg
(16) Leg of mutton cut	Lower neck and upper belly	Braising and stewing	Bone out, remove excess fat and sinew, and use as required	11 kg
(17) Shank	Lower front leg	Consommé, beef tea	Bone out, remove excess fat and sinew, and use as required	6 kg

Quality points of beef

- The lean meat should be bright red, with small flecks of white fat (marbled).
- The fat should be firm, brittle in texture, creamy white in colour and odourless. Older animals and dairy breeds usually have fat that is a deeper-yellow colour.

ACTIVITY

1   Name two joints from a hindquarter of beef and two from a forequarter.
2   List joints from a beef carcass that are traditionally roasted.
3   What is meant by marbling?
4   A sirloin of beef may be cut into steaks. Name these steaks.
5   What is a chateaubriand?

Joints and cuts of veal

Veal is a by-product of the dairy industry, as male calves cannot be used for milk production. Veal is obtained from good-quality carcasses weighing around 100 kg. This quality of veal is required for first class cookery and is produced from calves slaughtered at between 12 and 24 weeks. The average weight of English or Dutch milk-fed veal calves is 18 kg.

The joints of veal are shown in Table 5.3.

Veal is included as it appears on many restaurant menus for consistency of chef training but doesn't form part of the technical diploma in professional cookery, it is also a bi-product of the dairy industry.

Table 5.3 Joints, uses and weights of veal (numbers in left-hand column refer to those used in the picture of a side of veal)

Joint	Uses	Preparation	Approximate weight
(1) Knuckle	Osso buco, sauté, stock	Stewing (on the bone) (osso buco): cut and saw into 2–4-cm thick slices through the knuckle. Sauté: bone out and trim, then cut into even 25 g pieces.	2 kg
(2) Leg	Roasting, braising, escalopes, sauté	Braising or roasting whole: remove the aitch bone, clean and trim 4 cm off the knuckle bone. Trim off the excess sinew. Braising or roasting the nut: remove all the sinew; if there is insufficient fat on the joint then bard thinly (cover with bacon fat) and secure with string. Escalopes: remove all the sinew, cut against the grain into large 50–75 g slices and bat out thinly. Sauté: remove all the sinew and cut into 25 g pieces.	5 kg

Joint	Uses	Preparation	Approximate weight
(3) Loin	Roasting, frying, grilling	Roasting: bone out and trim the flap, roll out and secure with string. This joint may be stuffed before rolling. Frying: trim and cut into cutlets.	3.5 kg
(4) Best end	Roasting, frying, grilling	Roasting: bone out and trim the flap, roll out and secure with string. This joint may be stuffed before rolling. Frying: trim and cut into cutlets.	3 kg
(5) Shoulder	Braising, stewing	Braising: bone out as for lamb; usually stuffed. Stewing: bone out, remove all the sinew and cut into 25-g pieces.	2 kg
(6) Neck end	Stewing, saute	Stewing and sautéing: bone out and remove all the sinew; cut into approximately 25 g pieces.	2.5 kg
(7) Scrag	Stewing, stock	Stewing and sautéing: bone out and remove all the sinew; cut into approximately 25 g pieces.	1.5 kg
(8) Breast	Stewing, roasting	Stewing: as for neck end. Roasting: bone out, season, stuff and roll up, then tie with string.	2.5 kg

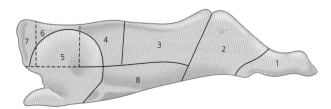

Side of veal

Quality points of veal

- Veal is available all year round.
- The flesh should be pale pink in colour and firm in texture – not soft or flabby.
- Cut surfaces should be slightly moist, not dry.
- Bones, as in all young animals, should be pinkish white and porous, with some blood in their structure.
- The fat should be firm and pinkish white.
- The kidney should be firm and well covered with fat.

Table 5.4 Joints of the leg (see the diagram showing dissection of veal)

Cut	Corresponding beef joint	Weight	Proportion	Uses
Cushion or nut	Topside	2.75 kg	15%	Escalopes, roasting, braising, sauté
Under cushion or under nut	Silverside	3 kg	17%	Escalopes, roasting, braising, sauté
Thick flank	Thick flank	2.5 kg	14%	Escalopes, roasting, braising, sauté
Knuckle (whole)	-	2.5 kg	14%	Osso buco, sauté
Bones (thigh and aitch)	-	2.5 kg	14%	Stock, jus lié, sauces
Useable trimmings	-	2 kg	11%	Pies, stewing

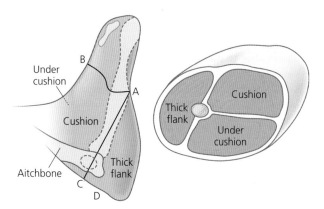

Joints and cuts of lamb and mutton

Lamb is the meat from a sheep under a year old; above that age the animal is called a **hogget** and its meat becomes **mutton**.

> **KEY TERMS**
>
> **Lamb:** a sheep under a year old
>
> **Hogget:** a sheep over a year old.
>
> **Mutton:** meat from a mature sheep.

There is greater demand for lamb than mutton as the lamb carcass provides smaller cuts of more tender meat. Mutton needs to be well ripened by long hanging before cooking and, as it is usually fatty, needs a good deal of trimming. The flesh of a younger lamb is usually more tender.

A good way to judge age is through weight, especially with legs of lamb: the highest quality weighs about 2.3 kilograms and never more than 4 kilograms. Smaller chops are also more tender and therefore more expensive. Mutton is rarely available to buy; when it is, it is much less expensive than lamb.

As a guide, when ordering lamb allow approximately 100 grams meat off the bone per portion and 150 grams on the bone per portion. However, these weights are only approximate and will vary according to the quality of the meat and what it will be used for. For example, a chef will often cut up a carcass differently from a shop butcher because a chef needs to consider the presentation of the particular joint, while the butcher is more often concerned with being economical.

In general, bones only need to be removed when preparing joints, to make carving easier. The bones are used for stock and the excess fat can be rendered down for second-class dripping.

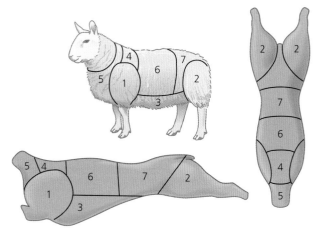

Joints of lamb

Loin chops

Lamb loin chops

Skin the loin, remove the excess fat and sinew, then cut into chops approximately 100–150 grams in weight.

A first-class loin chop should have a piece of kidney skewered in the centre.

Table 5.5 Joints, uses and weights (numbers in left-hand column refer to the pictures of joints of lamb)

Joint	Uses	Preparation	Approximate weight lamb	Approximate weight mutton
Whole carcass			16 kg	25 kg
(1) Shoulder (two)	Roasting, stewing	Boning: remove the blade bone and upper arm bone, then tie with string; the shoulder may be stuffed before tying. Cutting for stews: bone out the meat and cut into even 25–50-g pieces.	3 kg	4.5 kg
(2) Leg (two)	Roasting (mutton boiled)	Boning: remove the pelvic or aitchbone; trim the knuckle, cleaning 3 cm of bone; trim off excess fat and tie with string if necessary.	3.5 kg	5.5 kg
(3) Breast (two)	Roasting, stewing	Remove excess fat and skin. Roasting: bone; stuff and roll; tie with string. Stewing: bone and then cut into even 25–50-g pieces.	1.5 kg	2.5 kg
(4) Middle neck	Stewing	Remove excess fat, excess bone and gristle; cut into even 50-g pieces. When butchered correctly, this joint can give good uncovered second-class cutlets.	2 kg	3 kg
(5) Scrag end	Stewing, broth	Chop down the centre, remove the excess bone, fat and gristle, and cut into even 50-g pieces, or bone out and cut into pieces.	0.5 kg	1 kg
(6) Best end rack (two)	Roasting, grilling, frying	Remove the skin from head to tail and from breast to back. Remove the sinew and the tip of the blade bone. Clean the sinew from between the rib bones and trim the bones. Score the fat neatly to approximately 2 mm deep. Trim the overall length of the rib bones to two and a half times the length of the nut of meat.	2 kg	3 kg
(7) Saddle	Roasting, grilling, frying	For large banquets it is sometimes better to remove the chumps and use short saddles: remove the skin, starting from head to tail and from breast to back. Split down the centre of the backbone to produce two loins. Each loin can be roasted whole. Roasting: skin and remove the kidney. Trim the excess fat and sinew. Cut off the flaps, leaving about 15 cm each side so that they meet in the middle under the saddle. Remove the aitch or pelvic bone. Score neatly and tie with string. For presentation the tail may be left on, protected with foil and tied back. The saddle can also be completely boned, stuffed and tied. Grilling and frying: cut into loin and chump chops (see below).	3.5 kg	5.5 kg

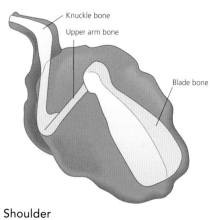

Shoulder

Best ends of lamb (racks, French trimmed)

Saddle of lamb

Double loin chop (also known as a Barnsley chop)

Barnsley chops

These are cut approximately two centimetres across a saddle on the bone.

When trimmed they are secured with a skewer and may include a piece of kidney in the centre of each chop.

Chump chops

These are cut from the chump end of the loin.

Cut into approximately 150-gram chops and trim where necessary.

Noisette

This is a cut from a boned-out loin.

Cut slantwise into approximately two-centimetre thick slices, bat out slightly and trim into a cutlet shape.

Rosette

Lamb rosettes

This is a cut from a boned-out loin approximately two centimetres thick. It is shaped round and tied with string.

Cutlets

Lamb cutlets

Prepare as for roasting, excluding the scoring, and divide evenly between the bones. Alternatively, the cutlets can be cut from the best end and prepared separately. A double cutlet consists of two bones, so a six-bone best end yields six single or three double cutlets.

Step by step: tunnel boning and trimming a leg of lamb

1 Use a boning knife to scrape back the flesh and expose the knuckle bone. Cut flesh away from the bone up to the joint, then cut through the joint.

2 Remove the knuckle bone, reshape the joint and tie it neatly.

Quality points of lamb and mutton

- A good-quality animal should be compact and evenly fleshed.
- The lean flesh should be firm and a pleasing dull red colour with a fine texture or grain.
- There should be an even distribution of surface fat, which should have a hard, brittle, flaky texture and a clear white colour.
- In a young animal the bones should be pink and porous, so that when they are cut a little blood can be seen inside them. As animals grow older, their bones become hard, dense and white, and are more likely to splinter when chopped.
- Good-quality lamb should have fine, white fat, with pink flesh when freshly cut.
- In mutton, the flesh is a deeper colour.
- Lamb has a very thin, parchment-like covering on the carcass, known as the 'fell', which is usually left on roasts to help them maintain their shape during cooking. It should, however, be removed from chops.

Joints and cuts of pork

The keeping quality of pork is less than that of other meat; therefore it must be handled, prepared and cooked with great care. Pork should always be well cooked.

At five to six weeks old, a piglet is known as a sucking or suckling pig. Its weight is then between 5 kilograms and 10 kilograms.

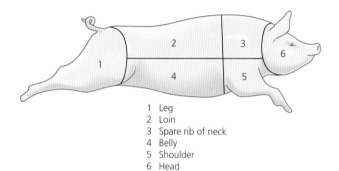

1 Leg
2 Loin
3 Spare rib of neck
4 Belly
5 Shoulder
6 Head

Pig carcass

Table 5.6 Cuts, uses and weights of pork (numbers in left-hand column refer to the drawing of a pig carcass)

Joint	Uses	Preparation	Approximate weight
(1) Leg	Roasting and boiling	Roasting: remove the pelvic or aitch bone, trim and score the rind neatly – that is, with a sharp-pointed knife, make a series of 3-mm deep incisions approximately 2 cm apart all over the skin of the joint; trim and clean the knuckle bone. Boiling: it is usual to pickle the joint either by rubbing dry salt and saltpetre into the meat, or by soaking in a brine solution; then remove the pelvic bone, trim and secure with string if necessary.	5 kg
(2) Loin	Roasting, frying and grilling	Roasting (on the bone): saw down the chine bone in order to facilitate carving; trim the excess fat and sinew and score the rind in the direction that the joint will be carved; season and secure with string. Roasting (boned out): remove the fillets and bone out carefully; trim off the excess fat and sinew, score the rind and neaten the flap, season, replace the filet mignon, roll up and secure the string; this joint is sometimes stuffed. Grilling or frying chops: remove the skin, excess fat and sinew, then cut and saw or chop through the loin in approximately 1-cm slices; remove the excess bone and trim neatly.	6 kg
(3) Spare rib	Roasting, pies	Roasting: remove the excess fat, bone and sinew, and trim neatly. Pies: remove the excess fat and sinew, bone out and cut as required.	1.5 kg
(4) Belly	Pickling, boiling, stuffed, rolled and roasted	Remove all the small rib bones, season with salt, pepper and chopped sage, roll and secure with string. This joint may be stuffed.	2 kg
(5) Shoulder	Roasting, sausages, pies	Roasting: the shoulder is usually boned out, the excess fat and sinew removed, seasoned, scored and rolled with string; it may be stuffed and can also be divided into two smaller joints. Sausages and pies: skin, bone out and remove the excess fat and sinew; cut into even pieces or mince.	3 kg
(6) Head (whole)	Brawn		4 kg

Leg of pork

Loin of pork

Quality points of pork

- Lean flesh should be pale pink, firm and of a fine texture.
- Fat should be white, firm, smooth and not excessive.
- Bones should be small, fine and pinkish.
- Skin or rind should be smooth.

ACTIVITY

1   Which joint of pork is most suitable for escalopes of pork?
2   Which joint of pork is suitable for roasting?

Bacon

Bacon is the cured flesh of a bacon-weight pig (60–75 kilograms) that is specifically reared for bacon because its shape and size yield economic bacon joints.

Bacon is cured either by dry salting and then smoking or by soaking in brine followed by smoking. Green bacon is brine-cured but not smoked; it has a milder flavour but does not keep for as long as smoked bacon. Depending on the degree of salting during the curing process, bacon joints may need to be soaked in cold water for a few hours before being cooked.

> **PROFESSIONAL TIP**
> Do not confuse gammon with ham. Both gammon and ham are from the hind legs of a pig. Gammon is the raw cured joint or steak while ham is the cooked meat.

Quality points of bacon

- There should be no sign of stickiness.
- There should be a pleasant smell.
- Rind should be thin, smooth and free from wrinkles.
- Fat should be white, smooth and not excessive in proportion to the lean.
- The lean should be a deep-pink colour and firm.

ACTIVITY

1   What is the difference between a ham and a gammon?
2   Which part of the pig does streaky bacon come from?

Table 5.7 Cuts, uses and weights of bacon (numbers in left-hand column refer to the drawing of a side of bacon)

Joint	Uses	Preparation	Approximate weight
(1) Collar	Boiling, grilling	Boiling: remove bone (if any) and tie with string. Grilling: remove the rind, trim off the outside surface and cut into thin slices (rashers), across the joint.	4.5 kg
(2) Hock	Boiling, grilling	Boiling: leave whole or bone out and secure with string.	4.5 kg
(3) Back	Grilling, frying	Grilling: remove all bones and rind, and cut into thin rashers. Frying: remove the rind, trim off the outside surface, and cut into rashers or chops of the required thickness.	9 kg
(4) Streaky	Grilling, frying	Grilling: remove all bones and rind, and cut into thin rashers. Frying: remove the rind, trim off the outside surface, and cut into rashers or chops of the required thickness.	4.5 kg
(5) Gammon	Boiling, grilling, frying	Frying and grilling: fairly thick slices are cut from the middle of the gammon; they are then trimmed and the rind removed.	7.5 kg

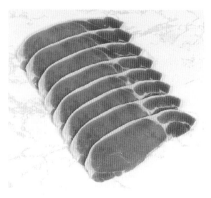

Back gammon

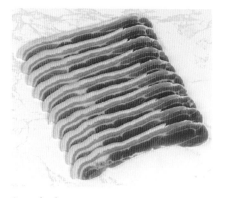

Streaky bacon

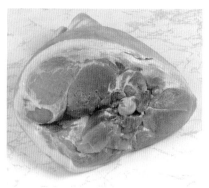

Gammon

Offal and other edible parts of the carcass

Offal is the edible parts taken from the inside of a carcass of meat. It includes liver, kidneys, heart and sweetbreads. Tripe, brains, tongue, head and oxtail are also sometimes included under this term.

Liver

Liver is a good source of protein and iron. It contains vitamins A and D, and is low in fat.

- Calf's liver – considered to be the most tender and tasty. It is also the most expensive.
- Lamb's liver – mild in flavour, light in colour and tender.
- Sheep's liver – from an older animal, so is firmer, deeper in colour and has a stronger flavour.
- Ox or beef liver – the cheapest type and, if taken from an older animal, can be coarse and have a strong flavour. It is usually braised.
- Pig's liver – has a strong, full flavour and is used mainly for pâté recipes.

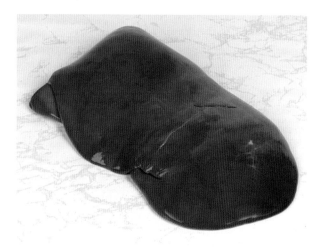

Calf's liver

Quality points:
- Liver should look fresh, moist and smooth, with a pleasant colour and no unpleasant smell.
- Liver should not be dry or contain an excessive number of tubes.

Kidneys

- Lamb's kidneys – light in colour, delicate in flavour and ideal for grilling and frying.
- Sheep's kidneys – darker in colour and stronger in flavour.
- Calf's kidneys – light in colour, delicate in flavour and used in a variety of dishes.
- Ox kidney – dark in colour, strong in flavour, and is either braised or used in pies and puddings (mixed with beef).
- Pig's kidneys – smooth, long and flat and have a strong flavour.

Quality points:
- Suet, the saturated fat in which kidneys are encased, should be left on until they are used, otherwise the kidneys will dry out. The suet should be removed when kidneys are being prepared for cooking.
- Both suet and kidneys should be moist and have no unpleasant smell. The food value is similar to that of liver.

Sweetbreads

These are the pancreas and thymus glands, known as heart breads and neck. The heart bread is round, plump and a better quality than the neck bread, which is long and uneven. Sweetbreads are an easily digested source of protein, which makes them valuable for use in special diets for people who are ill.

Calf's heart bread, considered the best, weighs up to 600 grams; lamb's heart bread weighs up to 100 grams.

Table 5.8 Other types of offal

Offal	Description	Quality points
Hearts	Hearts are a good source of protein. Most need slow braising to tenderise them. Lamb's hearts, sheep's hearts, ox or beef's heart, calf's hearts.	Hearts should not be too fatty and should not contain too many tubes. When cut they should be moist, not sticky, and with no unpleasant smell.
Tripe	Tripe is the stomach lining or white muscle of the ox, consisting of the rumen or paunch and the honeycomb tripe (considered the best); sheep tripe, darker in colour, is available in some areas. Tripe contains protein, is low in fat and high in calcium.	Tripe should be fresh, with no signs of stickiness or unpleasant smell.
Tongues	Ox, lamb and sheep tongues are those most used in cooking. Ox tongues are usually salted then soaked before being cooked. Lamb tongues are cooked fresh.	Tongues must be fresh and have no unpleasant smell.
Oxtail	Oxtails usually weigh 1–2 kg. Oxtail should be lean, without much fat	There should be no signs of stickiness or unpleasant smell.

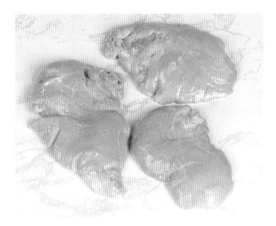

Veal sweetbreads

Quality points:
- Heart and neck breads should be fleshy and a good size.
- They should be creamy white and have no unpleasant smell.

Trotters

Pig's trotters are used as a garnish or in stocks to add flavour.

Pig's trotters

Quality points:
- Check they are clean and undamaged prior to use.

1.2 How meat composition influences processes and preparation methods

In order to preserve meat and offal to extend shelf life, it is important to reduce the micro-organisms that can cause food spoilage. The following methods are used to help preserve meat.

- Salting – meat can be pickled in brine. The salt draws out moisture and creates an environment inhospitable to bacteria. This method of preservation may be applied to silverside, brisket and ox tongues. Salting is used in the production of bacon, before the sides of pork are smoked, and for hams.
- Smoking – mainly used to enhance flavour and colour of the product; as a preservation method its effect is limited to the surface of the product. There are two main types of smoking used: hot smoking, which cooks the meat during the process, and cold smoking, which is part of a preservation method. Substitutes are used in some kitchens, which add flavour and the characteristics of smoked food to dishes. These add flavour without the carcinogenic substances associated with smoke.
- Pickling – using acidic foods such as yoghurt, vinegar or wine raises acidity levels and helps reduce the potential for food spoilage. This is used in Tandoori dishes.

- Chilling – meat is kept at a temperature just above freezing point in a controlled atmosphere. Chilled meat cannot be kept in the usual type of cold room for more than a few days, although sufficient time must be allowed for the meat to hang, enabling it to become tender.
- Vacuum-packing – small cuts of meat can be vacuum-packed prior to chilling. This removes contact with air and therefore may extend its shelf life. This process is also associated with the sous vide style of cooking, in which cuts of meat are slowly cooked, chilled and stored, or cook chill.
- Freezing – small carcasses, such as lamb and mutton, can be frozen; their quality is not usually affected by freezing. They can be kept frozen until required and then thawed out before being used. Some beef is frozen, but it is not such good quality as chilled beef as it takes on ice crystals which, when thawed, affect the quality of the meat.
- Canning – large quantities of meat are canned. Corned beef, for example, has a very high protein content. Pork is used for tinned luncheon meat.
- Marinating – steeping meat in a marinade mix to tenderise prior to cooking, normally contains something acidic, spicy and/or oil.

Advantages of preserving meat are that foods are ready prepared, saving time and labour. It is also easier to apply portion control and manage costs. The preservation process helps to extend seasons or the availability of some types of meat, as well as making storage easier while maintaining quality.

Disadvantages of preservation include the loss of some of the nutrients and a reduction in skills required, especially where meat and offal are bought in already prepared and preserved.

1.3 Using skills and techniques to prepare meat

Preparation

Meat from specific parts of an animal may be cut and cooked according to local custom (for more information on preparation methods, see the sections for each type of meat below). Some religions specify how meat should be butchered, especially in Jewish kosher and Islamic halal butchery, which stipulates the animal must be killed by an authorised person of the religion, all of the blood must be removed by draining, the meat must be soaked and salted and consumed within 72 hours. Kosher dietary laws also state that only the forequarters of permitted animals – goats, sheep, deer and cattle – may be used.

Step by step: preparing a wing rib

1 Remove the chine bone from the joint.

2 Cut away the chine.

3 Use a saw to finish removing the chine.

4 Trim and clean the top bone.

5 Remove the elastin.

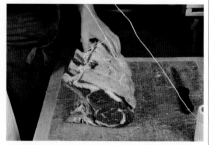

6 Tie the bone back – it will act as a trivet.

Step by step: preparing a best end of lamb (skinning, scoring, trimming and tying)

1 Remove the bark/skin, leaving as much fat as possible on the joint.

2 Mark/score 2 cm from the end of the bone.

3 Score down the middle of the back of the bone, scoring the cartilage.

4 Pull the skin fat and meat from the bone (to bring out the bone ends – this is an alternative to scraping them).

5 Remove the elastin.

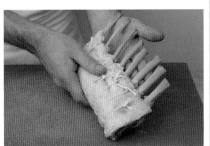

6 Tie the joint.

Step by step: preparing pork belly for roasting

1 Pork belly before it has been prepared.

2 Remove all the small rib bones.

3 Cut the fat.

4 Score the fat.

5 Season with salt, pepper and chopped sage.

6 Ready for roasting.

Step by step: preparation of liver

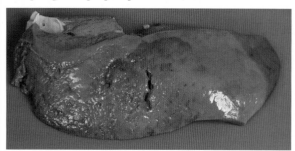

1 Remove skin, gristle and tubes.

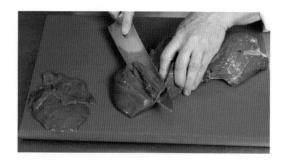

2 Cut into thin slices on the slant. Ox liver may be marinated in milk to tenderise.

Step by step: preparing kidneys

1 Whole kidneys.

2 Slice into kidney to open it up.

3 Lift up and cut out the white sinew from the centre.

4 Before, during and after preparation.

HEALTH AND SAFETY

When preparing uncooked meat or poultry followed by cooked food, or changing from one type of meat or poultry to another, all equipment, working areas and utensils must be thoroughly cleaned or changed. Wash all work surfaces with a bactericidal detergent to kill bacteria. This is particularly important when handling poultry and pork. If colour coded boards are used, it is essential to always use the correct colour coded boards for the preparation of foods and different ones for cooked foods.

HEALTH AND SAFETY

When using boning knives, wear a safety apron as protection; if a lot of boning is being done then it is also a good idea to wear protective gloves.

Preparation of sweetbreads

1 Soak in several changes of cold salted water to remove blood, which will darken the sweetbreads during cooking (over 2–3 hours).
2 Blanch and refresh.
3 Peel off the membrane and connective tissues.
4 The sweetbreads can then be pressed between two trays with a weight on top and refrigerated.

Preparation of tongue

1 Remove the bone and gristle from the throat end.
2 Soak in cold water for 2–4 hours.
3 If salted, soak for 3–4 hours.

Preparation of oxtail

1 Cut between the natural joints, trim off excess fat.
2 The large pieces may be split in two.

Preparation of trotters

1 Boil in water for a few minutes
2 Scrape with the back of a knife to remove the hairs,
3 Wash off in cold water and split in half.

Portion sizes and weights

Portion control is about the size and quantity of food served to an individual customer. This is based on the type of customer or establishment, quality of products and cost in relation to the selling price of the dish or commodity; the type of menu or number of courses offered will also influence portion size. A chef's knowledge should enable them to work out the number of portions that can be expected from a specific commodity. For example, steaks are often bought and sold by weight per cut, for example, sirloin steaks at 200 grams or fillet steaks at 150 grams per serving/steak. For roast joints or prepared meats, this is around 6–8 portions per boneless kilograms, therefore 125–200 grams per portion.

When preparing joints from a carcass this may follow the natural seam of a muscle formation, known as 'seam boning'.

If the bone is removed without breaking the seams through a tunnel it is known as tunnel boning', as in a shoulder of lamb.

Dish requirements will also cover the most effective preparation methods to ensure that the correct yield (amount of portions) will be presented. This could include the way meat is trimmed, marinated or tenderised before cooking.

1.4 Storage

Storing meat

Hang and store meat between 0 °C and 2 °C.

Meat and offal should be stored covered with cling film, labelled with appropriate information about the product and the use by date. This will also help to ensure good stock rotation.

Store uncooked meat on trays to prevent dripping, covered in separate refrigerators at a temperature of 3–5 °C, preferably at the lower temperature. When separate refrigerators are not available, store raw meat separately at the bottom of the refrigerator to prevent cross-contamination and away from cooked meats to prevent cross-contamination.

- Beef can be stored for up to three weeks.
- Veal can be stored for one to three weeks.
- Lamb can be stored for 10–15 days.
- Pork can be stored for 7–14 days.

Storing offal

Fresh offal (unfrozen) should be purchased as required and can be refrigerated under hygienic conditions at a temperature of –1 °C, at a relative humidity of 90 per cent, for up to seven days.

Frozen offal must be kept in a deep freeze and defrosted in a refrigerator as required. Liver and kidney dishes may traditionally have been served undercooked or lightly cooked, but it is now advised that they are cooked thoroughly all the way through to avoid possible food poisoning risks.

Produce meat and offal dishes

2.1 Cooking requirements for meat and offal

Meat is an extremely versatile product that can be cooked in many ways and matched with practically any vegetable, fruit or herb. The cut (for example, shin, steak, brisket), the method of heating (for example, roasting, braising, grilling) and the time and temperature all affect the way the meat will taste.

Cooking methods are described as 'wet' – for example, stewing, pot roasting or braising, or 'dry' – for example, roasting or grilling. Using the correct cooking method is important to ensure that specific joints or cuts are served at their best. If cooked too quickly or at too high a temperature, the meat shrinks and tightens which means that portions are smaller or insufficient to meet customer expectations. This will also affect the potential overall costs and profit targets of the business.

2.2 Effect of cooking methods on meat and offal

Raw meat is difficult to chew because the muscle fibres contain collagen (see 'Structure of meat', page 229), which is softened only by mincing – as in steak tartare – or by cooking. When meat is cooked, the protein gradually **coagulates** as the internal temperature increases. At 77°C, coagulation is complete and the protein begins to harden, so any further cooking makes the meat tougher.

Since tenderness combined with flavour is the aim in meat cookery, time and temperature are the key concerns. In principle, slow cooking retains the juices and produces a more tender result than fast cooking at high temperatures. There are, of course, occasions when high temperatures are essential: for instance, a steak needs to be grilled under a hot flame for a very short time in order to get a crisp, brown surface and a pink, juicy interior – using a low temperature would not give the desired result. In potentially tough cuts (for example, breast of lamb), or where there is a lot of connective tissue (for example, in neck of lamb), slower cooking converts the tissues to **gelatine** and helps to make the meat more tender. Meat containing bone will take longer to cook because bone is a poor conductor of heat.

Nutritional value of fat isn't affected by the cooking process, some of which is rendered, melt, which adds flavour and aids to sealing meat. Some is absorbed into the dish, where it raises to the top of a dish (it is normal to skim during cooking). Fat from roast meat is often used to flavour other dishes as in roast vegetables.

> **KEY TERMS**
>
> **Coagulate**: the transformation of liquids or moisture in meat to a solid form.
>
> **Gelatine**: a nearly transparent, glutinous substance, obtained by boiling the bones, ligaments, etc., of animals, and used in making jellies.

2.3 Sauces and dressings

Finishing or assembly of dishes involves the chef ensuring any sauces and accompaniments are of the correct consistency, size and balance for the dish. For example, if turned vegetables or croutons are part of the dish, the chef will ensure there are sufficient for each portion served. A sauce should be of the correct consistency for the dish. The accompaniments and sauces should enhance the overall dish in terms of texture, colour or flavour, for example, watercress with steak adds a peppery flavour to the dish when eaten with the steak.

Service temperature needs to be correct for the dish being served – hot, warm or cold as per the dish specification.

Some of the traditional sauces that could be served with meat are included in Unit 303 Stock, soups and sauces. Examples include:
- Jus-lié – a thickened gravy made with stock
- Roast gravy – made by de-glazing the roasting pan and reducing to intensify flavour
- Apple sauce with pork
- Mint sauce with lamb
- Béarnaise – an emulsified sauce served with grilled meats.

2.4 Producing dishes to specification using advanced skills

- Tough or coarse cuts of meat should be cooked using wet methods of cookery such as braising, pot roasting or stewing. These longer, slower methods of cooking dissolve the collagen, forming gelatine and making the meat more tender and tasty. An example is pot-roast shoulder of lamb with potatoes boulangère.

- Prime cuts, such as beef fillets, contain little collagen and do not require long cooking to tenderise the meat. Although most chefs would start the prime cuts at a high temperature for a short period, this does not always give a perfect result. These are often cooked using dry methods such as roasting, grilling and frying. An example is fried lambs' liver and bacon (*foie d'agneau au lard*).
- The four traditional stages of cooking a steak:
  - Blue – sealed on the outside but still raw on the inside, internal temperature around 29 °C.
  - Rare – seared on both sides and cooked until inside temperature is around 40–45 °C.
  - Medium – seared on both sides, cooked for around four minutes each side, internal temperature around 65 °C.
  - Well done – seared on both sides, cooked for around six minutes each side, internal temperature above 75 °C.
- **Searing** meat in hot fat or in a hot oven before roasting or stewing helps to produce a crisp exterior by coagulating the protein. However, it does not seal in the juices. Also, if the temperature is too high and the meat is cooked for too long, rapid evaporation of the juices and contraction of the meat will cause much of the juices and fat to be lost, making the meat tougher, drier and less tasty. This is particularly true for prime cuts, as they do not contain much fat or collagen to begin with. A lower temperature and longer in the oven will produce a better result.
- Sprinkling salt on meat before cooking will also speed up the loss of moisture because salt absorbs water.
- Marinating in a suitable marinade, such as wine and wine vinegar, helps to tenderise the meat and adds flavour.

PROFESSIONAL TIP
Meat bones are useful for giving flavour to soups and stocks, especially beef bones with plenty of marrow. Veal bones are gelatinous and help to enrich and thicken soups and sauces. Fat can be rendered down for frying or used as an ingredient (suet or lard).

Slow cooking meat

When cooking meats at low temperatures, there is one obvious flaw: the meat will not be exposed to the high cooking temperatures that develop that beautiful roasted flavour. This chemical reaction of browning is called the **Maillard reaction**. This occurs at 140 °C and above, when the wonderful roasted meat flavours begin to be released.

When slow cooking meats at lower temperatures they need to be started very quickly on a hot pan on the stove to start the Maillard reaction. In some cases you will need to return the meat quickly to the pan to re-caramelise the outside; alternatively, if the joint is dense and large, remove it from the low oven and increase the temperature to 190–200 °C. When the oven is up to temperature, put the joint back in for a short while to crisp the outside. The density of the meat and size of the joint will ensure that there is very little secondary cooking or residual heat left to cook through to the core.

KEY TERMS

Searing: browning or colouring the outer surface of meat

Maillard reaction: the chemical reaction that occurs when heat is applied to meat, causing browning.

As already mentioned, the collagen that makes up connective tissue requires long cooking at a moderate temperature to convert it into gelatine. This provides a form of secondary or internal basting. When **basting** the outside of the meat, take care not to raise the internal temperature of the meat too much as this will destroy the secondary basting properties of the collagen – at temperatures above 88 °C the collagen will dissolve rapidly into the braising medium, making the meat less tender and moist. Therefore, the traditional braising method of bringing a casserole to a simmer and placing it in the oven at 140 °C could, in theory, make the meat dry. The more modern approach to braising is to have the cooking medium at between 80 °C and 85 °C, and this can be controlled best on the top of the stove. Alternatively, set your oven at approximately 90 °C and check the cooking medium once in a while.

KEY TERM

Basting: moistening meat periodically, especially while cooking, using a liquid such as melted butter, or a sauce.

When slow cooking prime joints, generally the temperature of cooking is reduced as, in some cases, shrinkage can occur from 59 °C, up to 65 °C for sirloin of beef. A steak of sirloin beef has more collagen than a fillet and is generally cooked on a high heat, either roasted or pan fried. This will make the sirloin extremely tender and moist, with a roasted outer and the flavoursome roasted meat taste that people enjoy. An average sirloin joint for roasting can weigh 2–5 kilograms whole off the bone. The method for cooking this is to seal the meat on the outside and then place it into a pre-heated oven at 180 °C, cook at 180 °C for 10 minutes, then reduce the temperature to 64 °C (the oven door will need to be open at this stage). Once the oven has come down to 64 °C, close the door and cook for a further 1 hour 50 minutes. This will give you an extremely tender piece of sirloin.

Checking and finishing

During the cooking process a chef will check that meat is cooking correctly, without losing too much moisture, and is reaching the required colour or glaze before allowing the meat to rest before carving or serving. To maintain the quality and safety of meat and poultry dishes once cooked, it is advisable to check internal temperatures using a probe. The recommended temperatures are shown in Table 5.8.

Table 5.8 Recommended internal temperatures*

Meat	Recommended internal temperatures
Beef	Rare: 45–47 °C; medium: 50–52 °C; well done: 64–70 °C
Lamb	Pink: 55 °C; well done: 62 °C
Pork	60–65 °C
Veal	60 °C

* These temperatures only apply to complete pieces of meat. If they have been boned, rolled, layered or minced, higher temperatures may be required

Further guidance on checking and finishing dishes is provided in 'Preparing for assessment' on page 630.

2.5 Evaluation

Once a dish has been prepared, cooked and presented, the chef must ensure that it meets the dish specification and therefore the customer need by an evaluative process. Some parts of this evaluation happen during the cooking stage, for example when a joint is basted and monitored while cooking.

Overall evaluation is about ensuring that portion sizes are even, garnishes enhance the presentation of the dish and colour is appropriate, for instance breadcrumbs on a dish are golden brown with no black or burnt crumbs from a previous dish cooked in the pan. Taste in terms of seasoning and flavour must be checked and the finished temperature must be correct, along with the consistency of any sauces served with the dish.

Faults that may be rectified before the dish is served could relate to colour, appearance, cooking, temperature and aroma. One of the most common faults associated with cooking meat in many restaurants is the degree of cooking for steaks.

TEST YOURSELF

1 Give **three** examples of joints of beef from:
   a) the hindquarter
   b) the forequarter.
2 List **two** quality points to look for when buying beef.
3 Explain **three** reasons why portion control should be considered when preparing meat and offal.
4 Where are cutlets of lamb cut from?
5 Describe the stages in the preparation of a best end of lamb.
6 Describe **two** advantages of freezing as a preservation method for lamb.
7 Which part of the veal carcass are escalopes cut from?
8 Give **two** reasons why braising is a suitable cooking method for neck of lamb.
9 State the difference between pickling and salting as a preservation method for meat.
10 Describe the four stages in cooking steaks, including recommend temperatures.
11 List **four** quality points to look for when buying pork.
12 Why is braising the most appropriate cooking method for hearts?
13 What are the **two** types of sweetbread?
14 List the offal that would be used from a whole carcass of pork.
15 Which preservation methods could be used for offal?

1 Chateaubriand with Roquefort butter

Energy	Calories	Fat	Saturated fat	Carbohydrates	Sugar	Protein	Fibre
2368 kJ	566 kcal	43.3 g	19.9 g	0.0 g	0.0 g	44.0 g	0.0 g

	2–3 portions
Olive oil	
1 chateaubriand	500 g
Salt, pepper	
Roquefort cheese	25 g
Unsalted butter	40 g
Ground black pepper	½ tsp

Cooking

1 Heat the olive oil in a hot frying pan and brown the chateaubriand all over.

2 Season the chateaubriand and place in the oven at 190 °C; the timing depends on the degree of cooking required.

3 Remove from oven, allow to rest, then carve into thick slices.

4 Mash together the Roquefort cheese, butter and black pepper. Form into a roll using aluminium foil or cling film. Refrigerate.

5 Place a slice of Roquefort and butter on each slice of chateaubriand.

SERVING SUGGESTION

Serve with deep-fried potatoes and a tossed green salad.

PROFESSIONAL TIP

For the best texture and flavour, brown the chateaubriand all over before roasting, and leave it slightly underdone.

ON A BUDGET

Using cut steaks such as rump or sirloin could also be served this way, as it is prepared for single portions,

2 Roast wing rib of beef

Energy	Calories	Fat	Saturated fat	Carbohydrates	Sugar	Protein	Fibre
3185 kJ	758 kcal	31.04g	13.0g	31.6g	4.5g	90.0g	1.6g

	10 portions
Wing rib of beef	1 × 2 kg
Beef dripping	25 g
Gravy	
Carrots (for the mirepoix)	50 g
Onion (for the mirepoix)	50 g
Red wine	200 ml
Plain flour	30 g
Beef stock	300 ml
Yorkshire puddings (see recipe 3), prepared English mustard or horseradish sauce, to serve	

Mise en place

1 Place the dripping in a heavy roasting tray and heat on the stove top.

Cooking

1 Place the beef in the tray and brown well on all sides.

2 Place in the oven on 195 °C for 15 minutes then turn down to 75 °C for 2 hours.

3 Remove and allow to rest before carving.

For the gravy:

1 Remove the beef. Place the tray with the fat, sediment and the juice back on the stove.

2 Add the mirepoix and brown well.

3 Add the red wine and reduce by two-thirds.

4 Mix the flour and a little stock together to form a viscous batter-like mix.

5 Add the stock to the roasting tray and bring to the boil.

6 Pour in the flour mix and whisk into the liquid in the tray.

7 Bring to the boil, simmer and correct the seasoning.

8 Pass through a sieve and retain for service

To complete

1 Slice the beef and warm the Yorkshire puddings (see recipe 3), serve with the gravy, offer horseradish and mustard.

> **PROFESSIONAL TIP**
> This dish would work well with most vegetables or potatoes. As an alternative, add slightly blanched root vegetables to the roasting tray at the start of the beef cooking, remove and reheat for service. They will obtain maximum flavour from the beef and juices.

> **VARIATION**
> ● Other roasting joints could be used, for instance rolled sirloin.

3 Yorkshire pudding

Energy	Calories	Fat	Saturated fat	Carbohydrates	Sugar	Protein	Fibre
730 kJ	174 kcal	9.3 g	2.1 g	17.5 g	1.3 g	6.3 g	0.9 g

	4 portions	10 portions
Flour	85 g	215 g
Eggs	2	5
Milk	85 ml	215 ml
Water	40 ml	100 ml
Dripping or oil	20 g	50 g

Mise en place

1 Place the flour and eggs into a mixing bowl and mix to a smooth paste.

2 Gradually whisk in the milk and water and place in the refrigerator for 1 hour.

Cooking

1 Heat the pudding trays in the oven with a little dripping or oil in each well.

2 Carefully ladle the mixture in, up to about two-thirds full.

3 Place in the oven at 190 °C and slowly close the door (if you have a glass-fronted door it will be easy to monitor progress; if not, after about 30 minutes check the puddings). The myth about opening the door during cooking has an element of truth in it – however, it is slamming and the speed at which the door is opened that have most effect, so have just a small, careful peek to check and see if they are ready.

4 For the last 10 minutes of cooking, invert the puddings (take out and turn upside down in the tray) to dry out the base.

5 Serve immediately.

> **PROFESSIONAL TIP**
> The oven, and the oil, must be very hot before the mixture is placed into the pudding tray; if they are not hot enough, the puddings will not rise.

4 Sirloin steak with red wine jus (*entrecôte bordelaise*)

Energy	Calories	Fat	Saturated fat	Carbohydrates	Sugar	Protein	Fibre	
3013 KJ	717 kcal	62.2 g	21.6 g	6.0 g	3.0 g	26.1 g	1.4 g	*

* Using sunflower oil and 150 g raw steak per portion. Using sunflower oil and 200 g raw steak per portion provides: 3584 kJ/853 kcal Energy; 73.6 g Fat; 26.2 g Saturated fat; 6.0 g Carbohydrates; 3.0 g Sugar; 34.4 g Protein; 1.4 g Fibre

	4 portions	10 portions
Butter or oil	50 g	125 g
Sirloin steaks (approximately 150–200 g each)	4	10
Red wine	60 ml	150 ml
Red wine jus (see page 98)	¼ litre	½ litre
Parsley, chopped		

Cooking

1 Heat the butter or oil in a sauté pan.

2 Lightly season the steaks on both sides with salt and pepper.

3 Fry the steaks quickly on both sides, keeping them underdone.

4 Dress the steaks on a serving dish.

5 Pour off the fat from the pan.

6 Deglaze with the red wine. Reduce by half and strain.

7 Add the red wine jus, bring to the boil and correct the seasoning.

8 Coat the steaks with the sauce.

9 Sprinkle with chopped parsley and serve.

NOTES

Traditionally, two slices of beef bone marrow, poached in stock for 2–3 minutes, would be placed on each steak.

PROFESSIONAL TIP

Cook the steaks to order (and make the sauce by deglazing the pan and reducing), not in advance.

Healthy Eating Tips

- Use little or no salt to season the steaks.
- Fry in a small amount of an unsaturated oil and drain off all excess fat after frying.
- Serve with plenty of boiled new potatoes or a jacket potato and a selection of vegetables.

5 Boiled silverside, carrots and dumplings

Energy	Calories	Fat	Saturated fat	Carbohydrates	Sugar	Protein	Fibre
1068 kJ	254 kcal	10.17 g	4.6 g	15.5 g	5.5 g	26.3 g	2.6 g

3 Divide the suet paste into even pieces and lightly mould into balls (dumplings).

4 Add the dumplings and simmer for a further 15–20 minutes.

5 Serve by carving the meat across the grain, garnish with carrots, onions and dumplings, and moisten with a little of the cooking liquor.

NOTES

A large joint of silverside is approximately 6 kg; for this size of joint, soak it overnight and allow 25 minutes cooking time per 0.5 kg plus 25 minutes.

The beef is salted because this gives the desired flavour. The meat is usually salted before it is delivered to the kitchen.

	4 portions	**10 portions**
Silverside, pre-soaked in brine	400 g	1.25 kg
Onions	200 g	500 g
Carrots	200 g	500 g
Suet paste	100 g	250 g

Mise en place

Soak the meat in cold water for 1–2 hours to remove excess brine.

Cooking

1 Place in a saucepan and cover with cold water, bring to the boil, skim and simmer for 45 minutes.

2 Add the whole prepared onions and carrots and simmer until cooked.

Healthy Eating Tip

Adding carrots, onions, boiled potatoes and a green vegetable will give a healthy balance.

VARIATION

- Herbs can be added to the dumplings.
- Boiled brisket and tongue can be served with the silverside.
- French-style boiled beef is prepared using unsalted thin flank or brisket with onions, carrots, leeks, celery, cabbage and a bouquet garni, all cooked and served together, accompanied with pickled gherkins and coarse salt.

6 Boeuf bourguignonne

Energy	Calories	Fat	Saturated fat	Carbohydrates	Sugar	Protein	Fibre
2838 kJ	681 kcal	33.3 g	9.1 g	20.1 g	7.2 g	44.9 g	8.0 g

	4 portions	10 portions
Beef		
Beef shin pre-soaked in red wine (see below) for 12 hours	600 g	1.5 kg
Olive oil	50 ml	125 ml
Bottle of inexpensive red Bordeaux wine	1	2
Onion	100 g	250 g
Carrot	100 g	250 g
Celery sticks	75 g	180 g
Leek	100 g	250 g
Cloves of garlic	2	5
Sprig fresh thyme	1	2
Bay leaf	1	2
Seasoning		
Veal/brown stock to cover		
Garnish		
Button onions, cooked	150 g	300 g
Cooked bacon lardons	150 g	300 g
Button mushrooms, cooked	150 g	300 g
Parsley, chopped	2 tsp	5 tsp
To finish		
Mashed potato	300 g	750 g
Washed, picked spinach	300 g	750g
Cooked green beans	250 g	625 g

Mise en place

1 Trim the beef shin of all fat and sinew, and cut into 2½ cm-thick rondelles.

Cooking

1 Heat a little oil in a thick-bottomed pan and seal/brown the skin. Place in a large ovenproof dish.

2 Meanwhile, reduce the red wine by half.

3 Peel and trim the vegetables as appropriate, then add them to the pan that the beef has just come out of and gently brown the edges. Then place this, along with the garlic and herbs, in the ovenproof dish with the meat.

4 Add the reduced red wine to the casserole, then pour in enough stock to cover the meat and vegetables. Bring to the boil, then cook in the oven pre-heated to 180 °C for 40 minutes; after that, turn the oven down to 90–95 °C and cook for a further 4 hours until tender.

5 Remove from the oven and allow the meat to cool in the liquor. When cold, remove any fat. Reheat gently at the same temperature to serve.

6 Heat the garnish elements separately and sprinkle over each portion. Serve with a mound of mashed potato, wilted spinach and buttered green beans. Finish the whole dish with chopped parsley.

> **VARIATION**
> - Other joints of beef can be used here; beef or veal cheek can be used, reducing the time for the veal, or modernise the dish by using the slow cooked fillet preparation and serving the same garnish.

> **PROFESSIONAL TIP**
> Shallow fry the beef in hot oil to brown it all over, but do not let it boil in the oil. Then allow it to stew gently in the red wine.

7 Beef olives (*paupiettes de boeuf*)

Energy	Calories	Fat	Saturated fat	Carbohydrates	Sugar	Protein	Fibre	*
1134 kJ	271 kcal	13.1 g	2.6 g	13.6 g	5.0 g	25.4 g	1.5 g	

* Using 625 ml stock

	4 portions	10 portions
Stuffing		
White or wholemeal breadcrumbs	50 g	125 g
Parsley, chopped	1 tbsp	3 tbsp
Thyme, small pinch		
Suet, prepared and chopped	5 g	25 g
Onion, finely chopped and lightly sweated in oil	25 g	60 g
Salt		
Egg	½	1
Olives		
Lean beef (topside)	400 g	1.25 kg
Salt, pepper		
Dripping or oil	35 g	100 g
Carrot	100 g	250 g
Onion	100 g	250 g
Flour, browned in the oven	25 g	60 g
Tomato puree	25 g	60 g
Brown stock	500–750 ml	1.25–1.5 litres
Bouquet garni		

Mise en place

1 Combine all the stuffing ingredients and mix thoroughly.
2 Cut the meat into thin slices across the grain and bat out.

3 Trim to approximately 10 x 8 cm, chop the trimmings finely and add to the stuffing.
4 Season the slices of meat lightly with salt and pepper, and spread a quarter of the stuffing down the centre of each slice.
5 Roll up neatly and secure with string.

Cooking

1 Fry off the meat to a light brown colour, add the vegetables and continue cooking to a golden colour.
2 Drain off the fat into a clean pan. Make up to 25 g fat if there is not enough (increase the amount for 10 portions). Mix in the flour.
3 Mix in the tomato puree, cool and then mix in the boiling stock.
4 Bring to the boil, skim, season and pour on to the meat.
5 Add the bouquet garni.
6 Cover and simmer gently, preferably in the oven, for approximately 1½–2 hours.
7 Remove the string from the meat.
8 Skim, correct the sauce and pass on to the meat.

PROFESSIONAL TIP

When you have stuffed and shaped the olives, tie them with string so that they will keep their shape during cooking.

Healthy Eating Tips
- Use little or no salt to season the steaks.
- To avoid the use of allergenic ingredients, use gluten-free flour mix and crumbs as an alternative
- Fry in a small amount of an unsaturated oil and drain off all excess fat after frying.
- Serve with a large portion of potatoes and vegetables

SERVING SUGGESTION

Serve with braised vegetables and a herb mash, would look good in a large rim bowl for presentation to customer as finished dish

8 Beef stroganoff (*sauté de boeuf stroganoff*)

Energy	Calories	Fat	Saturated fat	Carbohydrates	Sugar	Protein	Fibre	*
1364 kJ	325 kcal	23.7 g	7.9 g	1.7 g	1.7 g	21.2 g	0.3 g	

* Using sunflower oil

	4 portions	10 portions
Fillet of beef (tail end)	400 g	1.5 kg
Butter or oil	50 g	125 g
Salt, pepper		
Shallots, finely chopped	25 g	60 g
Dry white wine	125 ml	300 ml
Cream	125 ml	300 ml
Lemon, juice of	¼	½
Parsley, chopped		

Mise en place

1 Cut the meat into strips approximately 1 × 5 cm.

Cooking

1 Place the butter or oil in a sauteuse over a fierce heat.

2 Add the beef strips, lightly season with salt and pepper, and allow to cook rapidly for a few seconds. The beef should be brown but underdone.

3 Drain the beef into a colander. Pour the butter back into the pan.

4 Add the shallots, cover with a lid and allow to cook gently until tender.

5 Drain off the fat, add the wine and reduce to one-third.

6 Add the cream and reduce by a quarter.

7 Add the lemon juice and the beef strips; do not reboil. Correct the seasoning.

SERVING SUGGESTION

Serve lightly sprinkled with chopped parsley. Accompany with rice pilaff (see page 406).

Healthy Eating Tips

- Use little or no salt to season the meat.
- Fry in a small amount of an unsaturated oil.
- Serve with a large portion of rice and a salad.

VARIATION

- Serve with fresh strips of pasta

FAULTS

Cooking temperature and time are important with this dish as overcooking the thin strips of meat will affect the finished dish.

ON A BUDGET

Flat iron steak, or flank steak could be used for this dish which are cheaper cuts.

9 Steak pudding

Energy	Calories	Fat	Saturated fat	Carbohydrates	Sugar	Protein	Fibre
1369 kJ	326 kcal	17.3 g	7.8 g	20.6 g	1.0 g	23.0 g	1.1 g

	4 portions	10 portions
Suet paste	200 g	500 g
Prepared stewing beef (chuck steak)	400 g	1.5 kg
Worcester sauce	1 tsp	3 tsp
Parsley, chopped	1 tsp	2½ tsp
Salt, pepper		
Onion, chopped (optional)	50–100 g	200 g
Water	125 ml approximately	300 ml approximately

Mise en place

1 Line a greased ½-litre basin with three-quarters of the suet paste and retain one-quarter for the top.

2 Mix all the other ingredients, except the water, together.

Cooking

1 Place in the basin with the water to within 1 cm of the top.

2 Moisten the edge of the suet paste, cover with the top and seal firmly.

3 Cover with greased greaseproof paper and also, if possible, foil or a pudding cloth tied securely with string.

4 Cook in a steamer for at least 3½ hours.

5 Serve with the paper and cloth removed, clean the basin, place on a round flat dish and fasten a napkin round the basin.

NOTES

Extra gravy should be served separately. If the gravy in the pudding is to be thickened, the meat can be lightly floured.

1 Line the basin with suet paste.

2 Fill the basin, then cover the top with paste.

3 Make a foil cover for the basin, with a fold to allow it to expand.

4 Tie the cover over the basin securely.

Healthy Eating Tips

- Use little or no salt as the Worcester sauce contains salt.
- Trim off as much fat as possible from the raw stewing beef.

HEALTH AND SAFETY

Worcester sauce is highlighted as an allergen as it contains allergens in its ingredients, chefs should be aware of contents for all elements of a recipe.

VARIATION

Try:

- Adding 50–100 g ox or sheep's kidneys cut in pieces with skin and gristle removed.
- Adding 50–100 g sliced or quartered mushrooms.
- Making the steak pudding with a cooked filling; in which case, simmer the meat until cooked in brown stock with onions, parsley, Worcester sauce and seasoning; cool quickly and proceed as above, steaming for 1–1½ hours.

10 Steak pie

Energy	Calories	Fat	Saturated fat	Carbohydrates	Sugar	Protein	Fibre
1442 kJ	346 kcal	22.2 g	2.9 g	13.6 g	1.8 g	24.3 g	0.4 g

* Using puff pastry (McCance data)

	4 portions	10 portions
Prepared stewing beef (chuck steak)	400 g	1.5 kg
Oil or dripping	50 ml	125 ml
Onion, chopped (optional)	100 g	250 g
Worcester sauce, few drops		
Parsley, chopped	1 tsp	3 tsp
Water, stock, red wine or dark beer	125 ml	300 ml
Salt, pepper		
Cornflour	10 g	25 g
Short, puff or rough puff paste (see pages 515, 517–518)	100 g	250 g

Mise en place

1 Cut the meat into 2-cm strips then cut into squares.

Cooking

1 Heat the oil in a frying pan until smoking, add the meat and quickly brown on all sides.

2 Drain the meat off in a colander.

3 Lightly fry the onion.

4 Place the meat, onion, Worcester sauce, parsley and the liquid in a pan, season lightly with salt and pepper.

5 Bring to the boil, skim, then allow to simmer gently until the meat is tender.

6 Dilute the cornflour with a little water, stir into the simmering mixture, reboil and correct seasoning.

7 Place the mixture into a pie dish and allow to cool.

8 Cover with the paste, eggwash and bake at 200 °C for approximately 30–45 minutes.

1 Dice the steak evenly.

2 Coat the meat in cornflour before frying it.

3 Cover the top with paste and trim off the excess.

4 Press gently to ensure a seal between the paste and the dish.

Healthy Eating Tips
- Use little or no salt as the Worcester sauce contains salt.
- There will be less fat in the dish if short paste is used.

VARIATION
- Try adding 50–100 g ox or sheep's kidneys with skin and gristle removed and cut into neat pieces
- Add 50–100 g sliced or quartered mushrooms
- Try adding 1 heaped tsp tomato purée and some mixed herbs
- In place of cornflour, the meat can be tossed in flour before frying off.
- In place of plain flour, 25–50 per cent wholemeal flour may be used in the pastry.

11 Carbonnade of beef (*carbonnade de boeuf*)

Energy	Calories	Fat	Saturated fat	Carbohydrates	Sugar	Protein	Fibre
1037 kJ	247 kcal	9.1 g	1.8 g	14.0 g	8.1 g	24.7 g	1.1 g

	4 portions	10 portions
Lean beef (topside)	400 g	1.25 kg
Salt, pepper		
Flour (white or wholemeal)	25 g	60 g
Dripping or oil	25 g	60 g
Onions, sliced	200 g	500 g
Beer	250 ml	625 ml
Caster sugar	10 g	25 g
Tomato purée	25 g	60 g
Brown stock		

Mise en place

1 Cut the meat into thin slices.
2 Season with salt and pepper and pass through the flour.

Cooking

1 Quickly colour on both sides in hot fat and place in a casserole.

2 Fry the onions to a light brown colour. Add to the meat.

3 Add the beer, sugar and tomato purée and sufficient brown stock to cover the meat.

4 Cover with a tight-fitting lid and simmer gently in a moderate oven at 150–200 °C until the meat is tender (approximately 2 hours).

5 Skim, correct the seasoning and serve.

1 Slice the meat thinly.

2 Pour the liquid over the browned meat and onions.

3 Pass each slice through the flour.

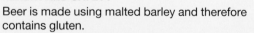

HEALTH AND SAFETY
Beer is made using malted barley and therefore contains gluten.

Chefs should ensure that information about dishes is accurate and verifiable.

12 Hamburg or Vienna steak (*bitok*)

Energy	Calories	Fat	Saturated fat	Carbohydrates	Sugar	Protein	Fibre
681 kJ	162 kcal	6.7 g	1.9 g	12.7 g	1.0 g	13.8 g	1.0 g

3 Divide into even pieces and, using a little flour, make into balls, flatten and shape round.

4 Shallow-fry in hot fat on both sides, reducing the heat after the first few minutes, making certain they are cooked right through.

SERVING SUGGESTION

Serve with a light sauce, such as piquant sauce (see page 106).

The 'steaks' may be garnished with French-fried onions and sometimes with a fried egg.

Serve with plenty of starchy carbohydrate and vegetables.

Healthy Eating Tips

- Use a small amount of an unsaturated oil to cook the onion and to shallow-fry the meat.
- The minced beef will produce more fat, which should be drained off.

	4 portions	10 portions
Onion, finely chopped	25 g	60 g
Butter or oil	10 g	25 g
Lean minced beef	200 g	500 g
Small egg	1	2–3
Breadcrumbs	100 g	250 g
Cold water or milk	2 tbsp (approximately)	60 ml (approximately)

HEALTH AND SAFETY

The use of temperature probe is the surest way of ensuring a core temperature of over 75°C.

Cooking

1 Cook the onion in the fat without colour, then allow to cool.

2 Add to the rest of the ingredients and mix in well.

13 Goulash (*goulash de boeuf*)

Energy	Calories	Fat	Saturated fat	Carbohydrates	Sugar	Protein	Fibre
1625 kJ	389 kcal	20.4 g	6.0 g	26.1 g	3.9 g	26.9 g	1.7 g

	4 portions	10 portions
Prepared stewing beef	400 g	1.25 kg
Lard or oil	35 g	100 g
Onions, chopped	100 g	250 g
Flour	25 g	60 g
Paprika	10–25 g	25–60 g
Tomato purée	25 g	60 g
Stock or water	750 ml (approximately)	2 litres (approximately)
Turned potatoes or small new potatoes	8	20
Choux paste (see page 517)	125 ml	300 ml

Mise en place

1 Remove excess fat from the beef. Cut into 2-cm square pieces.

Cooking

1 Season and fry in the hot fat until slightly coloured. Add the chopped onion.

2 Cover with a lid and sweat gently for 3–4 minutes.

3 Add the flour and paprika and mix in with a wooden spoon.

4 Cook out in the oven or on top of the stove. Add the tomato purée, mix in.

5 Gradually add the stock, stir to the boil, skim, season and cover.

6 Allow to simmer, preferably in the oven, for approximately 1½–2 hours until the meat is tender.

7 Add the potatoes and check that they are covered with the sauce. (Add more stock if required.)

8 Re-cover with the lid and cook gently until the potatoes are cooked.

9 Skim and correct the seasoning and consistency. A little cream or yoghurt may be added at the last moment.

SERVING SUGGESTION

Serve sprinkled with a few gnocchis made from choux paste (see page 517), reheated in hot salted water or lightly tossed in butter or margarine.

Healthy Eating Tips

- Trim off as much fat as possible before frying and drain all surplus fat after frying.
- Use the minimum amount of salt to season the meat.
- Serve with a large side salad.

VARIATION

- The use of gluten-free flour and potato-based gnocchi could be an alternative to enable the dish to meet specific customer requirements.

14 Best end or rack of lamb with breadcrumbs and parsley

Cooking

1 Roast the best end.

2 Ten minutes before cooking is completed cover the fat surface of the meat with a mixture of 25–50 g of fresh white breadcrumbs mixed with plenty of chopped parsley, an egg and 25–50 g melted butter or margarine.

3 Return to the oven to complete the cooking, browning carefully.

15 Roast leg of lamb with mint, lemon and cumin

Energy	Calories	Fat	Saturated fat	Carbohydrates	Sugar	Protein	Fibre	*
2192 KJ	524 kcal	39.1 g	13.7 g	0.3 g	0.3 g	29.7 g	0.0 g	

* Using a 225 g portion of lamb

Mise en place

1 Place the mint, lemon juice, cumin and olive oil in a food processor. Carefully blend the mint, lemon juice, cumin and olive oil to give maximum flavour.

2 Rub the mixture into the lamb and place in a suitable roasting tray.

Cooking

1 Roast the lamb in the normal way.

	4 portions	10 portions
Mint	25 g	62.5 g
Lemons, juice of	2	5
Cumin	2 tsp	5 tsp
Olive oil	4 tbsp	10 tbsp
Leg of lamb	3.5 kg	2 × 3.5 kg

SERVING SUGGESTION

Serve on a bed of boulangère potatoes or dauphinoise potatoes and a suitable green vegetable, e.g. leaf spinach with toasted pine nuts, or with a couscous salad.

Ingredients for roast leg of lamb with mint, lemon and cumin.

Place the leg of lamb in a roasting tray and rub with the mint and lemon mixture.

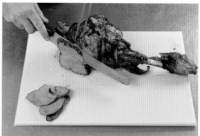

Carving a leg of lamb.

16 Pot roast shoulder of lamb

Energy	Calories	Fat	Saturated fat	Carbohydrates	Sugar	Protein	Fibre
3636 kJ	874 kcal	64 g	29 g	16.6 g	4.5 g	58.8 g	3.6 g

	10 portions
Lamb shoulder (boned)	3 kg
Gratin forcemeat (see recipe 18)	300 g
Salt	1 tsp
Pepper	½ tsp
Wholewheat flour	3 tbsp
Oil	3 tbsp
Brown stock	500 ml
Potatoes (turned)	10
Carrots (turned)	10
Button onions	10
Worcestershire sauce	1 tsp

See recipe 18 for how to prepare the gratin forcemeat.

Mise en place

1 Preheat the oven to 160 °C.

2 Stuff shoulder with forcemeat and tie.

3 Dredge the lamb with seasoned flour.

Cooking

1 Heat oil in a large roast pot over medium high heat.

2 Add the floured lamb to hot pan and brown evenly on all sides.

3 Pour stock to cover lamb.

4 Cover the pot and place in the preheated oven. Slow cook the lamb for 2½–3 hours.

5 During last 45 minutes, add the vegetables to the pot. Cover and continue to cook until meat and vegetables are tender.

6 Remove the cooked lamb and vegetables from the pot to a serving platter.

7 Strain the cooking liquor and adjust the consistency; this can be thickened if desired.

8 Finish with Worcestershire sauce and season to taste.

17 Sautéed forcemeat (*gratin forcemeat*)

Energy	Calories	Fat	Saturated fat	Carbohydrates	Sugar	Protein	Fibre	Sodium
2539 kJ	614 kcal	51.0 g	28.7 g	3.1 g	2.4 g	0.0 g	0.9 g	2,439.0 g

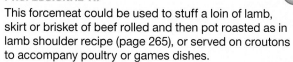

	300 g
Chicken liver or calves' liver (free from skin and sinews)	150 g
Pork fat, diced	5 g
Mushroom trimmings	50 g
Sautéed shallots	25 g
Butter	50 g
Spice salt	5 g
Pinch thyme and marjoram	
Brandy	10 ml

Cooking

1 Brown the pork fat in the butter, to set firm. Remove from pan and set to one side.

2 Sauté the liver in the same pan. Return pork to the pan.

3 Add mushrooms, sautéed shallot and seasonings.

4 Sauté all together, then flame with brandy.

5 Remove from heat and allow to cool.

6 Place all ingredients in food processor and blend. The forcemeat is then ready to use.

> **PROFESSIONAL TIP**
>
> This forcemeat could be used to stuff a loin of lamb, skirt or brisket of beef rolled and then pot roasted as in lamb shoulder recipe (page 265), or served on croutons to accompany poultry or games dishes.

18 Grilled lamb cutlets (*côtelettes d'agneau grillées*)

Energy	Calories	Fat	Saturated fat	Carbohydrates	Sugar	Protein	Fibre
1493 kJ	357 kcal	20.7 g	9.8 g	0.0 g	0.0 g	42.8 g	0.0 g

Cooking

1 If the grill is heated from below, place the prepared cutlet on the greased, preheated bars. Cook for approximately 5 minutes, turn and complete the cooking.

2 If using a salamander, place the cutlets on a greased tray, cook for approximately 5 minutes, turn and complete the cooking.

3 Serve dressed, garnished with a deep-fried potato and watercress. A compound butter (e.g. parsley, herb or garlic) may also be served.

4 Each cutlet bone may be capped with a cutlet frill.

Healthy Eating Tip
When served with boiled new potatoes and boiled or steamed vegetables, the dish becomes more 'balanced'.

Mise en place

1 Season the cutlets lightly with salt and mill pepper. Brush with oil or fat.

19 Grilled loin or chump chops, or noisettes of lamb

Energy	Calories	Fat	Saturated fat	Carbohydrates	Sugar	Protein	Fibre
1383 kJ	320 kcal	16.1 g	7.4 g	0.0 g	0.0 g	43.8 g	0.1 g

Mise en place

1 Season the chops or noisettes lightly with salt and mill pepper.

Cooking

1 Brush with oil and place on hot greased grill bars or place on a greased baking tray.

2 Cook quickly for the first 2–3 minutes on each side, in order to seal the pores of the meat.

3 Continue cooking steadily, allowing approximately 12–15 minutes in all.

SERVING SUGGESTION
A compound butter may also be served, together with deep-fried potatoes.

VARIATION
Try:
● Sprigs of rosemary or other herbs may be laid on the chops during the last few minutes of grilling to impart flavour.

20 Lamb kebabs (shish kebab)

Energy	Calories	Fat	Saturated fat	Carbohydrates	Sugar	Protein	Fibre
1544 kJ	378 kcal	25.1 g	9.9 g	7.2 g	5.9 g	29.6 g	20.6 g

NOTE

Kebabs, a dish of Turkish origin, are pieces of food impaled and cooked on skewers over a grill or barbecue. There are many variations and different flavours can be added by marinating the kebabs in oil, wine, vinegar or lemon juice with spices and herbs for 1–2 hours before cooking.

Kebabs can be made using tender cuts, or mince of lamb and beef, pork, liver, kidney, bacon, ham, sausage and chicken, using either the meats individually or combining two or three. Vegetables and fruit can also be added (e.g. onion, apple, pineapple, peppers, tomatoes, aubergine). Kebabs can be made using vegetables exclusively (e.g. peppers, onion, aubergine, tomatoes). Kebabs are usually served with a pilaff rice (see page 408).

The ideal cuts of lamb are the nut of the lean meat of the loin, best end or boned-out meat from a young shoulder of lamb.

Mise en place

1 Cut the meat into cubes and place on skewers with squares of green pepper, tomato, onion and bay leaves in between. The pieces of lamb and vegetables must be cut evenly so that they will cook evenly.

Cooking

1 Sprinkle with chopped thyme and cook over a hot grill.

SERVING SUGGESTION

Serve with pilaff rice, or with chickpeas and finely sliced raw onion.

VARIATION

Try:
- Miniature kebabs (one mouthful) can be made, impaled on cocktail sticks, grilled and served as a hot snack at receptions.
- Fish kebabs can be made using a firm fish, such as monkfish or tuna, and marinating in olive oil, lemon or lime juice, chopped fennel or dill, garlic and a dash of Tabasco or Worcester sauce.

21 Irish stew

Energy	Calories	Fat	Saturated fat	Carbohydrates	Sugar	Protein	Fibre
1544 kJ	378 kcal	25.1 g	9.9 g	7.2 g	5.9 g	29.6 g	20.6 g

	4 portions	10 portions
Stewing lamb	500 g	1.5 kg
Salt, pepper		
Bouquet garni		
Potatoes	400 g	1 kg
Onions	100 g	250 g
Celery	100 g	250 g
Savoy cabbage	100 g	250 g
Leeks	100 g	250 g
Button onions	100 g	250 g
Parsley, chopped		

Mise en place

1 Trim the meat and cut into even pieces. Blanch and refresh.

Cooking

1 Place in a shallow saucepan, cover with water, bring to the boil, season with salt and skim. If tough meat is being used, allow ½–1 hour stewing time before adding any vegetables.

2 Add the bouquet garni. Turn the potatoes into barrel shapes.

3 Cut the potato trimmings, onions, celery, cabbage and leeks into small neat pieces and add to the meat; simmer for 30 minutes.

4 Add the button onions and simmer for a further 30 minutes.

5 Add the potatoes and simmer gently with a lid on the pan until cooked.

6 Correct the seasoning and skim off all fat.

7 Serve sprinkled with chopped parsley.

NOTE
Keep the meat and vegetables covered with liquid during cooking, to keep the dish consistent and tasty.

VARIATION
Try:
● Alternatively, a more modern approach is to cook the meat for 1½–2 hours until almost tender, then add the vegetables and cook until all are tender. Optional accompaniments include Worcester sauce and/or pickled red cabbage (see page 190).

Healthy Eating Tips
● Trim as much fat as possible from the stewing lamb.
● Use the minimum amount of salt.
● Serve with colourful seasonal vegetables to create a 'healthy' dish.

Ingredients for Irish stew

Boil the meat.

Add the vegetables.

22 White lamb stew (*blanquette d'agneau*)

Energy	Calories	Fat	Saturated fat	Carbohydrates	Sugar	Protein	Fibre	*
1181 kJ	283 kcal	15.5 g	7.8 g	9.2 g	3.9 g	27.3 g	0.0 g	

* Using butter and 2 tbsp low-fat yoghurt

	4 portions	10 portions
Stewing lamb	500 g	1.5 kg
White stock	750 ml	1.5 litres
Onion, studded	50 g	125 g
Carrot	50 g	125 g
Bouquet garni		
Salt, pepper		
Butter, margarine or oil	25 g	60 g
Flour	25 g	60 g
Cream, yoghurt or quark	2–3 tbsp	5 tbsp
Parsley, chopped		

Mise en place

1 Trim the meat and cut into even pieces. Blanch and refresh.

Cooking

1 Place in a saucepan and cover with cold water.

2 Bring to the boil then place under running cold water until all the scum has been washed away.

3 Drain and place in a clean saucepan and cover with stock, bring to the boil and skim.

4 Add whole onion and carrot, and bouquet garni, season lightly with salt and simmer until tender, approximately 1–1½ hours.

5 Meanwhile prepare a blond roux with the butter and flour and make into a velouté with the cooking liquor. Cook out for approximately 20 minutes.

6 Correct the seasoning and consistency, and pass through a fine strainer on to the meat, which has been placed in a clean pan.

7 Reheat, mix in the cream and serve, finished with chopped parsley.

> **PROFESSIONAL TIP**
> To enrich this dish a liaison of yolks and cream is sometimes added at the last moment to the boiling sauce, which must not be allowed to reboil, otherwise the eggs will scramble and the sauce will curdle.

1 Ingredients for white lamb stew.

2 Cook the meat on the stove.

3 Make a roux.

4 Make a velouté by combining the roux with the liquid.

> **NOTE**
> Blanquette is cooked in white stock and then a sauce is made to coat the meat, as in this dish. Fricassee is cooked in a white sauce made as part of the cooking process, as in Chicken fricassee (page 310). Both are white stews which work well with chicken, lamb and veal.

23 Braised lamb shanks

Energy	Calories	Fat	Saturated fat	Carbohydrates	Sugar	Protein	Fibre
3098 kJ	742 kcal	24.2 g	7.4 g	10.5 g	8.1 g	98.3 g	4 g

	4 portions	10 portions
Lamb shanks	4	10
Olive oil for braising (to fill casserole about 1 cm deep)		
Leeks, roughly chopped	1	2
Celery sticks, roughly chopped	2	5
Carrots, roughly chopped	2	5
Onions, roughly chopped	2	5
Garlic head, broken into cloves (unpeeled)	1	2
Bay leaf	1	3
Thyme sprig	1	2
Rosemary sprig	1	2
Red wine	375 ml	1 litre
Chicken stock	600 ml	1.5 litres

Cooking

1 Take a casserole (or oven-proof dish) and place on the hob over a high heat. Pour in the olive oil and, when hot, add the lamb shanks, turning occasionally until brown.

2 Once browned, remove the lamb from the pot and tip in the leek, celery, carrot, onion and garlic cloves. Stir them all together and add the bay leaf, thyme and rosemary. These ingredients will all add flavour to the dish but won't be served at the end.

3 Once the vegetables are lightly browned, place the lamb back into the pot, allowing it to rest on top of the vegetables.

4 Pour in the red wine and chicken stock and bring to the boil.

5 Cover the pot with a lid or kitchen foil and place in the oven at 150 °C to braise for 2 hours 30 minutes, or up to 5 hours depending on the amount of lamb being used. When the meat is cooked, the bone can easily be turned out of the meat (if you would like to present the bone, only give it a small turn to check if the lamb is ready, as it will be difficult to re-insert the bone once removed).

6 Pass the cooking stock through a fine sieve and reduce to the correct consistency (coats the back of a spoon).

7 Serve the lamb shanks with the cooking juices poured over the top.

SERVING SUGGESTION
The lamb can be served with mashed potato and roast vegetables.

24 Roast saddle of lamb with rosemary mash

Energy	Calories	Fat	Saturated fat	Carbohydrates	Sugar	Protein	Fibre
3215 kJ	770 kcal	45.7 g	22.1 g	58.9 g	4.7 g	34.4 g	5.6 g

Cooking

1 Bone the saddle of lamb and roast in the normal way.

2 Bring the milk to the boil with the rosemary. Remove from the heat, cover and leave to infuse for 10–15 minutes.

3 To make the rosemary mash, prepare a potato purée using the milk infused with rosemary.

SERVING SUGGESTION
Allow the meat to rest before slicing, serving on a bed of herb mash with minted peas and glazed carrots

	4 portions	10 portions
Saddle of lamb, boned		
Milk	250 ml	625 ml
Rosemary	2 sprigs	5 sprigs
Potatoes, mashed	1.3 kg	3.2 kg

25 Mixed grill

Energy	Calories	Fat	Saturated fat	Carbohydrates	Sugar	Protein	Fibre	*
2050 kJ	488 kcal	40.8 g	19.3 g	0.0 g	0.0 g	30.4 g	0.0 g	

* 1 portion (2 cutlets). With deep-fried potatoes, parsley and watercress, 1 portion provides: 3050 kJ/726 kcal energy; 59.2 g fat; 26.6 g Saturated fat; 20.2 g Carbohydrates; 2.5 g sugar; 29.5 g protein; 4.9 g fibre

NOTE
These are the usually accepted items for a mixed grill, but it will be found that there are many variations to this list (e.g. steaks, liver, a Welsh rarebit and fried egg).

	4 portions	10 portions
Sausages	4	10
Cutlets	4	10
Kidneys	4	10
Tomatoes	4	10
Mushrooms	4	10
Streaky bacon, rashers	4	10
Deep-fried potato, to serve		
Watercress, to serve		
Parsley butter, to serve		

PROFESSIONAL TIP
The items must be cooked in the order listed above, so that they are all fully and evenly cooked at the end.

Healthy Eating Tips
Add only a small amount of compound butter and serve with plenty of potatoes and vegetables.

Cooking

1 Grill in the order listed above.

2 Dress neatly on an oval flat dish or plates.

26 Brown lamb or mutton stew (*navarin d'agneau*)

Energy	Calories	Fat	Saturated fat	Carbohydrates	Sugar	Protein	Fibre	*
1320 kJ	314 kcal	18.7 g	6.2 g	9.4 g	3.2 g	27.9 g	1.3 g	

* Using sunflower oil

Video: Making brown lamb stew

http://bit.ly/2oA0amN

	4 portions	10 portions
Stewing lamb	500 g	1.5 kg
Oil	2 tbsp	5 tbsp
Salt, pepper		
Carrot, chopped	100 g	250 g
Onion, chopped	100 g	250 g
Clove of garlic (if desired)	1	3
Flour (white or wholemeal)	25 g	60 g
Tomato purée	1 level tbsp	2¼ level tbsp
Brown stock (mutton stock or water)	500 g	1.25 litre
Bouquet garni		
Parsley, chopped, to serve		

Mise en place

1 Trim the meat and cut into even pieces.

Cooking

1 Partly fry off the seasoned meat in the oil, then add the carrot, onion and garlic, and continue frying.

2 Drain off the surplus fat, add the flour and mix.

3 Singe in the oven or brown on top of the stove for a few minutes, or add previously browned flour.

4 Add the tomato purée and stir with a wooden spoon.

5 Add the stock and season.

6 Add the bouquet garni, bring to the boil, skim and cover with a lid.

7 Simmer gently until cooked (preferably in the oven) for approximately 1–2 hours, until the lamb is tender.

8 When cooked, place the meat in a clean pan.

9 Correct the sauce and pass it on to the meat.

10 Serve sprinkled with chopped parsley.

Fry the lamb, onions and carrots.   Mix in the flour.

Add the stock and bring to the boil.

PROFESSIONAL TIP

Make sure the oil is hot before placing the meat in the pan to brown quickly all over.

Do not allow the meat to boil in the oil, because this will spoil the flavour and texture.

VARIATION

- Garnish with vegetables (glazed carrots and turnips, glazed button onions, potatoes, peas and diamonds of French beans, which may be cooked separately or in the stew).

27 Hotpot of lamb or mutton

Energy	Calories	Fat	Saturated fat	Carbohydrates	Sugar	Protein	Fibre	*
1505 KJ	360 kcal	17.0 g	6.4 g	22.0 g	1.8 g	29.0 g	2.5 g	

* Using sunflower oil

> ### NOTE
> Neck chops or neck fillet make a succulent dish.

When the dish is ready, the top layer of potatoes should be golden brown.

> ### VARIATION
> Try:
> - using leek in place of onion
> - adding 200 g lambs' kidneys
> - quickly frying off the meat and sweating the onions before putting in the pot
> - adding 100–200 g sliced mushrooms
> - adding a small tin of baked beans or a layer of thickly sliced tomatoes before adding the potatoes
> - using sausages in place of lamb.

	4 portions	10 portions
Stewing lamb	500 g	1.25 kg
Salt, pepper		
Onions, thinly sliced	100 g	250 g
Potatoes, thinly sliced	400 g	1.25 kg
Brown stock	1 litre	2.5 litres
Oil (optional)	25 g	60 g
Parsley, chopped		

Mise en place

1 Trim the meat and cut into even pieces.
2 Place in a deep earthenware dish. Season with salt and pepper.

Cooking

1 Lightly sauté the onions in the oil, if desired. Mix the onion and approximately three-quarters of the potatoes together.
2 Season and place on top of the meat; cover three parts with stock.
3 Neatly arrange an overlapping layer of the remaining potatoes on top, sliced about 2 mm thick.
4 Thoroughly clean the edges of the dish and place to cook in a hot oven at 230–250 °C until lightly coloured.
5 Reduce the heat and continue cooking for approximately 1½–2 hours.
6 Press the potatoes down occasionally during cooking.
7 Serve with the potatoes brushed with butter or margarine and sprinkle with the chopped parsley.

Chop the lamb into even-sized pieces.   Layer the potatoes over the lamb.

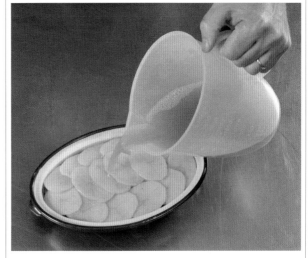

Pour over the stock.

28 Shepherd's pie (cottage pie)

Energy	Calories	Fat	Saturated fat	Carbohydrates	Sugar	Protein	Fibre	*
1744 kJ	415 kcal	25.3 g	9.1 g	22.1 g	2.5 g	26.3 g	1.6 g	

* Using sunflower oil, with hard margarine in topping

NOTE

This dish prepared with cooked beef is known as cottage pie.

Pipe the potato carefully so that the meat is completely covered.

When using reheated meats, care must be taken to heat thoroughly and quickly.

Healthy Eating Tips
- Use an oil rich in unsaturates (olive or sunflower) to lightly oil the pan.
- Drain off any excess fat after the lamb has been fried.
- Try replacing some of the meat with baked beans or lentils, and add tomatoes and/or mushrooms to the dish.
- When served with a large portion of green vegetables, a healthy balance is created.

	4 portions	10 portions
Onions, chopped	100 g	250 g
Oil	35 ml	100 ml
Lamb or mutton (minced), cooked	400 g	1.25 kg
Salt, pepper		
Worcester sauce	2–3 drops	5 drops
Potatoes, cooked	400 g	1.25 kg
Butter or margarine	25 g	60 g
Milk or eggwash		
Jus-lié or demi-glace	125–250 ml	300–600 ml

VARIATION

Try:
- adding 100–200 g sliced mushrooms
- adding a layer of thickly sliced tomatoes, then sprinkling with rosemary
- mixing a tin of baked beans in with the meat
- sprinkling with grated cheese and browning under a salamander
- varying the flavour of the mince by adding herbs or spices
- varying the potato topping by mixing in grated cheese, chopped spring onions or herbs, or by using duchess potato mixture
- serving lightly sprinkled with garam masala and with grilled pitta bread

Cooking

1 Cook the onion in the oil without colouring.

2 Add the cooked meat from which all fat and gristle has been removed.

3 Season and add Worcester sauce (sufficient to bind).

4 Bring to the boil; simmer for 10–15 minutes.

5 Place in an earthenware or pie dish.

6 Prepare the potatoes – mix with the butter or margarine, then mash and pipe, or arrange neatly on top.

7 Brush with the milk or eggwash.

8 Colour lightly under a salamander or in a hot oven.

9 Serve accompanied with a sauceboat of jus-lié.

29 Samosas

Energy	Calories	Fat	Saturated fat	Carbohydrates	Sugar	Protein	Fibre
237 kJ	57 kcal	3.6 g	1.3 g	5.5 g	1.5 g	0.9 g	0.8 g

Using short paste and lamb filling:

Samosa pastry	16–20 pasties	40–50 pasties
Short pastry made from ghee fat and fairly strong flour (the dough must be fairly elastic)	400 g	1 kg

Mise en place

1 Take a small piece of dough, roll into a ball 2 cm in diameter. Keep the rest of the dough covered with either a wet cloth, cling film or plastic, otherwise a skin will form on the dough.
2 Roll the ball into a circle about 9 cm round on a lightly floured surface. Cut the circle in half.
3 Moisten the straight edge with eggwash or water.
4 Shape the semicircle into a cone. Fill the cone with approximately 1½ tsp of filling, moisten the top edges with beaten egg white, flour paste or eggwash and press together well.
5 The samosas may be made in advance, covered with cling film or plastic, and refrigerated before being deep-fried.

Cooking

1 Deep-fry at 180 °C until golden brown; remove from fryer and drain well.

Filling 1: potato

	4 portions	10 portions
Potatoes, peeled	200 g	500 g
Vegetable oil	1½ tsp	3¾ tsp
Black mustard seeds	½ tsp	1¼ tsp
Onions, finely chopped	50 g	125 g
Fresh ginger, finely chopped	12 g	30 g
Fennel seeds	1 tsp	2½ tsp
Cumin seeds	¼ tsp	1 tsp
Turmeric	¼ tsp	1 tsp
Frozen peas	75 g	187 g
Salt, to taste		
Water	2½ tsp	6¼ tsp
Fresh coriander, finely chopped	1 tsp	2½ tsp
Garam masala	½ tsp	2½ tsp
Pinch of cayenne pepper		

Cooking

1 Cut the potatoes into 0.5-cm dice; cook in water until only just cooked.
2 Heat the oil in a suitable pan, add the mustard seeds and cook until they pop.
3 Add the onions and ginger. Fry for 7–8 minutes, stirring continuously until golden brown.
4 Stir in the fennel, cumin and turmeric, add the potatoes, peas, salt and water.
5 Reduce to a low heat, cover the pan and cook for 5 minutes.
6 Stir in the coriander; cook for a further 5 minutes.
7 Remove from the heat, stir in the garam masala and the cayenne seasoning.
8 Remove from the pan, place into a suitable bowl to cool before using.

Filling 2: lamb

	4 portions	10 portions
Saffron	½ tsp	1¼ tsp
Boiling water	2½ tsp	6¼ tsp
Vegetable oil	3 tsp	7½ tsp
Fresh ginger, finely chopped	12 g	30 g
Cloves of garlic, crushed and chopped	2	5
Onions, finely chopped	50 g	125 g
Salt, to taste		
Lean lamb, minced	400 g	1 kg
Pinch of cayenne pepper		
Garam masala	1 tsp	2½ tsp

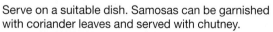
SERVING SUGGESTION
Serve on a suitable dish. Samosas can be garnished with coriander leaves and served with chutney.

Cooking

1 Infuse the saffron in the boiling water; allow to stand for 10 minutes.

2 Heat the vegetable oil in a suitable pan. Add the ginger, garlic, onions and salt, stirring continuously. Fry for 7–8 minutes, until the onions are soft and golden brown.

3 Stir in the lamb, add the saffron with the water. Cook, stirring the lamb until it is cooked.

4 Add the cayenne and garam masala, reduce the heat and allow to cook gently for a further 10 minutes.

5 The mixture should be fairly tight with very little moisture.

6 Transfer to a bowl and allow to cool before using.

1 Cut the paste into strips approx. 18 x 7 cm. Fold over the end of the pastry.

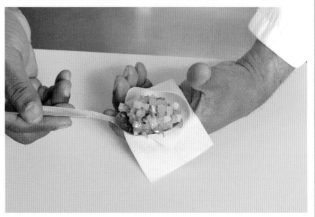

2 Flip over the pocket, ease it open and fill it.

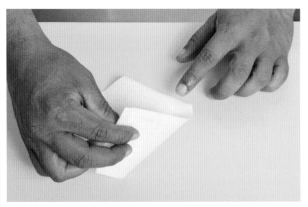

3 Eggwash the upper side of the pastry, then fold over the top part to form a pocket.

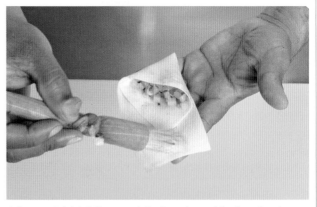

4 Once completely full, eggwash the top edge and the flap of pastry.

5 Fold over the pastry and wrap it round to seal the samosa. Deep fry until golden brown and crisp.

30 Grilled veal cutlet (*côte de veau grille*)

Energy	Calories	Fat	Saturated fat	Carbohydrates	Sugar	Protein	Fibre
990 kJ	236 kcal	9.7 g	2.6 g	0.0 g	0.0 g	36.9 g	0.0 g

Cooking

1 Place chop on previously heated grill bars.
2 Cook on both sides for 8–10 minutes in all.
3 Brush occasionally to prevent the meat from drying.

SERVING SUGGESTION

Serve with watercress, a deep-fried potato and a suitable sauce or butter, such as bearnaise or compound butter, or on a bed of plain or flavoured mashed potato (see page 177). Sprigs of rosemary may be added to the chop before or halfway through cooking.

Healthy eating tips
- Use the minimum amount of salt to season the meat.
- Serve with a small amount of sauce or butter and a large portion of mixed vegetables.

Mise en place

1 Lightly season the prepared chop with salt and mill pepper.
2 Brush with oil.

31 Braised shin of veal (osso buco)

Energy	Calories	Fat	Saturated fat	Carbohydrates	Sugar	Protein	Fibre
1732 kJ	413 kcal	19.7g	8.5g	12.5g	3.9g	47.3g	1.5g

	4 portions	10 portions
Salt and ground pepper		
Plain flour	45 g	112 g
Thick slices of veal shin on the bone	4 × 200 g	10 × 200 g
Butter	50 g	125 g
Oil	2 tbsp	5 tbsp
White wine	150 ml	375 ml
Plum tomatoes	450 g	1.125 kg
Light veal or chicken stock	300 ml	750 ml
Sprigs of parsley and thyme		
Bay leaf	1	1

Mise en place

1 Season the flour and use to coat the meat well on both sides.

Cooking

1 Heat the butter and oil in a casserole, add the veal and fry, turning once, until browned on both sides. Add the wine and cook, uncovered, for 10 minutes. Blanch, peel and chop the tomatoes, and add along with the stock and herbs.

2 Cover and cook in the centre of the oven until the meat is very tender and falls away from the marrow bone in the middle.

SERVING SUGGESTION

Delicious served with sauté potatoes or with a *risotto alla Milanese*.

NOTE

Part of the attraction of this dish is the marrow found in the bones. Although very rich, it is a special treat. Traditionally, osso buco is served with a gremolata, which is a combination of chopped parsley, garlic and lemon zest that is added to the dish at the very end. It has been omitted from this recipe, offering you a simple base.

32 Escalope of veal (*escalope de veau*)

Energy	Calories	Fat	Saturated fat	Carbohydrates	Sugar	Protein	Fibre
2079 KJ	495 kcal	39.8 g	11.4 g	10.3 g	0.5 g	24.7 g	1.0 g

* Fried in sunflower oil, using butter to finish

3 An optional finish is to pour over 50 g beurre noisette (nut-brown butter) and finish with a cordon of jus-lié.

NOTE

The escalopes need to be batted out thinly. Excess breadcrumbs must be shaken off before the escalopes are placed into the hot fat.
Pork or turkey escalopes could be used in the same way.

	4 portions	10 portions
Nut or cushion of veal	400 g	1.25 kg
Seasoned flour	25 g	60 g
Egg	1	2
Breadcrumbs	50 g	125 g
Oil, for frying	50 g	125 g
Butter, for frying	50 g	125 g
Beurre noisette (optional)	50 g	125 g
Jus-lié (page 99)	60 ml	150 ml

Mise en place

1 Trim and remove all sinew from the veal.

2 Cut into four even slices and bat out thinly using a little water.

3 Flour, egg and crumb. Shake off surplus crumbs. Mark with a palette knife.

Cooking

1 Place the escalopes into shallow hot fat and cook quickly for a few minutes on each side.

2 Dress on a serving dish or plate.

Healthy eating tips

- Use an unsaturated oil to fry the veal.
- Make sure the fat is hot so that less will be absorbed into the crumb.
- Drain the cooked escalope on kitchen paper.
- Use the minimum amount of salt.
- Serve with plenty of starchy carbohydrate and vegetables.

VARIATION

- **Escalope of veal Viennoise**: as for this recipe, but garnish the dish with chopped yolk and white of egg and chopped parsley; on top of each escalope place a slice of peeled lemon decorated with chopped egg yolk, egg white and parsley, an anchovy fillet and a stoned olive; finish with a little lemon juice and nut-brown butter.
- **Veal escalope Holstein**: prepare and cook the escalopes as for this recipe; add an egg fried in butter or oil, and place two neat fillets of anchovy criss-crossed on each egg; serve.
- **Escalope of veal with spaghetti and tomato sauce**: prepare escalopes as for this recipe, then garnish with spaghetti with tomato sauce (page 114), allowing 10 g spaghetti per portion.

33 Veal escalopes with Parma ham and mozzarella cheese (*involtini di vitello*)

Energy	Calories	Fat	Saturated fat	Carbohydrates	Sugar	Protein	Fibre
1642 kJ	394 kcal	26.1 g	15.5 g	0.1 g	0.1 g	39.8 g	0.0 g

	4 portions	10 portions
Small, thin veal escalopes	400 g (8 in total)	1.25 kg (20 in total)
Flour (for dusting)		
Parma ham, thinly sliced	100 g	250 g
Mozzarella cheese, thinly sliced	200 g	500 g
Fresh sage leaves, or dried sage	8	20
	1 tsp	2½ tsp
Salt, pepper		
Butter or oil	50 g	125 g
Parmesan cheese, grated		

Mise en place

1 Sprinkle each slice of veal lightly with flour and flatten.

2 Place a slice of Parma ham on each escalope.

3 Add several slices of mozzarella cheese to each.

4 Add a sage leaf or a light sprinkling of dried sage.

5 Season, roll up each escalope and secure with a toothpick or cocktail stick. Make sure the ham and cheese are well sealed within the escalope before cooking.

Cooking

1 Melt the butter in a sauté pan, add the escalopes and brown on all sides.

2 Transfer the escalopes and butter to a suitably sized ovenproof dish.

3 Sprinkle generously with grated Parmesan cheese and bake in a moderately hot oven at 190 °C for 10 minutes.

4 Clean the edges of the dish and serve.

Layer the veal, ham and cheese, then roll them up.

Transfer the fried escalopes to an ovenproof dish and sprinkle with cheese.

Healthy eating tips

● Use a small amount of oil to fry the escalopes and drain the cooked escalopes on kitchen paper.
● No added salt is necessary as there is plenty of salt in the cheese.
● Serve with plenty of vegetables.

34 Roast leg of pork

Energy	Calories	Fat	Saturated fat	Carbohydrates	Sugar	Protein	Fibre	*
1357 kJ	323 kcal	22.4 g	8.9 g	0.0 g	0.0 g	30.4 g	0.0 g	

* 113 g portion

Mise en place

1 Prepare leg for roasting (see page 238).
2 Moisten with water, oil, cider, wine or butter and lard, then sprinkle with salt, rubbing it well into the cracks of the skin. This will make the crackling crisp.
3 Place on a trivet in a roasting tin with a little oil or dripping on top.

Cooking

1 Start to cook in a hot oven at 230–250 °C, basting frequently.
2 reduce the heat to 180–185 °C, allowing approximately 25 minutes per 0.5 kg plus another 25 minutes. Pork must always be well cooked. If using a probe, the minimum temperature should be 75 °C for 2 minutes.
3 When cooked, remove from the pan and prepare a roast gravy from the sediment (see page 98).
4 Remove the crackling and cut into even pieces for serving.

SERVING SUGGESTION
Serve the joint garnished with picked watercress and accompanied by roast gravy, sage and onion dressing and apple sauce. If to be carved, proceed as for roast lamb (see page 264).

VARIATION
● Other joints can also be used for roasting (e.g. loin, shoulder and spare rib).

35 Sage and onion dressing for pork

Energy	Calories	Fat	Saturated fat	Carbohydrates	Sugar	Protein	Fibre
865 kJ	204 kcal	12.9 g	6.3 g	20.6 g	1.9 g	2.7 g	0.2 g

	4 portions	10 portions
Onion, chopped	50 g	125 g
Pork dripping	50 g	125 g
White breadcrumbs	100 g	250 g
Chopped parsley, pinch		
Powdered sage, good pinch		
Salt, pepper		

Cooking

1 Cook the onion in the dripping without colour.
2 Combine all the ingredients. Dressing is usually served separately.

HEALTHY EATING TIPS
- Use a small amount of unsaturated oil instead of dripping.
- Add the minimum amount of salt.

PROFESSIONAL TIP

Modern practice is to refer to this as a dressing if served separately to the meat, but as stuffing if used to stuff the meat.

36 Roast pork belly with shallots and champ potatoes

Energy	Calories	Fat	Saturated fat	Carbohydrates	Sugar	Protein	Fibre
2680 kJ	645 kcal	50.5 g	18.3 g	0.0 g	0.0 g	47.8 g	0.0 g

	4 portions	10 portions
Pork belly	1.2 kg	2.5 kg
Salt, pepper		
Olive oil	1 tbsp	3 tbsp
Shallots	20	50
Butter	70 g	175 g
Potatoes, peeled and chopped	1 kg	25 kg
Spring onions, chopped	8	20
Double cream	4 tbsp	10 tbsp

Mise en place

1 Place the pork on a rack in a roasting tray; season and oil.

Cooking

1 Roast in the oven for 10 minutes at 200 °C and then for 3–3½ hours at 140 °C.

2 Peel the shallots, fry gently in half the butter until caramelised. Keep warm.

3 Purée the potatoes.

4 Melt the remaining butter in a pan and sauté the spring onions until soft. Add the spring onion and the butter to the potato purée.

5 Add the cream and mix well.

6 Serve the pork with the caramelised shallots and potato.

7 Serve with a reduced brown stock flavoured with cider; alternatively, a red wine sauce may be served.

SERVING SUGGESTION

Serve accompanied with roasted parsnip and apple crisp.

37 Spare ribs of pork in barbecue sauce

Energy	Calories	Fat	Saturated fat	Carbohydrates	Sugar	Protein	Fibre	*
6151 kJ	1465 kcal	12.6 g	37.3 g	20.3 g	17.1 g	63.5 g	0.3 g	

* Using sunflower oil

	4 portions	10 portions
Onion, finely chopped	100 g	250 g
Clove of garlic, chopped	1	2
Oil	60 ml	150 ml
Vinegar	60 ml	150 ml
Tomato purée	150 g	375 g
Honey	60 ml	150 ml
Brown stock	250 ml	625 ml
Worcester sauce	4 tbsp	10 tbsp
Dry mustard	1 tsp	2 tsp
Pinch thyme		
Salt		
Spare ribs of pork	2 kg	5 kg

Mise en place

1 Sweat the onion and garlic in the oil without colour.
2 Mix in the vinegar, tomato purée, honey, stock, Worcester sauce, mustard and thyme, and season with salt.

Cooking

1 Allow the barbecue sauce to simmer for 10–15 minutes.
2 Place the prepared spare ribs fat side up on a trivet in a roasting tin.
3 Brush the spare ribs liberally with the barbecue sauce.
4 Place in a moderately hot oven: 180–200 °C. Cook for ¾–1 hour.
5 Baste generously with the barbecue sauce every 10–15 minutes.
6 The cooked spare ribs should be brown and crisp.
7 Cut the spare ribs into individual portions and serve.

PROFESSIONAL TIP
Apply plenty of barbecue sauce before and during cooking, to give the ribs a good flavour.

Healthy Eating Tips
● Sweat the onion and garlic in a little unsaturated oil.
● No added salt is necessary as the Worcester sauce is salty.

38 Pork loin chops with pesto and mozzarella

Energy	Calories	Fat	Saturated fat	Carbohydrates	Sugar	Protein	Fibre	*
3531 kJ	844 kcal	66.3 g	25.6 g	0.5 g	0.4 g	61.2 g	0.0 g	

* Using a 225 g portion of pork and 60 g of mozzarella

Cooking

1 Season, then shallow-fry or grill the chops until almost cooked.

2 Spread the pesto on top of each chop and top each with a slice of mozzarella.

3 Finish under the grill for approximately 1 minute until the cheese is golden and just cooked through.

SERVING SUGGESTION
Serve with a suitable pasta, e.g. butttered noodles and a green vegetable or tossed salad.

	4 portions	10 portions
Salt, pepper		
Loin chops	4	10
Olive oil (if frying)	2 tbsp	5 tbsp
Pesto	4 tsp	10 tsp
Mozzarella	4 slices	10 slices

39 Pork escalopes with Calvados sauce

Energy	Calories	Fat	Saturated fat	Carbohydrates	Sugar	Protein	Fibre	*
1856 kJ	447 kcal	34.2 g	20.3 g	12.7 g	12.5 g	22.8 g	1.1 g	

* Using lean meat only, and double cream

	4 portions	10 portions
Crisp eating apples (e.g. russet)	2	5
Cinnamon	¼ tsp	¾ tsp
Lemon juice	1 tbsp	2½ tbsp
Brown sugar	2 tsp	5 tsp
Butter, melted	25 g	70 g
Pork escalopes	4 x 100 g	10 x 100 g
Butter or oil	50 g	125 g
Shallots or onions, finely chopped	50 g	125 g
Calvados	30 ml	75 ml
Double cream or natural yoghurt	125 ml	300 ml
Salt, cayenne pepper		
Basil, sage or rosemary, chopped		

Mise en place

1 Core and peel the apples.
2 Cut into ½-cm-thick rings and sprinkle with a little cinnamon and a few drops of lemon juice.

Cooking

1 Place on a baking sheet, sprinkle with brown sugar and a little melted butter, and caramelise under the salamander or in the top of a hot oven.
2 Lightly sauté the escalopes on both sides in the butter.
3 Remove from the pan and keep warm.
4 Add the chopped shallots to the same pan, cover with a lid and cook gently without colouring (use a little more butter if necessary).
5 Strain off the fat, leaving the shallots in the pan, and deglaze with the Calvados.
6 Reduce by a half and then add the cream or yoghurt, seasoning and herbs.
7 Reboil, correct the seasoning and consistency, and pass through a fine strainer onto the meat.
8 Garnish with slices of caramelised apple.

NOTE
Special care must be taken not to overheat if using yoghurt, otherwise the sauce will curdle.

VARIATION
- Calvados can be replaced with twice the amount of cider and reduced by three-quarters as an alternative.
- Add a crushed clove of garlic and 1 tablespoon of continental mustard (2–3 cloves and 2½ tablespoons for 10 portions).

Healthy Eating Tips
- Try using yoghurt stabilised with a little cornflour, or half cream and half yoghurt.

40 Raised pork pie

Energy	Calories	Fat	Saturated fat	Carbohydrates	Sugar	Protein	Fibre
3005 kJ	721 kcal	47.7 g	19.3 g	49.2 g	1.7 g	26.7 g	2.7 g

Bread, soaked in milk	50 g	125 g
Stock or water	2 tbsp	5 tbsp
Eggwash		
Stock, hot	125 ml	375 ml
Gelatine	5 g	12.5 g
Picked watercress and salad to serve		

Mise en place

1 Cut the pork and bacon into small even pieces and combine with the rest of the main ingredients.

2 Keep one-quarter of the paste warm and covered.

3 Roll out the remaining three-quarters and carefully line a well-greased raised pie mould. Ensure that there is a thick rim of pastry.

4 Add the filling and press down firmly.

5 Roll out the remaining pastry for the lid and eggwash the edges of the pie.

6 Add the lid, seal firmly, neaten the edges, cut off any surplus paste; decorate if desired.

7 Make a hole 1 cm in diameter in the centre of the pie; brush all over with eggwash.

Cooking

1 Bake in a hot oven (230–250 °C) for approximately 20 minutes.

2 Reduce the heat to moderate (150–200 °C) and cook for 1½–2 hours in all.

3 If the pie colours too quickly, cover with greaseproof paper or foil. Remove from the oven and carefully remove tin. Eggwash the pie all over and return to the oven for a few minutes.

4 Remove from the oven and fill with approximately 125 ml of good hot stock in which 5 g of gelatine has been dissolved.

SERVING SUGGESTION
Serve cold, garnished with picked watercress and offer a suitable salad.

Hot water paste:

	4 portions	10 portions
Strong plain flour	250 g	500 g
Salt		
Lard or margarine (alternatively use 100 g lard and 25 g butter or margarine)	125 g	300 g
Water	125 ml	300 ml

Mise en place

1 Sift the flour and salt into a basin. Make a well in the centre.

Cooking

1 Boil the fat with the water and pour immediately into the flour.

2 Mix with a wooden spoon until cool enough to handle.

3 Mix to a smooth paste and use while still warm.

Raised pork pie:

	4 portions	10 portions
Shoulder of pork, without bone	300 g	1 kg
Bacon	100 g	250 g
Allspice (or mixed spice) and chopped sage	½ tsp	1½ tsp
Salt, pepper		

41 Stir-fried pork fillet

Energy	Calories	Fat	Saturated fat	Carbohydrates	Sugar	Protein	Fibre
831 kJ	199 kcal	9.8 g	2.2 g	5.1 g	4.2 g	22.9 g	0.8 g

Cooking

1 Gently fry the shallots, garlic and sliced mushrooms in a little oil in a frying pan or wok.

2 Add the pork cut into strips, stir well, increase the heat, season and add the Chinese five-spice powder; cook for 3–4 minutes then reduce the heat.

3 Add the soy sauce, honey and wine, and reduce for 2–3 minutes.

4 Correct the seasoning and serve.

	4 portions	10 portions
Shallots, finely chopped	2	6
Clove of garlic (optional), chopped	1	2
Button mushrooms, sliced	200 g	400 g
Olive oil		
Pork fillet	400 g	2 kg
Chinese five-spice powder	1 pinch	2 pinches
Soy sauce	1 tbsp	2 tbsp
Clear honey	2 tsp	3 tsp
Dry white wine	2 tbsp	5 tbsp
Salt, pepper		

Healthy Eating Tips
- No extra salt is needed, as soy sauce is added.
- Adding more vegetables and a large portion of rice or noodles can reduce the overall fat content.

VARIATION
- This dish could be prepared using chicken or turkey strips in the same quantities

42 Sweet and sour pork

Energy	Calories	Fat	Saturated fat	Carbohydrates	Sugar	Protein	Fibre
3067 kJ	730 kcal	43.9 g	9.2 g	69.7 g	54.7 g	13.4 g	1.6 g

	4 portions	10 portions
Loin of pork, boned	250 g	600 g
Sugar	12 g	30 g
Dry sherry	70 ml	180 ml
Soy sauce	70 ml	180 ml
Cornflour	50 g	125 g
Vegetable oil, for frying	70 ml	180 ml
Oil	2 tbsp	5 tbsp
Clove of garlic	1	2
Fresh root ginger	50 g	125 g
Onion, chopped	75 g	180 g
Green pepper, in 1-cm dice	1	2½
Chillies, chopped	2	5
Sweet and sour sauce (see page 114)	210 ml	500 ml
Pineapple rings (fresh or canned)	1	3
Spring onions	2	5

Mise en place

1 Cut the boned loin of pork into 2-cm pieces.

2 Marinate the pork for 30 minutes in the sugar, sherry and soy sauce.

3 Pass the pork through cornflour, pressing the cornflour in well.

Cooking

1 Deep-fry the pork pieces in oil at 190 °C until golden brown, then drain. Add the tablespoons of oil to a sauté pan.

2 Add the garlic and ginger, and fry until fragrant.

3 Add the onion, pepper and chillies, sauté for a few minutes.

4 Stir in the sweet and sour sauce, bring to the boil.

5 Add the pineapple cut into small chunks, thicken slightly with diluted cornflour. Simmer for 2 minutes.

6 Deep-fry the pork again until crisp. Drain, mix into the vegetables and sauce or serve separately.

7 Serve garnished with rings of spring onions or button onions.

> **PROFESSIONAL TIP**
> It is important to allow the pork enough time to marinate.
>
> This recipe would also work with chicken. It would not need as long to marinade or cook as the flesh is more tender.

43 Boiled bacon (hock, collar or gammon)

Energy	Calories	Fat	Saturated fat	Carbohydrates	Sugar	Protein	Fibre
1543 kJ	367 kcal	30.5 g	12.2 g	0.0 g	0.0 g	23.1 g	0.0 g

* Using 113 g per portion

Mise en place

1 Soak the bacon in cold water for 24 hours before cooking. Change the water.

Cooking

1 Bring to the boil, skim and simmer gently (approximately 25 minutes per 0.5 kg, plus another 25 minutes). Allow to cool in the liquid.

2 Remove the rind and brown skin; carve.

3 Serve with a little of the cooking liquor.

SERVING SUGGESTION
Boiled bacon may be served with pease pudding and a suitable sauce such as parsley (see page 101). It may also be served cold, or used as an ingredient in other dishes.

44 Griddled gammon with apricot salsa

Energy	Calories	Fat	Saturated fat	Carbohydrates	Sugar	Protein	Fibre
1112kJ	266 kcal	14.2 g	4.2 g	8.1 g	7.4 g	26.9 g	1.0 g

	4 portions	10 portions
Gammon steaks	4 × 150 g	10 × 150 g
Oil		
Apricot salsa		
Fresh apricots or dried, reconstituted, stoned and chopped	200 g	500 g
Lime, grated rind and juice	1	3
Fresh root ginger, grated	2 tsp	5 tsp
Clear honey	2 tsp	5 tsp
Olive oil	1 tbsp	2 ½ tsp
Sage, chopped, fresh	1 tbsp	2 ½ tbsp
Spring onions, chopped	4	10
Salt, pepper		

Cooking

1 Heat the griddle pan, lightly oil it then cook the gammon steaks.

2 Make the salsa: mix together in a processor the apricots, lime rind and juice, ginger, honey, olive oil and sage.

3 Add the finely chopped spring onions, correct the seasoning then mix well.

NOTE
The texture should be the consistency of thick cream but coarse. A little extra olive oil or some apricot juice may be required.

Healthy Eating Tips
- Use more juice and less oil in the salsa to reduce the fat.
- Gammon is a salty meat, so no extra salt is needed.

---

45 Potted meats

Energy	Calories	Fat	Saturated fat	Carbohydrates	Sugar	Protein	Fibre	*
1160 kJ	280 kcal	23.8 g	14.5 g	0.2 g	0.2 g	16.4 g	0.0 g	

* 1 of 4 portions

	1 pot
Cooked meat, e.g. beef, salt beef, tongue, venison, chicken or a combination	200 g
Salt, pepper and mace	
Clarified butter	100 g

Preparing the dish

1 Using an electric blender or chopper, reduce the meat, seasoning and 85 g of the butter to a paste.

2 Pack firmly into an earthenware or china pot and refrigerate until firm.

3 Cover with 1 cm of clarified butter and refrigerate.

SERVING SUGGESTION
Serve with a small tossed green salad and hot toast.

PROFESSIONAL TIP
Carefully purée the meat to give the desired texture.

Healthy Eating Tips
- Keep added salt to a minimum.
- Serve with plenty of salad vegetables and butter or toast (optional butter or spread).

46 Braised oxtail (*ragoût de queue de boeuf*)

Energy	Calories	Fat	Saturated fat	Carbohydrates	Sugar	Protein	Fibre
2481 kJ	595 kcal	38.0 g	12.0 g	12.3 g	4.7 g	51.6 g	1.4 g

Mise en place

1 Cut the oxtail into sections. Remove the excess fat.

Cooking

1 Fry on all sides in hot fat.

2 Place in a braising pan or casserole.

3 Roughly cut the onion and carrot. Fry them, then add them to the braising pan.

4 Mix in the flour.

5 Add tomato purée, brown stock, bouquet garni and garlic, and season lightly.

6 Bring to the boil, then skim.

7 Cover with a lid and simmer in the oven until tender (approximately 3 hours).

8 Remove the meat from the sauce, place in a clean pan.

9 Correct the sauce, pass on to the meat and reboil.

10 Serve sprinkled with chopped parsley.

	4 portions	10 portions
Oxtail	1 kg	2.5 kg
Dripping or oil	50 g	125 g
Onion	100 g	250 g
Carrot	100 g	250 g
Flour, browned in the oven	35 g	100 g
Tomato purée	25 g	60 g
Brown stock	1 litre	2.5 litres
Bouquet garni		
Clove of garlic	1	2
Salt, pepper		
Parsley, chopped		

SERVING SUGGESTION

This is usually garnished with glazed turned or neatly cut carrots and turnips, button onions, peas and diamonds of beans.

NOTE

Oxtail must be very well cooked so that the meat comes away from the bone easily.

HEALTHY EATING TIPS

- Keep added salt to a minimum.
- Fry in a small amount of an unsaturated oil and drain off all excess fat after frying.
- Serve with mashed potato and additional green vegetables.

VARIATION

- Haricot oxtail can be made using the same recipe with the addition of 100 g (250 g for 10 portions) cooked haricot beans, added approximately a ½ hour before the oxtail has completed cooking.

Brown the pieces of oxtail.

Mix in the flour.

47 Braised lambs' hearts (*coeurs d'agneau braisés*)

Energy	Calories	Fat	Saturated fat	Carbohydrates	Sugar	Protein	Fibre	*
614 kJ	147kcal	10.3 g	5.9 g	0.1 g	0.1 g	13.7 g	0.0 g	

* Using sunflower oil

	4 portions	10 portions
Lambs' hearts	4	10
Salt, pepper		
Fat or oil	25 g	60 g
Onions	100 g	250 g
Carrots	100 g	250 g
Brown stock	500 ml	1.25 litre
Bouquet garni		
Tomato puree	10 g	25 g
Demi-glace or jus-lié	250 ml	ml

Mise en place

1 Remove tubes and excess fat from the hearts.

Cooking

1 Season and colour quickly on all sides in hot fat to seal the pores.

2 Place into a small braising pan (any pan with a tight-fitting lid that may be placed in the oven) or in a casserole.

3 Place the hearts on the lightly fried sliced vegetables.

4 Add the stock, which should come two-thirds of the way up the meat; season lightly.

5 Add the bouquet garni and tomato purée and, if available, add a few mushroom trimmings.

6 Bring to the boil, skim, cover with a lid and cook in a moderate oven at 150–200°C.

7 After 1½ hours add the demi-glace or jus-lié, reboil, skim and strain.

8 Continue cooking until tender.

9 Remove the hearts and correct the seasoning, colour and consistency of the sauce.

10 Pass the sauce on to the sliced hearts and serve.

PROFESSIONAL TIP

Shallow frying the hearts in hot oil before braising gives the finished dish its attractive golden brown colour.

Healthy Eating Tips
- Lightly oil the pan using an unsaturated oil (olive or sunflower).
- Drain off any excess fat and skim all fat from the finished dish.
- Keep added salt to a minimum.

VARIATION
- The hearts can be prepared and cooked as above and, prior to cooking, the tube cavities can be filled with a firm stuffing.

48 Grilled lambs' kidneys (*rognons grillés*)

Energy	Calories	Fat	Saturated fat	Carbohydrates	Sugar	Protein	Fibre
614 kJ	147 kcal	10.3 g	5.9 g	0.1 g	0.1 g	13.7 g	0.0 g

Mise en place

1 Season the prepared skewered kidneys.

Cooking

1 Brush with melted butter, margarine or oil.

2 Place on preheated greased grill bars or on a greased baking tray.

3 Grill fairly quickly on both sides (approximately 5–10 minutes depending on size).

> **SERVING SUGGESTION**
> Serve with parsley butter, picked watercress and straw potatoes.

> **PROFESSIONAL TIP**
> Cook kidneys to order so that they are fresh and tasty.

49 Kidney sauté (*rognons sautés*)

Energy	Calories	Fat	Saturated fat	Carbohydrates	Sugar	Protein	Fibre	*
1680 kJ	400 kcal	28.3 g	4.3 g	15.5 g	3.7 g	21.8 g	1.8 g	

* Using sunflower oil

Cooking

1 Fry quickly in a frying pan using the butter, margarine or oil for approximately 4–5 minutes.

2 Place in a colander to drain, then discard the drained liquid.

3 Deglaze pan with demi-glace or jus, correct the seasoning and add the kidneys.

4 Do not reboil before serving as kidneys will toughen.

	4 portions	10 portions
Sheep's kidneys	8	20
Butter, margarine or oil	50 g	125 g
Demi-glace, lamb jus or jus-lié	250 ml	625 ml

Mise en place

1 Skin and halve the kidneys. Remove the sinews.

2 Cut each half into 3 or 5 pieces and season.

> **VARIATION**
> ● After draining the kidneys, the pan may be deglazed with white wine, sherry or port. As an alternative, a Suprême sauce (see page 104) may be used in place of demi-glace.
> ● An alternative recipe is **kidney sauté Turbigo**. Cook as for kidney sauté, then add 100 g small button mushrooms cooked in a little butter, margarine or oil, and eight small 2-cm-long grilled or fried chipolatas. Serve with the kidneys in an entrée dish, garnished with heart-shaped croutons. (Double these amounts for 10 portions.)

50 Fried lambs' liver and bacon (*foie d'agneau au lard*)

Energy	Calories	Fat	Saturated fat	Carbohydrates	Sugar	Protein	Fibre *
1039 kJ	250 kcal	20.1 g	3.8 g	0.1 g	0.1 g	17.2 g	0.0 g

* Using oil, jus-lié and reduced stock

	4 portions	**10 portions**
Liver	300 g	1 kg
Butter, margarine or oil, for frying	50 g	125 g
Streaky bacon	50 g (approximately 4 rashers)	125 g (approximately 10 rashers)
Jus-lié, reduced lamb stock or red wine jus	125 ml	300 ml

Mise en place

1 Skin the liver and remove the gristle. Cut into thin slices on the slant.

2 Pass the slices of liver through seasoned flour. Shake off the excess flour.

Cooking

1 Fry quickly on both sides in hot fat. (Liver is often served still pink in the centre but it is safer to cook it to a higher core temperature.)

2 Remove the rind and bone from the bacon and grill on both sides.

3 Serve the liver and bacon with a cordon and a sauceboat of jus or reduced stock.

Healthy Eating Tips
- Keep added salt to a minimum.
- Use a small amount of an unsaturated oil to fry the liver.
- Serve with plenty of potatoes and vegetables.

51 Braised veal sweetbreads (white) (*ris de veau braisé – à blanc*)

Energy	Calories	Fat	Saturated fat	Carbohydrates	Sugar	Protein	Fibre	*
1103 kJ	263 kcal	20.7 g	4.4 g	1.7 g	0.0 g	17.6 g	0.0 g	

* Using hard margarine and sunflower oil

	4 portions	10 portions
Heart-shaped sweetbreads	8	20
Salt, pepper		
Onion	100 g	250 g
Carrot	100 g	250 g
Oil, margarine or butter	50 g	125 g
Bouquet garni		
Veal stock	250 ml	625 ml

Mise en place

1 Wash, blanch, refresh and trim the sweetbreads (see page 245).

Cooking

1 Season and place in a casserole or sauté pan on a bed of roots smeared with the oil, margarine or butter.

2 Add the bouquet garni and stock.

3 Cover with buttered greaseproof paper and a lid.

4 Cook in a moderate oven at 150–200 °C for approximately 45 minutes.

5 Remove the lid and baste occasionally with cooking liquor to glaze.

6 Serve with some of the cooking liquor, thickened with diluted arrowroot if necessary, and passed on to the sweetbreads.

NOTE

Sweetbreads are glands, and two types are used for cooking. The thymus glands (throat) are usually long in shape and are of inferior quality. The pancreatic glands (stomach) are heart-shaped and of superior quality.

VARIATION

- **Braised veal sweetbreads (brown)**: prepare as in this recipe, but place on a lightly browned bed of roots. Barely cover with brown veal stock, or half-brown veal stock and half jus-lié. Cook in a moderate oven at 150–200 °C without a lid (for approximately 1 hour), basting frequently. Cover with the corrected, strained sauce to serve. (If veal stock is used, thicken with arrowroot.)
- **Braised veal sweetbreads with vegetables**: braise white with a julienne of vegetables in place of the bed of roots, the julienne served in the sauce.

52 Grilled veal sweetbreads

Energy	Calories	Fat	Saturated fat	Carbohydrates	Sugar	Protein	Fibre	Sodium
444 kJ	106 kcal	2.6 g	0.8 g	0.0 g	0.0 g	19.3 g	0.0 g	0.0 g

Cooking

1 Blanch, braise, cool and press the sweetbreads.

2 Cut in halves crosswise, pass through melted butter and grill gently on both sides.

3 Serve with a sauce and garnish as indicated.

SERVING SUGGESTION

In some recipes, the sweetbreads may be passed through butter and crumbs before being grilled, and garnished with noisette potatoes, buttered carrots, purée of peas and béarnaise sauce.

306 Poultry

Learning outcomes

In this unit you will be able to:

1 Prepare poultry dishes, including:
   - knowing different types of poultry: duck, chicken, poussin, turkey, goose, guinea fowl
   - knowing a range of quality points and understanding how they affect cooking methods
   - understanding how different factors affect the quality points
   - knowing the cuts and joints of poultry
   - understanding how texture, structure, fat content and colour influences the choice of processes and preparation methods needed to meet dish requirements
   - knowing how and being able to use preparation techniques on poultry, which are appropriate to dish requirements
   - being able to interpret a dish specification in order to determine the methodology, precise quantities, timings, presentation and balance of ingredients required

   - understanding the correct storage procedures to us throughout production and on completion of the final products.

2 Produce poultry dishes, including:
   - understanding how texture, fat content, muscle structure and colour of poultry affect the cooking requirements
   - understanding the effect of cooking methods on fat, muscle tissues, connective tissues and nutritional value
   - knowing how to and being able to use flavourings, coatings and stuffings to enhance or finish a poultry dish to meet dish specifications
   - understanding and being able to use cooking methods for poultry to meet dish requirements
   - being able to use tools and equipment with precision and speed
   - being able to measure and evaluate against quality standards throughout preparation and cooking
   - being able to evaluate products against dish requirements and production standards and recognise any faults.

Recipes included in this chapter

No.	Name
Chicken	
1	Chicken sauté chasseur
2	Fricassee of chicken
3	Chicken in red wine
4	Chicken spatchcock
5	Grilled chicken
6	Chicken Kiev
7	Deep fried chicken
8	Crumbed breast of chicken with asparagus
9	Steamed ballotine of chicken with herb stuffing and red wine jus
10	Tandoori chicken
11	Chicken tikka
12	Terrine of chicken and vegetables

No.	Name
Turkey	
13	Roast turkey
14	Turkey escalopes
Duck	
15	Roast duck or duckling
16	Sage and onion dressing for duck
17	Duckling with orange sauce
18	Confit duck leg with red cabbage and green Beans
Guinea fowl	
19	Suprêmes of guinea fowl with pepper and basil coulis
Goose	
20	Roast goose

Prepare poultry dishes

The word 'poultry' refers to all domestic fowl (birds) bred for food. It includes chickens, poussin, turkeys, ducks and geese.

Poultry is the most popular meat from animal or bird, with chicken as the type most commonly used in cooking.

The lighter or white breast meat is more tender than the dark or leg meat, but all flesh of poultry is more easily digested than that of butchers' meat. It contains protein, so it is useful for building and repairing body tissues and providing heat and energy. The fat content is low and contains a high percentage of unsaturated fat.

Flavour and texture of the meat is influenced by the way the birds are reared (fully domesticated, free range or even wild) and what they are fed upon. For instance, 'corn-fed chicken' will have a yellow tinge to the fat and flesh. The type or breed of bird will affect its characteristics and influence the cooking technique that a chef may use to produce the dish.

Birds which are reared intensively will be slaughtered within 10 weeks. These may have a higher water content which means the flesh shrinks quicker when cooked, as well as having less flavour.

There are two main methods for slaughtering poultry in preparation for the table:

- Electrical stunning – birds are shackled and the head then dipped in water with an electrical current passing through, before passing through a mechanical neck cutter.
- Controlled atmosphere – birds are placed caged in a chamber and gassed until dead, this method is considered more humane and less stressful for the bird.

Where religious requirements need to be met then there is agreement that poultry will not be stunned before slaughter.

1.1 Quality points for different types of poultry

Chicken

Originally chickens were classified according to size and cooking method by specific names, as shown in Table 6.1.

There is approximately 15–20 per cent bone in whole poultry.

	Description	Uses	Weight	Number of portions
Single baby chicken (poussin)	These are spring chickens that are 4–6 weeks old.	Roasting or grilling as whole bird or split	0.3–0.5 kg	1
Double baby chicken (poussin)	These are spring chickens that are 4–6 weeks old.	Roasting or grilling as whole bird or split	0.5–0.75 kg	2
Small roasting bird (broiler chickens)	3–4 months old	Roasting or grilling	0.75–1 kg	3–4
Medium roasting bird	Fully grown, tender prime birds	Roasting, grilling, sautéing, casseroles, suprêmes and pies	1–2 kg	4–6
Large roasting or boiling chicken		Roasting, boiling and casseroles	2–3 kg	6 8
Capon	Large, specially bred cock birds	Roasting	3–4.5 kg	8–12
Old boiling fowl		Stocks and soups	2.5–4 kg	

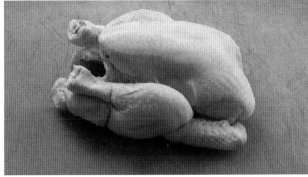

Poussin

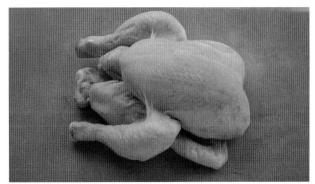

Corn-fed chicken

Quality points

Good-quality chickens have the following features:

- The breast should be plump and firm.
- The wishbone (breastbone) should be easy to bend between fingers and thumb.
- The skin should be white and unbroken. Broiler chickens have a faint bluish tint.
- Corn-fed chickens are yellow. Free range chickens have more colour, a firmer texture and more flavour.
- Bresse chickens are specially bred in France and are highly regarded.
- Old birds have coarse scales, large spurs on the legs and long hairs on the skin.

Turkey

Turkeys are a domesticated bird with a number of specific named breeds based on colour of the main plumage, for example, White, Bronze or Norfolk Black. These can vary in weight from 3.5 to 20 kilograms.

Turkeys are trussed in the same way as a chicken. The wishbone should always be removed before trussing, to make carving easier. The sinews should be drawn out of the legs. Allow 200 grams per portion raw weight.

When cooking a large turkey, the legs may be removed, boned, rolled, tied and roasted separately from the remainder of the bird, which is known as the crown, and includes both breasts still on the bone for roasting. This will reduce the cooking time, and help the legs and breast to cook more evenly (see recipe 14).

Stuffing may be rolled in foil, steamed or baked and thickly sliced. If a firmer stuffing is required, mix in one or two raw eggs before cooking. If stuffing a whole bird prior to cooking, this should only be in the neck cavity to ensure the stuffing cooks thorough in the roasting process.

Escalope's can be taken from the whole breast as per recipe 15 or cut into strips to use in a stir fry dish (see Unit 305, recipe 47, replacing the pork with turkey).

Quality points

Good-quality turkeys have the following features:

- Large, full breast with undamaged skin and no signs of stickiness.
- Smooth legs with supple feet and a short spur. As birds age the legs turn scaly and the feet harden.
- Bronze birds may have residue dark stubs from feathers visible.

Duck, duckling

Ducks are water-based fowl with webbed feet. They tend to have a higher fat content than other poultry, therefore it is normal when roasting whole to roast on a trivet to allow the fat to run from the bird.

The type and breed of duck should be taken into account in relation to the dish requirements. Ducks are usually slaughtered at 6–16 weeks.

Approximate sizes are as follows:

- duck: 3–4 kilograms
- duckling: 1–2 kilograms.

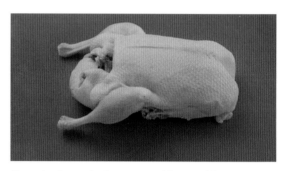

Gressingham duck prepared for cooking

Goose, gosling

These are water-based fowl with webbed feet. They tend to have rich dark flesh and are well covered with fat, therefore it is normal when roasting whole to roast on a trivet to allow the fat to run from the bird.

Geese tend to be a seasonal bird in the autumn and winter, which is why they are often an alternative for a Christmas roast.

Geese are usually aged around 24–28 weeks at slaughter. They have an approximate size of 2.5–12 kilograms, with the average bird around 6 kilograms.

Goose and duck fat is often used for its flavour in roasting vegetables and frying.

Quality points for duck and goose

Good-quality birds have the following features:

- Plump breasts with pliable breast bones.
- Moist but not sticky to the touch.
- Webbed feet that tear easily.
- A lower back that bends easily.
- Feet and bill should be yellow.
- Wild duck and geese will feature under game, and tend to be darker fleshed and have a richer flavour.

Guinea fowl

A grey and white feathered bird that, once plucked, resembles a chicken, although it has darker flesh and scales on the lower leg which enable it to be identified. Normally around the size of a medium chicken, the quality points for selecting guinea fowl are as for chicken.

Younger birds are known as squabs and are normally slaughtered at around 10–15 weeks. Guinea fowl can be prepared and cooked as per chicken recipes.

Quality points

The quality points described for chicken can be used for buying and using guinea fowl, which is why it is used as an alternative in some recipes for chicken.

General quality points for purchasing poultry

- Check for damage to packaging or flesh of the bird on delivery.
- Check for any inappropriate aroma or blemishes.
- Check temperature of the product to ensure it is safe to receive.

ACTIVITY

Cook two chicken breasts in the same way but using two different types of chicken, then evaluate the following differences:

1 weight loss to check yield
2 moisture loss in cooking process
3 flavour of the meat
4 texture.

1.2 Joints and cuts of poultry

When preparing poultry for cooking, recipes will specify the parts of the bird or cuts which will meet the dish specification. These will also influence the cooking method used (see Section 1.4 Using skills and techniques to prepare poultry for examples of the different cuts or joints). In some kitchens poultry is bought in prepared, for instance breasts or suprêmes, crowns of turkey or duck legs, while in other kitchens whole birds are bought and chefs then butcher the meat according to dish requirements. This means that bones (carcass) is available to be used for stocks and so on.

Cuts of chicken

The pieces of cut chicken are as follows (numbers refer to the diagram):

1 wing
2 breast
3 thigh
4 drumstick
5 winglet
6 carcass.

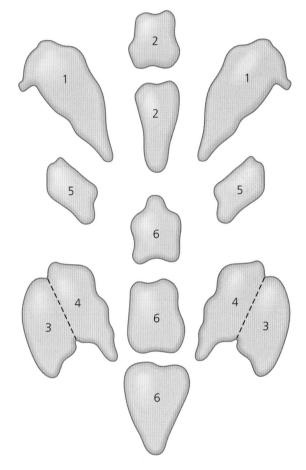

Cuts of chicken

1.3 How meat composition influences processes and preparation methods

Different cuts of poultry and joints may require separate process and preparation methods to maximise flavour and cooking techniques. As poultry is low in fat some dishes may need the addition of a layer of fat. **Barding** is laying a slice of fat or fatty bacon over or around the meat to help add moisture and flavour. Some breasts are split and a flavoured butter added and then sealed inside, as in Chicken Kiev. **Marinating** the darker or leg meat will add flavour, as well as helping to tenderise the flesh before cooking for a shorter period as in grilled chicken. Barbeque chicken is often marinated before cooking.

Escalopes may be tenderised by **batting** out with a meat mallet to help shape and gain an even thickness prior to cooking.

1.4 Using skills and techniques to prepare poultry

Step by step: trussing a chicken for roasting or boiling

1 Place the bird on its back.

2 To facilitate carving, remove the wishbone.

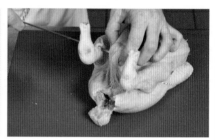

3 Insert a trussing needle through the bird, midway between the leg joints.

4 Thread the needle through the tendon and breast.

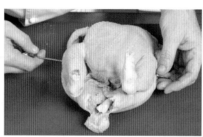

5 The needle comes out by the wing.

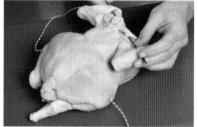

6 Turn over the chicken.

7 Run the needle under the skin through the wing, crown and other wing.

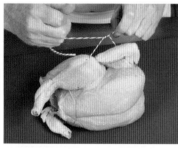

8 The string ends at the same place it started. Tie the ends of the string securely.

9 The trussed bird will hold its shape during roasting.

Step by step: cutting for sauté, fricassee and pies

1 Using a large, sharp knife, remove the wishbone.

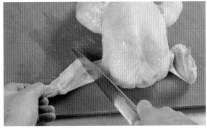

2 Remove the winglets and trim.

3 Remove the legs from the carcass, cutting around the oyster.

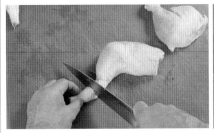

4 Cut off the feet.

5 Separate the thighs from the drumsticks.

6 Trim the drumsticks neatly.

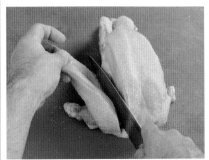

7 Remove each breast.

8 Separate the wings from the breasts and trim them.

9 Cut each breast in half.

10 Cut the cavity, splitting the carcass (this may be used for stock).

11 Chicken cut for sauté, on the bone, clockwise from top left: thighs, wings, breast pieces, drumsticks and winglets.

Step by step: preparation of a chicken for grilling (spatchcock)

1 Remove the wishbone.
2 Cut off the feet at the first joint.

3 Insert a large knife through the neck end and out of the vent (the gap at the other end, just under the parson's nose). Cut through the backbone and open out the bird.

4 Place the bird on its back, remove the back and rib bones.

5 Chicken spatchcock with the rib bones removed, ready for cooking. This process would also work well with poussin.

Step by step: preparation for suprêmes

A suprême is the wing and half the breast of a chicken with the trimmed wing bone attached; one chicken yields two suprêmes. Use a chicken weighing 1.25–1.5 kg.

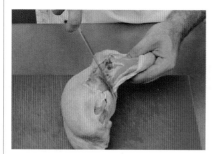

1 Cut off both the legs.
2 Remove the skin from the breasts.

3 Scrape the wing bone bare where it meets the breast and cut off the winglets near the joints, leaving 1.5–2 cm of bare bone attached to the breasts. Then remove the wishbone where it meets the breast and cut off the winglets near the joints, leaving 1.5–2 cm of bare bone attached to the breasts. Then remove the wishbone.

4 Cut the breasts close to the breastbone and follow the bone down to the wing joint.

5 Cut through the joint and pull off the suprêmes, using the knife to help.

6 Lift the fillets (the small, tender part that is slightly separate from the rest of the breast) from the suprêmes and remove the sinew from both. Make an incision lengthways, along the thick side of the suprêmes (do not cut all the way through); open this and place the fillets inside. Close, lightly flatten with a bat moistened with water, and trim if necessary.

Step by step: preparation for ballotines

A ballotine is a boned, stuffed leg of bird.

1 Using a small, sharp knife, cut the leg open around the bone

2 Start to separate the knuckle from the meat (scrape the flesh off the bone) of the drumstick towards the claw joint.

3 Lift the knuckle away (tunnel boning).

4 Cut off the drumstick bone, leaving approximately 2–3 cm at the claw joint end.

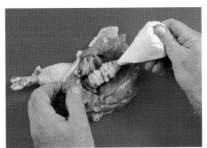

5 Fill the cavities in both the drumstick and thigh with a savoury stuffing.

6 Neaten the shape and secure with string using a trussing needle. Ballotines of chicken may be cooked and served using any of the recipes for sautéed chicken presented in this chapter.

Chicken mousse, mousselines and quenelles

Mousse, mousselines and quenelles are smooth light dishes that are easy to digest. They are made from a mixture known as forcemeat or farce using minced meat, which are then mixed with cream, egg white and seasoned. Used as a dish on their own, or as a garnish or as a stuffing for other dishes as in suprême or ballotine dishes.

Cutting cooked chicken (roasted or boiled)

1 Remove the legs and cut in two (drumstick and thigh).
2 Remove the wings.
3 Separate the breast from the carcass and divide in two.
4 Serve a drumstick with a wing and a thigh with a breast.

ACTIVITY

1 What is another name for suprême of chicken?
2 List the quality points for fresh chicken.
3 Name the food poisoning bacteria associated with fresh chicken.

Step by step: preparing a turkey for roasting

1 Remove the legs; cooking them separately will reduce the cooking time and enable the legs and breast to cook more evenly.

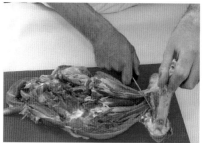

2 Remove the leg bones.

3 Stuff and roll each leg if required.

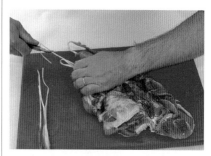

4 Tie the stuffed legs securely.

5 Cut the remainder of the bird in half, again to reduce the cooking time.

6 The two halves, ready for roasting, with one leg left whole and the other boned, stuffed and rolled.

Roasting duck

Preparation is the same as for chicken. The only difference is that these birds have a gizzard, which should not be split but should be trimmed off with a knife.

Roast birds can be served whole and carved at the table, or can be carved up before they are served, in which case:

- remove the legs and cut into two (drumstick and thigh)
- remove the wings and divide the breast into two
- serve a drumstick with a wing and a thigh piece with a piece of breast.

For more details on preparing poultry, see the recipe section at the end of this unit.

1.5 Interpret dish specifications

For more on interpreting dish specifications, see Section 2.5 Evaluation on page 308.

1.6 Storage

The correct storage of poultry is important for food safety. Uncooked poultry should be stored on trays to prevent dripping, in a separate fridge whenever possible. If a separate refrigerator is not available, store covered, labelled and dated and place in the bottom of a multi-use fridge away from other items.

- Chilled birds should be stored in a refrigerator between 1 °C and 4 °C.
- Oven-ready birds are eviscerated (gutted) and should be stored in a refrigerator.
- Frozen birds must be kept in a deep freeze until required, but must be completely thawed, preferably in a refrigerator, before being cooked. This procedure is essential to reduce the risk of food poisoning: chickens are potential carriers of campylobacter and salmonella and if birds are cooked from frozen there is a risk that the degree of heat required to kill off campylobacter and salmonella may not reach the centre of the bird.

When using frozen poultry, check that:

- the packaging is undamaged
- there are no signs of freezer burns, which are indicated by white patches on the skin.

Preservation of poultry

Poultry can be preserved or shelf life extended through the use of appropriate preservation methods.

- Chilling, vacuum-packing or freezing – place the poultry in a bag, use a vacuum to remove air and seal it. This can be completed for small joints or portions that can then be stored at the correct temperature.
- Smoking – small cuts or parts of poultry are normally smoked. This enhances the flavour and colour of the flesh. This process is often used in cold dishes, where a hot smoking process has been used once chilled.
- Canning – preservation of cooked poultry either as pieces or as a product (for example, goose liver or confit duck legs).

Further information on the advantages and disadvantages of preserving poultry can be found in Unit 301 Legal and social responsibilities in the professional kitchen.

Produce poultry dishes

2.1 Cooking requirements for poultry

When preparing uncooked poultry followed by cooked food, or changing from one type of meat or poultry to another, all equipment, working areas and utensils must be thoroughly cleaned or changed. Unhygienic equipment, utensils and preparation areas increase the risk of cross-contamination and danger to health. Wash all work surfaces with a bactericidal detergent to kill bacteria. This is particularly important when handling poultry. If colour-coded boards are used, it is essential to always use the correct boards for the preparation of raw or cooked poultry.

All poultry must be cooked until it reaches a core temperature of 75 °C, even if this is not stated in the recipe.

> **PROFESSIONAL TIP**
> Chicken is a delicate meat as most birds are under six months old when slaughtered. Chefs need to be aware of this when cooking, especially the breast, as a high heat will render the fibres tough and dry. The correct method of cookery is important to ensure the quality of the finished dish.

FAULTS

This chicken looks done on the outside, but when it is cut open, you can see it is not cooked through.

Video: Buying poultry
http://bit.ly/2qa7IKo

2.2 Effect of cooking methods on poultry

The application of heat to poultry cooks the meat and makes it tender and safe to eat.

The use of the correct cooking method ensures that the nutritional value of the dish is maintained and this is linked to the time and temperature taken to complete the process, as the proteins within poultry are cooked out at around 75 °C. It is important not to overcook poultry as the flesh then toughens or becomes stringy to eat and appears dried out in the final presentation of the dish.

2.3 Flavourings, coatings and stuffings

The addition of flavourings, coatings and stuffing are used to enhance dishes. These include regional recipes and traditional techniques and help to add moisture:

- Flavours – includes marinades, seasonings, herbs.
- Coatings – breadcrumbs, batters.
- Stuffing – herb butter, sage and onion.

For more on flavourings, coatings and stuffings, see the recipe section at the end of this unit.

2.4 Producing dishes to specification

Much depends upon the age and type of poultry and the cut being prepared. Young birds can be roasted, fried and grilled, whereas older birds and perhaps legs of larger birds tend to be pot roasted or braised, for example, confit duck legs or coq au vin.

The finishing or assembly of dishes is where the chef ensures any sauces and accompaniments are of the correct consistency, size and balance for the dish. For example, if grilled vegetables or croutons are part of the dish, the chef will ensure there are sufficient for each portion served. A sauce should be of the correct consistency for the dish. The accompaniments and sauces should enhance the overall dish in terms of texture, colour and flavour. Service temperature needs to be correct for the dish being served – hot, warm or cold as per the dish specification.

See the recipe section at the end of this unit for details of producing different types of poultry dishes.

2.5 Evaluation

Once a dish has been prepared, cooked and presented, the chef will have ensured that it meets the dish specification and therefore the customer need by an evaluative process. Some parts of this process happen during the cooking stage, for example when a bird is basted and monitored while cooking.

The overall evaluation involves ensuring that portion sizes are even, garnishes enhance the presentation of the dish and colour is appropriate, for instance breadcrumbs on a dish are golden brown with no black or burnt crumbs from a previous dish cooked in the pan. Taste in terms of seasoning and flavour is checked and the finished temperature is correct along with the consistency of any sauces served with the dish.

Faults that may be rectified before the dish is served could be about colour, appearance, cooking, temperature and aroma. One of the most common faults with cooking poultry is the undercooking of chicken – it must be returned to the kitchen if it is pink inside. The recommendation is to use a temperature probe to ensure thorough cooking.

TEST YOURSELF

1 List the **five** birds that are termed as poultry.
2 List **five** quality points for a good quality chicken.
3 Describe each of the following cuts:
   a) spatchcock chicken.
   b) suprême
   c) chicken cut for sauté.
4 State **three** considerations when roasting a chicken.
5 What is the safe internal temperature required for cooked poultry?
6 Describe how vacuum-packing is used to preserve poultry.
7 Describe **three** advantages of canning when used to preserve poultry.
8 Explain why shallow frying is an appropriate cooking method for turkey escalopes.
9 Briefly describe how to truss a turkey.
10 State why you would baste a whole duck when roasting.

1 Chicken sauté chasseur (*poulet sauté chasseur*)

Energy	Calories	Fat	Saturated fat	Carbohydrates	Sugar	Protein	Fibre	*
2430 kJ	579 kcal	45.8 g	20.7 g	2.1 g	1.6 g	37.6 g	1.5 g	

* Using butter

	4 portions	10 portions
Butter or oil	50 g	125 g
Salt, pepper		
Chicken, 1.25–1.5 kg, cut for sauté	1	2½
Shallots, chopped	10 g	25 g
Button mushrooms, washed and sliced	100 g	250 g
Dry white wine	3 tbsp	8 tbsp
Jus-lié, demi-glace or reduced brown stock	250 ml	625 ml
Tomato concassé	200 g	500 g
Parsley and tarragon, chopped		

PROFESSIONAL TIP

The leg meat takes longer to cook than the breast meat, which is why the drumsticks and thighs should be added first.

Add the soft herbs and tomatoes just before serving.

NOTE

Demi-glaze is a traditional brown sauce made with a flour-based roux, so gluten in the dish should be listed for allergen information.

Cooking

1 Place the butter or oil in a sauté pan on a fairly hot stove.

2 Season the pieces of chicken and place in the pan in the following order: drumsticks, thighs, wings and breast.

3 Cook to a golden brown on both sides.

4 Cover with a lid and cook on the stove or in the oven until tender. Dress neatly in a suitable dish.

5 Add the shallots to the sauté pan, rubbing them into the pan sediment to extract the flavour. Cover with a lid and cook on a gentle heat for 1–2 minutes.

6 Add the washed, sliced mushrooms and cover with a lid. Cook gently for 3–4 minutes, without colour. Drain off the fat.

7 Add the white wine and reduce by half. Add the jus-lié, demi-glace or reduced stock.

8 Add the tomatoes. Simmer for 5 minutes.

9 Correct the seasoning and pour over the chicken.

10 Sprinkle with chopped parsley and tarragon and serve.

11 Ballotines of chicken chasseur can be prepared as above or lightly braised (as shown).

Healthy Eating Tips

- Use a minimum amount of salt to season the chicken.
- The fat content can be reduced if the skin is removed from the chicken.
- Use a little unsaturated oil to cook the chicken, and drain off all excess fat from the cooked chicken.
- Serve with a large portion of new potatoes and seasonal vegetables.

Place the chicken into the hot pan.

Cook the shallots in the pan that was used for the chicken – they will pick up the sediment and flavour.

2 Fricassee of chicken (*fricassée de volaille*)

Energy	Calories	Fat	Saturated fat	Carbohydrates	Sugar	Protein	Fibre	*
2699 kJ	643 kcal	51.3 g	23.3 g	7.4 g	0.6 g	38.2 g	0.4 g	

* Using butter

	4 portions	10 portions
Chicken, 1.25–1.5 kg	1	2–3
Salt, pepper		
Butter, margarine or oil	50 g	125 g
Flour	35 g	100 g
Chicken stock	0.5 litre	1.75 litres
Egg yolks	1–2	5
Cream or non-dairy cream	4 tbsp	10 tbsp
Parsley, chopped		

Mise en place

1 Cut the chicken as for sauté (page 303); season with salt and pepper.

Cooking

1 Place the butter in a sauté pan. Heat gently.

2 Add the pieces of chicken. Cover with a lid.

3 Cook gently on both sides without colouring. Mix in the flour.

4 Cook out carefully without colouring. Gradually mix in the stock.

5 Bring to the boil and skim. Allow to simmer gently until cooked.

6 Mix the yolks and cream in a basin (liaison).

7 Pick out the chicken into a clean pan.

8 Pour a little boiling sauce on to the yolks and cream and mix well.

9 Pour all back into the sauce, combine thoroughly but do not reboil.

10 Correct the seasoning and pass through a fine strainer.

11 Pour over the chicken, reheat without boiling.

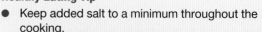

SERVING SUGGESTION

Serve sprinkled with chopped parsley and garnish with heart-shaped croutons, fried in butter, if desired.

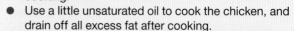

VARIATION

- **Fricassée de volaille à l'ancienne**: a fricassee of chicken with button onions and mushrooms can be made in a similar way, with the addition of 50–100 g button onions and 50–100 g button mushrooms. They are peeled and the mushrooms left whole, turned or quartered depending on size and quality. The onions are added to the chicken as soon as it comes to the boil and the mushrooms 15 minutes later. Heart-shaped croutons may be used to garnish. This is a classic dish.

PROFESSIONAL TIP

Sauté the chicken lightly. Add the liaison of yolks and cream carefully. Do not allow the sauce to come back to the boil once the liaison has been added, or it will curdle.

Healthy Eating Tip

- Keep added salt to a minimum throughout the cooking.
- Use a little unsaturated oil to cook the chicken, and drain off all excess fat after cooking.
- Try oven-baking the croutons brushed with olive oil.
- The sauce is high in fat, so serve with plenty of starchy carbohydrate and vegetables.

NOTE

Gluten-free flour mix can be used as an alternative in this recipe.

3 Chicken in red wine (coq au vin)

Energy	Calories	Fat	Saturated fat	Carbohydrates	Sugar	Protein	Fibre	*
4794 kJ	1141 kcal	95.7 g	32.9 g	16.6 g	2.3 g	49.0 g	1.7 g	

* Using sunflower oil and hard margarine

	4 portions	10 portions
Roasting chicken, 1.5 kg	1	2–3
Lardons	50 g	125 g
Small chipolatas	4	10
Button mushrooms	50 g	125 g
Butter	50 g	125 g
Sunflower oil	3 tbsp	7 tbsp
Small button onions	12	30
Red wine	500 ml	900 ml
Beurre manié		
Butter	25 g	60 g
Flour	10 g	25 g

Mise en place

1 Cut the chicken as for sauté (see page 303). Blanch the lardons.
2 If the chipolatas are large divide into two.
3 Wash and cut the mushrooms into quarters.

Cooking

1 Sauté the lardons, mushrooms and chipolatas in a mixture of butter and oil. Remove when cooked.
2 Lightly season the pieces of chicken and place in the pan in the correct order (see recipe 1) with button onions. Sauté until almost cooked.
3 Place in a casserole with the mushrooms and lardons.
4 Drain off the fat from the sauté pan. Deglaze with the red wine and stock; bring to the boil.
5 Transfer the liquid to the casserole (just covering the chicken); cover with a lid and finish cooking.
6 Remove the chicken and onions, place into a clean pan.
7 Lightly thicken the liquor with a beurre manié of butter and flour by whisking small pieces of it into the simmering liquid.
8 Pass the sauce over the chicken and onions, add the mushrooms, chipolatas and lardons. Correct the seasoning and reheat.
9 Add the beurre manié slowly, mixing well, to create a thick, smooth sauce.

Sauté the other ingredients before the chicken.

Sauté the chicken pieces and onion in the same pan.

Add a beurre manié.

4 Chicken spatchcock (*poulet grillé à la crapaudine*)

Energy	Calories	Fat	Saturated fat	Carbohydrates	Sugar	Protein	Fibre	*
1560 kJ	372 kcal	24.1 g	8.0 g	0.0 g	0.0 g	38.9 g	0.0 g	

* Using sunflower oil and hard margarine

	4 portions	10 portions
Chicken, 1.25–1.5 kg	1	2½

Mise en place

1 Cut horizontally from below the point of the breast over the top of the legs down to the wing joints, without removing the breasts. Fold back the breasts.

2 Snap and reverse the backbone into the opposite direction so that the point of the breast now extends forward.

3 Flatten slightly. Remove any small bones.

4 Skewer the wings and legs in position.

5 Season with salt and mill pepper.

6 Brush with oil or melted butter.

Cooking

1 Place on preheated grill bars or on a flat tray under a salamander.

2 Brush frequently with melted fat or oil during cooking and allow approximately 15–20 minutes on each side.

3 Test if cooked by piercing the drumstick with a needle or skewer – there should be no sign of blood.

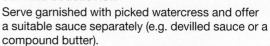

SERVING SUGGESTION
Serve garnished with picked watercress and offer a suitable sauce separately (e.g. devilled sauce or a compound butter).

HEALTH AND SAFETY
The use of a temperature probe to ensure the core temperature of over 75 °C is the surest method of checking a whole grilled chicken on the bone.

VARIATION
● This dish could be made using poussin. Depending upon the size this would be 1 or 2 portions from one bird.

Healthy Eating Tip
● Use a well-seasoned pan to dry-fry the lardons and chipolatas, then add the mushrooms.

5   Grilled chicken (*poulet grillé*)

Energy	Calories	Fat	Saturated fat	Carbohydrates	Sugar	Protein	Fibre	*
975 kJ	234 kcal	15.7 g	4.3 g	0.0 g	0.0 g	23.3 g	0.0 g	

* Based on chicken with bone, wing and leg quarters

Mise en place

1   Season the chicken with salt and mill pepper, and prepare as for grilling (see page 304).

Cooking

1   Brush with oil or melted butter or margarine, and place on preheated greased grill bars or on a barbecue or a flat baking tray under a salamander.

2   Brush frequently with melted fat during cooking; allow approximately 15–20 minutes each side.

3   Test if cooked by piercing the drumstick with a skewer or trussing needle; there should be no sign of blood issuing from the leg.

4   Serve garnished with picked watercress and offer a suitable sauce separately.

VARIATION

● Grilled chicken is frequently garnished with streaky bacon, tomatoes and mushrooms.

● The chicken may be marinated for 2–3 hours before grilling, in a mixture of oil, lemon juice, spices, herbs, freshly grated ginger, finely chopped garlic, salt and pepper. Chicken or turkey portions can also be grilled and marinated beforehand if wished (breasts or boned-out lightly battened thighs of chicken).

Healthy Eating Tips

● Use a minimum of salt and an unsaturated oil.

● Garnish with grilled tomatoes and mushrooms.

● Serve with Delmonico potatoes and green vegetables.

6 Chicken Kiev

Energy	Calories	Fat	Saturated fat	Carbohydrates	Sugar	Protein	Fibre
2094 kJ	500 kcal	26.1 g	14.4 g	24.4 g	0.9 g	43.4 g	1.0 g

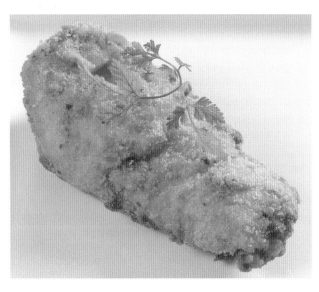

	4 portions	10 portions
Suprêmes of chicken	4 × 150 g	10 × 150 g
Butter	100 g	250 g
Seasoned flour	25 g	65 g
Eggs	2	5
Breadcrumbs	100 g	250 g

Mise en place

1 Make an incision along the thick sides of the suprêmes. Insert 25 g cold butter into each. Season.

2 Pass through seasoned flour, eggwash and crumbs, ensuring complete coverage. Eggwash and crumb twice if necessary.

> **VARIATION**
>
> Additional ingredients may be added to the butter before insertion:
> - chopped garlic and parsley
> - fine herbs such as tarragon or chives
> - liver pâté.

> **PROFESSIONAL TIP**
>
> The butter must be pushed well into the supreme. The incision must be sealed or the butter will leak out during cooking.

Cooking

1 Deep-fry until completely cooked. When the chicken is cooked, a probe in the thickest part will read 75 °C+ and the juices will run clear when the chicken is pierced. Drain and serve.

Carefully make an incision in the top of the suprême.

Stuff with softened butter.

Dip the chicken in egg and then coat with breadcrumbs.

7 Crumbed breast of chicken with asparagus (*suprême de volaille aux pointes d'asperges*)

Energy	Calories	Fat	Saturated fat	Carbohydrates	Sugar	Protein	Fibre
1831 kJ	439 kcal	26.4 g	8.9 g	15.7 g	1.5 g	35.5 g	1.3 g

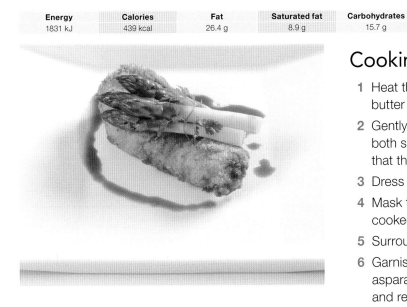

Cooking

1 Heat the oil and 50 g (125 g for 10 portions) of the butter or margarine in a sauté pan.

2 Gently fry the suprêmes to a golden brown on both sides (6–8 minutes). Use a probe to check that the centre has reached 75 °C.

3 Dress the suprêmes on a flat dish and keep warm.

4 Mask the suprêmes with the remaining butter cooked to the nut-brown stage.

5 Surround the suprêmes with a cordon of jus-lié.

6 Garnish each suprême with a neat bundle of asparagus points (previously cooked, refreshed and reheated with a little butter).

	4 portions	10 portions
Suprêmes of chicken (page 304)	4 × 125 g	10 × 125 g
Egg	1	2
Breadcrumbs (white or wholemeal)	50 g	125 g
Oil	50 g	125 g
Butter or margarine	100 g	250 g
Jus-lié	60 ml	150 ml
Asparagus	200 g	500 g

Healthy Eating Tips

● Use a minimum amount of salt.
● Remove the skin from the suprêmes and fry in a little unsaturated vegetable oil. Drain on kitchen paper.
● Try omitting the additional cooked butter.
● Serve with plenty of boiled new potatoes and vegetables.

Mise en place

1 Pané the chicken suprêmes. Shake off all surplus crumbs.

2 Neaten and mark on one side with a palette knife.

8 Deep-fried chicken

Energy	Calories	Fat	Saturated fat	Carbohydrates	Sugar	Protein	Fibre
1754 kJ	421 kcal	28.6 g	6.1 g	14.5 g	0.4 g	27.2 g	0.5 g

Mise en place

1 Cut the chicken as for sauté. It is advisable to remove the bones from chicken that will be deep-fried.

2 Coat with flour, egg and crumbs (pané), or pass them through a light batter (see page 357) to which herbs can be added.

HEALTH AND SAFETY

When deep-frying meat always place in fat in a motion away from your body; try not to drop as this may splash and cause burns.

VARIATION
To create a traditional southern fried chicken dish, add the following to the crumb mix:
- 1 level teaspoon cayenne pepper
- 1 level teaspoon hot smoked paprika
- 1 level teaspoon onion powder
- 1 level teaspoon garlic powder.

Cooking

1 Deep-fry in hot fat (approx. 170–180 °C) until golden brown and cooked through – about 5 minutes. When the chicken is cooked, a probe in the thickest part will read 75 °C+, and the juices will run clear when the chicken is pierced.

2 For suprêmes, make an incision, stuff with a compound butter, flour, egg and crumb, and deep-fry as in Chicken Kiev (recipe 7).

Healthy Eating Tip
- The fat content can be reduced if the skin is removed from the chicken.

9 Steamed ballotine of chicken with herb stuffing and red wine jus

Energy	Calories	Fat	Saturated fat	Carbohydrates	Sugar	Protein	Fibre	Sodium
1626 kJ	390 kcal	24.9 g	8.5 g	1.8 g	0.9 g	33.4 g	0.5 g	1.1 g

	4 portions	10 portions
Chicken legs	4	10
Forcemeat		
Minced chicken	120 g	300 g
Parsley	1 tsp	2½ tsp
Tarragon	1 tsp	2½ tsp
Chives	1 tsp	2½ tsp
Chervil	1 tsp	2½ tsp
Salt and pepper		
Red wine jus		
Shallots, sliced	75 g	190 g
Butter	25 g	60 g
Garlic, finely chopped	5 g	10 g
Red wine	175 ml	440 ml
Chicken stock	175 ml	440 ml
Veal or beef jus	125 ml	310 ml
Bay leaves	1	2½
Thyme, sprigs	½	1

Mise en place

1 Remove the thighbone and the knuckle joint from the chicken legs.

2 Prepare the forcemeat by mixing all of the herbs with the minced chicken. Season the mix and divide it evenly between the chicken legs.

3 Season the legs then stuff them with the forcemeat. Roll the leg into a neat cylinder and either tie the leg in place or roll tightly in cling film and tie off the ends.

Cooking

1 Steam the legs for 30–45 minutes or until the core temperature reaches 84 °C.

2 Allow the chicken legs to rest so they will stay in shape.

3 When the chicken legs have rested and cooled slightly, remove the string or cling film and dry them.

4 Heat a sauté pan with oil and butter, add the chicken legs and brown the skin. Slice and serve with the sauce.

For the red wine jus:

1 Caramelise the shallots in the butter until golden, adding the garlic at the end.

2 Strain off any excess butter, deglaze the pan with the red wine and reduce by half.

3 Add the chicken stock and veal jus and reduce to a sauce consistency, adding the aromats for the last 5 minutes.

4 Pass through a muslin cloth and serve over the sliced ballotines.

> **VARIATION**
>
> Instead of the traditional chicken legs, ballotines can be made using:
> - chicken suprêmes, batted out, stuffed and rolled
> - chicken breasts: make an incision in the thicker part and fill this with forcemeat.
>
> These ballotines are cooked in the same way.

10 Tandoori chicken

Energy	Calories	Fat	Saturated fat	Carbohydrates	Sugar	Protein	Fibre	*
1436 kJ	342 kcal	14.1 g	4.6 g	10.1 g	8.6 g	44.6 g	0.3 g	

* Estimated edible meat used; vegetable oil used

	4 portions	10 portions
Chicken, cut as for sauté (page 303)	1.25–1.5 kg	3–4 kg
Salt	1 tsp	2½ tsp
Lemon, juice of	1	2½
Natural yoghurt	300 ml	800 ml
Small onion, chopped	1	3
Clove of garlic, peeled	1	3
Ginger, piece of, peeled and quartered	5 cm	12 cm
Fresh hot green chilli, sliced	½	1
Garam masala	2 tsp	5 tsp
Ground cumin	1 tsp	2½ tsp
Red and yellow colouring, few drops each		

Mise en place

1 Cut slits bone-deep in the chicken pieces.
2 Sprinkle the salt and lemon juice on both sides of the pieces, lightly rubbing into the slits; leave for 20 minutes.

3 Combine the remaining ingredients in a blender or food processor.
4 Brush the chicken pieces on both sides, ensuring the marinade goes into the slits. Cover and refrigerate for 6–24 hours.

Cooking

1 Preheat the oven to the maximum temperature.
2 Shake off as much of the marinade as possible from the chicken pieces; place on skewers and bake for 15–20 minutes or until cooked.

SERVING SUGGESTION
Serve with red onion rings and lime or lemon wedges.

PROFESSIONAL TIP
If cooking in a tandoori oven, make sure the chicken is secure on the skewer, so that it cannot slip off during cooking.

Healthy Eating Tips
- Skin the chicken and reduce the salt by half.
- Serve with rice and vegetables.

HEALTH AND SAFETY
Check ingredients of gram masala to ensure that no allergens have been included.

11 Chicken tikka

Energy	Calories	Fat	Saturated fat	Carbohydrates	Sugar	Protein	Fibre	*
1780 kJ	427 kcal	27.1 g	5.2 g	5.5 g	5.1 g	41.3 g	0.6 g	

* Estimated edible meat used

Mise en place

1 Place the chicken pieces into a suitable dish.

2 Mix together the yoghurt, seasoning, spices, garlic, lemon juice and tomato purée.

3 Pour this over the chicken, mix well and leave to marinate for at least 3 hours.

Cooking

1 In a suitable shallow tray, add the chopped onion and half the oil.

2 Lay the chicken pieces on top and grill under the salamander, turning the pieces over once or gently cook in a moderate oven at 180 °C for 20–30 minutes.

3 Baste with the remaining oil.

SERVING SUGGESTION
Serve on a bed of lettuce garnished with wedges of lemon.

Healthy Eating Tips
- Skin the chicken and keep the added salt to a minimum.
- Use half the amount of unsaturated oil.
- Serve with rice and a vegetable dish.

PROFESSIONAL TIP
Baste the chicken during grilling so that it does not become too dry.

	4 portions	10 portions
Chicken, cut for sauté	1 × 1.5 kg	2.5 × 1.5 kg
Natural yoghurt	125 ml	250 ml
Grated ginger	1 tsp	2 ½ tsp
Ground coriander	1 tsp	2 ½ tsp
Ground cumin	1 tsp	2 ½ tsp
Chilli powder	1 tsp	2 ½ tsp
Clove garlic, crushed and chopped	1	2–3
Lemon, juice of	½	1
Tomato puree	50 g	125 g
Onion, finely chopped	50 g	125 g
Oil	60 ml	150 ml
Lemon, wedges of	4	10
Seasoning		

Mix together the spices.

Mix the chicken pieces into the marinade.

Griddle the chicken in a shallow tray.

12 Terrine of chicken and vegetables

Energy	Calories	Fat	Saturated fat	Carbohydrates	Sugar	Protein	Fibre
930 kJ	226 kcal	17.3 g	9.4 g	2.0 g	1.8 g	15.5 g	0.9 g

	8–10 portions
Carrots, turnips and swedes, peeled and cut into 7-mm dice	50 g of each
Broccoli, small florets	50 g
Baby corn, cut into 7-mm rounds	50 g
French beans, cut into 7-mm lengths	50 g
Chicken (white meat only), minced	400 g
Egg whites	2
Double cream	200 ml
Salt, mill pepper	

Mise en place

1 Blanch all the vegetables individually in boiling salted water, ensuring that they remain firm. Refresh in cold water, and drain well.

2 Blend the chicken and egg whites in a food processor until smooth. Turn out into a large mixing bowl and gradually beat in the double cream. Don't over mix in the food processor as it could start the cooking process.

3 Season with salt and mill pepper and fold in the vegetables.

4 Line a lightly greased 1-litre terrine with cling film.

Cooking

1 Spoon the farce into the mould and overlap the cling film.

2 Cover with foil, put the lid on and cook in a bain-marie in a moderate oven for about 45 minutes. Use a temperature probe to check that the centre has reached 70 °C.

3 When cooked, remove the lid and leave to cool overnight.

Healthy Eating Tips
- Keep the added salt to a minimum.
- Serve with plenty of salad vegetables and bread or toast (optional butter or spread).

VARIATION
- Asparagus in whole spears could be included as vegetables in the terrine

FAULTS
If the terrine falls apart when sliced, it could be that the mixture was too wet when prepared or water entered the dish during cooking. Do not allow the bain marie to boil or splash over the top of the terrine.

Cut the vegetables into neat dice, rounds and florets.

Line the terrine with cling film and spoon in the mixture.

Cover the mixture with cling film and press down gently.

13 Roast turkey

Energy	Calories	Fat	Saturated fat	Carbohydrates	Sugar	Protein	Fibre
1076 kJ	257 kcal	9.6g	3.2g	0.0g	0.0g	42.0g	0.0g

	4 portions	10 portions
Turkey, with legs on	1 small hen	4–5 kg
Sea salt and freshly ground black pepper		
Unsalted butter, melted	250 g	625 g

Mise en place

1 Adjust an oven rack to its lowest position and remove the other racks in the oven. Pre-heat to 165 °C.

2 Remove turkey parts from neck and breast cavities and reserve for other uses, if desired. Dry bird well with paper towels, inside and out. Salt and pepper inside the breast cavity.

Cooking

1 Set the bird on a roasting rack in a roasting pan, breast-side up, brush generously with half the butter and season with salt and pepper. Tent the bird with foil.

2 Roast the turkey for 2 hours. Remove the foil and baste with the remaining butter. Increase the oven temperature to 220 °C and continue to roast until an instant-read thermometer registers 74 °C in the thigh of the bird, about 45 minutes more.

3 Remove turkey from the oven and set aside to rest for 15 minutes before carving. Carve and serve with roast gravy, cranberry sauce, bread sauce, and either or both sausage meat and chestnut dressing and parsley and thyme dressing. Chipolata sausages and rolled rashers of grilled bacon may also be served.

PROFESSIONAL TIP
The secret to keeping turkey moist is to baste as much as you can and, when the turkey is cooked, place it on its breast, breast-side down, allowing all the cavity juices to penetrate the meat.

VARIATION
● Use oil or fat to roast the turkey rather than the butter.

14 Turkey escalopes

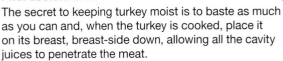

Energy	Calories	Fat	Saturated fat	Carbohydrates	Sugar	Protein	Fibre
1712 kJ	414 kcal	37.0 g	14.0 g	5.9 g	0.2 g	14.8 g	0.3 g

Cooking

1 Cut 100-g slices from boned-out turkey breast and lightly flour. Gently cook on both sides in butter or oil with a minimum of colour; alternatively flour, egg and crumb the slices and shallow fry.

2 Serve with a suitable sauce and/or garnish (e.g. pan-fried turkey escalope cooked with oyster mushrooms and finished with white wine and cream).

PROFESSIONAL TIP
The oil or fat must be hot enough before the escalopes are placed in the pan. If it is too cool, the breadcrumbs will absorb the fat and the dish will be greasy.

15 Roast duck or duckling (*canard ou caneton rôti*)

Energy	Calories	Fat	Saturated fat	Carbohydrates	Sugar	Protein	Fibre	*
3083 kJ	734 kcal	60.5 g	16.9 g	8.2 g	7.8 g	40.0 g	1.4g	

* With apple sauce and watercress

	4 portions	10 portions
Duck	1	2–3
Salt		
Oil		
Brown stock	0.25 litres	600 ml
Salt, pepper		
Watercress, bunch	1	2
Apple sauce (page 109)	125 ml	300 ml

Mise en place

1 Lightly season the duck inside and out with salt.
2 Truss and brush lightly with oil.

Cooking

1 Place duck on its side in a roasting tin, with a few drops of water.
2 Place in a hot oven for 20–25 minutes.
3 Turn on to the other side.
4 Cook for a further 20–25 minutes. Baste frequently.

5 To test if cooked, pierce with a fork between the drumstick and thigh and hold over a plate. The juice issuing from the duck should not show any signs of blood. If using a probe, the temperature should be 62 °C. If the duck is required pink, the temperature should be 57 °C.

6 Prepare the roast gravy with the stock and the sediment in the roasting tray. Correct the seasoning, remove the surface fat.

SERVING SUGGESTION

Serve garnished with picked watercress. Accompany with a sauce boat of hot apple sauce, a sauceboat of gravy, and game chips. Also serve a sauceboat of sage and onion dressing (recipe 17).

NOTE

Arrange the duck to cook sitting on one leg, then the other leg and then the breast, so the whole bird cooks evenly.

The temperatures in this recipe reflect industry standards for cooking duck. An environmental health officer may advise higher temperatures.

Healthy Eating Tips

- Use the minimum amount of salt to season the duck and the roast gravy.
- Take care to remove all the fat from the roasting tray before making the gravy.
- This dish is high in fat and should be served with plenty of boiled new potatoes and a variety of vegetables.

Video: Preparing a duck
http://bit.ly/2oMFljw

16 Sage and onion dressing for duck

Cooking

1 Gently cook the onion in the fat without colour. Add the chopped liver (if required) and fry until cooked.

2 Add the herbs and seasoning. Mix in the crumbs. Form into thick sausage shapes, in foil.

3 Place in a tray and finish in a hot oven at 180 °C for approximately 5–10 minutes. Check with a probe that the centre has reached 75 °C.

SERVING SUGGESTION
Serve separately with roast duck or with other poultry.

	4 portions	10 portions
Onion, chopped	100 g	250 g
Duck fat or butter	100 g	250 g
Powdered sage	¼ tsp	½ tsp
Parsley, chopped	¼ tsp	½ tsp
Salt, pepper		
White or wholemeal breadcrumbs	100 g	250 g
Duck liver (optional), chopped	50 g	125 g

17 Duckling with orange sauce (*caneton bigarade*)

Energy	Calories	Fat	Saturated fat	Carbohydrates	Sugar	Protein	Fibre	*
3125 kJ	744 kcal	60.1 g	17.1 g	11.8 g	9.3 g	39.9 g	0.1 g	

* Using butter

	4 portions	10 portions
Duckling, 2 kg	1	2–3
Butter	50 g	125 g
Carrots	50 g	125 g
Onions	50 g	125 g
Celery	25 g	60 g
Bay leaf	1	2–3
Small sprig thyme	1	2–3
Salt, pepper		
Brown stock	250 ml	625 ml
Arrowroot	10 g	25 g
Oranges	2	5
Lemon	1	2
Vinegar	2 tbsp	5 tbsp
Sugar	25 g	60 g

Mise en place

1 Clean and truss the duck. Use a fifth of the butter to grease a deep pan. Add the mirepoix (vegetables and herbs).

2 Season the duck. Place the duck on the mirepoix.

3 Coat the duck with the remaining butter.

Cooking

1 Cover the pan with a tight-fitting lid. Place the pan in the oven at 200–230 °C.

2 Baste occasionally; cook for approximately 1 hour.

3 Remove the lid and continue cooking the duck, basting frequently until tender (about a further 30 minutes).

4 Remove the duck, cut out the string and keep the duck in a warm place. Drain off all the fat from the pan.

5 Deglaze with the stock, bring to the boil and allow to simmer for a few minutes. Thicken by adding the arrowroot diluted in a little cold water.

6 Reboil, correct the seasoning, degrease and pass through a fine strainer.

7 Thinly remove the zest from half the oranges and the lemon(s), and cut into fine julienne.

8 Blanch the julienne of zest for 3–4 minutes, then refresh.

9 Place the vinegar and sugar in a small sauteuse and cook to a light caramel stage.

10 Add the juice of the oranges and lemon(s).

11 Add the sauce and bring to the boil.

12 Correct the seasoning and pass through a fine strainer.

13 Add the julienne to the sauce; keep warm.

14 Remove the legs from the duck, bone out and cut into thin slices.

15 Carve the duck breasts into thin slices and dress neatly.

16 Coat with the sauce and serve.

An alternative method of service is to cut the duck into eight pieces, which may then be either left on the bone or the bones removed.

PROFESSIONAL TIP

Baste the duck during cooking; the butter will give it flavour.

Healthy Eating Tips

● Use the minimum amount of salt to season the duck and the final sauce.
● Take care to remove all the fat from the roasting tray before deglazing with the stock.
● Reduce the fat by removing the skin from the duck and 'balance' this fatty dish with a large portion of boiled potatoes and vegetables.

VARIATION

● Use duck breasts to enable the chef to manage costs and aid portion control. Score skin, pan fry breast to seal, then finish in the oven with a lid on the pan for 6–8 minutes. Allow to rest for 8–10 minutes in a warm place. Serve sliced with the sauce and garnish with watercress.

18 Confit duck leg with red cabbage and green beans

Energy	Calories	Fat	Saturated fat	Carbohydrates	Sugar	Protein	Fibre
3859 kJ	9332 kcal	83 g	28 g	7.7 g	6.4 g	39.2 g	4.2 g

	4 portions	10 portions
Confit oil	1 litre	2.5 litres
Garlic cloves	4	10
Bay leaf	1	3
Sprig of thyme	1	2
Duck legs	4 × 200g	10 × 200g
Butter	50g	125g
Green beans, cooked and trimmed	300g	750g
Braised red cabbage	250g	625g
Seasoning		

Cooking

1 Gently heat the confit oil, add the garlic, bay leaf and thyme.

2 Put the duck legs in the oil and place on a medium to low heat, ensuring the legs are covered.

3 Cook gently for 4–4½ hours (if using goose, 5–6½ hours may be needed).

4 To test if the legs are cooked, squeeze the flesh on the thigh bone and it should just fall away.

5 When cooked, remove the legs carefully and place on a draining tray.

6 When drained, put the confit leg on a baking tray and place in a pre-heated oven at 210 °C; remove when the skin is golden brown (approximately 9–10 minutes), taking care as the meat is delicate.

7 Heat the butter in a medium sauté pan and reheat the green beans.

8 Place the braised cabbage in a small pan and reheat slowly.

9 Place the duck leg in a serving dish or plate along with the red cabbage and green beans.

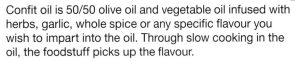

PROFESSIONAL TIPS

Confit oil is 50/50 olive oil and vegetable oil infused with herbs, garlic, whole spice or any specific flavour you wish to impart into the oil. Through slow cooking in the oil, the foodstuff picks up the flavour.

Confit duck legs can be prepared up to three or four days in advance. Remove them carefully from the fat they are stored in, clean off any excess fat and place directly into the oven. This is a great timesaver in a busy service.

Healthy Eating Tip

● Confit is a traditional dish that is cooked in duck fat. This recipe has been developed to highlight how reducing the use of saturated fat by replacing it with oil can still produce a dish of quality.

19 Suprêmes of guinea fowl with a pepper and basil coulis

Energy	Calories	Fat	Saturated fat	Carbohydrates	Sugar	Protein	Fibre	*
904 kJ	216 kcal	4.7 g	1.3 g	7.8 g	7.3 g	35.8 g	0.5 g	

* Using chicken instead of guinea fowl

	4 portions	10 portions
Red peppers	3	7
Olive oil	150 ml	375 ml
Fresh basil, chopped	2 tbsp	5 tbsp
Salt, pepper		
Guinea fowl suprêmes (approximately 150 g each)	4	10

Cooking

1 Skin the peppers by brushing with oil and gently scorching in the oven or under the grill. Alternatively, use a blowtorch with great care. Once scorched, peel the skin from the peppers, cut in half and deseed.

2 Place the skinned and deseeded peppers in a food processor, blend with the olive oil and pass through a strainer.

3 Add the chopped basil and season.

4 Season the guinea fowl and either shallow-fry or grill.

5 Pour the coulis onto individual plates. Place the guinea fowl on top and serve immediately.

20 Roast goose

Energy	Calories	Fat	Saturated fat	Carbohydrates	Sugar	Protein	Fibre
9541 kJ	2305 kcal	210.5 g	60.0 g	0.0 g	0.0 g	103.1 g	0.0 g

The average weight of a goose is 5–6 kg.

Mise en place

1 Clean and truss the goose as for a chicken (see page 302).

Cooking

1 Roast the goose using the same procedure for roasting a duck (recipe 16), with the oven at 200–230 °C. Turn oven temperature down to 180 °C after 20 minutes. Allow 15–20 minutes per 0.5 kg.

307 Fish and shellfish

Learning outcomes

In this unit you will be able to:

1 Prepare fish and shellfish dishes and products, including:
   - knowing different types of fish and shellfish
   - knowing quality points and understanding how they affect preparation and cooking methods
   - knowing the different types of folds and products that can be produced from whole fish and which species would be suitable to meet dish requirements
   - knowing how to use preservation methods on different fish and shellfish, including picking, chilling, freezing (conventional and blast)
   - understanding the effects of preservation methods
   - being able to use preparation techniques on fish and shellfish
   - understanding the correct storage procedures to use throughout production and on completion of the final products.

2 Produce fish and shellfish dishes and products, including:
   - understand how to produce different types of sauces and dressings
   - understand how sauces and dressings are used to enhance fish and shellfish dishes
   - being able to use cooking techniques on different types of fish and shellfish
   - being able to cook dishes to the required specification
   - being able to measure and evaluate against quality standards throughout preparation and cooking
   - being able to evaluate products against dish requirements and production standards and recognise any faults.

Recipes included in this chapter

No.	Recipe
Fish	
1	Breadcrumbed deep-fried fish fillets
2	Goujons of plaice
3	Fish meunière
4	Délice of flat white fish Dugléré
5	Fillets of fish in white wine sauce
6	Grilled fillets of sole, plaice or haddock
7	Frying batters for fish
8	Deep-fried fish in batter
9	Whitebait
10	Pan-fried fillets of sea bass with rosemary mash and mushrooms
11	Poached smoked haddock
12	Steamed fish with garlic, spring onions and ginger
13	Griddled monkfish with leeks and Parmesan
14	Grilled round fish
15	Poached salmon
16	Grilled swordfish and somen noodle salad with coriander vinaigrette
17	Roast fillet of sea bass with vanilla and fennel
18	Baked cod with a herb crust

No.	Recipe
19	Red mullet ceviche with organic leaves
20	Sardines with tapenade
21	Fish kedgeree
22	Fish pie
23	Soused herring or mackerel
24	Haddock and smoked salmon terrine
25	Salmon fishcakes
26	Baked salmon salad with sea vegetables
27	Fish en papillote
Shellfish	
28	Scallops with caramelised cauliflower
29	Oysters in their own shells
30	Oyster tempura
31	Crab cakes with rocket salad and lemon dressing
32	Mussels in white wine sauce
33	Scallops and bacon
34	Prawns with chilli and garlic
35	Seafood stir-fry
36	Seafood in puff pastry
37	Sauté squid with white wine, garlic and chilli

Prepare fish dishes and products

Introduction

Marine and freshwater fish have been a crucial part of the human diet worldwide for many centuries. Fish is a good source of essential protein, minerals and vitamins and easier to digest than meat.

There are more than 20,000 species of fish in the world's seas, yet we use only a fraction of these. This may be because certain types are neither edible nor ethical. Using certain types of fish might be considered wrong if, for example, there was a danger of the species becoming extinct or greatly reduced in numbers.

More people are choosing to eat fish in preference to meat; for some it is for health reasons and others a matter of choice, so fish consumption is steadily increasing. This popularity has resulted in a far greater selection becoming available and, because of swift and efficient refrigerated transport, well over 200 types of fish are on sale throughout the year. However, demand has also led to overfishing, causing a steep decline in the stocks of some species. To meet increasing demand, some species of fish such as trout, salmon, cod, sea bass and turbot are reared in fish farms to supplement natural resources.

> **HEALTH AND SAFETY**
> Fish and shellfish are highly perishable items and, if not stored and cooked properly, could cause food poisoning. Fish and shellfish also appear on the list of 14 allergens for which customers must be provided with full information.

Those preparing fish and shellfish must also be aware of relevant health and safety issues and legislation:

- Food Safety Act 1990 and Food Hygiene Regulations 2006
- Food Information Regulations 2014
- Health and Safety at Work Act (1974) (updated in 1994)
- Control of Substances Hazardous to Health (COSHH) 2002
- Management of Health and Safety at Work Regulations 1999
- Manual Handling Operations Regulations 1992
- Personal Protective Equipment at work Regulations 1992

For more details on these regulations see Unit 301.

Types of fish

Fish are vertebrates (animals with a backbone) and are split into two main groups: flat and round.

White fish can be **round** (for example, cod, whiting and hake) or **flat** (for example, plaice, sole and turbot). White fish are categorised as **demersal** fish and live at or near the bottom of the sea.

> **Healthy Eating Tip**
> - The flesh of white fish does not contain any fat. Vitamins A and D are only present in the liver and used in cod liver or halibut liver oil.

White, round fish: cod

White, flat fish: turbot

Oily fish are round in shape (examples include herring, mackerel, salmon, tuna and sardines). These are categorised as **pelagic** fish and swim in mid-depth water.

Oily fish: salmon

Healthy Eating Tip
- Oily fish contain fat-soluble vitamins A and D in their flesh and omega-3 fatty acids (the unsaturated fatty acids that are essential for good health). It is recommended that we eat more oily fish. However, owing to its fat content, oily fish is not as easily digestible as white fish.

KEY TERMS

Flat fish: have a flatter profile and always have white flesh because the oils are stored in the liver. They include sole, plaice, dabs, turbot, brill, bream, flounder and halibut.

Round fish: can vary greatly in size from small sardines to very large tuna. They can either have white flesh, such as bass, grouper, mullet, haddock and cod, or darker, oily flesh such as tuna, mackerel, herring, trout and salmon.

Oily fish: are always round and, because the fish oils are dispersed through the flesh (rather than stored in the liver as in white fish), the flesh is darker. These include mackerel, salmon, sardines, trout, herring and tuna.

Demersal fish: live in deep water and feed from the bottom of the sea; they are almost always white fleshed fish and can be round or flat.

Pelagic fish: live in more shallow or mid-depth waters and are usually round, oily fish such as mackerel, herring and sardines.

Seasonality and availability of fish

The quality of fish can vary due to climatic and environmental conditions. Generally, all fish spawn (release their eggs) over a period of four to six weeks. During spawning, they use up a lot of their reserves of fat and protein in the production of eggs. This has the effect of making their flesh watery and soft. Fish in this condition are termed 'spent fish'. This takes anything between one and two months, depending on local environmental conditions.

Weather conditions also have an enormous effect on fishing activities. The full range of species may not always be available during stormy weather, for instance. Table 7.1 shows when different types of fish are available.

Table 7.1 Seasonality of fish

	Jan	Feb	Mar	Apr	May	Jun	Jul	Aug	Sep	Oct	Nov	Dec
Bream					*	*						
Brill			*	*	*							
Cod												
Eel												
Mullet (grey)			*	*								
Gurnard												
Haddock												
Hake			*	*	*							
Halibut				*	*							
Herring												
John Dory												
Mackerel												
Monk fish												
Plaice			*	*								
Red mullet												
Salmon (farmed)												
Salmon (wild)												

	Jan	Feb	Mar	Apr	May	Jun	Jul	Aug	Sep	Oct	Nov	Dec
Sardines												
Sea bass				*	*							
Sea trout								*	*	*	*	
Skate												
Squid												
Sole (Dover)	*											
Sole (lemon)												
Trout												
Tuna												
Turbot				*	*	*						
Whiting												

Key

Available	At best

1.1 Quality points

Whole fish

All fish should be fresh and undamaged with no unpleasant smell. Whole fresh fish must have:

- clear, bright eyes, which are not sunken
- bright red gills
- no missing scales; scales should also be firmly attached to the skin
- moist skin with a fresh sea slime (fresh fish feels slightly slippery)
- shiny skin with bright, natural colouring
- a stiff tail and fins
- firm textured flesh
- a fresh sea smell and no odour of ammonia.

Fillets

Fillets should be:

- neatly cut and trimmed with firm flesh
- neatly packed, close together
- a translucent white colour if from a white fish, with no discolouration.

Smoked fish

Smoked fish should have:

- a bright, glossy surface
- firm flesh (sticky or soggy flesh means the fish may have been of poor quality or undersmoked)
- a pleasant, smoky smell

Frozen fish

Frozen fish should:

- be frozen hard with no signs of thawing (defrosting)
- be in packaging that is not damaged

* Spawning and roeing – this can deprive the flesh of nutrients and will decrease quality and yield.

- show no evidence of freezer burn; this shows as dull, white, dry patches.

1.2 Folds, cuts and products

While some fish are used whole, for example, trout, sea bass, sardines and bream, many are cut or prepared in some way:

- Darnes – thick slices through the bone of round fish (for example, salmon and cod).
- Tronçons – thick slices of large flat fish through the bone (for example, turbot and halibut). These are likely to be large, so tronçons may be further cut in half or quarters.
- Fillets – cuts that remove the flesh from the backbone and rib bones, leaving fish fillets with no bones. Fillets from a round fish are taken either side of the backbone so there will be two long fillets. With a flat fish, two fillets are taken from either side of the backbone on both sides of the fish, which gives four fillets.
- Suprêmes – prime cuts without bone or skin and cut at an angle from large fillets of fish such as salmon, turbot or halibut.
- Goujons – prepared from skinless filleted fish such as sole or plaice, cut into strips approximately 8 × 0.5 centimetres.
- Délice – a neatly folded, skinless, boneless fillet from a flat fish.
- Paupiettes – fillets of fish such as sole, plaice or whiting, often spread with a suitable stuffing or mousseline and rolled before cooking.
- En tresse – neatly plaited strips of fish, for example, sole fillets cut into three even pieces lengthwise to within one centimetre of the top, and neatly plaited.
- Mousseline – a light mousse made with blended fish and the addition of cream or egg whites. It may be used as a mousse in terrines or as a stuffing for paupiettes.

Fish is now frequently ordered from the fishmonger in portion sizes of specified weight and by number (see below). When preparing portions or cuts of fish in the kitchen it is important to consider neat preparation, even size and weight of cuts wherever possible. Store prepared cuts on a clean tray, cover with cling film label, date, refrigerate and use as soon as possible.

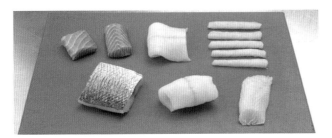

Cuts of fish

Portion sizes and weights

Fresh fish can be bought by the kilogram, by the number of fillets or whole fish of the weight that is required. For example, 30 kilograms of salmon could be ordered as 2 × 15 kilograms, 3 × 10 kilograms or 6 × 5 kilograms, the number of whole fish, fillets, suprêmes or darnes. Frozen fish can be purchased whole (gutted), filleted, cut or prepared in a wide variety of different ways.

There are a number of ways that can be used to achieve good portion control with fish; for example:
- the specific size and weight of a whole fish such as trout, sea bass or Dover sole served as a portion
- specified cut sizes of darnes, tronçons and suprêmes used as a portion
- number of fillets from small fish, for example, two fillets from a plaice as a portion
- number of shellfish items such as six king prawns, four scallops or half a lobster making a portion
- number and size of paupiettes or délice offered as a portion
- dividing a fish pie into equal-sized portions
- filling an individual pie dish, serving dish or pastry case
- using a ladle to measure the portion size of a fish stew or soup.

Unless otherwise stated, as a guide allow 100 grams fish off the bone and approximately 300 grams on the bone for a portion.

Products

Increasing numbers of fish products become available each year. Many are intended for fast food and convenience markets but 'high-end' quality products are available too. Fish products are often pre-portioned and sometimes pre-cooked, but many do need some simple cooking or finishing. Fish products are available coated, marinated, in a sauce, as pies, on skewers or as canapes, and a range of products are chilled and ready to serve. Some fish products are sold as individual units and others sold in multi-portion packs.

1.3 Preservation methods

Fish deteriorates quickly after it has been caught. Over many years ways have been sought to preserve fish for longer to maximise its use. The following are some of the ways that fish is preserved.

Chilling

Chilling fish between 0°C and 4°C and preferably in a specific fish refrigerator running below 2°C will help it to keep a little longer. Remove fish from delivery boxes, place on clean trays, cover with cling film, date and label. Fish may also be vacuum-packed before chilling.

Vacuum-packing

The cleaned fish is placed in a plastic pouch and using a vacuum-packaging machine all the surrounding air is removed and the pouch is tightly sealed. The vacuum-packed fish will also need refrigeration or freezing.

Freezing

Fish is either frozen at sea or as soon as possible after reaching port. The freezing process should be fast, as in blast freezing, because the longer it takes to freeze, the larger and more angular the ice crystals become, which break the protein strands within the fish. This results in excess liquid leaking out when it is defrosted, leaving an inferior product. When fish is frozen quickly, the ice crystals are small and there is less leakage and less damage.

Most fish should be defrosted before being cooked but some prepared frozen fish products, such as breadcrumbed scampi or plaice, can be cooked from frozen. Plaice, halibut, turbot, haddock, pollack, sole, cod, trout, salmon, herring, whiting, scampi, smoked haddock and kippers as well as a very wide range of prepared products are available frozen. Fish and fish products can also be frozen 'in house'.

Frozen fish should be checked to ensure:
- no evidence of freezer burn (very low temperatures can damage the fish, leaving white patches and a change in texture and flavour. It can be avoided by wrapping or packaging the fish well and not storing for too long)

- packaging is undamaged
- there is minimum fluid loss during defrosting
- the flesh is still firm after defrosting.

Frozen fish should be stored at –18 °C to –23 °C and defrosted at the bottom of a refrigerator in a covered deep tray to catch any drips. It should *not* be defrosted in water as this spoils the taste and texture of the fish and valuable water-soluble nutrients are lost. Fish should not be re-frozen as this will impair its taste and texture as well as being a food safety risk.

Canning

Oily fish such as sardines, salmon, anchovies, pilchards, tuna, herring and herring roe (eggs) are often canned. They can be canned either in their own juice (as with salmon) or in oil, a sauce, brine or spring water. In some countries, bottling is popular for fish such as herrings and this is a similar process to canning.

Canned fish

> **Healthy Eating Tips**
> - If the small bones in sardines, whitebait and canned salmon are eaten they provide calcium and phosphorus.

Salting

Salting of fish is often accompanied by a smoking process (see below). Cured herrings are packed in salt. Caviar is the roe (unfertilised eggs) of the sturgeon and is slightly salted then sieved, canned and refrigerated. In some Caribbean countries, salted dried fish, especially cod, has been popular for many years and this is now available in numerous other countries. It needs to be soaked in plenty of cold water for several hours before use.

Pickling

Herrings pickled in vinegar are filleted, rolled and skewered, and are known as rollmops.

Smoking

Fish that is to be smoked may be gutted or left whole. It is then soaked in a strong salt solution (brine) and in some cases a dye is used to add colour (although use of dyes has become less popular). After soaking, the fish are drained, hung on racks in a kiln and exposed to smoke for five to six hours.

- Cold smoking takes place at no more than 33 °C, to avoid cooking the flesh. Therefore, all cold-smoked fish is actually raw and, with the exception of smoked salmon and trout, is usually cooked before being eaten.
- Hot-smoked fish is cured between 70 °C and 80 °C to cook the flesh at the same time, so does not require further cooking. Smoked fish should be wrapped up well and kept separate from other fish to prevent the smell and dye penetrating other foods.

> **PROFESSIONAL TIP**
> There is a high salt content in salted, pickled and smoked fish. There is no need to add salt when cooking.

Table 7.2 Advantages and disadvantages of preservation methods for fish

Methods	Advantages	Disadvantages
Chilling	Chilling temperatures slow down deterioration and fish will keep a little longer. No change in flavour texture or colour. Use as for any fresh fish.	Only a short-term method of preservation – fish should still be used as quickly as possible.
Freezing	Very low temperatures completely stop any deterioration and when done properly, quality, colour and flavour is not lost. A fairly long-term preservation method (up to 6 months). Once defrosted, cook quickly and use as for fresh fish.	If not done with care, damage such as freezer burn can occur (see above). If not frozen quickly the ice crystals can damage the flavour, texture and appearance of the fish.
Canning	Offers long-term preservation without refrigeration or freezing.	The texture, colour, appearance and flavour of the fish can be changed in the canning process, but some consumers actually prefer this, especially for sandwiches.

Methods	Advantages	Disadvantages
.Pickling/ smoking	Adds a different flavour texture and sometimes colour.	Only gives short-term preservation on its own – to extend the life other preservation methods such as chilling, freezing or bottling may be used.
Drying	Once completely dry offers long-term preservation without refrigeration or freezing.	Fish needs lengthy soaking before use to re-hydrate and to remove some of the salt used in the drying process. The texture, colour and flavour are different to fresh fish.

1.4 Techniques to prepare fish

Tools and equipment for preparing fish

Tools and equipment may vary slightly for each of type of fish preparation. However, the basic equipment that will be needed includes:

- knives – mainly a rigid blade chef's knife and a flexible blade fish filleting knife
- a large chopping knife – for removal of fish heads
- fish tweezers – for removal of pin bones
- scissors – for trimming and removal of fins
- colour-coded boards – blue for raw fish and yellow for cooked fish
- trays – preferably different colours or types for raw and cooked fish
- bowls – for trimmings, debris and additional ingredients
- kitchen paper and colour-coded or single-use disposable cloths.

All fish should be washed under running cold water before and after preparation.

> **HEALTH AND SAFETY**
> Unhygienic equipment, utensils and preparation areas increase the risk of cross-contamination and danger to health. Use equipment reserved just for raw fish. If this is not possible, wash and sanitise equipment before and immediately after each use. Clean small equipment and chopping boards (usually blue) thoroughly with colour-coded or single-use cloths, along with detergent/disinfectant or sanitiser, or by putting them through a dishwasher. Clean surfaces and larger equipment with hot water and detergent, then disinfect or use hot water and sanitiser.

> **HEALTH AND SAFETY**
> Thorough, regular and effective hand washing is essential before and after handling fish.

Trimming

Whole fish are trimmed to remove the scales, fins and the head using fish scissors and a knife. If the head is to be left on (as in the case of a salmon for a cold buffet), the gills and the eyes are removed.

Gutting and scaling

If the fish needs to be gutted, the following procedure should be used.

1. Cut from the vent (hole) to two-thirds along the fish.
2. Draw out the intestines with the fingers or, in the case of a large fish, use the hook handle of a ladle.
3. Ensure that the blood lying along the main bone is removed, then wash and drain thoroughly.
4. If the fish is to be stuffed then it may be gutted by removing the innards through the gill slits, thus leaving the stomach skin intact, forming a pouch in which to put the stuffing. When this method is used, care must be taken to ensure that the inside of the fish is clean and clear of all traces of blood.

To scale a fish such as salmon, trout, sea bass or bream, hold the fish at the tail end, scrape the back of the blade of a knife blade against the grain of the scales towards the head until the scales are removed. Wash the fish thoroughly to remove any loose scales.

Gutting a red mullet

Step by step: filleting round fish (salmon)

1 Using a chef's knife with a rigid blade, remove the head and clean thoroughly.

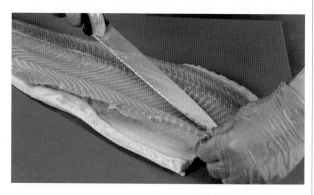

2 Using a fish filleting knife, remove the first fillet by cutting along the backbone from head to tail. Keeping the knife close to the bone, remove the fillet.

3 Reverse the fish and remove the second fillet in the same way, this time cutting from tail to head.

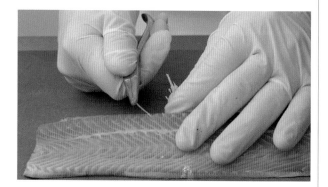

4 Trim the fillets to neaten them and, if required, skin them using the filleting knife.

Step by step: filleting and skinning flat fish (with the exception of Dover sole)

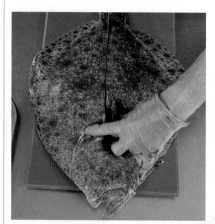

1 Using a filleting knife, make an incision from the head to tail down the line of the backbone.

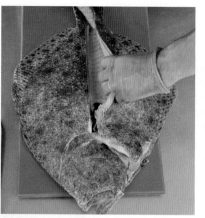

2 Remove the first fillet, holding the knife almost parallel to the work surface and keeping the knife close to the bone.

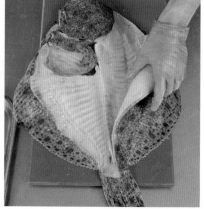

3 Repeat for the second fillet.

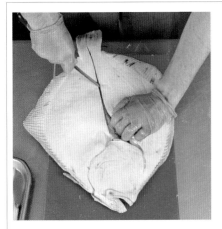

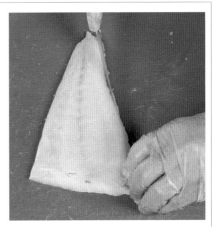

4 Turn the fish over and repeat, removing the last two fillets.

5 Hold the fillet firmly at the tail end. Cut the flesh as close to the tail as possible, as far as the skin. Keep the knife parallel to the work surface, grip the skin firmly and move the knife from side to side to remove the skin.

6 Trim the fillets to neaten them.

PROFESSIONAL TIP

If you find it difficult to grip the tail end of a fillet because it is slippery, place a little salt on the fingers and this will help you to hold it firmly. Alternatively, grip using kitchen paper.

FAULTS

When filleting a whole round or flat fish it is important to keep the blade of the knife very close to the bone so the flesh of the fish is removed cleanly and neatly. If too much flesh is left on the bones it is very wasteful, looks unprofessional and will affect the quality of the finished dish.

Step by step: preparation of whole Dover sole

1 Score the skin just above the tail.

2 Hold the tail of the fish firmly, then cut and scrape the skin until you have lifted enough to grip.

3 Gently pull the skin away from the tail to the head. Both black and white skins may be removed in this way.

4 Trim the tail and side fins with fish scissors, remove the eyes, and clean and wash the fish thoroughly.

Step by step: stuffing a whole round fish

1 Scale, trim and gut the fish.

2 Remove the spine, snipping the top with scissors. Clean the cavity.

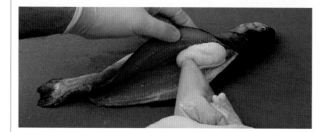

3 Using a piping bag, fill the cavity with the required stuffing.

4 Secure the cut edges in place before cooking.

Wrapping

Thick fillets, suprêmes and items such as monkfish tails are sometimes wrapped before cooking. The wrapping may hold another ingredient in place, such as a fish mousseline or stuffing, duxelle, julienne of vegetables or shellfish. Suitable wrappings include filo pastry, thinly sliced cured ham or pig's caul (a lacy, fatty membrane surrounding the internal organs of a pig).

Before cooking, fish can be enclosed in a suitable pastry such as shortcrust, puff or a suet pastry for dumplings.

>
> **HEALTH AND SAFETY**
>
> Keep working areas clean and tidy – always clean as you go, do not allow debris to build up.

Flavourings and coatings

Additional flavours may be added by the use of seasonings, herbs, spices or spice pastes and various purées. Marinades can be used with most fish and the ingredients for these will vary, depending on the type of fish, method of cooking and the flavours required. Popular marinade mixture ingredients include wine, vinegar, oils, herbs, spices, yoghurt, lemon or lime juice and prepared sauces such as soy sauce, tabasco, Worcestershire sauce or teriyaki sauce.

Coatings such as flour, egg and breadcrumbs (pané), milk and flour and batter are often added to fish before cooking (for more information see Section 2.2 Cooking techniques).

Pickling and marinating fish

Pickling is used as a way of preserving raw or cooked fish by immersing it into an acidic mixture, which may also be salted or flavoured. Pickling alone will not fully preserve fish and if not for immediate consumption another preservation method such as chilling or bottling may be used. The flavour, texture and appearance of the fish may be changed by the pickling process. The essential ingredient of a pickling liquid is vinegar or citrus juice, often accompanied by salt. The pickling liquid can also be flavoured with sugar, herbs, spices or citrus juices and a variety of other flavourings. In some regions or countries (for example, Scandinavian countries), pickled fish is a traditional dish.

Marinating is used for a variety of fish and has a number of benefits. As well as adding the desired flavour, a marinade can prevent fish and shellfish from drying out during cooking and can also stop fish sticking to grill bars and frying pans. A marinade is a mixture of ingredients as described above. If using a highly acidic marinade, such as one made with vinegar or citrus, this can actually 'cook' the fish so it is best not to leave in the marinade too long.

1.5 Storage

Spoilage and deterioration of fish is mainly caused by the actions of natural enzymes and bacteria. The enzymes in the gut of the living fish help convert its food to tissue and energy. When the fish dies, these enzymes carry on working and, along with bacteria in the digestive system, start breaking down the flesh itself. Warmer temperatures speed up this process. Bacteria also exist on the skin and in the fish intestine causing deterioration. Although these bacteria are harmless to humans, they reduce the eating quality of the fish and can result in unpleasant odours.

To ensure quality, fish should always be stored correctly:

- Once caught, fish has a shelf life of 10–12 days if properly refrigerated at a temperature between 0 °C and 4 °C.
- If the fish is delivered whole with the innards still intact, it should be gutted and the cavity washed well before it is stored.
- Use the fish as soon as possible after delivery.
- Remove fish from the delivery containers, rinse, pat dry, place on a clean tray (this could be blue), cover with cling film and store in a refrigerator just for fish (running at a temperature of 1–2 °C) or at the bottom of a multi-use refrigerator (running at 1–4 °C). Make sure the fish is well covered or wrapped, labelled and dated. Some fish refrigerators are also humidity controlled to keep fish in the best possible condition.
- For fish products that do not need to be used immediately, use in rotation (use older items before newly delivered items), observe use-by dates on fresh and chilled fish products and best before dates on longer-life products such as canned fish.
- Ready-to-eat cooked fish, such as 'hot' smoked mackerel, cooked prawns and crab, should be stored on shelves above other raw food to avoid cross-contamination.
- Frozen fish should be well wrapped or packaged and stored in a freezer at −18 °C to −23 °C. Frozen fish needs to be defrosted in a covered container at the bottom of the refrigerator.
- Smoked fish should be stored and refrigerated as for other fish, but also very well wrapped in cling film or in a covered container to avoid the strong smell permeating other items in the refrigerator.

> **HEALTH AND SAFETY**
> Fish offal and bones present a high risk of contamination and must not be mixed or stored with raw prepared fish.

> **ACTIVITY**
>
> 1 Name three dishes using three different types of flat fish that could be used for a hotel restaurant menu.
> 2 Name three types of round fish that could be used to produce dishes for a buffet with hot and cold food.
> 3 List five of the quality points to be considered when receiving a delivery of fresh fish. What would you do if the fish did not meet these quality requirements?
> 4 In a large restaurant what measures are necessary to ensure that raw fish and fish dishes are at the required temperature? What should these temperatures be?
> 5 What would the order to your fish supplier state to provide correct portion sizes for anticipated orders of 30 each of trout, salmon fillet, smoked haddock and sea bass?
> 6 Select three fried fish dishes from the recipe section. Using the same fish, design different dishes that would provide a healthy option.

Produce fish dishes and products

2.1 Sauces and dressings

Cooked fish can be served with a wide variety of flavourings and sauces. Sauces can enhance individual fish dishes by adding moisture, colour, flavour, a variation in texture and interest to the dish. However, it is important to remember that sauces should be added to enhance the fish, not overwhelm it, so there should not be too much sauce and it should not be too strongly flavoured. The sauces selected will depend on the type of fish, the cooking method and flavourings, coatings, toppings or finishes already used. It is possible that no further sauce or dressing is needed. However, the following may be served with various types of fish:

- flavoured or herb or spice infused oils
- flavoured or herb butter
- fish glazes
- roux-based sauces such as béchamel and velouté, or jus lié-based sauces
- emulsified butter sauces such as hollandaise sauce
- reduction sauces with stock, butter or cream

- butter or olive oil monter sauces and sabayon
- mayonnaise and mayonnaise-based sauces such as tartare sauce
- cold dressings such as vinaigrette, raita, various chutneys, pickles and jellies
- salsas, tapanade and pesto
- commercially prepared sauces such as ketchups, Worcestershire sauce, sweet chilli sauce.

For more information on sauces see Unit 303 Soups, stocks and sauces.

2.2 Cooking techniques

When cooked, fish loses its translucent look and becomes opaque. It will also flake easily so should be handled with care because it is delicate and can break up.

Fish cooks quickly and can easily become dry and lose its flavour if overcooked, so it is important to consider methods of cooking carefully.

To maintain the quality and food safety of fish dishes it is advisable to check the internal temperature using a temperature probe. It is recommended that all fish should be cooked to a minimum internal temperature of 63 °C. This is for whole fish and cuts of fish. Made up fish dishes such as fish pie or fish cakes should be cooked to a minimum of 75 °C.

> **HEALTH AND SAFETY**
> Environmental health officers may require higher internal temperatures for whole fish and fillets, especially if vulnerable groups of people are being served.

The main methods used for cooking fish are described below.

Frying

Frying is probably the most popular method of cooking fish.

- Shallow frying – this method is suitable for small whole fish, cuts or fillets cooked in oil or fat in a frying pan. The fish should be seasoned and lightly coated with flour before frying to protect it. The fish can also be lightly coated with semolina, matzo meal, oatmeal or breadcrumbs before frying. If using butter as the frying medium, it must be clarified (see butter sauces in Unit 303) to reduce the risk of the fish burning. Oil is probably best for cooking fish and a little butter may be added for flavour. Turn the fish only once during cooking, to avoid breaking it.

- Deep frying – this method is suitable for small whole white fish, cuts fillets and goujons, as well as made up items such as fishcakes. All white fish are suitable for deep-frying in batter, including cod, pollack, haddock and skate. Depending on size, the fish may be left whole, be portioned or filleted. The fish should be seasoned and coated before frying, usually with a batter or an egg and breadcrumb (**pané**) mixture; milk and flour can also be used. This prevents penetration of the cooking fat or oil into the fish. Where possible always use a thermostatically controlled deep fryer only half filled with oil to the level indicated on the fryer. Heat the oil to 175 °C and place the fish into the oil carefully and cook until evenly coloured. Remove carefully and drain the fish on absorbent paper after cooking.
- Stir-frying – this is a very fast and popular method of cooking. The fish is cooked along with suitable vegetables and often noodles, bean shoots or rice. Spices, flavourings and suitable sauces may also be added. Use a wok or deep frying pan and a high cooking temperature. Food should be cut into thin strips and all prepared before cooking begins. This method is well suited to cooking firm-fleshed fish cut into strips or shellfish.

Grilling

Grilling, or griddling, is a fast and convenient method suitable for fillets or small whole fish. When grilling whole fish, trim and prepare neatly and descale. Cut through the flesh at the thickest part of the fish to allow even cooking. Lightly oil and season fish or fillets; to avoid breaking do not turn more than once.

> **FAULT**
> All fish needs to be carefully cooked and most fish cooks quickly. Overcooking impairs the quality of the finished fish dish, leaving it dry, lacking in flavour and misshapen.

Poaching

Poaching is suitable for whole fish such as salmon, trout or bass; certain cuts on the bone such as salmon, turbot, brill, halibut, cod, skate; and fillets, suprêmes or fish prepared into délice or paupiette.

The prepared fish should be completely immersed in the cooking liquid, which can be water, water and milk, fish stock (for white fish) or a court bouillon for oily fish. Most kinds of fish can be cooked in this way and should be poached gently for 5–8 minutes, depending on the thickness of the fish. Whole fish are covered with a cold liquid and brought to the boil then cooked just below simmering point. Cut fish are usually placed in the liquid at simmering point then cooked just below simmering point. The resulting liquid is ideal for use in sauces and soups.

When poaching smoked fish, place in cold, unsalted water and bring to a steady simmer. This liquid will be salty and may not be suitable for use in stocks and sauces.

En sous vide

This has become a popular way to cook fish. The method involves sealing the fish in a vacuum-pack and cooking it in a hot water bath. Sealing the fish first helps to reduce moisture loss and also allows the fish to cook with a marinade or sauce if required.

Boiling

Boiling is used mainly for shellfish and, as many shellfish are sold already cooked, this is often done at sea or soon after landing because shellfish can deteriorate quickly. Shellfish are often boiled from live and include lobsters, crabs, oysters, langoustines, crayfish, clams, cockles, mussels, prawns and shrimps.

Roasting

Roast fish now frequently appears on restaurant menus but because fish is much more delicate than meat, only whole fish or larger, firmer cuts are suitable to roast and care must be taken not to overcook. The fish may be brushed with oil or clarified butter before cooking and may be raised a little from the base of the pan using vegetables. The fish is usually basted with the pan juices as it cooks.

When the fish is cooked and removed, the tray can be **deglazed** with a suitable wine (usually a dry white) and fish stock to form the base of an accompanying sauce. Examples of this method of cooking include roast cod on garlic mash and roast sea bass flavoured with fennel.

KEY TERMS

Pané: coating the fish with a light coating of seasoned flour, beaten egg then breadcrumbs.

Deglaze: to add wine stock or water to the pan in which the fish was cooked to use the sediments and flavours left in the pan.

Baking

Similar to roasting, whole, portioned or filleted fish may be oven baked. To retain their natural moisture it is necessary to protect the fish from direct heat and this may be done by coating or wrapping. Prepared fish dishes such as fish pies may also be finished by baking.

Before baking whole, fish may be stuffed with items such as a duxelle-based mixture, flavoured breadcrumbs, fish mousses, herbs or vegetables such as onion or fennel. They can be wrapped in pastry (puff or filo), coated with a crumb mixture or completely covered with a thick coating of dampened sea salt. A prepared dish such as a fish pie may be topped or finished with puff pastry, mashed potato or a sauce.

En papilotte

Fish and accompanying ingredients are oven cooked together in a sealed package made from parchment paper or foil. The package expands in the oven, creating steam that cooks the fish and other ingredients. The method is excellent for preserving flavour and is considered a very healthy method of cooking. The package is often opened in front of the customer releasing the aromas at the table.

Steaming

Most fish are suitable for steaming but generally cuts of fish such as darnes, tronçons, fillets and suprêmes are used. Steaming is also a good method for fish fillets made into délice or paupiettes, as they are or with a filling. It is a popular method to cook fish for use in other dishes such as fish cakes or fish pie.

Preparation is usually simple and any fish that can be poached or boiled may also be cooked by steaming. The fish is seasoned then placed in a steamer tray lined with greaseproof paper or in a small steamer over simmering water covered with a lid. Most items are cooked for 10–15 minutes, depending on the thickness of the fish or the fillets. Steaming can also be done by putting the food between two plates over a pan of simmering water.

Steaming is an easy method of cooking; because it is quick, it conserves flavour, colour and nutrients. It is also suitable for large-scale cookery.

Fish may be steamed then a suitable sauce used to finish the dish or be offered with it. Any cooking liquor from the steamed fish may be strained off, reduced and incorporated into the sauce. Preparation can also include adding finely cut ingredients such as ginger, spring onions, garlic, mushrooms and soft herbs, lemon juice and dry white wine, either to the fish on the steamer dish before cooking or when the fish is served. Fish is sometimes marinated before steaming to add extra flavour.

Braising and stewing

These tend to be less popular methods for cooking fish but generally whole fish or larger cuts of fish can be braised in a closed container covering or half covering the fish with a suitable liquid such as those used for poaching. Vegetables such as onions, shallots and fennel may be added, along with herbs, spices, lemon or lime.

Stewing of fish may be similar to braising but in many countries there are traditional fish stews using a variety of fish and shellfish, along with onions, shallots, tomatoes, garlic and a variety of other ingredients. Examples of these stews are bouillabaisse from France, cataplana from Portugal and cioppino from Italy.

Tables 7.3 and 7.4 give examples of fish types that are available and suitable cooking methods.

Table 7.3 Examples of white fish and suitable cooking methods

	Baking	Boiling	Deep-frying	Grilling	Poaching	Roasting	Shallow-frying	Steaming	Stir-frying
Cod	■	■	■	■	■	■	■	■	
Coley		■	■	■		■	■	■	■
Dover sole			■		■	■	■		■
Grouper			■		■	■	■	■	■
Haddock	■	■	■	■	■	■	■	■	■
Hake	■	■	■	■	■	■	■	■	■
Halibut	■		■	■	■	■	■	■	■
Huss		■	■	■	■	■	■	■	■
John Dory	■	■	■	■	■	■	■	■	■
Lemon sole		■	■		■	■	■	■	■
Monkfish	■	■	■	■	■		■	■	
Plaice	■	■	■		■	■		■	■
Sea bass	■	■	■	■	■		■	■	■
Shark	■	■	■		■	■	■	■	■
Skate wings		■	■		■	■	■		■
Swordfish	■		■	■	■	■	■		■
Turbot	■	■	■	■	■	■	■		■

Table 7.4 Examples of oily fish and suitable cooking methods

	Baking	Boiling	Deep-frying	Grilling	Poaching	Roasting	Shallow-frying	Steaming	Stir-frying
Barracuda				■			■		
Dorade (red sea bream)	■			■			■		■
Emperor bream	■			■			■		
Herring				■			■		
Mackerel	■			■			■		
Marlin	■			■			■		
Red mullet	■			■		■	■		
Red snapper	■			■			■		

	Baking	Boiling	Deep-frying	Grilling	Poaching	Roasting	Shallow-frying	Steaming	Stir-frying
Salmon									
Sardines									
Trout									
Tuna									
Whitebait									

2.3 Evaluation

See the evaluation section on page 349 for evaluation of fish and shellfish dishes.

Prepare shellfish dishes and products

Introduction

Shellfish such as lobsters and crabs are all invertebrates, which means they do not have an internal skeleton. They are split into two main groups: **molluscs** and **crustaceans**.

Molluscs have either:

- an external, hinged double shell such as scallops and mussels (these are called **bivalves**)
- a single spiral shell such as winkles or whelks (these are called **univalves**)
- soft bodies with an internal shell such as squid and octopus (these are also called **cephalopods**).

Molluscs (mussels)

Crustaceans have tough outer shells and also have flexible joints to allow quick movement, for example, crab and lobster.

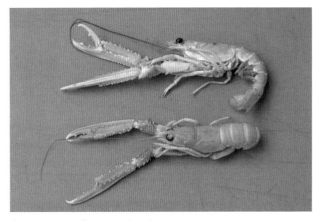

Crustaceans (langoustines)

KEY TERMS

Molluscs: shellfish with either a hinged double shell, such as mussels, or a spiral shell, such as winkles.

Crustaceans: shellfish with tough outer shells and flexible joints to allow quick movement; for example, crab and lobster.

Bivalves: molluscs with an external hinged double shell; for example, scallops and mussels.

Univalves: molluscs with a single spiral shell; for example, winkles or whelks.

Cephalopods: molluscs with a soft body and an internal shell; for example, squid and octopus.

Video: Buying shellfish
http://bit.ly/2plu4bJ

1.1 Quality points

Shellfish have tender, fine-textured flesh, and can be prepared in a variety of ways, but are prone to rapid spoilage. The reason for this is that they contain quantities of certain amino acids (the 'building blocks' of proteins), which encourage bacterial growth.

To ensure freshness and best flavour, choose live shellfish and cook them yourself. This is possible with the expansion of globalisation and air freight, which has created a faster trade in live shellfish with greater availability.

Choosing shellfish

- Shells should not be cracked or broken.
- The shells of mussels and oysters should be tightly shut; open shells that do not close when tapped sharply should be discarded.
- Lobsters, crabs and prawns should have a good colour and feel heavy for their size.
- Lobsters and crabs should have all their limbs.

1.2 Folds, cuts and products

This learning outcome does not include shellfish. For details of shellfish products, see Section 1.4 below and the recipe section at the end of this unit.

1.3 Preservation methods

Because shellfish deteriorate so quickly after they are caught, preservation methods similar to those used for fish are frequently used. Shellfish are often boiled soon after they are caught which slows down deterioration and therefore some shellfish on sale are 'ready cooked'. They may also be chilled or frozen while still at sea. Popular preservation methods are cooking, chilling, freezing, pickling, vacuum-packaging, canning and smoking.

Table 7.5 Seasonality of shellfish

	Jan	Feb	Mar	Apr	May	Jun	Jul	Aug	Sep	Oct	Nov	Dec
Crab (brown cock)	■	■	■	■								
Crab (spider)	■	■	■	■	■	■	■	■	■			
Crab (brown hen)	■	■			■	■	■	■	■	■	■	■
Clams	■	■	■	■					■	■	■	■
Cockles	■	■	■	■					■	■	■	■
Crayfish (signal)					■	■	■	■	■	■		
Lobster								■	■	■		
Langoustines			■	■	■	■	■					
Mussels									■	■	■	■
Oysters (rock)	■	■	■	■					■	■	■	■
Oysters (native)	■	■	■	■					■	■	■	■
Prawns	■	■	■	■	■	■	■	■	■	■	■	
Scallops	■	■	■	■	■	■	■	■	■	■		■

Key

Available	At best
	■

1.4 Techniques to prepare shellfish

The flesh of fish and shellfish is different from meat: the connective tissue is very fragile, the muscle fibres are shorter and the fat content is relatively low. Shellfish should be cooked as little as possible – to the point that the protein in the muscle groups just coagulates. Beyond this point the flesh tends to dry out, making it tougher and drier. Shellfish are known for their dramatic colour change, from blue-grey to a vibrant orange colour. This is because they contain red and yellow pigments called carotenoids, bound to molecules of protein. Once heat is applied, the bonds are broken and the bright colour is revealed.

Shrimps and prawns

Prawns are crustaceans that come in a variety of sizes and have a firm, meaty flesh. Cold water prawns are often cooked and sometimes peeled while still at sea; the tropical warm-water prawns, such as king and tiger prawns, are often reared in fish farms.

Smell is a good guide to freshness. Shrimps and prawns can be used for garnishes, decorating fish dishes, cocktails, sauces, salads, hors d'oeuvres, omelettes and snack and savoury dishes. They can also be used for a variety of hot dishes such as stir-fries, risotto and curries.

King prawns are a larger variety, which can be used in any of the ways listed above and are an ingredient of many Asian dishes.

Step by step: preparing prawns (removing head, carapace, legs and tail and dark intestinal vein)

1 Remove any loose parts; then, holding the prawn in one hand, remove the head by twisting it off the body.

2 Turn the prawn over and pull the shell open along the length of the belly, starting at the head end and pulling it open so that you can pull the prawn from the shell. Sometimes the very end of the tail is left on for appearance reasons.

3 Once the shell is off, check if there is a black line running down the back (the intestinal tract). It is not harmful to eat, but looks better without it. Using a small, sharp knife, make a shallow cut along the length of the black line.

4 Lift out the intestinal tract (this is called 'deveining'). Wash before using.

Langoustines, crayfish, scampi, Dublin Bay prawns

Langoustine, crayfish, scampi and Dublin Bay prawns are also known as Norway lobster. They are succulent, white shellfish related to the lobster, though are more the size of a large prawn and are sold fresh, frozen, raw or cooked. Their tails are prepared like prawns and they are used in a variety of ways: salads, rice dishes, stir-fries, deep-fried, poached and served with a number of different sauces.

They are also used as garnishes for hot and cold fish dishes, especially on buffets.

Freshwater crayfish are also known as *écrevisse*. These are small freshwater crustaceans with claws, found in lakes and lowland streams. They are prepared and cooked like shrimps and prawns and used in many dishes, including soup. They are often used whole to garnish hot and cold fish dishes.

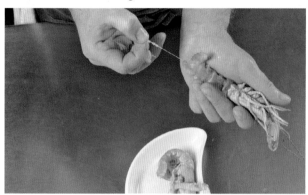

Remove the cord from langoustine before cooking.

Lobster

Although there are many different lobster varieties around the world, the two main ones are the American lobster, which tends to be the largest, and the European lobster, which is usually smaller. Lobsters grow very slowly and can live up to 100 years. As they grow they shed their hard shell and form another one. Maturity is reached at about five years old and at a length of 18–20 centimetres. They can weigh up to nine kilograms but are usually around two to three kilograms.

Lobsters are served cold in cocktails, lobster mayonnaise, hors d'oeuvres, salads, sandwiches and in halves on cold buffets. Used hot, they are served simply, usually cut and served in half, in the shell with clarified butter or a sauce or as one of the classic hot lobster dishes such as Lobster Thermador. Lobster is also used in soups, sauces, rice dishes, stir-fry dishes and as a filling for ravioli or cannelloni.

Check points for buying lobster

- Purchase alive, with both claws attached, to ensure freshness.
- Lobsters should be heavy in proportion to their size.
- The coral (the roe) of the hen (female) lobster is necessary to give the required colour for certain soups, sauces and lobster dishes.
- Hen lobsters are distinguished from cock lobsters by their broader tails.

Step by step: preparing lobster

1 Remove lobster claws and legs.

2 Cut lobster in half.

3 Remove meat from a cleaned lobster.

Crawfish

These are sometimes referred to as 'spiny lobsters', but unlike lobsters they have no claws and their meat is solely in the tail. Crawfish vary considerably in size from one to three kilograms; they are cooked in the same way as lobsters and the tail meat can be used in any of the lobster recipes. Because of their impressive appearance, crawfish dressed whole are sometimes used on cold buffets.

Crab

There are large numbers of edible species of crab (estimated at around 4,000). The meat from crabs is very different in the claws and the main body. Crab claws have a sweet, dense white meat similar to lobster, while the flesh from under the main body shell is soft, rich and brown. The male crabs tend to have larger claws and more white meat. Whole crab can be served simply in the shell with salad and bread or prepared as dressed crab when the white and brown meats are arranged separately back into the shell. Crab meat can be used cold for hors d'oeuvres, cocktails, salads and sandwiches. Served hot, it can be used in a variety of shellfish dishes, covered with a suitable sauce and served with rice, pasta, gnocchi or in Chinese-style dumplings in *bouchées* or pancakes, or made into crab fishcakes.

- European brown crab is available all year, reaches 20–25 cm across, has large front claws with dark pincers, a red or brown shell and red legs.
- Atlantic blue crab has a blue-brown shell and tends to be smaller than the European crab at 10–15 centimetres. These crabs regularly shed their hard shell and are sometimes caught with their newly forming soft shell. These can be cooked in the shell and eaten whole – soft shell crab.
- Dungeness crab can reach up to 20 centimetres and has plenty of good, white, dense meat in the claws. The meat in the shell tends to be different from other crabs and is pale grey-green in colour.
- Spider crabs resemble a large spider. The meat has a good flavour but it has no large claws so no white claw meat.

Crab

Check points for buying crab
- Buy alive where possible to ensure freshness.
- Ensure that both claws are attached.
- Crabs should be heavy in relation to size.

Preparing crab
1 Cook crab in the same way as lobsters.
2 With the crab on its back, twist off the legs and claws.
3 Grip where the body meets the shell and pull to separate the body from the shell.
4 Pull out and discard the gills (dead man's fingers).
5 Scoop out the brown meat from the shell and place in a bowl.
6 Using a lobster pick or skewer, pick the white meat out of the cavities in the body and put in a separate bowl.
7 Crack the claws and legs using a hammer, mallet or steel.

8 Remove the meat, discarding the cartilage. Add the meat to the other white meat. Use all the crab meat as required.

Cockles
Cockles are enclosed in small, attractive, cream-coloured shells. As they live in sand, it is essential to purge them by washing well under running cold water and leaving them in cold salted water (changed frequently) until no traces of sand remain.

Cockles can be cooked either by steaming, boiling in unsalted water, on a preheated griddle or as for any mussel recipe. They should be cooked only until the shells open.

They can be used in soups, sauces, salads, stir-fries and rice dishes and as garnish for fish dishes.

Mussels
Mussels are bivalves. They can be from either the sea or freshwater (rivers and lakes), but the sea varieties are by far the most widely used. Mussels are now extensively 'farmed' – cultivated on wooden frames in the sea, producing tender, delicately flavoured plump flesh.

They are produced off British coasts and also imported from France, Holland and Belgium. French mussels are small; Dutch and Belgian mussels are plumper. The production of mussels is considered to be ecologically sound which means the species is not threatened or damaging to the environment.

Mussels can be used for soups, sauces and salads and cooked in a wide variety of hot dishes, including the popular and traditional *moules marinière* (mussels with white wine).

Mussels

Check points for buying mussels
- Shells must be tightly closed or close when tapped, indicating that the mussels are alive.

- There should not be an excessive number of barnacles attached.
- They should smell fresh.

Mussels should be kept in containers, covered with cling film or damp cloths, and stored in a cold room or fish refrigerator.

Preparing mussels

1 Discard any mussels with broken shells.
2 Scrape the shells with a small knife to remove any barnacles.
3 Remove the byssus threads (these are sometimes called the beard, which gives rise to the term 'de-bearding').
4 Wash in several changes of cold water to clean the mussels then drain well in a colander.

Scallops

Scallops are bivalves with a fan-shaped shell. They vary in size from 15 centimetres for great scallops to around 8 centimetres for bay scallops and queen scallops that are the size of cockles. Inside they have round white flesh and the orange coral, which is the roe and is often discarded. Scallops are popular and prices tend to remain high, especially for 'hand-dived' scallops rather than those caught by a dredging trawler. They are prepared by prising the two halves of the shell apart and removing the white flesh and the orange roe if required.

Scallops are found on the seabed and are therefore dirty or sandy, so it is advisable to purchase them ready cleaned. If scallops are bought in their shells, the shells should be tightly shut, which indicates they are alive and fresh. The orange roe should be bright and moist. Scallops in their shells should be covered with damp cloths or cling film and kept in a refrigerated cold room or fish refrigerator.

To remove scallops from their shells, insert a sharp knife between the two halves of the shell and prise apart. The flesh can then be removed with a knife. Scallops should then be well washed.

Scallops

Whelks

The common whelk is familiar around the coast of Britain. Whelks can often be bought in jars, pickled in vinegar and brine.

British winkles

Winkles are small, black shellfish that look like snails and grow up to three centimetres in size. The British winkle can be readily identified on rocky shores.

Oysters

Oysters are highly regarded saltwater bivalves found near the bottom of the sea, with a number of different species available. The upper shell (valve) is flattish and attached by a ligament hinge to the lower, bowl-shaped shell. Those available around British coastlines are native, flat or rock oysters with Colchester and Whitstable in Kent being significant oyster areas.

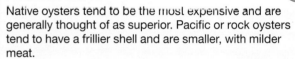

> **PROFESSIONAL TIP**
> Native oysters tend to be the most expensive and are generally thought of as superior. Pacific or rock oysters tend to have a frillier shell and are smaller, with milder meat.

When buying oysters, the shells should be clean, bright, tightly closed and unbroken. Traditionally oysters are eaten raw so it is essential they are very fresh and cleaned or purged well. Preparation usually involves prising the two halves of the shell apart with a small, pointed oyster knife. The oysters are then served with lemon and red pepper.

Store oysters in their delivery boxes, in the refrigerator, covered with damp cloths or kitchen paper. The shells should be tightly closed or should close when tapped. Discard any that do not close. Unopened live oysters can be kept in the fish refrigerator for two to three days. Do not store in an airtight container or under fresh water as this will cause them to die. Shelled (shucked) oysters can be kept refrigerated in a sealed container for four to five days.

Squid

Squid is traditionally popular in Mediterranean cuisine and has increased in popularity throughout the rest of Europe. It is available all year round, either fresh or frozen. Squid vary in size, from 7 centimetres in length up to 25 centimetres. The ink sac, 'beak' and transparent cartilage need to be removed. The main body is usually cut into rings for cooking. Squid needs to be cooked very quickly or by a slow moist method, otherwise it can be tough.

Step by step: preparing squid

1   Pull the head away from the body, together with the innards and remove the beak.

2   Taking care not to break the ink bag, remove the long transparent blade of cartilage (the backbone or quill).

3   Cut the tentacles just below the eye and remove the small round cartilage at the base of the tentacles.

4   Scrape or peel off the reddish membrane that covers the pouch, rub with salt and wash under cold water.

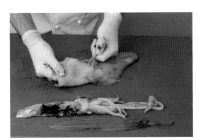

5   Discard the head, innards and pieces of cartilage. Cut up the squid as required.

1.5 Storage

Unless live, shellfish will start to spoil as soon as they have been removed from their natural environment; therefore, the longer shellfish are stored, the more they will deteriorate due to the bacteria present. Best practice is to cook immediately and store in the same way as cooked fish. Shellfish can be blanched quickly to remove the shell and membrane (especially in lobsters), but they will still need to be stored as a raw product as they will require further cooking.

Bear in mind the following quality, purchasing and storage points:

- When possible, all shellfish should be purchased live so as to ensure freshness.
- Shellfish should be kept in suitable containers, covered with damp cloths or cling film, and stored in a cold room or fish refrigerator.
- Shellfish should be cooked as soon as possible after purchasing.

Produce shellfish dishes and products

2.1 Sauces and dressings

For examples of sauces to use with shellfish, see the recipe section at the end of this chapter and Unit 303 Stocks, soups and sauces.

2.2 Cooking techniques

Lobster

To cook lobster:

1   Wash then plunge the lobster into a pan of boiling salted water containing 60 millilitres of vinegar to 1 litre of water.
2   Cover with a lid, re-boil, then allow to simmer for 15–20 minutes according to size.
3   Overcooking can cause the tail flesh to toughen and the claw meat to become hard and fibrous.
4   Allow to cool in the cooking liquid when possible.

To clean a cooked lobster:

1   Remove the claws and the pincers from the claws.
2   Crack the claws and joints and remove the meat.

3 Cut the lobster in half by inserting the point of a large knife 2 centimetres above the tail on the natural central line.

4 Cut through the tail firmly.

5 Turn the lobster around and cut through the upper shell (carapace).

6 Remove the halves of the sac (which contains grit). This is situated at the top, near the head.

7 Using a small knife, remove the intestinal vein from the tail and wash if necessary.

Mussels

There are various ways the mussels can be cooked, but the most popular and traditional way is described below.

1 Prepare a cooking liquid of shallots, butter, garlic and herbs along with fish stock.

2 Cook the mussels in the liquid in a large pan with a tightly fitting lid until they open – this usually takes 4–5 minutes. Remove the mussels and check they have all opened; throw away any that are still closed.

3 Strain the liquid, whisk in cream and serve with the mussels still in their shells.

Alternatively:

1 In a thick-bottomed pan with a tight-fitting lid, place 25 grams of chopped shallot or onion for each litre of mussels.

2 Add the mussels, cover with a lid and cook on a fierce heat for 4–5 minutes until the shells open completely.

3 Remove the mussels from the shells.

4 Retain the carefully strained liquid to make a sauce to serve with the mussels.

Scallops

Scallops should be lightly cooked; if overcooked they shrink and toughen.

● Poach gently for 2–3 minutes in dry white wine with a little onion, carrot, thyme, bay leaf and parsley. Serve with a suitable sauce such as white wine or mornay sauce.

● Lightly pan fry on both sides for a few seconds in butter or oil in a very hot pan (if the scallops are very thick they can be cut in half sideways). Serve with a suitable garnish such as sliced mushrooms or a fine brunoise of vegetables and tomato, and a liquid that need not be thickened such as white wine and fish stock, or cream- or butter-thickened sauce. Fried scallops can also be served hot on a base of salad leaves.

● Deep-fry, either coated in egg and bread crumbs or passed through a light batter, and served with segments of lemon and a suitable sauce such as tartare sauce.

● Wrap in thin streaky bacon or cured ham and place on skewers for grilling or barbecuing.

Whelks

To cook:

1 Place the whelks in a saucepan and add water to cover them by 1 centimetre.

2 Add a sprig of thyme and a bay leaf.

3 Bring to a boil and cook at a gentle simmer.

Whelks require only a very short cooking time otherwise the flesh has a tendency to become tough.

British winkles

To remove the winkles easily from their shells they need to be cooked.

1 Place the winkles in a saucepan and cover with water. Add salt and a bay leaf.

2 Turn on the heat, wait for the first sign of boiling and allow to boil 2 to 3 minutes longer – no more.

3 Take out one winkle to test if cooked; if uncooked it will resist removal from the shell.

4 When the winkles are cooked, run them quickly under cold water, otherwise they will toughen.

5 Serve simply in melted butter.

2.3 Evaluation

See the evaluation section below for evaluation of fish and shellfish.

ACTIVITY

1 Explain the difference between molluscs and crustaceans and give examples of two of each.

2 What are the different types of scallops available? Find prices for each and compare.

3 How could you provide information about shellfish on your menu for a customer allergic to shellfish.

4 List the food safety precautions that must be in place when storing, preparing and cooking shellfish.

5 Suggest some of the ways that shellfish could be preserved to make them last longer. Provide a recipe for a shellfish dish using a preserved shellfish item.

6 Suggest a selection of shellfish you would include on a mixed shellfish platter for a buffet.

2.3 Evaluation

Assembling fish dishes

As shown in the fish recipe pages, there are numerous ways to serve and present fish using skill and imagination.

Each establishment or business will have their own production and dish standards and it is important to follow these.

- Portions – should be of an even size and weight and similar in shape
- Colour and harmony – all items on the plate or serving dish should be considered and planned. Different items should be placed carefully and balanced. The colour of the actual fish must be appropriate to its type. Variations in colour in a finished dish may be approached by use of toppings, coatings, sauces, vegetables and garnishes.
- Texture and consistency – these are likely to be determined by the dish itself. For example, a poached fish dish with a sauce will be soft and moist but different textures and variations in consistency can be added with vegetables and accompaniments. It is very important not to overcook fish as this will affect texture and the eating quality will be spoiled.
- Taste – properly cooked fresh fish will have its own distinct taste. Cooking procedures may change the overall taste and other flavours will be added with coatings, toppings, marinades, sauces and accompaniments. It is important for chefs to actually taste these items and decide if there are any faults or if adjustments are needed.
- Temperature – ensure that the fish, accompaniments, garnishes and sauces are at the right temperature for the dish requirements: hot, warm or cold. Dishes to be served cold should remain below 5°C until required and generally for whole pieces of fish temperatures can be slightly lower than usual cooking temperature requirements as long as this is above 63°C. However, for prepared or made up fish dishes such as fishcakes the usual 75°C + is required. Follow your individual dish requirements. Serving dishes and plates must be clean and also at the required temperature.

- Aroma – the dish must smell appetising and appropriate to dish requirements. Never serve any fish dish if you have concerns about unpleasant aromas.

Presenting fish dishes

Presentation of any dish is a very important consideration, complying with dish requirements and standards.

Dishes must be presented as attractively as possible with different components of the dish in harmony with each other. Accompaniments and garnishes should be appropriate in flavour, colour, texture and size; they must not be not so large that they overpower the dish and need to enhance the presentation of the entire dish.

The plate should be the right size for requirements and appear to be well balanced with the food.

Sauces

A number of different effects can be achieved with an accompanying sauce. Popular techniques include:

- flooding – sauce is ladled or spooned across the plate before or after the fish is plated
- masking – carefully spooning a sauce over the fish to thinly coat it
- drizzling – a thin 'drizzle' of sauce is added to the plate or the plated fish
- glazing – usually refers to finishing the fish with a light sauce and finishing under a salamander
- cordon – a line or ribbon of sauce used across the plate before or after plating the fish.

> **HEALTH AND SAFETY**
> Take care that if garnishes and sauces introduce an allergen that was not previously present in the fish dish, clear information is given to customers to indicate this.

ACTIVITY

1   The term 'sustainable fish' is frequently seen in the press and on some restaurant menus. What is meant by 'sustainable fish'? Select six varieties of fish that are considered to be sustainable.

2   Produce the fish section for the menu of a large restaurant using sustainable fish cooked by at least three different methods. Describe how each dish would be served, the sauces that could be used and how the dishes would be garnished. Highlight two of the dishes as a healthy eating option.

3   Produce two of the dishes (you can use recipes from this chapter) and write comments on your finished dishes.

TAKE IT FURTHER

A coastal hotel has decided to open a new seafood restaurant. Design a new menu for the restaurant to include:

● eight first course dishes and ten main course dishes
● at least five different fish cooking methods
● at least five shellfish
● a menu description for each dish.

For the above, design a notice to be included in the menu stating how you would inform customers with allergies about the content of your menu.

TEST YOURSELF

1 Describe **three** types of oily fish and how each may be cooked and presented.

2 What is the difference in production procedure for hot-smoked salmon and cold-smoked salmon?

3 At what temperatures should fresh fish and frozen fish be stored? Which preservation methods allow fish to be stored at room temperature?

4 How is removing skin from a Dover sole different to removing skin from plaice?

5 State **three** types of shellfish that may be purchased alive. How would you store them until they are ready to use?

6 Describe **three** quality points to look for when taking delivery of lobsters and mussels.

7 Suggest **five** suitable ways of cooking salmon? Would you include a sauce with any of these?

8 What are the quality points you would look for in a 10-kilogram box of chilled haddock fillets?

9 Describe **four** preservation methods commonly used for fish.

10 Suggest **two** fish that would be suitable for adding a stuffing. How would you cook these fish?

1 Breadcrumbed fried fish fillets

Energy	Calories	Fat	Saturated fat	Carbohydrates	Sugar	Protein	Fibre
1736 kJ	415 kcal	14.0 g	1.8 g	41.6 g	35 g	33.2 g	2.1 g

SERVING SUGGESTION
Serve with either lemon quarters or tartare sauce.

NOTE
Suitable for cod, haddock, pollack, coley and flat fish fillets such as plaice and sole.

Mise en place

1 Pass the fillets through flour, beaten egg and fresh white breadcrumbs. (Pat the surfaces well to avoid loose crumbs falling into the oil, burning and spoiling both the fat and the fish.)

Cooking

1 Deep-fry at 175 °C, until the fish turns a golden-brown. Remove and drain well.

Pané the fish, passing it through flour, egg and breadcrumbs in turn.

Video: Deep-frying fish
http://bit.ly/2oPBAuc

2 Goujons of plaice

Energy	Calories	Fat	Saturated fat	Carbohydrates	Sugar	Protein	Fibre
994 kJ	261 kcal	13.8 g	0.0 g	13.8 g	0.0 g	21.3 g	0.6 g

Mise en place

1 Prepare bowls of seasoned flour, egg-wash and breadcrumbs

2 Cut fillets of plaice into strips approximately 8 × 0.5 cm. Wash and dry well.

3 Pass through flour, beaten egg and fresh white breadcrumbs. Pat the surfaces well so that there are no loose crumbs which could fall into the fat and burn.

Cooking

1 Deep-fry at 175 °C, then drain well.

PROFESSIONAL TIP

Keep the coating ingredients (flour, egg and breadcrumbs) separate. Shake off any excess flour and egg before dipping the fish into the breadcrumbs.

Other fish, such as sole or salmon, may be used instead of plaice.

SERVING SUGGESTION

Serve with lemon quarters and a suitable sauce (e.g. tartare).

HEALTH AND SAFETY

There are three of the highlighted allergens in this dish according to Food Information for Consumers Regulation No. 1169/201: fish, eggs and gluten.

Video: Cutting a fillet of fish into goujons

http://bit.ly/2pps8B2

VARIATION

● Other white fish can be used.

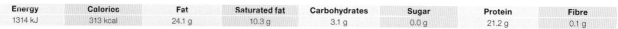

3 Fish meunière

Energy	Calories	Fat	Saturated fat	Carbohydrates	Sugar	Protein	Fibre
1314 kJ	313 kcal	24.1 g	10.3 g	3.1 g	0.0 g	21.2 g	0.1 g

Mise en place

1 Prepare a bowl of seasoned flour. Clarify the butter.
2 Prepare and clean the fish, wash and drain.
3 Pass through seasoned flour, shake off all surplus.

Cooking

1 Shallow-fry on both sides, presentation side first, in hot clarified butter or oil.
2 Dress neatly on an oval flat dish or plate/plates.
3 Peel a lemon, removing the peel, white pith and pips.
4 Cut the lemon into slices and place one on each portion.
5 Squeeze some lemon juice on the fish.

6 Allow 10–25 g butter per portion and colour in a clean frying pan to the nut-brown stage (*beurre noisette*).
7 Pour over the fish.
8 Sprinkle with chopped parsley and serve.

NOTE

Many fish, whole or filleted, may be cooked by this method, for example, sole, sea bass, bream, fillets of plaice, trout, brill, cod, turbot, herring and scampi.

PROFESSIONAL TIP

When the butter has browned, try adding a squeeze of lemon juice or a splash of white wine for extra flavour.

FAULTS

The skinned fish fillets are delicate and can break up easily. Turn them carefully to avoid breaking them.

Healthy Eating Tips

● Use a small amount of unsaturated oil to fry the fish instead of the butter.
● Use less *beurre noisette* per portion. Some customers will prefer the finished dish without the additional fat.

Making fish meunière

http://bit.ly/2qj39fO

4 Délice of flat white fish Duglére

Energy	Calories	Fat	Saturated fat	Carbohydrates	Sugar	Protein	Fibre	Sodium
1974 kJ	475 kcal	38 g	23 g	3.5 g	3.2 g	28 g	0.7 g	0.58 g

	4 portions	10 portions
Fillets of flat white fish (e.g. Sole, plaice)	400–600 g	1–1.5 kg
Fish stock for poaching	approx. 200 ml	approx. 500 ml
For the sauce		
Butter	25 g	60 g
Shallots, finely chopped	20 g	50 g
Fish stock	60 ml	150 ml
Dry white wine	60 ml	150 ml
Whipping cream	200 ml	500 ml
Butter, sliced and kept cold on ice	50 g	125 g
Tomatoes, skinned and neatly diced (concassé)	2	5
Parsley, chopped finely	10 g	20 g

Mise en place

1 Skin the fish, trim and wash.

2 Fold the fillets neatly, ensuring the skinned side is facing inwards (délice).

Cooking

1 Using a wide, shallow pan, poach the délice gently in fish stock for 4–6 minutes (depending on the thickness of the fillet).

2 To make the sauce, sweat the finely chopped shallots with the butter in a saucepan, until translucent.

3 Add the fish stock and reduce by one-third.

4 Add the dry white wine and reduce again by half.

5 Add the whipping cream and reduce by one-third, until the cream starts to thicken the sauce to a coating consistency.

6 Add the cold, sliced butter and ripple the sauce over the butter until the butter has emulsified into the sauce (monté). Check the seasoning and adjust accordingly. Do not allow the sauce to reboil as the butter will split and the sauce will become greasy.

7 Add the neatly cut tomato concassé and finely chopped parsley to the sauce.

8 To serve, drain the fish well and place neatly on a plate; carefully coat each délice with the sauce and serve.

Sweat the finely chopped shallots.

Coat each délice with sauce.

> **PROFESSIONAL TIP**
> Completely coat the cooked fish fillets neatly with the sauce (nappe). Do not use too much sauce.

> **HEALTH AND SAFETY**
> Clear away all the fish preparation items, sanitise surfaces and wash hands thoroughly before beginning the cooking process.

> **SERVING SUGGESTION**
> Could be served in shallow, individual dishes.

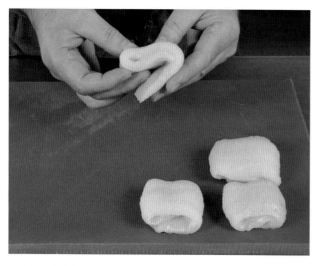

Fold each fillet into a délice.

5 Fillets of fish in white wine sauce (*filets de poisson vin blanc*)

Energy	Calories	Fat	Saturated fat	Carbohydrates	Sugar	Protein	Fibre
1421 kJ	342 kcal	24.0 g	12.8 g	5.8 g	0.9 g	25.9 g	0.2 g

	4 portions	10 portions
Fillets of white fish	400–600 g	1–1.5 kg
Butter, for dish and greaseproof paper		
Shallots, finely chopped and sweated	10 g	25 g
Fish stock	60 ml	150 ml
Dry white wine	60 ml	150 ml
Lemon, juice of	¼	½
Fish velouté	250 ml	625 ml
Butter	50 g	125 g
Cream, lightly whipped	2 tbsp	5 tbsp

Mise en place

1 Skin and fillet the fish, trim and wash.
2 Butter and season an earthenware dish.

Cooking

1 Sprinkle with the sweated chopped shallots and add the fillets of sole.
2 Season, add the fish stock, wine and lemon juice.
3 Cover with buttered greaseproof paper.
4 Poach in a moderate oven at 150–200 °C for 7–10 minutes.
5 Drain the fish well; dress neatly on a flat dish or clean earthenware dish.
6 Bring the cooking liquor to the boil with the velouté.
7 Correct the seasoning and consistency and pass through double muslin or a fine strainer.

8 Mix in the butter then, finally, add the cream.
9 Coat the fillets with the sauce. Garnish with *fleurons* (puff paste crescents).

PROFESSIONAL TIP

In this recipe, the shallots should be sweated before use; however, if they are very finely chopped, they could be added raw.

Healthy Eating Tips

- Keep the added salt to a minimum.
- Reduce the amount of butter and cream added to finish the sauce.
- Less sauce could be added, plus a large portion of potatoes and vegetables.

VARIATION

Fish bonne femme: add the following to the fish before cooking: 100 g thinly sliced white button mushrooms and chopped parsley.

Fish bréval: as for *bonne-femme* plus 100 g diced, peeled and deseeded tomatoes.

Video: Filleting a flat fish

http://bit.ly/2pCTtjX

HEALTH AND SAFETY

Fish deteriorates and spoils very quickly. Keep refrigerated until needed.

FAULTS

Remove the cooked fish from the cooking dish carefully. The cooked fish fillets are delicate and could easily break up, making the dish look untidy and unprofessional.

ON A BUDGET

Less expensive skinned white fish can be used for this dish.

6 Grilled fillets of sole, plaice or haddock

Energy	Calories	Fat	Saturated fat	Carbohydrates	Sugar	Protein	Fibre
802 kJ	191 kcal	7.8 g	1.0 g	3.9 g	0.1 g	26.6 g	0.2 g

> **PROFESSIONAL TIP**
> Oil the grill bars well, so that the fish does not stick.

> **NOTE**
> Grilling is a healthy method of cooking so do not use too much oil for brushing.

> **HEALTH AND SAFETY**
> Take care when using a salamander as without due care arms can easily be burned.

> **FAULTS**
> Fish fillets cook very quickly, so take care not to overcook or the fish will break up as well losing texture and flavour.

Mise en place

1 Remove the skin from sole and plaice. Wash the fillets and dry them well.

2 Pass through flour, shake off surplus and brush with oil.

Cooking

1 Place on hot grill bars, a griddle or a greased baking sheet if grilling under a salamander. Brush occasionally with oil. Turn the fish carefully and grill on both sides. Do not overcook.

2 Serve with lemon quarters and a suitable sauce (e.g. compound butter or salsa).

7 Frying batters (pâtes à frire) for fish

Recipe A:

	6–8 portions	10 portions
Flour	200 g	500 g
Salt		
Yeast	10 g	25 g
Water or milk	250 ml	625 ml

Preparing the dish

1 Sift the flour and salt into a basin.
2 Dissolve the yeast in a little of the water.
3 Make a well in the flour. Add the yeast and the liquid.
4 Gradually incorporate the flour and beat to a smooth mixture.
5 Allow to rest for at least 1 hour before using.

Recipe C:

	6–8 portions	10 portions
Flour	200 g	500 g
Salt		
Water or milk	250 ml	625 ml
Oil	2 tbsp	5 tbsp
Egg whites, stiffly beaten	2	5 tbsp

1 As for Recipe B, but fold in the egg whites just before using.

> **VARIATION**
> ● Other ingredients can be added to batter (e.g. chopped fresh herbs, grated ginger, garam masala, beer).

> **HEALTH AND SAFETY**
> Wash hands thoroughly before and after handling raw eggs.

> **FAULTS**
> Follow the recipe carefully so the batter coats the fish well but is not too thick.

Recipe B:

	6–8 portions	10 portions
Flour	200 g	500 g
Salt		
Egg	1	2–3
Water or milk	250 ml	625 ml
Oil	2 tbsp	5 tbsp

1 Sift the flour and salt into a basin. Make a well. Add the egg and the liquid.
2 Gradually incorporate the flour and beat to a smooth mixture.
3 Mix in the oil. Allow to rest before using.

8 Deep-fried fish in batter

Energy	Calories	Fat	Saturated fat	Carbohydrates	Sugar	Protein	Fibre
736 kJ	415 kcal	14.0 g	1.8 g	41.6 g	3.5 g	33.2 g	2.1 g

NOTE
Most suitable for cod, haddock, pollock and coley.

HEALTH AND SAFETY
Fried fish in batter will include some of the identified allergens under the Food Information for Consumers Regulation No. 1169/201: fish, gluten, eggs and milk. Customers may need to be informed that these allergens are present.

Healthy Eating Tip
● Drain fried fish well then place on kitchen paper to remove excess oil.

FAULTS
The frying oil needs to be at the correct temperature (around 175 °C). If lower than this, the finished product will be pale and oily and if too high it could be dark and the fish may not be properly cooked in the middle.

ON A BUDGET
A variety of lower cost white fish can be deep fried.

Mise en place

1 Pass the prepared, washed and well-dried fish through flour, shake off the surplus and pass through the batter.

Cooking

1 Place carefully away from you into the hot deep-fryer at 175 °C until the fish turns a golden-brown. Remove and drain well.

SERVING SUGGESTION
Serve with either lemon quarters or tartare sauce.

PROFESSIONAL TIP
Remove any excess batter before frying; too much batter will make the dish too heavy.

Pass the prepared fish through the batter.

Shake off any excess and then lower carefully into the fryer.

9 Whitebait (*blanchailles*)

Energy	Calories	Fat	Saturated fat	Carbohydrates	Sugar	Protein	Fibre
2174 kJ	525 kcal	47.5 g	0.0 g	5.3 g	0.1 g	19.5 g	0.2 g

Allow 100 grams per portion.

Mise en place

1 Pick over the whitebait, wash carefully and drain well.

2 Pass through milk and seasoned flour.

Cooking

1 Shake off surplus flour in a wide-mesh sieve and place fish into a frying-basket.

2 Plunge into very hot oil, around (190 °C).

3 Cook until brown and crisp (approximately 1 minute).

4 Drain well.

5 Season lightly with salt and cayenne pepper.

SERVING SUGGESTION

Serve garnished with fried or picked parsley and quarters of lemon.

PROFESSIONAL TIP

It is important to shake off any excess flour before frying; too much flour will cause the fish to stick together.

HEALTH AND SAFETY

Take extra care when frying in very hot oil. Keep hands well clear of the oil and handle the frying basket carefully. Have everything ready before you start.

Healthy Eating Tip

● Serve with a crisp green salad.

VARIATION

● Serve immediately while crisp and hot.

10 Pan-fried fillets of sea bass with rosemary mash and mushrooms

Energy	Calories	Fat	Saturated fat	Carbohydrates	Sugar	Protein	Fibre	*
1183 kJ	282 kcal	11.6 g	3.6 g	5.3 g	0.6 g	28.2 g	1.3 g	

* Using lemon sole in place of sea bass, and button mushrooms

	4 portions	10 portions
Vegetable oil	10 ml	25 ml
Salt, pepper		
Sea bass portions, skin on	4 × 100 g	10 × 100 g
Wild mushrooms, sliced	200 g	600 g
Extra virgin olive oil	10 ml	25 ml
Rosemary, chopped	pinch	1 tsp
Mashed potato (see page 177)	400 g	1 kg

Cooking

1 Heat the vegetable oil in a non-stick pan, season and then fry the sea bass, skin side first, until it has a golden colour and is crisp.

2 Turn the fish over and gently seal without colouring.

3 Remove from the pan (skin side up) and keep warm.

4 Quickly and lightly fry the mushrooms in extra virgin olive oil.

5 Mix the rosemary into the mashed potato.

6 Arrange the potato in the centre of hot plates.

7 Place the fish on top, skin side up.

8 Garnish with the mushrooms and olive oil and serve.

Healthy Eating Tips

- Using an unsaturated oil (sunflower or olive), lightly oil the non-stick pan to fry the sole.
- Use a little olive oil to fry the mushrooms.
- Keep added salt to a minimum.
- Serve with plenty of seasonal vegetables and new potatoes.

PROFESSIONAL TIP

Do not cook too many portions at once. This would cool the pan and it would be difficult to crisp the skin.

HEALTH AND SAFETY

The handle of the frying pan will get hot so have a thick cloth to hand. The oil may also 'spit' when the fish is placed in the pan so stand back a little.

11 Poached smoked haddock

Energy	Calories	Fat	Saturated fat	Carbohydrates	Sugar	Protein	Fibre
702 kJ	168 kcal	7.6 g	4.6 g	1.4 g	1.4 g	23.4 g	0 g

PROFESSIONAL TIP

The illustration shows undyed smoked haddock which is now much preferred. Previously dyed smoked haddock which was a bright yellow colour may have been used but the use of dyes on fish has now fallen from favour.

NOTE

Smoked haddock has a strong smell. Keep well wrapped when in the refrigerator when stored with other items as this can be passed to other foods.

	4 portions	10 portions
Smoked haddock fillets	400–600 g	1.2 kg
Milk and water, mixed		

HEALTH AND SAFETY

Smoking and salting does preserve fish slightly but refrigerate and treat as other fresh fish.

Mise en place

1 Skin and cut fillets into even portions.

Cooking

1 Place fillets into a shallow pan and just cover with half milk and water.

2 Poach gently for a few minutes until cooked.

3 Drain well and serve.

Healthy Eating Tip

● Poaching is a healthy way to cook fish as no fat is used.

VARIATION

● This is a popular breakfast dish and is also served as a lunch and a snack dish.
● When cooked, garnish with slices of peeled tomato or tomato concassé, lightly coat with cream, flash under the salamander and serve.
● Top with a poached egg.
● When cooked, lightly coat with Welsh rarebit mixture, brown under the salamander and garnish with peeled slices of tomato or tomato concassé.

12 Steamed fish with garlic, spring onions and ginger

Energy	Calories	Fat	Saturated fat	Carbohydrates	Sugar	Protein	Fibre *
468 kJ	112 kcal	3.5 g	0.7 g	1.2 g	0.4 g	18.7 g	0.1 g

* Using 2 cloves of garlic

	4 portions	10 portions
White fish fillets, e.g. cod, sole	400 g	1.5 kg
Salt		
Fresh ginger, freshly chopped	1 tbsp	2½ tbsp
Spring onions, finely chopped	2 tbsp	5 tbsp
Light soy sauce	1 tbsp	2½ tbsp
Cloves of garlic, peeled and thinly sliced	4	10
Oil	1 tbsp	2½ tbsp

Mise en place

1 Wash and dry the fish well; rub very lightly with salt on both sides.

2 Put the fish on to plates, scatter the ginger evenly on top. Place another plate on top or place on steamer trays and cover with greaseproof paper.

Cooking

1 Put the plates into the steamer and steam gently until just cooked (5–15 minutes, according to the thickness of the fish).

2 Remove the plates, sprinkle on the spring onions and soy sauce.

3 Brown the garlic slices in the hot oil and pour over the dish.

NOTE
Garlic and ginger have intense flavours, so they must be chopped very finely.

Healthy Eating Tips
- Steaming is a healthy way of cooking.
- Serve with a large portion of rice or noodles and stir-fried vegetables.

VARIATION
- This is a Chinese-themed recipe that can be adapted in many ways – for example, replace the spring onions and garlic and use thinly sliced mushrooms, diced tomato (skinned and deseeded), finely chopped shallots, lemon juice, white wine, chopped parsley, dill or chervil.

PROFESSIONAL TIP
Steamed fish is delicate and can break easily so handle carefully.

HEALTH AND SAFETY
Take care when using a steamer; released steam is very hot and can burn the hands and face. Use thick cloths to remove items and when opening the door of a steaming oven open gradually and stand back.

13 Griddled monkfish with leeks and Parmesan

Energy	Calories	Fat	Saturated fat	Carbohydrates	Sugar	Protein	Fibre
1009 kJ	239 kcal	8.3 g	5.0 g	1.0 g	0.8 g	40.3 g	0.6 g

	4 portions	10 portions
Oil		
Leeks, finely sliced	100 g	250 g
Parmesan, grated	100 g	250 g
Salt, pepper		
Prepared monkfish fillets	750 g	1.8 kg
Egg white, lightly beaten	2	5
Lemons	1	3
Mixed salad leaves, to serve		

Mise en place

1 Heat the griddle pan and oil lightly.

2 Cut the leeks into fine julienne with the Parmesan, then season.

3 Cut the monkfish into 1.5-cm-thick slices. Dry, dip in the beaten egg white, then in the leek and Parmesan.

Cooking

1 Place the monkfish on the griddle to cook (approximately 3–4 minutes). Turn gently to cook the other side.

2 Garnish with lemon wedges and mixed leaves.

Healthy Eating Tips
- There is no need to add salt as there is plenty in the cheese.
- Serve with a large portion of potatoes and vegetables or salad.

PROFESSIONAL TIP
Handle the fish gently on the griddle as it can easily break.

HEALTH AND SAFETY
Griddles can become very hot. Make sure that you have a thick cloth to hand and keep jacket sleeves rolled down to protect the arms.

ON A BUDGET
Monkfish can be expensive. You use any thick fillets of fish.

14 Grilled round fish (herring, mackerel, bass, mullet, trout)

Energy	Calories	Fat	Saturated fat	Carbohydrates	Sugar	Protein	Fibre
2238 kJ	536 kcal	33.2 g	8.3 g	0 g	0 g	59.5 g	0 g

Mise en place

1 Descale fish where necessary, using the back of a knife working from tail to head of fish.

2 Remove heads if required clean out intestines, trim off all fins and tails using fish scissors. Leave herring roes in the fish.

3 Wash and dry well.

Cooking

1 Make three light incisions 2-mm deep on either side of the fish. This is known as 'scoring' and helps the heat to penetrate the fish.

2 Pass through flour, shake off surplus and brush with oil.

3 Place on hot grill bars, a griddle or a greased baking sheet if grilling under a salamander. Brush occasionally with oil.

4 Turn the fish carefully and grill on both sides. Do not overcook.

> **SERVING SUGGESTION**
> Serve with lemon quarters and a suitable sauce (e.g. compound butter or salsa).
>
> Herrings are traditionally served with a mustard sauce. Mackerel may be butterfly filleted and grilled. Trout is sometimes filled with a suitable stuffing before cooking.

---

15 Poached salmon

Energy	Calories	Fat	Saturated fat	Carbohydrates	Sugar	Protein	Fibre
1284 kJ	309 kcal	20.6 g	5.5 g	0.3 g	0.3 g	30.4 g	0.1 g

> **SERVING SUGGESTION**
> Serve with a suitable sauce (e.g. hollandaise) or melted herb butter, and thinly sliced cucumber.
>
> Depending on the size of the salmon, either a whole or half a darne would be served as a portion.

Video: Cutting fish into darnes

http://bit.ly/2ppsKqj

> **HEALTH AND SAFETY**
> Take care to avoid burns and scalds when placing the salmon into the hot court bouillon.

> **Healthy Eating Tip**
> ● Salmon is an oily fish and including this in the diet is considered beneficial to health. Poaching is a healthy method of cookery because no fat is used

> **ON A BUDGET**
> Most fish and fish fillets can be poached.

Mise en place

1 Prepare the court bouillon and bring it to boiling point. Have a fish slice ready to remove the fish.

Cooking

1 Place the prepared and washed darnes of salmon in a barely simmering court bouillon for approximately 5 minutes.

2 Drain well and carefully remove the centre bone. Ensure that the fish is cleaned of any cooked blood.

16 Grilled swordfish and somen noodle salad with coriander vinaigrette

Energy	Calories	Fat	Saturated fat	Carbohydrates	Sugar	Protein	Fibre *
2599 kJ	623 kcal	42.6 g	6.1 g	40.3 g	2.7 g	21.2 g	2.5 g

* Using pasta for buckwheat pasta

	4 portions	10 portions
Swordfish fillets	4 × 75 g	10 × 75 g
Buckwheat somen noodles	200 g	500 g
Romaine lettuce	1/4	1/2
Celery, finely diced	50 g	125 g
Chopped onion (red)	50 g	125 g
Vinaigrette		
Olive oil	70 ml	170 ml
Sesame oil	35 ml	85 ml
Vegetable oil	35 ml	85 ml
Rice wine	2 tbsp	4 ½ tbsp
Lemon, juice of	½	1
Lime, juice of	1	2
Fresh ginger, chopped	2 tbsp	5 tbsp
Garlic, chopped	1 clove	3 cloves
Coriander leaves	1 tbsp	2 ½ tbsp
Sesame seeds (black)	2 tbsp	5 tbsp
Seasoning		

Mise en place

1 Prepare the vinaigrette by placing all the ingredients in a liquidiser, purée until smooth. Season to taste.

Cooking

1 Season the swordfish fillets, brush with the vinaigrette. Grill for 2–3 minutes on each side until cooked. Allow to cool.

2 Cook the noodles in boiling water, refresh and drain. Allow to cool.

3 Shred the lettuce finely, mix with the diced celery and finely chopped red onion. Season with the vinaigrette.

4 Serve by arranging the noodles in the centre of each plate.

5 Place the swordfish fillets on the noodles, season with vinaigrette.

6 Garnish with romaine lettuce. Freshly ground black pepper may be used to finish the dish.

Healthy Eating Tips
- Keep added salt to a minimum.
- Adding less vinaigrette to the finished dish can reduce the fat content.

PROFESSIONAL TIP
Use a separate board for preparing the vegetables. These will not be cooked and must not be contaminated by bacteria from the raw fish.

HEALTH AND SAFETY
Remember that fish are one of the identified allergens. It also deteriorates rapidly so must be kept under refrigeration until needed.

FAULTS
If overcooked swordfish can become tough and unpalatable.

17 Roast fillet of sea bass with vanilla and fennel

Energy	Calories	Fat	Saturated fat	Carbohydrates	Sugar	Protein	Fibre
2322 kJ	559 kcal	40 g	10.1 g	8.5 g	7.6 g	34 g	9.3 g

	4 portions	10 portions
Sea bass fillets (approximately 160 g each, cut from a 2–3 kg fish) skin on, scaled and pin-boned	4	10
Seasoning		
Vegetable oil	50 ml	125 ml
Fennel		
Bulbs of baby fennel	8	20
Vegetable oil	50 ml	125 ml
Fish stock	500 ml	1 ¼ litres
Clove of garlic	1	3
Vanilla sauce		
Shallots, peeled and sliced	1	3
Fish stock	500 ml	1 ¼ litres
White wine	200 ml	500 ml
Vanilla pods	2	5
Butter	50 g	125 g
Chives, chopped	1 tsp	3 tsp
Tomato concassé	100 g	250 g

Mise en place

1 Trim the fennel bulbs well, ensuring they are free from blemishes and root.

Cooking

For the fennel:

1 Heat the oil in a pan, place in the fennel and slightly brown.

2 Add the stock and garlic, bring to the boil and cook until tender.

For the sauce:

1 Heat a small amount of vegetable oil in a pan.

2 Add the shallots and cook without colour, add the stock, white wine and split vanilla and reduce by two-thirds.

3 Pass through a chinois and reserve for serving.

4 Add the butter and chopped chives.

To finish:

1 Pre-heat the oven to 180 °C.

2 Heat the oil in a non-stick pan and place the seasoned sea bass fillets in skin side down.

3 Cook for 2 minutes on the stove and then place in the oven for 3 minutes (depending on thickness) still with the skin side down.

4 Meanwhile, reheat the fennel and add the tomato concassé to the sauce.

5 Remove the sea bass from the oven and turn in the pan, finishing the flesh side for 30 seconds to 1 minute.

6 Lay the fennel in the centre of the plate and place the sea bass on top.

7 Finish the dish with the sauce over the bass and around, serve immediately.

NOTE
A marriage of flavour: vanilla, bass and fennel are made for each other. This fish can also be steamed.

PROFESSIONAL TIP
Take note of how long the fish has been in the oven. Because it is out of sight it is easy to forget about it and overcook the fish.

HEALTH AND SAFETY
Remember that raw fish can be a contaminant so keep it covered and away from other foods. It is also one of the 14 identified allergens.

Healthy Eating Tip
● Reduce the amount of butter used in the sauce and increase the vegetable content for a healthier dish.

VARIATION
● Substitute the sea bass with any other fish fillets.

18 Baked cod with a herb crust

Energy	Calories	Fat	Saturated fat	Carbohydrates	Sugar	Protein	Fibre *
1882 kJ	452 kcal	30.8 g	18.7 g	12.7 g	0.8 g	31.7 g	0.4 g

* Using mustard powder (1 tsp) for herb mustard

	4 portions	10 portions
Cod fillets, 100–150 g each	4	10
Herb or English mustard	½ teaspoon	2 teaspoons
Fresh breadcrumbs	100 g	250 g
Butter, margarine or oil	100 g	250 g
Cheddar cheese, grated	100 g	250 g
Parsley, chopped	1 tsp	1 tbsp
Salt, pepper		

Mise en place

1 Line a suitable oven tray with baking parchment paper.

2 Place the prepared, washed and dried fish on a greased baking tray or ovenproof dish.

3 Combine the ingredients for the herb crust (the mustard, breadcrumbs, butter, margarine or oil, cheese, parsley and seasoning).

Cooking

1 Press the herb crust evenly over the fish.

2 Bake in the oven at 180 °C for approximately 15–20 minutes until cooked and the crust is a light golden-brown.

SERVING SUGGESTION

Serve either with lemon quarters or a suitable salsa (page 121) or sauce, e.g. hot tomato sauce.

PROFESSIONAL TIP

Add a little beaten egg to the breadcrumb mixture to help bind the mixture together.

Healthy Eating Tips

● Use a little sunflower oil when making the herb crust.
● Cheese is salty – no added salt is needed.
● Serve with a large portion of tomato or cucumber salsa and new potatoes.

HEALTH AND SAFETY

Although this is a simple dish it includes some identified allergens: fish, mustard, gluten and milk (in the cheese).

VARIATION

● Any other thick fish fillets may be used.

SERVING SUGGESTION

Serve with a selection of vegetables to add colour and interest to this dish.

19 Red mullet ceviche with organic leaves

Energy	Calories	Fat	Saturated fat	Carbohydrates	Sugar	Protein	Fibre
1245 kJ	299 kcal	20. 2 g	1.9 g	5 g	4.4 g	23.9 g	2.6 g

2 Place the red mullet in a container and cover with the liquid. Top with cling film to ensure all the air is kept out, capitalising on maximum curing. This will need to remain in the fridge for a minimum of 6 hours.

Dressing:

1 Add water and a pinch of saffron to a pan and bring to the boil.

2 Whisk in the vegetable oil and vinegar.

3 Season.

To finish:

1 Mix the dressed leaves lightly in vinaigrette.

2 Place the red mullet carefully on a plate with a little of the curing liquor, shallots and cucumber.

3 Top with the organic salad and finish with the saffron dressing and caviar (if using).

	4 portions	10 portions
Shallots, finely diced	2	5
Olive oil	1 tbsp	3 tbsp
White wine vinegar	50 ml	125 ml
Fish stock	1 lltre	2 ½ litres
Lemons, juice of	1	3
Cucumber, diced	2 tbsp	5 tbsp
Red mullet fillet spined and scaled (approximately 120 g each)	4	10
Mixed salad leaves	100 g	225 g
Vinaigrette	50 ml	125 ml
Caviar (optional)	50 g	125 g
Saffron dressing		
Water	10 ml	25 ml
Saffron	Pinch	Pinch
Vegetable oil	50 ml	125 ml
Vinegar	10 ml	25 ml

Cooking

1 Bring the shallots, olive oil, white wine vinegar and fish stock to the boil. Add the lemon and cucumber and allow to cool at room temperature.

PROFESSIONAL TIP

A cured dish always tastes of the true ingredients. Using red mullet, as here, the flavours are bold and earthy, and paired with the saffron it makes a perfect summer starter.

HEALTH AND SAFETY

This dish is cured rather than cooked. The curing process makes it safe to eat but always ensure that you use very fresh fish.

Healthy Eating Tip

● Serve with salad and maybe some wholemeal bread.

ON A BUDGET

Red mullet can be expensive. Try using cheaper fish cuts.

20 Sardines with tapenade

Energy	Calories	Fat	Saturated fat	Carbohydrates	Sugar	Protein	Fibre
3167 kJ	757 kcal	44.0 g	12.4 g	0.0 g	0.0 g	90.5 g	0.8 g

	4 portions	10 portions
Sardines	8–12	20–30
Tapenade		
Kalamata olives	20	50
Capers	1 tbsp	2½ tbsp
Lemon juice	1 tsp	2½ tsp
Olive oil	2 tsp	5 tsp
Anchovy paste (optional)	½ tsp	1¼ tsp
Freshly ground black pepper		

Mise en place

1 Prepare the tapenade by finely chopping the olives and capers (this can be done in a food processor). Add the lemon juice, olive oil, anchovy paste and black pepper. Mix well.

Cooking

1 Allow 2–3 sardines per portion. Clean the sardines and grill them on both sides, either on an open flame grill or under the salamander.

2 When cooked, smear the tapenade on top. The tapenade can be heated slightly before spreading.

VARIATION
- A little chopped basil and a crushed chopped clove of garlic may also be added to the tapenade.

PROFESSIONAL TIP
Sardines can also be filleted or purchased filleted. Cook in the same way but they will cook even faster.

Healthy Eating Tip
- Sardines are considered a healthy choice because they are an oily fish, so are beneficial to health. The tapenade, however, will add extra fat and salt so consider using it sparingly or not at all.

FAULTS
Sardines can overcook very quickly on a hot grill and become dry and lack flavour. Cook them quickly.

21 Fish kedgeree (*cadgery de poisson*)

Energy	Calories	Fat	Saturated fat	Carbohydrates	Sugar	Protein	Fibre
1974 kJ	472 kcal	28.2 g	15.3 g	29.3 g	4.7 g	25.7 g	1.2 g

* Using smoked haddock

SERVING SUGGESTION
Serve hot with a sauce boat of curry sauce.

NOTE
This dish is traditionally served for breakfast, lunch or supper. The fish used should be named on the menu (e.g. salmon kedgeree).

Healthy Eating Tips
- Reduce the amount of butter used to heat the rice, fish and eggs.
- Garnish with grilled tomatoes and serve with bread or toast.

HEALTH AND SAFETY
Using eggs, fish and rice makes this a high-risk dish. Once cooked serve immediately and avoid reheating.

VARIATION
- Seasoned boiled rice could be used in place of the pilaff rice.
- Add a few sautéed mushrooms or sliced sundried tomatoes just before serving.

ON A BUDGET
This is a fairly economical dish but could be made more so by reducing the amount of fish and increasing the egg.

	4 portions	10 portions
Fish (usually smoked haddock or fresh salmon)	400 g	1 kg
Milk for poaching		
Rice pilaff (see page 406)	200 g	500 g
Eggs, hard-boiled	2	5
Butter	50 g	125 g
Salt, pepper		
Chives, chopped	1 tsp	2 tsp
Curry sauce, to serve	250 ml	625 ml

Mise en place

1 Carefully check the fish on a blue board to make sure there are no small bones.
2 Heat the milk in a shallow pan.
3 Make the rice pilaff.

Cooking

1 Poach the fish in milk. Remove all skin and bone. Flake the fish.
2 Cook the rice pilaff. Cut the eggs into dice.
3 Combine the eggs, fish and rice and heat in the butter. Correct the seasoning and add the chives.

22 Fish pie

Energy	Calories	Fat	Saturated fat	Carbohydrates	Sugar	Protein	Fibre
879 kJ	209 kcal	12.0 g	5.3 g	11.9 g	3.2 g	14.1 g	0.9 g

2 Add the fish, mushrooms, egg and parsley. Correct the seasoning.

3 Place in a buttered pie dish.

4 Carefully spread or pipe the potato on top. Brush with eggwash or milk.

5 Brown in a hot oven or under the salamander and serve.

Healthy Eating Tips
- Keep the added salt to a minimum.
- This is a healthy main course dish, particularly when served with plenty of vegetables.

VARIATION

Many variations can be made to this recipe with the addition of:
- prawns or shrimps
- herbs such as dill, tarragon or fennel
- raw fish poached in white wine, the cooking liquor strained off, double cream added in place of béchamel and reduced to a light consistency.

PROFESSIONAL TIP

Multi-portion and individual fish pies are popular.

HEALTH AND SAFETY

As the fish in a fish pie may be reheated after the initial cooking, make sure it is heated very well to 75 °C+ to avoid the possibility of food poisoning.

ON A BUDGET

Lower cost fish types can be used in this dish.

	4 portions	10 portions
Béchamel (thin) (see page 100)	250 ml	625 ml
Cooked fish (free from skin and bone)	200 g	500 g
Mushrooms, cooked and diced	50 g	125 g
Egg, hard-boiled and chopped	1	3
Parsley, chopped		
Salt, pepper		
Potatoes, mashed or duchess	200 g	500 g
Eggwash or milk, to finish		

Mise en place

1 Prepare the béchamel, cook the fish and hard boil the eggs then chop them.

2 Boil the potatoes and mash them.

Cooking

1 Bring the béchamel to the boil.

1 Cook the fish, mushrooms and egg in the béchamel.

2 Pipe mashed potato over the top before baking.

23 Soused herring or mackerel

Energy	Calories	Fat	Saturated fat	Carbohydrates	Sugar	Protein	Fibre *
2419 kJ	576 kcal	44.5 g	9.4 g	3.0 g	3.0 g	41.0 g	1.1 g

* For 4 portions

	4 portions	10 portions
Herrings or mackerel	2	5
Salt, pepper		
Button onions	25 g	60 g
Carrots, peeled and fluted	25 g	60 g
Bay leaf	½	1 ½
Peppercorns	6	12
Thyme	1 sprig	2 sprigs
Vinegar	60 ml	150 ml

Mise en place

1 Clean, scale and fillet the fish.
2 Wash the fillets well and season with salt and pepper.
3 Roll up with the skin outside. Place in an earthenware dish.
4 Peel and wash the onion. Cut the onion and carrots into neat, thin rings.

Cooking

1 Blanch the vegetables for 2–3 minutes.
2 Add to the fish with the remainder of the ingredients.
3 Cover with greaseproof paper and cook in a moderate oven for 15–20 minutes.
4 Allow to cool, place in a dish with the onion and carrot.
5 Garnish with picked parsley, dill or chives.

Healthy Eating Tips
- Serve with plenty of salad vegetables and bread or toast (optional butter or spread).
- Keep added salt to a minimum.

HEALTH AND SAFETY
Soused fish is eaten cold. Cool the dish quickly and refrigerate until needed.

24 Haddock and smoked salmon terrine

Energy	Calories	Fat	Saturated fat	Carbohydrates	Sugar	Protein	Fibre
563 kJ	133 kcal	3.4 g	0.7 g	0.1 g	0.1 g	25.7 g	0.1 g

	4 portions	10 portions
Smoked salmon	140 g	350 g
Haddock, halibut or Arctic bass fillets, skinned	320 g	800 g
Salt, pepper		
Eggs, lightly beaten	1	2
Crème fraiche	40 ml	105 ml
Capers	3 tbsp	7 tbsp
Green or pink peppercorns	1 tbsp	2 tbsp

Mise en place

1 Line or grease the loaf tin; alternatively, line with cling film.

2 Cut the white fish into small pieces on a blue board using a chef's knife. Make sure all bones are removed.

Cooking

1 Line the tin with thin slices of smoked salmon. Let the ends overhang the mould. Reserve the remaining salmon until needed.

2 Cut two long slices of haddock the length of the tin and set aside.

3 Cut the rest of the haddock into small pieces. Season all the haddock with salt and pepper.

4 In a suitable basin, combine the eggs, crème fraiche, capers, and green or pink peppercorns. Add the pieces of haddock.

5 Spoon the mixture into the mould until one-third full. Smooth with a spatula.

6 Wrap the long haddock fillets in the reserved smoked salmon. Lay them on top of the layer of the fish mixture in the terrine.

7 Fill with the remainder of the haddock and crème fraiche mixture.

8 Smooth the surface and fold over the overhanging pieces of smoked salmon.

9 Cover with tin foil, secure well.

10 Cook in a water bath (bain-marie) of boiling water. Place in the oven at 200 °C for approximately 45 minutes, until set.

11 Remove from oven and bain-marie. Allow to cool. Do not remove foil cover.

12 Place heavy weights on top and leave in refrigerator for 24 hours.

13 When ready to serve, remove the weights and foil. Remove from mould.

SERVING SUGGESTION
Cut into thick slices, serve on suitable plates with a dill mayonnaise and garnished with salad leaves and fresh dill.

PROFESSIONAL TIP
Line the tin carefully so that the finished terrine will look neat and attractive.

Healthy Eating Tips
- Season with the minimum amount of salt.
- Offer the customer the mayonnaise separately.
- Serve with warm wholemeal bread or rolls (butter optional).

NOTE
Weights are placed on the terrine as it cools to ensure it is firm so will cut cleanly and neatly.

HEALTH AND SAFETY
Keep all ingredients under refrigeration until needed. As this is served cold, cool it quickly and refrigerate.

VARIATION
- Use any skinned white fish fillets in the filling.

25 Salmon fishcakes

Energy	Calories	Fat	Saturated fat	Carbohydrates	Sugar	Protein	Fibre
1821 kJ	438 kcal	30.2 g	13.6 g	18.0 g	1.5 g	24.5 g	1.6 g

	4 portions	10 portions
Salmon fillet – skinned and boneless	400 g	1 kg
Cooked mashed potato (make sure this is fairly dry)	225 g	560 g
Spring onions	3	8
Flat leaf parsley	10 g	25 g
Dill	Sprig	Sprig
Butter	50 g	125 g
Juice of lemon	1	2
Crème fraiche	1 tbsp	2½ tbsp
Thai fish sauce (nam pla) – optional	Few drops	1 tsp
Eggs	1	2
Pané		
Plain flour, seasoned		
Eggs		
Breadcrumbs		

Mise en place

1 Prepare the mashed potato and bowls of seasoned flour, eggwash and breadcrumbs.

Cooking

1 Place the salmon in an oiled roasting tin, season with salt and pepper, dot with butter and squeeze the lemon juice over.

2 Bake for approximately 7 minutes at 200 °C.

3 Allow the salmon to cool a little then flake into bite-sized pieces.

4 Chop the spring onions and herbs.

5 Add the salmon to the potato, herbs, spring onion, beaten egg and crème fraiche. Add a little nam pla if using and season.

6 Form into neat, even-sized cake shapes (a ring mould could be used), place on a tray lined with cling film and chill thoroughly for several hours.

7 Coat with seasoned flour, egg and breadcrumbs (pané), chill well again or the formed cakes can be frozen.

8 Cook in a deep-fryer as required, drain on kitchen paper.

SERVING SUGGESTION
Serve with a suitable sauce or salsa and/or mixed salad leaves.

HEALTH AND SAFETY
Take care when using the deep-fryer. Place the fishcakes in a frying basket and lower it slowly into the hot oil. Have a tray with kitchen paper ready for the fishcakes when they are cooked.

Healthy Eating Tip
● Serve with plenty of fresh vegetables or salad.

VARIATION
● Other types of fish could be used in place of the salmon.

FAULTS
If the potato mixture is too wet the fishcakes will not form properly and will fall apart on cooking.

SERVING SUGGESTION
Smaller fishcakes can be made and served as canapes.

26 Baked salmon salad with sea vegetables

Energy	Calories	Fat	Saturated fat	Carbohydrates	Sugar	Protein	Fibre
2548 kJ	613 kcal	45 g	6.9 g	18.3 g	1.8 g	37.5 g	2.8 g

	4 portions	**10 portions**
Fillet of salmon	4 × 150 g	10 × 150 g
Lime	2	5
Olive oil	2 tbsp	5 tbsp
Chopped basil leaves	1 tbsp	4 tbsp
Mayonnaise	4 tbsp	7 tbsp
Stale bread diced	100 g	250 g
Broad beans	200 g	500 g
Sea kale	100 g	250 g
Samphire	100 g	250 g
Salad leaves, rocket		
Lime wedges	4	10

Mise en place

1 Place the salmon on a suitable tray (on parchment paper or a non-stick tray), season with salt and pepper, sprinkle with olive oil and squeeze the juice of a lime on top.

2 Place the diced bread in a food processor with 2 tablespoons olive oil. Blitz until the bread becomes breadcrumbs.

Cooking

1 Cover with foil and gently bake in a pre-heated oven at 200 °C for approximately 15 minutes.

2 Place the breadcrumbs on a tray and bake until golden brown.

3 Add the basil leaves to the mayonnaise; add a squeeze of lime juice.

4 Blanch the broad beans and seakale in boiling water for 3–4 minutes.

5 Refresh in cold water and drain.

6 Place in a bowl, sprinkle with olive oil, lime juice and season.

7 Dress the cooked salmon on a suitable plate, mask thinly with the basil mayonnaise and sprinkle with breadcrumbs.

8 Garnish with the broad beans, sea kale and sapphire, finish with the salad leaves and lime wedges.

> **VARIATION**
> - As an alternative to breadcrumbs, use roasted flaked almonds.

> **HEALTH AND SAFETY**
> Prepare your cooking area. When taking hot trays from the oven, warn others that they are hot and make sure you have somewhere to put them down.

> **Healthy Eating Tip**
> - The mayonnaise could be replaced with a low fat fromage frais

27 Fish en papillote

This method of cookery is fresh-tasting and suitable for most fish or shellfish. This recipe outlines the technique.

Mise en place

1 Place a large piece of baking parchment or foil onto an ovenproof tray or dish. A loose, sealed pouch is made from the paper or foil once the ingredients are added.
2 The fish should be portioned, free from bones and may or may not be skinned.
3 Garnish the fish with a fine selection of vegetables chosen from carrots, leeks, celery, white mushrooms and wild mushrooms; a small amount of freshly chopped herbs may be added as desired.

Cooking

1 Moisten the fish with a little dry white wine, then seal the foil parcel and bake for 15–20 minutes (size and fish dependent).

SERVING SUGGESTION

Serve with an appropriate sauce (e.g. white wine).

Open the pouch at the table, releasing the fragrant aromas in front of the customer.

PROFESSIONAL TIP

This method cooks the fish and other ingredients in a sealed paper or foil pouch.

HEALTH AND SAFETY

Take care when opening the cooked fish in the pouch, as the trapped steam can burn.

Healthy Eating Tip

● This is a very healthy way to cook fish in its own steam.

VARIATION

Vary the types of fish used and vary the vegetables.

ON A BUDGET

This recipe would be suitable for cheaper fish fillets.

Seal the foil package.

Make sure it is tightly sealed on all sides.

28 Scallops with caramelised cauliflower

Energy	Calories	Fat	Saturated fat	Carbohydrates	Sugar	Protein	Fibre
968 kJ	230 kcal	7.9 g	4.3 g	21.2 g	18.5 g	19.7 g	1.9 g

	4 portions	10 portions
Raisins	100 g	250 g
Capers	100 g	250 g
Water	180 ml	450 ml
Sherry vinegar	1 tbsp	2 tbsp
Grated nutmeg		
Salt and cayenne pepper		
Butter	30 g	75 g
Head of cauliflower, sliced into ½ cm-thick pieces	½	1
Large hand-dived scallops (roe removed)	12	30

Mise en place

1 Slice the cauliflower.

2 Remove the roe from the scallops.

Cooking

1 In a small saucepan, cook the raisins and capers in the water until the raisins are plump – about 5 minutes.

2 Pour mixture into a blender and add the vinegar, nutmeg, salt and pepper. Blend just until smooth.

3 Set sauce aside.

4 In a sauté pan, heat butter and cook the cauliflower until golden on both sides. To prevent cauliflower from burning, if necessary, add about 1 tablespoon of water to pan during cooking. Set cauliflower aside.

5 In a separate pan, sauté the scallops in a little butter, about 1 minute on each side. To serve, place 3 scallops on each plate, top with cauliflower and finish with the caper-raisin emulsion.

PROFESSIONAL TIP

When using scallops, use hand-dived scallops as your first choice, as dredged scallops are sometimes unethically sourced. You will pay a little more for the hand-dived variety but the difference is certainly worth it.

NOTE

Be careful not to overcook the scallops, as they will become tough.

HEALTH AND SAFETY

Shellfish are high-risk food safety items, so keep refrigerated until needed, handle with care and cook as the recipe states.

Healthy Eating Tip

Reduce the amount of butter used or use a little sunflower oil instead.

SERVING SUGGESTION

There are a number of ways that this dish could be served attractively. Consider how this could be done.

29 Oysters in their own shell

Energy	Calories	Fat	Saturated fat	Carbohydrates	Sugar	Protein	Fibre
192 kJ	45 kcal	0.9 g	0.2 g	2.7 g	1 g	6.8 g	0.4 g

	4 portions
Rock or native oysters	24
Lemon	1
To accompany	
Brown bread and butter	
Tabasco or chilli sauce	

Mise en place

1 Select only those oysters that are tightly shut and have a fresh smell (category A is best, which means the waters they have grown in are clean).

2 Make sure you have a suitable oyster knife for opening oysters.

3 Check that the oysters are clean and closed.

Preparing the dish

1 To open an oyster, only the point of the rigid oyster knife is used. Hold the oyster with a thick oven cloth to protect your hand.

2 With the oyster in the palm of your hand, push the point of the knife about 1 cm deep into the 'hinge' between the 'lid' and the body of the oyster.

3 Once the lid has been penetrated, push down. The lid should pop open. Lift up the top shell, cutting the muscle attached to it.

4 Remove any splintered shell from the flesh and solid shell.

5 Return each oyster to its shell and serve on a bed of crushed ice with chilli sauce, brown bread and lemon.

NOTE

Make sure the oysters have been grown in or fished from clean waters, and take note of the famous rule only to use them when there is an 'r' in the month, although rock oysters are available throughout the year.

PROFESSIONAL TIP

Always purchase oysters from a reputable supplier and keep under refrigeration until needed.

HEALTH AND SAFETY

Oysters are often eaten raw so can be a food safety risk. This is why it is so important to purchase from a reliable source.

30 Oyster tempura

Energy	Calories	Fat	Saturated fat	Carbohydrates	Sugar	Protein	Fibre
858 kJ	205 kcal	12.2 g	1.6 g	10.8 g	0.1 g	13.7 g	0.4 g

The recipe for tempura batter is on page 219.

Mise en place

1 Open the fresh oysters.

2 Place each oyster on a tray on absorbent kitchen paper.

3 Lightly sprinkle with finely chopped herbs, parsley, basil and chives.

4 Pass each oyster through seasoned flour.

Cooking

1 Dip in tempura batter and fry in hot deep oil at 180 °C until golden brown.

2 Remove from the oil and drain well.

3 Serve immediately after cooking.

SERVING SUGGESTION

Serve as a garnish or as a hot hors d'oeuvre with a suitable dip, such as tomato chilli and lemon and garlic mayonnaise.

PROFESSIONAL TIP

Make sure the oil is at the right temperature. Cook the oysters quickly until crisp and golden brown.

HEALTH AND SAFETY

Oysters are a high-risk ingredient so keep refrigerated until needed. Oysters are also one of the identified allergen ingredients that customers must be informed of.

Healthy Eating Tip

● Drain well after frying and place on kitchen paper to absorb as much excess oil as possible.

31 Crab cakes with rocket salad and lemon dressing

Energy	Calories	Fat	Saturated fat	Carbohydrates	Sugar	Protein	Fibre
3002 kJ	719 kcal	43.9g	10.5g	41.9g	4.4g	41.4g	4.1g

	4 portions	10 portions
Crab cakes		
Shallots, finely chopped	25 g	60 g
Spring onions, finely chopped	4	10
Fish/shellfish glaze	75 ml	185 ml
Crab meat	400 g	1 kg
Mayonnaise	75 g	185 g
Lemons, juice of	1	3
Plum tomatoes skinned, cut into concassé	2	5
Wholegrain mustard	1 tsp	3 tsp
Seasoning		
Fresh white breadcrumbs	200 g	500 g
Eggs, beaten with 100 ml of milk	2	5
Salad and lemon dressing		
Vegetable oil	170 ml	425 ml
White wine vinegar	25 ml	60 ml
Lemons, juice of	1	3
Seasoning		
Rocket, washed and picked	250 g	625 g
Parmesan, shaved	100 g	250 g

Mise en place

1 Mix the shallots, spring onions and the fish glaze with the hand-picked crab meat.

2 Add the mayonnaise, lemon juice, tomato concassé and mustard, check and adjust the seasoning.

Cooking

Crab cakes:

1 Allow the crab cake mixture to rest for 30 minutes in the refrigerator.

2 Scale into 80–90 g balls and shape into discs 1½ cm high, place in the freezer for 30 minutes to harden.

3 When firm to the touch, coat in breadcrumbs using the flour, egg and breadcrumbs.

4 Allow to rest for a further 30 minutes.

5 Heat a little oil in a non-stick pan, carefully place the cakes in and cook on each side until golden brown.

For the salad and dressing:

1 Combine the oil, vinegar and lemon juice together, check the seasoning.

2 Place the rocket and Parmesan in a large bowl and add a little dressing, just to coat.

3 Place this in the centre of each plate, top with the crab cakes and serve.

> **PROFESSIONAL TIP**
> Any excess crab meat can be used up in this recipe which is a quick, classic dish. The crab can be exchanged for salmon or most fresh fish trimmings.

> **HEALTH AND SAFETY**
> Cooked crab is a high-risk item. To avoid possible food poisoning keep the crab refrigerated until needed and serve immediately after cooking.

> **Healthy Eating Tip**
> ● Serve the fishcakes with a mixed leaf salad or some lightly stir-fried vegetables.

> **VARIATION**
> ● Use other shellfish such as chopped prawns.

32 Mussels in white wine sauce (*moules marinière*)

Energy	Calories	Fat	Saturated fat	Carbohydrates	Sugar	Protein	Fibre
1900 kJ	452 kcal	14.3 g	5.3 g	18.1 g	0.9 g	61.4 g	0.6 g

3 Drain off all the cooking liquor in a colander set over a clean bowl to retain the cooking juices.

4 Carefully check the mussels and discard any that have not opened.

5 Place in a dish and cover to keep warm.

6 Make a roux from the flour and butter; pour over the cooking liquor, ensuring it is free from sand and stirring continuously to avoid lumps.

7 Correct the seasoning and garnish with more chopped parsley.

8 Pour over the mussels and serve.

	4 portions	10 portions
Shallots, chopped	50 g	125 g
Parsley, chopped	1 tbsp	2 tbsp
White wine	60 ml	150 ml
Strong fish stock	200 ml	500 ml
Mussels	2 kg	5 kg
Butter	25 g	60 g
Flour	25 g	60 g
Seasoning		

VARIATION

● For an Eastern influence, why not add a little red chilli and replace the parsley with coriander?

PROFESSIONAL TIP

This is a traditional French dish, also very popular in Belgium.

HEALTH AND SAFETY

Only use mussels that are tightly closed or that close when they are tapped. This ensures that they are live so have not deteriorated.

Mise en place

1 Clean the mussels in several changes of water.

2 Remove any barnacles attached to the shells and remove the 'beards'.

Cooking

1 Take a thick-bottomed pan and add the shallots, parsley, wine, fish stock and the cleaned mussels.

2 Cover with a tight-fitting lid and cook over a high heat until the shells open.

Healthy Eating Tips
● Serve with a large salad and plain French baguette

SERVING SUGGESTION

Serve in individual bowls with a further bowl for discarded shells.

33 Scallops and bacon

Energy	Calories	Fat	Saturated fat	Carbohydrates	Sugar	Protein	Fibre
1705 kJ	410 kcal	26.8 g	6.9 g	5.9 g	28 g	36.4 g	2.3 g

	4 portions	10 portions
Large scallops, shelled, roe and skirt removed, washed	12	30
Pancetta bacon rashers (rind off)	12	30
Olive oil	50 ml	125 ml
Lemon	1	2
Asparagus sticks, peeled, blanched for 1 minute and refreshed	16	40
Seasoning		

Mise en place

1 Open the scallop shells and remove the white scallop discarding the orange roe.

2 Slice each scallop.

3 Wrap the scallops in the pancetta, pin with a cocktail stick and season (be mindful that the pancetta is salty).

Cooking

1 Heat the oil in a non-stick pan, place the scallops in and cook until golden-brown. Squeeze the lemon over the scallops and allow the juice to evaporate slightly.

2 Remove from the pan and retain with all the pan juices.

3 Return the pan to the heat and add the asparagus, cooking for a further 2 minutes.

4 To serve, divide the asparagus onto plates. Top with the scallops, pour over the pan juices and serve.

NOTE
The orange-coloured roe inside the scallop shell is not used in this dish

HEALTH AND SAFETY
Only buy scallops from a reputable supplier and use quickly as scallops are high risk and are sometimes linked with food poisoning. They are also one of the 14 identified allergens.

SERVING SUGGESTION
Serve in individual portions.

34 Prawns with chilli and garlic

Energy	Calories	Fat	Saturated fat	Carbohydrates	Sugar	Protein	Fibre
728 kJ	173 kcal	3.8 g	0.6 g	7.5 g	6.3 g	28.6 g	0.1 g

Cooking

1 Remove the prawns from the marinade and heat a small amount of oil in a non-stick frying pan.

2 Place the prawns in the pan and cook until pink and cooked through, basting with any leftover marinade while cooking.

3 To serve, place warm garlic bread on plates and pile up the prawns on top, allowing the cooking juices to run into the bread. Drizzle with salsa verde and serve with a green salad.

To make garlic bread:

1 Take one baguette and cut into thin slices at an angle, for example oval shaped slices.

2 Mix 100 g of softened butter with 1 clove of finely pureed garlic, salt and black pepper and use to spread on one side of each slice of bread.

3 Bake the slices at 180°C for 10 minutes. When plating, drizzle with a little of the salsa.

	4 portions	10 portions
Clove of garlic, crushed	1	3
Lime, juice of	1	2
Lemon, juice of	½	1
Mild red chillies, deseeded and finely chopped	2	5
Olive oil	1 tbsp	3 tbsp
Honey	1 tbsp	3 tbsp
Extra-large prawns, raw, shells on, heads removed	32	80
Black pepper		
To serve		
Salsa verde (see page 121)	100 ml	250 ml
Garlic bread	4 slices	10 slices
Green salad		

PROFESSIONAL TIP

Marinating the prawns before cooking adds extra flavour.

HEALTH AND SAFETY

Keep the prawns under refrigeration until needed.

Healthy Eating Tip

● For the healthiest version serve with the salad and omit the bread or salsa.

SERVING SUGGESTION

Individual dishes as shown or in a larger serving dish for customers to share.

Mise en place

1 In a shallow dish, mix together the garlic, lime juice, lemon juice, chillies, olive oil and honey.

2 Make an incision (do not cut all the way through – leave the prawn intact) in the back of each prawn and remove the entrails. Wash and dry well.

3 Add the prawns to the oil and chilli mix, season with black pepper and marinate in the fridge for 30 minutes.

4 Prepare the green salad, salsa verde and garlic bread.

35 Seafood stir-fry

Energy	Calories	Fat	Saturated fat	Carbohydrates	Sugar	Protein	Fibre
724 kJ	172 kcal	4.9 g	0.8 g	9.1 g	4.9 g	23.2 g	2.1 g

	4 portions	10 portions
Small asparagus spears	100 g	250 g
Sunflower or groundnut oil	1 tbsp	2½ tbsp
Fresh ginger, grated	1 tsp	2½ tbsp
Leeks, cut into julienne	100 g	250 g
Carrots, cut into julienne	100 g	250 g
Sweetcorn	100 g	250 g
Light soy sauce	2 tbsp	5 tbsp
Oyster sauce	1 tbsp	2½ tbsp
Clear honey	1 tsp	2½ tbsp
Cooked assorted shellfish, e.g. prawns, mussels, scallops	400 g	1 kg
Garnish		
Large cooked prawns	4	10
Fresh chives	25 g	62 g

Mise en place

1 Make sure that all the ingredients are assembled before starting to stir fry.

2 Heat oil in a wok or large frying pan.

Cooking

1 Blanch the asparagus for 2 minutes in boiling water, refresh then drain.

2 Heat the oil in a wok, add the ginger, leek, carrots and sweetcorn. Stir-fry for 3 minutes without colour.

3 Add the soy sauce, oyster sauce and honey. Stir.

4 Stir in the cooked shellfish and continue to stir-fry for 2–3 minutes until the vegetables are just tender and the shellfish is thoroughly heated through.

5 Add the blanched asparagus and stir-fry for another 1 minute.

SERVING SUGGESTION
Serve with fresh cooked noodles garnished with large fresh prawns and chopped chives.

PROFESSIONAL TIP
If you blanch the vegetables before stir-frying them, they will keep their colour and flavour and can be fried more quickly.

Healthy Eating Tips
- No added salt is needed – soy sauce is high in sodium.
- Increasing the ratio of vegetables to seafood will improve the 'balance' of this dish.

Ingredients for seafood stir-fry.

Fry the ginger and vegetables.

Combine all the ingredients and continue to stir-fry.

HEALTH AND SAFETY
Stir-frying should be completed at a high temperature to cook the food quickly. Take care of possible burns from 'spitting' hot oil. Stand back a little.

VARIATION
- Vegetables and shellfish can be varied.

FAULTS
A stir-fry dish should be cooked quickly. If cooked slowly, for too long, the shellfish will be rubbery and the vegetables will lose colour.

SERVING SUGGESTION
Serve in individual bowls or a larger serving dish for several people.

36 Seafood in puff pastry (*bouchées de fruits de mer*)

Energy	Calories	Fat	Saturated fat	Carbohydrates	Sugar	Protein	Fibre *
1327 kJ	316 kcal	17.6 g	3.1 g	28.9 g	1.1 g	12.2 g	1.2 g

* Fried in peanut oil

	4 portions	10 portions
Button mushrooms	50 g	125 g
Butter	25 g	60 g
Lemon, juice of	¼	½
Cooked lobster, prawns, shrimps, mussels, scallops	200 g	500 g
White wine sauce (see page 111)	125 ml	300 ml
Chopped parsley		
Salt, pepper		
Puff pastry bouchée cases	4	10
Picked parsley, to garnish		

Mise en place

1 Either make or prepare ready-made bouchée cases.

2 Peel and wash the mushrooms, cut into neat dice. Heat a sauté pan with the butter to cook the mushrooms.

Cooking

1 Cook in butter with the lemon juice.

2 Add the cooked shellfish (mussels, prawns, shrimps left whole, the scallops and lobster cut into dice).

3 Cover the pan with a lid and heat through slowly for 3–4 minutes.

4 Add the white wine sauce and chopped parsley, and correct the seasoning.

5 Meanwhile warm the *bouchées* in the oven or hot plate.

6 Fill the *bouchées* with the mixture and place the lids on top.

7 Serve garnished with picked parsley.

NOTE
Vol-au-vents can be prepared and cooked in the same way as *bouchées*.

Healthy Eating Tips
- The white wine sauce is seasoned, so added salt is not required.
- Serve with a salad garnish.

Raw puff pastry *bouchées*, before baking.

Combine the shellfish with the mushrooms.

Fill the pastry cases.

PROFESSIONAL TIP
Bouchées are small, canape-sized filled puff pastry cases. Larger versions are called vol au vents.

HEALTH AND SAFETY
When using these as canapes or on a buffet, leave out for the shortest possible time as the fillings are considered food safety high risk ingredients.

VARIATION
- The seafood filling can be varied.

37 Sauté squid with white wine, garlic and chilli

Energy	Calories	Fat	Saturated fat	Carbohydrates	Sugar	Protein	Fibre
1084 kJ	260 kcal	17.6g	2.4g	2.3g	0.3g	23.4g	0.2g

PROFESSIONAL TIP
The texture of squid is unlike that of other species. The flesh is very high in protein and dense, giving it that 'rubbery' texture when overcooked. Cook quickly and over a high heat.

Cut the body of the squid in half.   Cut into strips.

Cook the squid with the other ingredients.

	4 portions	10 portions
Squid, cleaned	600 g	1½ kg
Vegetable oil	60 ml	150 ml
Garlic cloves, crushed	2	5
Sprigs of parsley, chopped	3–4	7–8
Red chilli pepper, seeds removed, finely chopped	1	3
White wine	60 ml	150 ml
Fish stock	60 ml	150 ml

Mise en place

1 Prepare the strips of squid.

2 Cut the squid in to halves and then into thick strips.

3 Heat the oil to cook the squid.

Cooking

1 Place the squid in the pan and sauté quickly (this will not take long – the squid will toughen if cooked for too long).

2 Add the garlic, chopped parsley and the chilli. Toss the squid around the pan, working in all the flavours.

3 Add the wine and stock, quickly bring to the boil, check the seasoning and serve.

HEALTH AND SAFETY
Keep the squid refrigerated until needed and serve quickly once cooked.

Healthy Eating Tip
● Serve with green salad.

FAULTS
Squid will become very tough and inedible if cooked for too long. Cook it quickly.

308 Farinaceous dishes

Learning outcomes

In this unit you will be able to:

1 Prepare and produce farinaceous dishes, including:
- knowing different types of farinaceous ingredients – rice, pasta, grain, gnocchi – and knowing the types of dishes they are used to produce
- knowing the quality points of ingredients and understanding how they affect the preparation and cooking of farinaceous dishes; ingredients must include rice, pasta, grains, gnocchi, as well as hen, duck and quail eggs
- understanding the correct storage procedures to use throughout production and on completion of the final products

- being able to use cooking techniques on different rice, pasta, grains, gnocchi and eggs
- understanding how to and being able to produce different types of sauces, dressings and fillings
- understanding how sauces, dressings and fillings are used to create and finish farinaceous dishes
- being able to measure and evaluate against quality standards throughout preparation and cooking
- being able to evaluate products against dish requirements and production standards and recognise any faults.

Recipes included in this chapter

No.	Name
	Rice
1	Braised or pilaff rice
2	Steamed rice
3	Plain boiled rice
4	Stir-fried rice
5	Paella
6	Risotto with Parmesan
7	Risotto with asparagus (amuse bouche)
	Pasta
8	Fresh egg pasta dough: method 1 and method 2
9	Lasagne
10	Pumpkin tortellini with brown butter and balsamic vinaigrette
11	Ravioli
12	Spaghetti bolognaise
13	Cannelloni
14	Macaroni cheese
15	Tagliatelle carbonara
	Grains
16	Crisp polenta and roasted Mediterranean vegetables

No.	Name
17	Couscous with chorizo sausage and chicken
18	Couscous fritters with feta
	Eggs
19	Scrambled eggs
20	Soft-boiled eggs in the shell
21	Hard-boiled eggs with cheese and tomato sauce
22	French-fried eggs
23	Fried eggs
24	Poached eggs with cheese sauce
25	Omelette
26	Egg white omelette
27	Spanish omelette
28	Feta, mint, lentil and pistachio omelette
29	Scotch eggs
30	Eggs in cocotte
31	Eggs sur le plat
	Gnocchi
32	Gnocchi parisienne (choux paste)
33	Potato gnocchi (gnocchi piemontese)
34	Gnocchi romaine (semolina)

Additional recipe available online at www.hoddereducation.co.uk/practical-cookery-resources

Prepare and produce rice dishes

Introduction

Rice is a type of grain. It is one of the world's most important crops, being the main food crop for about half the world's population. A hot, wet atmosphere is required to grow rice, and therefore it is grown mainly in India, the Far East, South America, Italy and the southern states of the USA. In order to grow, rice needs more water than any other cereal crop.

There are around 250 different varieties of rice. Rice is used in a variety of dishes, starters, main courses and desserts as well as a variety of buffet and takeaway food. Rice introduces texture, flavour and carbohydrate content to dishes. The finished result depends on the cooking method and ingredients used.

1.1 Types of rice

Long grain

White **long-grain rice** is a narrow, pointed grain that has had the full bran and most of the germ removed so that it is less fibrous than brown rice. Because of its firm structure (which helps to keep the grains separate when cooked), it is suitable for plain boiling, steaming, braising and savoury dishes such as kedgeree and curry.

Short grain

White **short-grain rice** is a short, rounded grain with a soft texture. It is suitable for sweet dishes and risotto, and is often used in rice desserts.

Brown rice

This is any rice that has had the outer husk removed, but retains its brown bran. As a result, it is more nutritious and contains more fibre than white rice. It takes longer to cook than long-grain rice because water and heat take longer to penetrate the bran layers. The nutty flavour of **brown rice** lends itself to some recipes, but does not substitute well in traditional dishes such as paella, risotto or puddings. Brown rice can be used for any other rice recipes, but allow extra cooking time to soften the grain (use one part grain to two parts water for 35–40 minutes).

Arborio rice (risotto rice)

Arborio rice is a medium- to long-grain rice and is used in risottos because it can absorb a good deal of cooking liquid without becoming too soft.

Long-grain rice

Short-grain rice

Brown rice

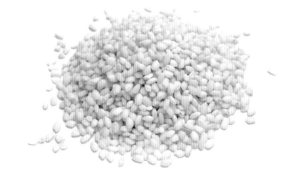
Arborio rice

Long-grain rice: a narrow, pointed grain that has had the full bran and most of the germ removed. Used for plain boiling and in savoury dishes.

Short-grain rice: a short, rounded grain with a soft texture. Suitable for sweet dishes and risotto.

Brown rice: any rice that has had the outer covering removed, retaining its bran.

Other types of rice

Many other types of rice are now available, which can add different colours and textures to dishes. Some of these are described below.

- Basmati – a narrow, long-grain rice with a distinctive aroma and flavour, suitable for serving with Indian dishes.
- Wholegrain rice – the whole unprocessed grain of the rice. Only the inedible outer husk is removed.
- Wild rice – this is not, in fact, rice but the seed of an aquatic grass. Difficulty in harvesting makes it expensive, but its colour (a purplish black) and subtle nutty flavour make it a good base for a special dish or rice salad. It can be economically mixed with other types of rice, but may need pre-cooking as it takes 45–50 minutes to cook.
- Red rice – an unmilled, short-grain rice from the Camargue region in France, with a brownish-red colour and a nutty flavour. It is slightly sticky when cooked and is particularly good in salads.
- Pre-cooked instant rice – par-boiled, ready cooked and vacuum-sealed cooked rice are popular convenience products.
- Ground rice – used for milk puddings and for thickening sauces and soups.
- Easy-cook rice – traditional rice is steamed under pressure before it is milled. This process hardens the grain and reduces the possibility of overcooking and helps prevent the rice grains sticking together. The uncooked rice is pale golden in colour and turns white during cooking. It is used in the same way as untreated rice.
- Aromatic rice – speciality rice which has a distinct flavour and aroma. Their quality and flavour can differ from one year to the next, based on the harvest and climate change. Examples include, jasmine rice (a soft, sticky rice often used when making sushi) and American aromatic rice such as Japonica.

Wild rice

Other rice products

Rice is used to produce many different rice products.

- Rice flakes (brown and white) – can be added to muesli or made into a milk pudding or porridge.
- Rice flour – used for thickening cream soups.
- Rice paper – edible paper made from milled rice; used in pastry work for nougat and macaroons, etc.
- Rice wine – made from fermented wine. Sake is the most famous rice wine and is drunk extensively in Japan. Other examples include mirin (a sweet rice wine used in rice dishes) and Shaoxing (a Chinese rice wine).
- Rice cakes – were best known in Japan and the countries of the Pacific. However, rice cakes are now popular around the world as a low fat, low calorie snack. Rice cakes are bland in taste so some have added flavour.
- Rice noodles – made from rice flour, which is blended with water and either rolled out, cut or extruded. They are usually dried and come in a range of shapes and sizes similar to pasta shapes. Fresh noodles are often cut into wide ribbons and used in soups and stir-fries. The noodles become slightly transparent when cooked and tend to be slightly more chewy than flour noodles. Sometimes ingredients such as mung bean flour may be added to change their consistency and appearance. Types of rice noodles include laska noodles, rice fluke noodles, rice sticks and river rice noodles.
- Rice vermicelli – thin rice noodles used in soups, salads and spring rolls.

1.2 Quality points

- Aroma – some rices, such as basmati or jasmine, do have a definite aroma. However, for most the aroma is more related to the actual dish produced with the rice. For example, risotto and paella will have the distinctive aroma of those dishes. Any rice dish with an unpleasant odour must not be used.
- Freshness – dry rice has a long shelf life but check best before dates on packaging. Use rice in the order of delivery, i.e. rotate the stock. Cooked rice should preferably be used immediately. If not for immediate use, cool and chill quickly, store below 5 °C and use within two days.
- Types – there are many types of rice available, allowing the chef to use the different characteristics of the rice to produce a range of interesting new and authentic dishes. The size of various rice grains will also differ before and after cooking and this again will add to the authenticity and characteristics of the dish.
- Texture – some rice dishes will require separate grains to be evident, for example, pilaff or stir-fried rice, while other dishes are soft with the rice sticking together in

the cooking liquid, such as risotto or rice pudding. The required texture of a finished dish must match the recipe or dish specification.

- Colour – the colour of rice will depend on its type and on any processing it has undergone, for example, the removal of bran layers from brown rice will result in white grains. Cooking changes the colour of rice as the grains absorb water. Some rice dishes are a distinctive colour because of other ingredients used – paella is yellow because of the addition of saffron.

1.3 Storage

Packaging and storing

- Most rice is sold pre-packaged in varying sizes of package and weight so it is possible to purchase the correct size according to needs.
- Dry rice can be stored in its original packaging or in a storage container in a dry, well ventilated food store.
- Humidity control is important as high humidity can cause spoilage of rice.
- Refer to best before dates on packaging.
- The shelf life of brown rice is shorter than that of white rice because the bran layers contain oil that can become rancid. Refrigeration is recommended for longer shelf life.

Refrigerated storage

- Cooked rice can be rapidly chilled then refrigerated below 5 °C for up to two days, or stored in the freezer for six months.
- When placing cooked rice in the refrigerator use a suitable clean container with a lid or cover with cling film.
- Label with the date and with a multi-use refrigerator position near the top to avoid cross contamination.

Damage

- Rice delivered in damaged packaging or with any visible damage to the actual rice should be rejected.
- Cooked rice dishes that have become damaged or contaminated in any way should also be discarded.

Temperature

- Because rice dishes are high risk in relation to food poisoning they should never fall below 63 °C once cooked and while being held for service. Cooked rice should be discarded after two hours.
- A refrigerated rice dish should be stored below 5 °C. If this is not done, the spores of *bacillus cereus* (a bacterium found in the soil) may revert to bacteria and multiply, which could cause food poisoning.

1.4 Cooking ingredients

Preparing rice

Washing

Most rice does not require washing. However, the washing of rice removes any excess starch, which tends to cloud the cooking liquid.

Using correct amounts of liquid

Rice absorbs liquid easily. Table 8.1 gives a guide on liquid absorption, to help you to achieve the correct texture and overall result for each type of rice dish.

Table 8.1 Proportions of liquid for cooking different types of rice

Type of rice	Proportion of liquid to rice
Boiling	3–4 times liquid to rice
Pilaff (braised)	2:1 liquid to rice
Risotto	3:1 liquid to rice
Paella	4:1 liquid to rice
Sushi	1:5 liquid to rice
Wild rice	4:1 liquid to rice

Cooking rice

As the rice grains cook, starch is released which naturally thickens a liquid. This is particularly important when making risotto and sushi. Liquids used for cooking rice include water, white stock, milk, or a combination of half white stock and half white wine.

Rice grains are porous so absorb water easily. Rice should be cooked so that it is **al dente** (to the bite), except if the rice is being moulded.

Rice can be cooked using a number of different cooking methods, including:
- boiling
- steaming
- frying
- stewing
- braising
- microwaving.

When boiling rice, cook in a large quantity of water and drain well.

KEY TERM

Al dente: cooked firm, to the bite.

Once cooked, keep hot (above 65°C for no longer than two hours), or cool quickly (within 90 minutes) and keep cool (below 5°C). If this is not done, the spores of *bacillus cereus* (a bacterium found in the soil) may revert to bacteria and multiply.

You will find more details on rice dishes in the recipe section at the end of this unit.

1.5 Sauces, dressings and fillings

For sauces and dressings to accompany rice, see the recipe section at the end of this unit and Unit 303 Stocks, soups and sauces.

1.6 Evaluation

See page 405 for evaluation of all types of farinaceous dishes.

Prepare and produce pasta dishes

Introduction

Pasta is the name for a type of flour paste originating from Italy and consisting of dough made from durum wheat and water. Fresh pasta usually has egg as the liquid content. Pasta can also have added colour and flavour in the form of tomato puree or spinach. The dough is stretched and flattened into various shapes and either used fresh or dried.

Durum wheat has a 15 per cent protein content, which makes it a good alternative to rice and potatoes for vegetarians. Further protein would be added by the use of eggs. Pasta also contains carbohydrates in the form of starch, which gives the body energy.

1.1 Types of pasta

There are basically four types of pasta, each of which may be plain or flavoured with added ingredients:

● dried durum wheat pasta
● fresh durum wheat egg pasta
● semolina pasta
● wholewheat pasta.

Dried pasta

Almost 90 per cent of the pasta consumed is dried. Dried pasta comes in numerous shapes and sizes; the most common in the UK include spaghetti, fettuccini (long, narrow ribbons), penne (short tubes cut diagonally), farfalle (bow tie or butterfly shaped) and fusilli (short spirals) but new pasta shapes appear frequently. Sheets of pasta are used for lasagne.

A variety of dried pasta shapes.

Fresh pasta

Fresh pasta is usually made from flour produced from durum wheat. This type of wheat produces a strong flour, making a dough that is less elastic than bread flour, which makes it suitable for forming pasta shapes. The flour is mixed with a little oil then made into a dough with either water or egg. Pasta dough is sometimes coloured with powdered spinach or tomato purée. Fresh pasta must be kept wrapped to prevent it drying out and stored below 5 °C; this is very important for pasta made with egg.

Stuffed pasta

Stuffed or filled pasta refers to pasta shapes stuffed with a variety of fillings. Shapes include cannelloni (tubes), ravioli and tortellini.

Examples of fillings include:
- mushroom duxelle
- minced beef or lamb flavoured with herbs
- ricotta cheese with basil
- fish mousseline
- couscous with chopped cooked vegetables
- minced lamb, beef or pork with spinach purée
- minced chicken with fine herbs (parsley, chervil and tarragon).

Tortellini is a type of stuffed pasta.

1.2 Quality points

- Aroma – pasta itself has very little aroma and the aroma of pasta dishes is taken from the ingredients used with it. Fresh pasta that has been stored incorrectly or for too long may have an unpleasant odour and must be thrown away. Always discard any pasta or pasta dish with an unpleasant aroma.
- Freshness – dried pasta has a long shelf life but check best before dates on packaging. Use dried pasta in the order of delivery, i.e. rotate the stock. On delivery the box or container should be sealed and the pasta unbroken. If delivered in a damaged state do not accept it and discard

any pasta that becomes damaged in storage. Dried pasta must be stored in a dry, humidity controlled area, with good ventilation, preferably in a sealed storage container with a lid to protect from pests. The pasta should have an even colour and no signs of infestation. Fresh pasta – because eggs are used to produce fresh pasta it has a short shelf life and needs to be refrigerated. On delivery fresh pasta should be below 5 °C, evenly coloured, intact and not torn or damaged. The pasta must easily separate. Refrigerated pasta must labelled with the date; once again stock rotation is important 'first in – first out'.
- Texture – dried pasta is dry and brittle and softens when it is cooked. Fresh pasta should be soft and of an even texture, not dry round the edges. Fresh and dried pasta should never be overcooked and should remain al dente – slightly resistant to the bite, but cooked through. The texture of a finished pasta dish will also be affected by the amount and type of sauce used. Follow dish specifications at all times.
- Temperatures – if a pasta dish is refrigerated for use later, for example, lasagne or cannelloni, it must be chilled rapidly and kept at temperatures below 5 °C until required for reheating. The dish should be reheated to 75 °C or higher and all hot pasta dishes being held for service should remain at or above 63 °C.

1.3 Storage

Cooked pasta

If cooked pasta is not to be used immediately, drain and rinse thoroughly with cold water. If it is left in water, it will continue to absorb water and become mushy. When the pasta is cool, drain and toss lightly with oil to prevent it from sticking and drying out. Cover tightly and refrigerate or freeze. Stuffed fresh pasta should be blanched, drained, frozen or chilled in small amounts which allows for easy separation.

- Cooked pasta may be stored in plastic bags in the refrigerator, labelled and dated. Refrigerate the pasta and sauce separately or the pasta will become soggy.
- Frozen pasta will keep for up to three months, but it is recommended that it is used within one month.

To reheat, put pasta in a colander and immerse in rapidly boiling water just long enough to heat through. Do not allow it to continue to cook. Pasta may also be reheated in a microwave.

It is advisable to serve pasta immediately after it is reheated and the dish is finished. If pasta is to be held on a buffet (for example, a tray of lasagne) then it must be kept at the required temperature of above 63 °C and served within two hours.

Fresh pasta

Fresh pasta will carry a use-by date. Once delivered, store fresh pasta above any raw items in the refrigerator below 5 °C. Shaped fresh pasta is best stored in large containers, sprinkled with semolina and layered with sheets of silicone paper or sheets of plastic wrapping.

Keep fresh pasta chilled until required. It may also be vacuum packed and chilled. Fresh pasta must have an even colour and not be darkening or brittle at the edges. Pasta must be the required shape, not torn, damaged or sticky, and should easily separate.

A pasta machine

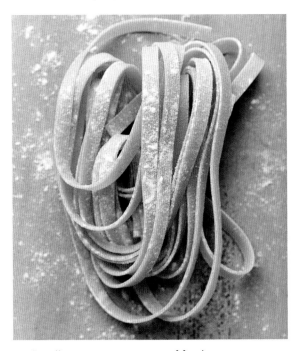

Tagliatelle is a common type of fresh pasta.

1.4 Cooking ingredients

Equipment for preparing pasta dishes

The following equipment is used in pasta making:
- Pasta machine – used for rolling out the pasta to the correct thickness and size.
- Pastry brushes – used to brush the pasta with oil and to brush the excess semolina from the pasta after rolling.
- Fluted cutting roller – used to cut and shape and divide the pasta.
- Palette knife – used to lift the pasta sheets and shapes.
- Ravioli tray – used to mould and shape the ravioli (though modern ravioli is more frequently shaped by use of cutters and finished by hand).

Cooking dried pasta

When using dried pasta, a usual portion weight is 80–100 grams as a first course. If larger portions are required, increase accordingly.

Most types of dried pasta that have been properly stored and are within their best before date will cook in less than 10 minutes.

To cook dried pasta:
1 Bring plenty of water (at least 4 litres for every 600 grams of dry pasta) to a rolling boil.
2 Add about 1 tablespoon of salt per 4 litres of water, if desired.
3 Add the pasta in small quantities to maintain the rolling boil.
4 Stir occasionally to prevent sticking. Do not cover the pan.
5 Follow package directions for cooking time. Do not overcook. Pasta should be al dente (slightly resistant to the bite, but cooked through).
6 Drain to stop the cooking action. Do not rinse unless the recipe says to do so. For salads, drain and rinse pasta with cold water.

Mixing cooked pasta with other ingredients

Always mix in a large bowl or saucepan, to enable the ingredients to be well distributed throughout the pasta. This is done with a large spoon or spatula.

> **PROFESSIONAL TIP**
> It is sometimes suggested that you add a little oil to the water when cooking pasta. However, this has no benefit; it is the constant movement of the water that will stop it sticking together.

PROFESSIONAL TIP

The water should always be at a rolling boil – the more water in proportion to pasta, the quicker it will return to the boil after the pasta is added. This means fast cooking and better-textured pasta.

Fresh egg pasta dough can be made using a food processor (see recipe 9). If using a pasta rolling machine, divide the dough into three or four pieces. Pass each section by hand through the machine, turning the rollers with the other hand. Repeat five or six times, adjusting the rollers each time to make the pasta thinner.

PROFESSIONAL TIP

After rolling fresh pasta allow it to dry a little by hanging on a pasta rack or over wooden spoon handles.

Straining

Always strain pasta carefully as it is a delicate product. Use a colander or a spider.

Blanching and refreshing

Pasta can be cooked for a few minutes and then refreshed under cold water to speed up the service. Only minimal finishing/reheating will be needed when orders are placed in the restaurant. This is done by plunging the pasta into boiling water for a short time.

Other cooking methods for pasta

- **Baking** – this technique is used for dishes such as cannelloni and pasta bakes. These dishes have usually been previously cooked by boiling and require baking to finish the product. This technique requires the dish to cook evenly and have a well distributed colour.
- **Poaching** – used for ravioli, where gentle cooking is needed.
- **Deep frying** – some pasta dishes are first boiled, then coated and deep fried. For example, stuffed ravioli can be dipped in egg wash or butter milk, coated in breadcrumbs and deep fried.

HEALTH AND SAFETY

When cooked, the core temperature of a pasta dish such as lasagne must be 75 °C or more. Always check the internal temperature using a temperature probe.

Step by step: making fresh egg pasta dough

To make 500 g of dough, use 400 g strong flour, 4 medium eggs (beaten), approximately 1 tbsp of olive oil (as required) and salt.

1   Sift the flour and salt.

2   Make a well in the flour. Pour the beaten eggs into the well.

3   Gradually incorporate the flour.

4   Mix until a dough is formed. Only add oil to adjust to required consistency. The amount of oil will vary according to the type of flour and the size of the eggs.

5   Pull and knead the dough until it is of a smooth elastic consistency and then cover and allow to rest in a cool place for 30 minutes.

6   After resting, knead the dough again. Roll out the dough on a well-floured surface to the thickness of 0.5 mm. Trim the sides and cut as required using a large knife.

Using a pasta machine

You will find more details on pasta dishes in the recipe section at the end of this unit.

1.5 Sauces, dressings and fillings

Examples of sauces to accompany pasta include:

- tomato-based sauces
- cream, butter or béchamel-based sauces
- reduction-based sauces
- rich meat sauce such as bologne sauce
- olive oil and garlic
- soft white or blue cheese sauces
- pesto.

For more details on how to prepare sauces for pasta, see the recipe section at the end of this unit and Unit 303 Stocks, soups and sauces.

Step by step: making stuffed pasta (tortellini)

1 Place the filling in the centre of a piece of pasta. Lightly brush the pasta with beaten egg.

2 Fold the pasta over the filling, creating a semi-circle, and press down around the filling to seal.

3 Trim the excess pasta with a pastry cutter.

4 Bring the two ends of the long edge together, egg wash the seam and firmly press the ends together to seal.

Examples of cheeses used in pasta cooking:

- Parmesan – the most popular hard cheese used with pasta, ideal for grating. The flavour is best when it is freshly grated. If it is bought ready grated, or if it is grated and stored, the flavour deteriorates.
- Pecorino – a strong ewes' milk cheese, sometimes studded with peppercorns. Used for strongly flavoured dishes, it can be grated or thinly sliced.
- Ricotta – creamy-white in colour and made from the discarded whey of other cheeses, ricotta is widely used in fillings for pasta such as cannelloni and ravioli, and for sauces.
- Mozzarella – traditionally made from the milk of the water buffalo, mozzarella is pure white and creamy, with a mild but distinctive flavour, and usually round or pear-shaped. It will keep for only a few days in a container half-filled with milk and water or in its delivery packaging containing liquid.

- Gorgonzola or dolcelatte – these are both distinctive blue cheeses that can be used in sauces.

1.6 Evaluation

See page 405 for evaluation of all types of farinaceous dishes.

ACTIVITY

1. Which type of flour is used for pasta dough?
2. Name four different pasta shapes and a dish you may produce using one of them.
3. What does 'al dente' mean?
4. What equipment might you need when making fresh pasta?
5. How should cooked pasta be stored?

Prepare and produce grain dishes

Introduction

Cereals are important crops and are the oldest farmed agricultural products. A cereal is the edible fruit of any grass which is used as food. Cereals are used in a range of dishes and food products, including breads, soups, stews, pilaffs, risotto and of course breakfast cereal.

Grain is the name given to the edible fruit of cereals. In many countries grains are a staple food as they are inexpensive and readily available.

The whole grain is made up of three parts of the grain kernel: the bran, endosperm and the germ.

- The **bran** is the outer layers of the cereal. Bran contains a high amount of dietary fibre, some of the B vitamins (for example, thiamine and niacin) and the minerals zinc, copper and iron. It also contains protein and other chemicals.
- The **endosperm** is the middle layer and is the largest section of the grain. It is the main energy supply of the plant and is rich in carbohydrates, protein and vitamin B.
- The **germ** is the smallest part of the grain. It is rich in nutrients, containing B vitamins, minerals, vitamin E and other chemicals.

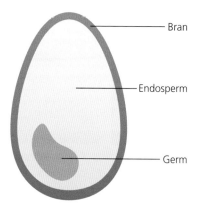

The structure of a grain

KEY TERMS

Cereal: the edible fruit of any grass used for food.

Grain: edible fruit of cereals.

Bran: outer layers of the cereal.

Endosperm: the middle layer of the grain.

Germ: the part from which the new plant would shoot.

1.1 Types of grain

Barley

Barley grows in a wider variety of climatic conditions than any other cereal. It is usually found in the shops as whole or pot barley or polished pearl barley, but you can also buy barley flakes and kernels. It can be cooked on its own (one part grain to three parts water for 45–60 minutes) and is used as an alternative to rice, pasta or potatoes, or added to stews. Malt extract is made from sprouted barley grains.

Barley must have its fibrous hull (outer shell) removed before it is eaten (hulled barley). Hulled barley still has its bran and germ and is considered a whole grain, making it a healthy food. Pearl barley is hulled barley that has been processed further to remove the bran.

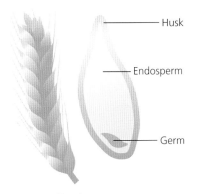

A grain of barley

Corn or maize

Fresh corn – available in the form of sweetcorn and corn on the cob – is eaten as a vegetable. The dried grain is most often eaten as cornflakes or popcorn. Tortillas are made from maize meal, as are numerous other snack foods.

The flour made from corn (cornmeal or maize meal) is used to make **polenta** and can be added to soups, pancakes and muffins. It is also a popular ingredient in Italian cakes. When cooking polenta (one part grain to three parts water, for 15–20 minutes), stir carefully to avoid lumps. After cooking it can be used like mashed potato but it is quite bland, so often strong flavoured ingredients such as Gorgonzola, Parmesan and fresh herbs are added. It can be pressed into a tray when cold and cut into slices or brushed with garlic and olive oil and grilled. Ready cooked polenta is also available.

Do not confuse cornmeal with refined corn starch or flour, which is used for thickening. Corn is gluten free.

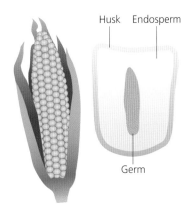

A grain of maize

Polenta

Wheat

This is a widely used and popular cereal. It is used for bread, cakes, biscuits, pastry, breakfast cereals and pasta. Wheat grains can be eaten whole and have a satisfying, chewy texture. (Cook one part grain to three parts water for 40–60 minutes.) Some different types of wheat are described below.

- Cracked or kibbled wheat is the dried whole grains cut by steel blades.
- Bulgar wheat is parboiled before cracking, has a light texture and only needs rehydrating by soaking in boiling water or stock.
- Semolina is a grainy yellow flour ground from durum or hard wheat and is the main ingredient of dried Italian pasta.
- Couscous is made from semolina grains that have been rolled, dampened and coated with finer wheat flour. Soak couscous in two parts of water or stock to rehydrate. Traditionally, couscous is steamed after soaking, often in a sieve over a simmering stew.

Cracked wheat

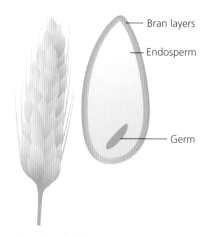

A grain of wheat

- Bran layers
- Endosperm
- Germ

Bulgar wheat

Buckwheat

When roasted, the seeds of buckwheat are dark reddish brown. Buckwheat can be cooked (one part grain to two parts water for six minutes; leave to stand for six minutes) and served like rice, or it can be added to stews and casseroles. Buckwheat flour can be added to cakes, muffins and pancakes, where it gives a distinctive flavour. Soba noodles, made from buckwheat, are an essential ingredient in Japanese cooking. Buckwheat is gluten free.

Buckwheat

Couscous

Flour

- Strong wheat flour (with a high gluten content) is required for yeasted bread making. Plain flour is used for general cooking, including cakes and pastry.
- Self-raising flour is plain flour with the addition of raising agents.
- Wheat flakes are used for porridge, muesli and flapjacks.

Millet

Millet is a group of small-seeded cereal crops or grains widely grown around the world for food. The main millet varieties are:

- pearl millet
- foxtail millet
- proso millet (also known as common millet, broom corn millet, hog millet or white millet)
- finger millet.

Millet

As none of the millet types is closely related to wheat, they can be eaten by those with **coeliac disease** or other forms of allergies or intolerances to wheat. Those with coeliac disease can replace certain cereal grains in their diets with millets in various forms, including breakfast cereals. Millet can also often be used in place of buckwheat, rice or quinoa.

> **KEY TERM**
>
> **Coeliac disease:** a disease in which a person is unable to digest the protein gluten (found in wheat and some other cereals). Gluten causes the person's immune system to attack its own tissues, specifically the lining of the small intestine. Symptoms include diarrhoea, bloating, abdominal pain, weight loss and malnutrition.

In western India, millet flour (called 'bajari' in Marathi) has been commonly used with 'jowar' (sorghum) flour for hundreds of years to make the local staple flat bread, 'bhakri'.

The protein content in millet is very close to that of wheat; both provide about 11 per cent protein by weight. Millet is rich in B vitamins (especially niacin, B6 and folacin), calcium, iron, potassium, and zinc. Millet contains no gluten, so cannot be used for a risen bread. However, when combined with wheat or xanthan gum (for those who have coeliac disease), millet can be used to make raised bread. Alone, they are suited to flatbread.

Millet can be used as an alternative to rice, but the tiny grains need to be cracked before they will absorb water easily. Before boiling, sauté with a little vegetable oil for 2–3 minutes until the grains crack, then add water carefully (one part grain to three parts water). Bring to the boil and simmer for 15–20 minutes until fluffy. Millet flakes can be made into porridge or added to muesli. Millet flour is also available and is sometimes used to make pasta.

Oats

There are various grades of oatmeal, rolled oats or jumbo oat flakes. All forms can be used to make porridge, combined with groundnuts to make a nut roast or added to stews. Oatmeal is low in gluten so cannot be used to make risen bread, but can be mixed with wheat flour to add flavour and texture to bread, muffins and pancakes. Oatmeal contains some oils and can become rancid, so observe the best before date.

Oatmeal is created by grinding oats into coarse powder; various grades are available depending on thoroughness of grinding (including coarse, pinhead and fine). The main uses of oats are:

- as an ingredient in baking
- in the manufacture of bannocks or oatcakes
- as a stuffing for poultry
- as a coating for some cheeses
- as an ingredient of black pudding
- for making traditional porridge (or 'porage').

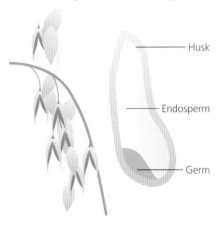

Husk

Endosperm

Germ

A grain of oats

Rye

Rye is one of the few cereals (along with wheat and barley) that has enough gluten to make yeasted bread. However, with less gluten than wheat, rye flour makes a denser, richer-flavoured bread. It is more usual to mix rye flour with wheat flour. Rye grains should be cooked using one part grain to three parts water for 45–60 minutes. Kibbled (cracked) rye is often added to granary-type loaves. Rye grains can be added to stews and rye flakes are good in muesli.

Rye

Spelt

Originating in the Middle East, spelt is closely related to common wheat and has been popular for decades in Eastern Europe. It has an intense nutty, wheat flavour. The flour is excellent for bread-making and spelt pasta is becoming more widely available.

Quinoa

Quinoa is an ancient crop that fed the native South American Aztecs for thousands of years and has recently been cultivated in the UK. It is a seed that is high in protein, making it useful for vegetarians.

The small, round grains look similar to millet, but are pale brown in colour. The taste is mild and the texture firm and slightly chewy. It can be cooked like millet and absorbs twice its volume in liquid.

Cook for 15 minutes (one part grain to three parts water); it is ready when all the grains have turned from white to transparent and the spiral-like germ has separated. It can be used in place of more common cereals, pasta or rice (risottos, pilaff), and is served in salads and used in stuffing.

Quinoa

1.2 Quality points

- Aroma and appearance – grains should always look and smell faintly sweet or have no aroma at all. If you detect a musty or oily scent, the grains have passed their peak and should not be used.
- Colour – this will vary between different types and may become lighter in colour when cooked.
- Packaging and storage – most grains will now be packaged by weight. Do not accept grains delivered in broken or damaged packaging and either store in the original packaging or in a sealed container. Use in rotation, first in, first out. Like pasta and rice, grains in their dry state have a long shelf life if kept in a dry, cool store room. However, observe the best before date.

- Size and texture – grains are usually small but different types will vary in size. In their dry state grains will be hard but will soften on cooking, giving a softer texture which will once again vary between the different types.

1.3 Storage

Whole grains must be stored more carefully than refined grains, since the healthy oils found largely in the germ of the whole grain can be negatively affected by heat, light and moisture. Because each grain has a different fat content, shelf life varies. All whole grains should be stored in the original packaging or in airtight containers with tight-fitting lids or closures.

- Whole intact grains – the shelf life of whole intact grains is generally longer than flours. If stored properly in airtight containers, intact grains will keep for up to six months in a cool, dry store, or up to one year in a freezer.
- Whole grain flours and meals – these spoil more quickly than intact grains because their protective bran layer has been broken up and oxygen can reach all parts of the grain, causing oxidation. If stored properly in airtight containers, most grain flours and meals will keep for one to three months in a cool, dry store or up to two to six months in a freezer.

Once cooked it is advisable to serve grains immediately. If they are to be held on a hot plate for service the temperature must be retained at 63°C or above for no longer than two hours. After such time they must be thrown away. Cooked grains that are to be served cold in salads should be cooled quickly, covered, labelled, dated and refrigerated below 5°C and stored above any raw foods.

1.4 Cooking ingredients

Grains should first be washed under cold water in a colander to remove any dust or foreign bodies. Look for any unusual infestation such as insects. The dry grain does not need to be soaked.

Cooking times will vary, depending on the type of grain. Whole grains take longer to cook than grains that have been processed.

Cooking methods

Grains are usually boiled or steamed, but may be braised or stewed, especially when part of another dish. Grains are starch based and when heated, they absorb water, swell and burst, cooking at between 60 °C and 70 °C.

Grains such as couscous, bulgar wheat and millet only usually require boiling water to be poured onto them, stirred and cooled, but may also be gently steamed.

Porridge oats and oat flakes, polenta and cornmeal are often soaked or paboiled, poured into a lined baking sheet (usually on a silpat mat) and baked in the oven. As the grains are baked, the water will cook the product, evaporate and the result is a dry baked grain.

Porridge is made by mixing oats with water/milk, bringing to the boil and simmering gently. Grains may be cooked in water or soaked in wine or fruit juice.

You will find more details on cooking methods using grains in the recipe section at the end of this unit.

1.5 Sauces, dressings and fillings

For sauces and dressings to use with grains, see the recipe section at the end of this unit and Unit 303 Stocks, soups and sauces.

1.6 Evaluation

See page 405 for evaluation of all types of grain dishes.

ACTIVITY

1 Name three different types of grain used in culinary work. How should they be stored before cooking?

2 State three different ways that wheat is used for menu items.

3 Describe how you would prepare and cook quinoa.

4 What is the nutritional value of including grains in a menu?

5 What is the difference between couscous and bulgar wheat?

Prepare and produce egg dishes

Introduction

Eggs have a wide range of culinary uses, both savoury and sweet. They are important in the diet when eaten alone but also as an ingredient, used for binding, enriching, colouring, emulsifying and aerating.

- The outer shell of an egg is mainly calcium carbonate lined with membranes and has a number of pores for gas exchange.
- These pores are covered by a wax-like layer, known as the cuticle. This cuticle partly protects the egg against microbial invasion and controls water loss.

- The egg white is divided into a thick layer around the yolk and a thinner layer next to the shell. The thick layer anchors the yolk, together with the chalazae, in the middle of the egg. The egg white has an important function as it possesses special antibacterial properties, which prevent the growth and multiplication of micro-organisms.

1.1 Types of egg

In cookery, hens' eggs are almost exclusively used, but eggs from turkeys, geese, ducks, guinea fowl, quail and gulls are also edible.

Quails' eggs are used in a variety of ways, for example, as a garnish to many hot and cold dishes, as a starter or main course, such as a salad of assorted leaves with hot wild mushrooms and poached quail eggs, or tartlet of quail eggs on chopped mushrooms coated with hollandaise sauce.

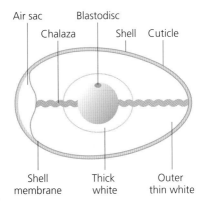

Air sac   Blastodisc
   Chalaza   Shell   Cuticle

Shell   Thick   Outer
membrane   white   thin white

The structure of an egg

Quiche Lorraine is made using eggs.

- Fresh eggs are used extensively in hors d'oeuvres, meat and poultry dishes, soup, pasta, egg dishes, salads, fish dishes, sweets and pastries, sauces and savouries. As well as buying eggs fresh, egg products are available in liquid, frozen or spray-dried form.
- Whole egg is used primarily for cake production, where its foaming and coagulation properties are required.
- Egg whites are used for meringues and light sponges where their foaming property is crucial.
- Eggs are useful as a main dish as they provide the energy, fat, minerals and vitamins needed for growth and repair of the body. The yolk contains protein, fats, B vitamins and minerals. The egg white consists of mainly protein and water.

> **PROFESSIONAL TIP**
> Liquid egg may be purchased in cartons as whole egg, whites or yolks. The egg is pasteurised before packaging so is safer for use where low temperatures are required or where the egg is left uncooked.

Egg production methods

Enriched colony cage

Enriched colony cage systems hold between 40 and 80 birds. They help to promote a bird's natural behaviour by providing a nest box, scratching area, perches and more space to move around and flap their wings.

Barn eggs

Hens are housed in buildings with one or more levels. The hens have space to move around freely, have litter for scratching, and are provided with nest boxes and perches. Hens can interact with each other, while still being protected from the weather and predators.

Free-range eggs

Free-range eggs are produced in ways similar to the barn system with nest boxes, perches and scratching areas provided and room for hens to move around freely. The birds also have access to the outdoors during daylight hours.

Organic eggs

Organic eggs come from hens that are free range on organically-farmed land and are kept in smaller numbers than other methods, with plenty of space for nest boxes, perches and scratch areas. They are fed on an organically-produced diet.

> **PROFESSIONAL TIP**
> 'Battery' cages for hens are now banned in the UK. All UK eggs are now produced in the enriched colony system, barn free range or organic systems.

All boxes of whole eggs sold in this country must state the method of production clearly, so you can check whether the eggs have come from hens kept in cages or whether they have come from higher welfare alternative systems such as barn, free range or organic.

Each egg is also stamped with a code for the method of production, country of origin and the originating farm:

- O = Organic
- 1 = Free range
- 2 = Barn
- 3 = Caged.

Sizes

Hens' eggs are graded in four sizes: small, medium, large and very large, as shown in Table 8.2.

Table 8.2 Hens' egg sizes

Very large	73 g	Size 0
		Size 1
Large	63–73 g	Size 1
		Size 2
		Size 3
Medium	53–63 g	Size 3
		Size 4
		Size 5
Small	53 g and under	Size 5
		Size 6
		Size 7

1.2 Quality points

- Appearance – eggshell should be clean, well-shaped, strong and slightly rough. Not cracked or broken. The yolk should be firm, round (not flattened) and of a good even colour. Over time, as eggs are kept, the yolk loses strength and begins to flatten, water evaporates from the egg and is replaced by air. There should be a high proportion of thick white to thin white. If eggs are stored for too long, the thick white gradually changes into thin white, and water passes from the white into the yolk.
- Colour – the colour of the eggshell and yolk does not affect its quality, it depends on the diet of the hen.
- Aroma – there should be no detectable unpleasant odour from a fresh egg, though cooked eggs, especially when hard boiled, do have a characteristic odour.
- Freshness – ensure eggs are purchased from a reputable supplier. Most eggs now carry a date stamp. On delivery do not accept dirty or damaged eggs or packaging. Store in the refrigerator away from strong-smelling foods. A number of procedures are now in place by producers to ensure good quality fresh eggs (see below).

Grading

The size of the eggs does not affect their quality, but it does affect their price. Eggs are tested for quality, then weighed and graded.

- Grade A – naturally clean, fresh eggs, internally perfect with intact shells and an air cell not exceeding six millimetres in depth.
- Grade B – eggs that have been downgraded because they have been cleaned or preserved or because they are internally imperfect, cracked or have an air cell exceeding six millimetres but not more than nine millimetres in depth. Grade B eggs are broken and pasteurised.
- Grade C – eggs that are fit for breaking for manufacturing purposes but cannot be sold in their shells to the consumer.

> **PROFESSIONAL TIP**
> For catering or restaurant use, eggs are packed onto trays of 30 eggs then into outer cases. These often contain 60, 180 or 360 eggs.

British Lion Quality Code of Practice

The Lion Quality mark on eggs and egg boxes means that the eggs have been produced to the highest standards of food safety, including a programme of vaccination against salmonella where the size of the flock is 350 or more.

The Code of Practice covers breeding flocks, hatcheries and laying birds and will include hygiene and animal welfare requirements, farm handling of eggs, distribution of eggs, packing centre procedures, advice to retailers, consumers and caterers and environmental policy and enforcement.

To guarantee traceability, every process involved in the production of Lion Quality eggs must be approved. All Lion Quality hen flocks must be accompanied by a passport certificate and all Lion Quality egg movement has to be fully traceable.

> **HEALTH AND SAFETY**
> Hens can pass salmonella bacteria into their eggs and therefore cause food poisoning. To reduce this risk, pasteurised eggs may be used where appropriate (for example, in omelettes, scrambled eggs and desserts).

> **PROFESSIONAL TIP**
> Always buy eggs from a reputable retailer where they will have been transported and stored under the correct conditions.

1.3 Storage

- Store eggs in a cool but not too dry place – 1 °C to 4 °C is ideal – where the humidity of the air and the amount of carbon dioxide present are controlled. Eggs will keep well under these conditions, but always follow the date advice on packaging.
- Because egg shells are porous, eggs will absorb any strong odours, so they should not be stored near strong-smelling foods, such as onions, fish and cheese.
- Eggs should be stored away from possible contaminants, such as raw meat or fish.
- Stock of eggs should be rotated: first in, first out.
- Pasteurised eggs are washed, sanitised and then broken into sterilised containers. After separating or combining the yolks and whites, they are strained, pasteurised (heated to 63 °C for approximately one minute), then rapidly cooled.
- Egg dishes should be consumed as soon as possible after preparation or, if not for immediate use, refrigerated.
- It is advisable to serve eggs immediately after they are cooked. It is not advisable to hold cooked eggs such as fried, scrambled or poached for too long, such as on a breakfast buffet.

- Poached eggs may be cooked in advance, chilled in ice water and reheated in a water bath when required.

1.4 Cooking ingredients

Eggs are very versatile. Fried, scrambled, poached and boiled eggs and omelettes are mainly served at breakfast. A variety of dishes may be served for lunch, high tea, supper and snacks (this may include scotch eggs, which are deep-fried). They are also used widely as an ingredient in baking and desserts.

> **HEALTH AND SAFETY**
> Hands should be washed before and after handling eggs. Preparation surfaces, utensils and containers should be cleaned regularly and always cleaned between preparation of different dishes.

1.5 Sauces, dressings and fillings

Eggs can be served in a variety of ways with different garnishes and sauces. Sauces can be hot (for example, tomato, cream sauce, curry sauce, cheese sauce) or cold (for example, mayonnaise and green sauce).

There are many different types of fillings for omelettes, such as mushrooms, ham, cheese, mixed herbs and smoked fish. Omelettes may also be coated with a sauce, for example, cheese sauce, cream sauce or tomato sauce.

Scrambled eggs may have additions such as smoked fish, mushrooms, bacon, ham or tomato. The egg mixture may be flavoured with mixed herbs and chilli.

1.6 Evaluation

See page 405 for evaluation of all types of egg dishes.

> **ACTIVITY**
> 1 What are the various production systems used for hens' eggs?
> 2 Name the main food poisoning bacteria associated with hens' eggs.
> 3 Name four basic egg dishes that could be used on a breakfast menu.
> 4 Describe how to make a perfect omelette. Give four examples of fillings that could be used with omelettes.
> 5 Describe how eggs should best be stored.

Prepare and produce gnocchi dishes

Introduction

Gnocchi are soft dumplings made from a range of ingredients including semolina, wheat flour, choux paste potato, cornmeal and occasionally other ingredients.

1.1 Types of gnocchi

Types of gnocchi include parisienne, made with a choux paste, potato gnocchi, also known as gnocchi piemontese and gnocchi romaine which is made with semolina.

1.2 Quality points

Traditionally gnocchi is made in-house then cooked and served quickly or chilled until required. It can now be purchased chilled or frozen and is frequently vacuum-packed to prevent it drying out. The different kinds of gnocchi have differing characteristics but generally gnocchi should be fairly firm without being hard or brittle. It should be well shaped, an even colour and smell fresh.

1.3 Storage

Gnocchi should be stored in the same way as fresh pasta or in the packaging in which it was delivered.

1.4 Cooking ingredients

Gnocchi dough may be rolled out and cut into boiling water, piped and cut into boiling water or shaped into small pieces by hand then boiled. Gnocchi may also be baked in one piece on a tray then cut into shapes. Occasionally a gnocchi dish can be braised and some gnocchi dishes can be deep-fried. Gnocchi is served in a similar way to pasta and is often finished with a sauce and maybe cheese.

1.5 Sauces, dressings and fillings

For sauces and dressings to use with gnocchi, see the recipe section at the end of this unit and Unit 303 Stocks, soups and sauces.

1.6 Evaluation

See below for evaluation of all types of farinaceous dishes.

1.6 Evaluation of farinaceous, egg and grain dishes

All chefs producing farinaceous, grain and egg dishes must be able to evaluate all items against quality requirements and the dish specification to ensure each dish meets the standard required. They must also know the procedure to follow if any faults or breaches in quality standards are identified throughout preparation and cooking.

- Portion size – portion size and control is of great importance to any food business and the success of the business and its profits may depend on portion control. Portion control can be based on many factors including:
  - the number of items in a portion, such as the number of eggs in an omelette, the number of pieces of cannelloni or ravioli to a portion
  - dishes used, such as an individual dish used for a pasta item
  - ladles, scoops, moulds and spoons for a portion of rice or grain salad
  - cutting into equal sized portions, for example, a large lasagne divided into squares
  - weighed portions of pasta, rice or grains.
- Colour – should be what is expected for that dish and according to the dish specification. For example, scrambled eggs should be a pleasant yellow colour but the colour of pasta may be changed by the addition of a tomato sauce.

- Taste – it is important for chefs to taste the dishes they produce to ensure the dish is what is expected and seasoning and flavouring levels are correct
- Texture and consistency – following a recipe or specification and using skill and experience will help to ensure that the correct texture and consistency has been achieved. For example, rice and pasta not overcooked, sauces not too thick or too thin and that the expected texture has been achieved – crisp, soft, aerated and etc so on.
- Harmony – all of the items in a meal should be in harmony with each other. Consider flavours, textures, colours, cooking methods, garnishes and sauces as they all need to work together.
- Presentation – this is very important and needs to be to an agreed specification so a particular dish is presented to the same standard by all involved. Ways of arranging and presenting individual items as well as garnishes will help to maintain quality of presentation.
- Temperature – for food safety reasons, food must be stored, cooked and held at specific hot or cold temperatures. For more information, see Unit 301 Legal and social responsibilities in the professional kitchen. Dishes presented to customers must be served at the temperature expected: hot, cold or warm. Serving dishes and plates also need to be at the correct temperature.
- Aroma – the aroma of food can be one of its most powerful selling points, with individual food and dishes having their own unique aroma. If you detect an aroma that you think is wrong for the dish or any unpleasant aromas do not serve the dish but speak to someone in charge.

TEST YOURSELF

1   State the difference between wholegrain and basmati rice.

2   How much liquid to rice would you use when cooking risotto?

3   What other ingredients would you add to a basic risotto?

4   State **three** sauces and three cheeses that could be used with pasta.

5   State **three** pieces of equipment you would use when preparing fresh pasta.

6   What is the minimum legal temperature for holding cooked lasagne hot for service?

7   State **one** menu use of each of the following grains:
a)   wheat   b)   oats   c)   rye.

8   What is an appropriate cooking method for millet? Why would bread made with millet not rise?

9   State **four** quality points to look for when buying fresh eggs and list the four sizing categories used for eggs.

10   Describe the process for cooking a perfect poached egg.

11   What are the ways quails' eggs could be used?

TAKE IT FURTHER

A college wants to include a pasta café with a salad bar as part of its catering provision for students. Suggest eight pasta dishes (you can include gnocchi) you could include on the menu. Provide cooking and serving guidelines for staff for **one** of the dishes.

Suggest a range of 10 salads for the salad bar to provide variety in colour, texture and flavour.

1 Braised or pilaff rice (*riz pilaff*)

Energy	Calories	Fat	Saturated fat	Carbohydrates	Sugar	Protein	Fibre	*
774 kJ	184 kcal	10.4 g	4.5 g	22.1 g	0.3 g	1.9 g	0.6 g	

* Using white rice and hard margarine. Using brown rice and hard margarine, 1 portion provides: 769 kJ/183 kcal energy; 10.9 g fat; 4.6 g saturated fat; 20.7 g carbohydrates; 0.7 g sugar; 1.9 g protein; 1.0 g fibre

	4 portions	10 portions
Butter or oil	50 g	125 g
Onion, chopped	25 g	60 g
Rice, long grain, white or brown	100 g	250 g
White stock (preferably chicken)	200 ml	500 ml
Salt, mill pepper		

Cooking

1 Place half the butter or oil into a small sauteuse. Add the onion.

2 Cook gently without colouring for 2–3 minutes. Add the rice.

3 Cook gently without colouring for 2–3 minutes.

4 Add twice the amount of stock to rice.

5 Season, cover with buttered paper, bring to the boil.

6 Place in a hot oven (230–250 °C) for approximately 15 minutes, until cooked.

7 Remove immediately into a cool sauteuse.

8 Carefully mix in the remaining butter or oil with a two-pronged fork.

9 Correct the seasoning and serve.

Cook the rice gently without colouring.

Add the stock.

Cover with buttered paper, with a small hole at the centre.

It is usual to use long-grain rice for pilaff because the grains are firm and there is less likelihood of them breaking up and becoming mushy. During cooking, the long-grain rice absorbs more liquid, loses less starch and retains its shape as it swells; short or medium grains may split at the ends and become less distinct in outline.

PROFESSIONAL TIP

Cook the rice for the exact time specified in the recipe. If it cooks for longer, it will be overcooked and the grains will not separate.

Healthy Eating Tips

- Use an unsaturated oil (sunflower or olive). Lightly oil the pan and drain off any excess after the frying is complete.
- Keep the added salt to a minimum.

VARIATION

Add wild mushrooms – about 50–100 g for 4 portions.

Pilaff may also be infused with herbs and spices such as cardamom.

2 Steamed rice

Energy	Calories	Fat	Saturated fat	Carbohydrates	Sugar	Protein	Fibre
1277 kJ	305 kcal	1.4 g	0.0 g	63.7 g	0.0 g	7.1 g	0.0 g

	4 portions
Rice (dry weight)	100 g

Mise en place

1 Wash the rice in a sieve and prepare the saucepan or steamer tray.

2 Place the washed rice into a saucepan and add water until the water level is 2.5 cm above the rice.

Cooking

1 Bring to the boil over a fierce heat until most of the water has evaporated.

2 Turn the heat down as low as possible, cover the pan with a lid and allow the rice to complete cooking in the steam.

3 Once cooked, the rice should be allowed to stand in the covered steamer for 10 minutes.

PROFESSIONAL TIP

Larger amounts of rice can be steamed by placing in a non-perforated steamer tray and placing in a commercial steamer oven. For even cooking cover the top with baking parchment paper

HEALTH AND SAFETY

Take care when using steaming pans or ovens. Remove the lid or open the door carefully and stand back. Escaping steam can burn the hands and face

Healthy Eating Tip

Steaming is a healthy way to cook rice because no fat or salt is added

FAULTS

Take care not to overcook the rice or it will become over-wet and soggy

SERVING SUGGESTION

Steamed rice can be served with a main course item or served separately

3 Plain boiled rice

Energy	Calories	Fat	Saturated fat	Carbohydrates	Sugar	Protein	Fibre
37 kJ	90 kcal	0.1 g	0.0 g	20.0 g	0.0 g	1.9 g	0.0 g

	4 portions
Rice (dry weight)	100 g

Cooking

1 Wash the long-grain rice. Add to plenty of boiling water.

2 Stir to the boil and simmer gently until tender (approximately 12–15 minutes). For brown rice simmer for 40–45 minutes until tender.

3 Pour into a sieve and rinse well under cold running water, then boiling water. Drain and leave in sieve, placed over a bowl and covered with a cloth.

4 Place on a tray in the hotplate and keep hot. See health and safety comment.

5 Serve separately in a vegetable dish.

PROFESSIONAL TIP

Brown rice takes longer to cook because the outer bran layers have been retained so the water takes longer to penetrate the grain

HEALTH AND SAFETY

Wherever possible use rice quickly after cooking and avoid reheating it as it is a high risk item that could lead to food poisoning if not used correctly. If keeping rice hot for service do not allow to fall below 63 °C. If cooling, cool quickly within 90 minutes and chill. This is a requirement under Food Hygiene Regulations 2006

Healthy Eating Tip

If adding salt add sparingly or leave it out altogether. There is no need to add fat such as butter to the cooked rice.

VARIATION

Use different types of rice but check the cooking time needed for each type.

FAULTS

Undercooked rice will be hard and gritty. Overcooked rice will be wet, sticky, soggy and will stick together. You will need to get it just right.

4 Stir-fried rice

Energy	Calories	Fat	Saturated fat	Carbohydrates	Sugar	Protein	Fibre
1423 kJ	338 kcal	10.2 g	1.9 g	30.6 g	0.6 g	32.7 g	0.6 g

* Using 125 g chicken (average dark and light meat) and 25 g mung beans per portion

Mise en place

1 Prepare and cook meat or poultry in fine shreds; dice and lightly cook any vegetables. Add bean sprouts just before the egg.

Cooking

1 Place a wok or thick-bottomed pan over fierce heat, add some oil and heat until smoking.

2 Add the cold rice and stir-fry for about 1 minute.

3 Add the other ingredients and continue to stir-fry over fierce heat for 4–5 minutes.

4 Add the beaten egg and continue cooking for a further 1–2 minutes.

5 Correct the seasoning and serve immediately.

Stir-fried rice dishes consist of a combination of cold pre-cooked rice and ingredients such as cooked meat or poultry, fish, vegetables or egg.

Healthy Eating Tip

- Use an unsaturated oil (sunflower or olive) to lightly oil the pan. Oil sprays are also useful.
- Soy sauce adds sodium (salt), so no added salt is needed.

PROFESSIONAL TIP
Make sure that the vegetables do not overcook or they will lose colour and texture

HEALTH AND SAFETY
Cooked rice and egg are both high-risk food safety items

VARIATION
Omit the chicken and the eggs for a vegetarian version. Vary the vegetables used, cooked brown rice could also be used

SERVING SUGGESTION
Serve in individual serving dishes or a serving bowl.

5 Paella (savoury rice with chicken, fish, vegetables and spices)

Energy	Calories	Fat	Saturated fat	Carbohydrates	Sugar	Protein	Fibre
3383 kJ	804 kcal	31.0 g	6.2 g	48.8 g	3.8 g	85.7 g	1.3 g

* Using edible chicken meat

	4 portions	10 portions
Lobster, cooked	400 g	1 kg
Squid	200 g	500 g
Gambas (Mediterranean prawns), cooked	400 g	1 kg
Mussels	400 g	1 kg
White stock	1 litre	2.5 litres
Pinch of saffron		
Onion, finely chopped	50 g	125 g
Clove of garlic, finely chopped	1	2–3
Red pepper, diced	50 g	125 g
Green pepper, diced	50 g	125 g
Roasting chicken, cut for sauté	1.5 kg	3.75 kg
Olive oil	60 ml	150 ml
Short-grain rice	200 g	500 g
Thyme, bay leaf, seasoning	200 g	500 g
Tomatoes, skinned, deseeded, diced		
Lemon wedges, to finish		

Mise en place

1 Prepare all of the shellfish and keep it refrigerated until you start cooking.

2 Prepare the stock (see recipe).

3 Finely chop the onion and dice the peppers.

4 Prepare the lobster: cut it in half, remove the claws and legs, discard the sac and trail. Remove the meat from the claws and cut the tail into 3–4 pieces, leaving the meat in the shell.

5 Clean the squid: pull the body and head apart. Extract the transparent 'pen' from the body. Rinse well, pulling off the thin purple membrane on the outside. Remove the ink sac. Cut the body into rings and the tentacles into 1-cm lengths.

6 Prepare the gambas by shelling the body.

Cooking

1 Boil the mussels in water or white stock until the shells open. Shell the mussels and retain the cooking liquid.

2 Boil the white stock and mussel liquor together, infused with saffron. Simmer for 5–10 minutes.

3 Sweat the finely chopped onion in a suitable pan, without colour. Add the garlic and the peppers.

4 Sauté the chicken in olive oil until cooked and golden brown, then drain.

5 Add the rice to the onions and garlic and sweat for 2 minutes.

6 Add about 200 ml white stock and mussel liquor.

7 Add the thyme, bay leaf and seasoning. Bring to the boil, then cover with a lightly oiled greaseproof paper and lid. Cook for 5–8 minutes, in a moderately hot oven at 180 °C.

8 Add the squid and cook for another 5 minutes.

9 Add the tomatoes, chicken and lobster pieces, mussels and gambas. Stir gently, cover with a lid and reheat the rice in the oven.

10 Correct the consistency of the rice if necessary by adding more stock, so that it looks sufficiently moist without being too wet. Correct the seasoning.

11 When all is reheated and cooked, place in a suitable serving dish, decorate with 4 (10) gambas and 4 (10) mussels halved and shelled. Finish with wedges of lemon.

NOTE

For a traditional paella, a raw lobster may be used, which should be prepared as follows. Remove the legs and claws and crack the claws. Cut the lobster in half crosswise, between the tail and the carapace. Cut the carapace in two lengthwise. Discard the sac. Cut across the tail in thick slices through the shell. Remove the trail, wash the lobster pieces and cook with the rice.

HEALTH AND SAFETY

As a variety of shellfish and rice are used in paella this is a high-risk dish. Keep the shellfish under refrigeration until it is needed and serve the finished dish immediately.

VARIATION

Vary the types of shellfish used.

Healthy Eating Tips

- To reduce the fat, skin the chicken and use a little unsaturated oil to sweat the onions and fry the chicken.
- No added salt is necessary.
- Serve with a large green salad.

FAULTS

Cook the shellfish thoroughly to make it safe to eat but do not overcook or shellfish will become rubbery and lose flavour.

PROFESSIONAL TIP

Completely prepare all the ingredients before starting cooking.

ON A BUDGET

A budget version could be made using some less expensive fish instead of shellfish in the recipe.

SERVING SUGGESTION

Serve in individual shallow serving dish or a multi-portion paella dish.

6 Risotto with Parmesan (*risotto con Parmigiano*)

Energy	Calories	Fat	Saturated fat	Carbohydrates	Sugar	Protein	Fibre
2598 kJ	621 kcal	36.2 g	14.1 g	49.9 g	4.2 g	23.0 g	0.3 g

Mise en place

1 Prepare all the ingredients ready to cook.

2 Finely chop the onion.

3 Place the stock in a saucepan and bring to boiling point.

5 Melt half of the butter in a heavy pan.

Cooking

1 Bring the stock to a simmer, next to where you will cook the risotto. Take a wide, heavy-bottomed pan or casserole, put half the butter in over a medium heat and melt.

2 Add the onion and sweat until it softens and becomes slightly translucent.

3 Add the rice and stir with a heat-resistant spatula until it is thoroughly coated in butter (about 2 minutes). Then take a soup ladle of hot stock and pour it into the rice.

	4 portions	10 portions
Chicken stock	1.2 litres	3 litres
Butter	80 g	200 g
Onion, peeled and finely chopped	½	1
Arborio rice	240 g	600 g
Parmesan, freshly grated	75 g	180 g
Salt, pepper		

4 Continue to cook and stir until this liquid addition is completely absorbed (about 3 minutes).

5 Repeat this procedure several times until the rice has swollen and is nearly tender. The rice should not be soft but neither should it be chalky. Taste and wait: if it is undercooked, it will leave a gritty, chalky residue in your mouth.

6 Normally the rice is ready about 20 minutes after the first addition of stock.

7 Add the other half of the butter and half the Parmesan off the heat. Stir these in, season and cover. Leave to rest and swell a little more for 3 minutes. Serve immediately after this in soup plates, with more Parmesan offered separately.

This is the classic risotto. With the addition of saffron and bone marrow it becomes risotto Milanese.

Healthy Eating Tips
- Use an unsaturated oil (sunflower or olive), instead of butter, to sweat the onion. Lightly oil the pan and drain off any excess after the frying is complete.
- Additional salt is not necessary.
- Serve with a large salad and tomato bread.

PROFESSIONAL TIP
Add the stock slowly, to give the rice time to absorb the liquid. Stir regularly during cooking.

VARIATION
Risotto variations include:
- saffron or Milanese-style – soak ¼ teaspoon saffron in a little hot stock and mix into the risotto near the end of the cooking time
- seafood – add any one or a mixture of cooked mussels, shrimp, prawns, etc., just before the rice is cooked; also use half fish stock, half chicken stock
- mushrooms.

HEALTH AND SAFETY
Serve risotto once cooked and avoid holding hot for service, but if this is done always hold above 63 °C and for no longer than two hours (Food Hygiene Regulations 2006).

FAULTS
Take care not to overcook the risotto as it will become over soft and sticky.

7 Risotto with asparagus (amuse bouche)

See risotto recipe (recipe 7).
Garnish the risotto with fresh cooked asparagus heads.

NOTE
Many other recipes are suitable for an amuse bouche, such as different risottos, ravioli with different fillings, different types of soups with a range of combinations and flavours, various fritters, vegetable, meat, fish and poultry. A common amuse bouche which is simple and quick is deep fried king prawns with a chilli sauce.

Remember: amuse bouche are bite-sized appetizers.

Risotto
http://bit.ly/2oAbm2O

PROFESSIONAL TIP
Serve only very small portions

8 Fresh egg pasta dough (method 1)

Energy	Calories	Fat	Saturated fat	Carbohydrates	Sugar	Protein	Fibre
1672 kJ	400 kcal	17.2 g	9.4 g	50.0 g	10.2 g	11.8 g	4.0 g

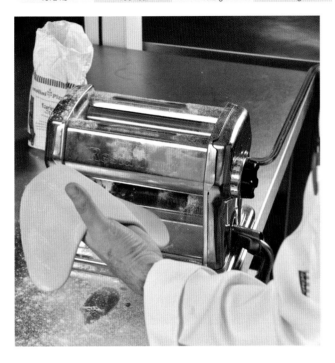

	4 portions	10 portions
Pasta flour	175g	350g
Whole eggs	1	2
Egg yolks	3	6

Mise en place

1. Weigh out and measure the ingredients exactly and place on an ingredients tray.
2. Assemble the food processor.
3. Prepare (and clamp down) a pasta rolling machine if using.

Preparing the dish

1. Place all the ingredients into a food processor and mix quickly until a wet crumb mix appears; this should take no more than 30–45 seconds.
2. Tip the mix out on to a clean surface; this is where the working of the pasta begins.
3. The pasta dough may feel wet at this stage, however the working of the gluten will take the moisture back in to the dry mass, leaving a velvety-smooth finish that is malleable and easy to work; most of this process should be carried out using a pasta machine.
4. Rest the dough for 30 minutes and then it is ready to use.
5. For a classical noodle shape, roll out to a thin rectangle 45 × 15 cm. Cut into 0.5-cm strips. Leave to dry.

Combine the ingredients into a wet crumb mix.

Work and roll out the pasta using a pasta machine.

A pasta machine with an attachment can be used to cut noodles.

PROFESSIONAL TIPS

Do not knead the dough too much, or it will become tough.

Let it rest before rolling it out.

The best way to roll out pasta dough at the correct thickness is to use a pasta machine.

Making fresh pasta
http://bit.ly/2plPHc1

Fresh egg pasta dough (method 2)

Sieve the flour and salt, shape into a well. Pour the beaten eggs into the well.

Gradually incorporate the flour and only add oil to adjust to required consistency. The amount of oil will vary according to the type of flour and the size of the eggs.

Pull and knead the dough until it is of a smooth elastic consistency. Cover the dough with a dampened cloth and allow to rest in a cool place for 30 minutes.

After resting, knead the dough again.

Roll out the dough on a well-floured surface to a thickness of ½ mm or use a pasta rolling machine.

Trim the side and cut the dough as required using a large knife.

HEALTH AND SAFETY
Because the pasta contains fresh eggs, refrigerate the dough once made.

Healthy Eating Tip

The sauces and finishes used with pasta will determine how healthy the finished dish is. Avoid sauces with significant amounts of oil or butter. Avoid too much cheese.

VARIATION
Fresh pasta can be coloured by addition of tomato puree or spinach powder to the dough.

FAULTS

If the pasta dough is too dry it will not roll out or shape. If it is too wet it will be difficult to handle, sticky and will not shape well. Follow the recipe carefully

SERVING SUGGESTION
Serve with a suitable sauce and other ingredients (see recipes).

9 Lasagne

Energy	Calories	Fat	Saturated fat	Carbohydrates	Sugar	Protein	Fibre
2416 kJ	575 kcal	28.7 g	11.4 g	56.1 g	10.0 g	26.7 g	5.8 g

	4 portions	10 portions
Lasagne	200 g	500 g
Oil	1 tbsp	3 tbsp
Streaky bacon, thin strips of	50 g	125 g
Onion, chopped	100 g	250 g
Carrot, chopped	50 g	125 g
Celery, chopped	50 g	125 g
Minced beef	200 g	500 g
Tomato purée	1 tbsp	2½ tbsp
Jus-lié or demi-glace	375 ml	1 litre
Clove of garlic	1	1½
Salt, pepper		
Marjoram	½ tsp	1½ tsp
Mushrooms, sliced	100 g	250 g
Béchamel sauce	250 ml	600 ml
Parmesan or Cheddar cheese, grated	25 g	125 g

Mise en place

1 Prepare and cut the pasta dough to the correct size.

2 Prepare a large pan of water and bring to the boil.

3 Cut the bacon into strips.

4 Chop the vegetables.

5 Prepare the béchamel sauce.

6 Prepare the noodle paste and roll out to 1 mm thick.

7 Cut into 6-cm squares.

8 Allow to rest in a cool place and dry slightly on a cloth dusted with flour.

Cooking

1 Whether using fresh or ready-bought lasagne, cook in gently simmering salted water for approximately 10 minutes.

2 Refresh in cold water, then drain on a cloth.

3 Gently heat the oil in a thick-bottomed pan, add the bacon and cook for 2–3 minutes.

4 Add the onion, carrot and celery, cover the pan with a lid and cook for 5 minutes.

6 Add the minced beef, increase the heat and stir until lightly brown.

7 Remove from the heat and mix in the tomato purée.

8 Return to the heat, mix in the jus-lié or demi-glace, stir to boil.

9 Add the garlic, salt, pepper and marjoram, and simmer for 15 minutes. Remove the garlic.

10 Mix in the mushrooms, re-boil for 2 minutes, then remove from the heat.

11 Butter an ovenproof dish and cover the bottom with a layer of the meat sauce.

12 Add a layer of lasagne and cover with meat sauce.

13 Add another layer of lasagne and cover with the remainder of the meat sauce.

14 Cover with the béchamel.

15 Sprinkle with cheese, cover with a lid and place in a moderately hot oven at 190°C for approximately 20 minutes.

16 Remove the lid, cook for a further 15 minutes and serve in the cleaned ovenproof dish.

This recipe can be made using 200 g of ready-bought lasagne or it can be prepared fresh using 200 g flour noodle paste. Wholemeal lasagne can be made using noodle paste made with 100 g wholemeal flour and 100 g strong flour.

Brown the minced beef in a thick-bottomed pan.

Place a layer of lasagne over a layer of meat sauce.

Cover the final layer with béchamel.

VARIATION

Traditionally, pasta dishes are substantial in quantity but because they are so popular they are also sometimes requested as lighter dishes. Obviously the portion size can be reduced but other variations can also be considered.

For example, freshly made pasta cut into 8–10-cm rounds or squares, rectangles or diamonds, lightly poached or steamed, well drained and placed on a light tasty mixture (e.g. a tablespoon of mousse of chicken or fish or shellfish, well-cooked dried spinach flavoured with toasted pine nuts and grated nutmeg or a duxelle mixture) using just the one piece of pasta on top or a piece top and bottom. A light sauce should be used (e.g. a measure of well-reduced chicken stock with a little skimmed milk, blitzed to a froth just before serving, pesto sauce, a drizzle of good-quality olive oil, or a light tomato sauce. The dish can be finished with a suitable garnish (e.g. lightly fried wild or cultivated sliced mushrooms).

Fillings for lasagne can be varied in many ways. Tomato sauce may be used instead of jus-lié.

Healthy Eating Tip

- Use an unsaturated oil (sunflower or olive). Lightly oil the pan and drain off any excess after the frying is complete. Skim the fat from the finished meat sauce.
- Season with the minimum amount of salt.
- The fat content can be proportionally reduced by increasing the ratio of pasta to sauce and thinning the béchamel.

PROFESSIONAL TIP

If using ready-made dry pasta, either soften the pasta as in the recipe or assemble the lasagne with the dry pasta but ensure that the sauces have enough liquid for the pasta to absorb. The assembled dish will also need to be cooked for long enough to soften the pasta (40–50 minutes).

NOTE

Bacon can be left out for a 'non pork' version.

HEALTH AND SAFETY

If making lasagne for use at another time, cool it quickly (within 90 minutes) before chilling. A blast chiller is best for this. When reheating do this thoroughly. Heat to a minimum of 75 °C – test with a clean food probe (Food Safety Act 1990 and the Food Hygiene Regulations 2006).

FAULTS

When layering the lasagne make sure that the pasta sheets are separated by a sauce otherwise they will stick together, making the finished dish heavy.

ON A BUDGET

Adding vegetables to the sauce or making a vegetable only sauce will reduce cost as well as making the dish healthier.

SERVING SUGGESTION

Lasagne can be made in individual dishes or in a large dish and portioned.

10 Pumpkin tortellini with brown butter balsamic vinaigrette

Energy	Calories	Fat	Saturated fat	Carbohydrates	Sugar	Protein	Fibre
1744 kJ	417 kcal	24.6 g	9.0 g	43.2 g	4.1 g	8.2 g	3.0 g

	4 portions	10 portions
Small pumpkin	1	$2\frac{1}{2}$
Olive oil	1 tbsp	$2\frac{1}{2}$ tbsp
Ground cinnamon	$\frac{1}{2}$ tsp	$1\frac{1}{4}$ tsp
Ground nutmeg	$\frac{1}{4}$ tsp	$\frac{3}{4}$ tsp
Caster sugar	1 tsp	$1\frac{1}{2}$ tsp
Seasoning		
Ravioli paste (recipe 2)		
Egg	1	2
Butter	50 g	125 g
Shallots, finely chopped	25 g	62 g
Balsamic vinegar	2 tbsp	5 tbsp
Spinach leaves	100 g	250 g
Sage, chopped	1 tbsp	$2\frac{1}{2}$ tbsp

Mise en place

1 Prepare the pasta dough.
2 Prepare the pumpkin for roasting.
3 Chop the shallots.
4 Wash the spinach leaves.

Cooking

1 First prepare the pumpkin filling. Cut in half and scoop out the seeds.

2 Place in a roasting tray. Sprinkle with olive oil, cinnamon and nutmeg. Add a little water to the pan. Roast the pumpkin at 200 °C for approximately 45 minutes, until tender.

3 Remove from the oven, allow to cool, scrape out the flesh. Puree the flesh with the sugar in a food processor until smooth, then season.

4 To make the tortellini, roll out the ravioli paste into 2-mm thick sheets. Cut the pasta sheets into 8-cm squares.

5 Place 1 teaspoon of the pumpkin filling in the centre of each square. Lightly brush two sides of the pasta with beaten egg and fold the pasta in half, creating a triangle. Join the two ends of the long side of the triangle to form the tortellini, egg wash the seam and firmly press the ends together to seal.

6 Cook the tortellini in boiling salted water for 3–4 minutes until al dente.

7 To prepare the vinaigrette, cook the butter until nut brown, remove from the heat, add the shallots and balsamic vinegar, then season.

8 Place the washed spinach leaves in a pan with one-third of the vinaigrette and quickly wilt the spinach. Season.

9 To serve, place the wilted spinach in the centre of the plates. Arrange the well-drained tortellini on the spinach. Spoon the vinaigrette around the plates. Finish with a sprinkling of fresh sage.

Healthy Eating Tips
- Lightly brush the pumpkin with olive oil when roasting.
- Keep the amount of added salt to a minimum throughout.

VARIATION
Other vegetables, fish, minced meat or poultry fillings could be used. See the fillings used in ravioli.

11 Ravioli

Energy	Calories	Fat	Saturated fat	Carbohydrates	Sugar	Protein	Fibre
1027 kJ	249 kcal	9.4 g	1.4 g	38.9 g	0.8 g	4.8 g	1.6 g

	4 portions	10 portions
Flour	200 g	500 g
Salt		
Olive oil	35 ml	150 ml
Water	105 ml	250 ml

Mise en place

1 Prepare the pasta dough.
2 Place required filling in a bowl.
3 Weigh the flour, add the salt and place in a mixing bowl.
4 Carefully measure the liquids.

Cooking

To make the dough and form the ravioli:

1 Sieve the flour and salt. Make a well. Add the liquid.
2 Knead to a smooth dough. Rest for at least 30 minutes in a cool place.
3 Roll out thinly to a rectangle: 30 cm × 45 cm.
4 Cut in half and egg wash.
5 Place the stuffing in a piping bag with a large plain tube.
6 Pipe out the filling in small pieces, each about the size of a cherry, approximately 4 cm apart, on to one half of the paste or pipe larger amounts spaced further apart for larger ravioli.

7 Carefully cover with the other half of the paste and seal, taking care to avoid air pockets.
8 Mark each with the back of a plain cutter.
9 Cut in between each line of filling, down and across with a serrated pastry wheel or with a suitable sized pastry cutter.
10 Separate on a tray sprinkled with semolina.
11 Poach in gently boiling salted water for approximately 10 minutes. Drain well.
12 Place in an earthenware serving dish.
13 Cover with 250 ml jus-lié, demi-glace or tomato sauce.
14 Sprinkle with 50 g grated cheese.
15 Brown under the salamander and serve.

> **NOTES**
> Fresh egg pasta dough (recipe 8, methods 1 or 2) can also be used. If you have prepared the dough, start this recipe at step 3.

Possible fillings

Here are some examples of stuffing for ravioli, tortellini and other pastas. Each recipe provides enough stuffing to use with 400 g pasta.

Chicken, cooked, minced	200 g
Ham, minced	100 g
Butter	25 g
2 yolks or 1 egg	
Cheese, grated	25 g
Nutmeg, grated	pinch
Salt and pepper	
Fresh white breadcrumbs	25 g

Dry spinach, cooked, pureed	200 g
Ricotta cheese	200 g
Butter	25 g
Nutmeg	
Salt and pepper	

Lean pork mince, cooked	200 g

Lean veal mince, cooked	200 g
Butter	25 g
Cheese, grated	25 g
2 yolks or 1 egg	
Fresh white breadcrumbs	25 g
Salt and pepper	

Marjoram, chopped	Pinch
Ricotta cheese	150 g
Parmesan, grated	75 g
Egg	1
Nutmeg	
Salt and pepper	

Beef or pork mince, cooked	200g
Spinach, cooked	100 g
Onion, chopped, cooked	50 g
Oregano	
Salt and pepper	

Fish, chopped, cooked	200 g
Mushrooms, chopped, cooked	100 g
Parsley, chopped	
Anchovy paste	

PROFESSIONAL TIP
Sprinkling tray with semolina rather than flour is better for stopping shaped pasta sticking together or sticking to the tray.

HEALTH AND SAFETY
If prepared ravioli is not going to be cooked immediately, place on a single layer on a tray sprinkled with semolina, cover with cling film and place in a refrigerator running between 1 °C and 4 °C. Egg pasta has two of the identified allergens (eggs and flour); check any fillings and sauces you use for possible allergens too (Food Information for Consumers Regulation No. 1169/201).

Healthy Eating Tip
Make low fat, low salt fillings and sauces, serve with a salad.

VARIATION
Vary with different fillings and sauces.

SERVING SUGGESTION
Serve in individual portions topped with herbs and parmesan

12 Spaghetti Bolognaise (*spaghetti alla Bolognese*)

Energy	Calories	Fat	Saturated fat	Carbohydrates	Sugar	Protein	Fibre	Sodium
1741 kJ	416 kcal	22.7 g	12 g	21.7 g	2.7 g	32.5 g	2 g	0.403 g

	4 portions	10 portions
Butter or oil	20 g	50 g
Onion, chopped	50 g	125 g
Clove of garlic, chopped	1	2
Good quality lean minced beef, pork or lamb	400 g	1 kg
Jus-lié	125 ml	300 ml
Tomato purée	1 tbsp	2½ tbsp
Marjoram or oregano	Pinch	½ tsp
Mushrooms, diced	100 g	250 g
Salt, mill pepper		
Spaghetti	400 g	1kg
Grated cheese, to serve		

Mise en place

1 Weigh, measure and prepare all ingredients.

Cooking

1 Place half the butter or oil in a sauteuse.

2 Add the chopped onion and garlic, and cook for 4–5 minutes without colour.

3 Add the beef and cook, colouring lightly.

4 Add the jus-lié, the tomato purée and the herbs.

5 Simmer until tender.

6 Add the mushrooms and simmer for 5 minutes. Correct the seasoning.

7 Meanwhile, cook the spaghetti in plenty of boiling salted water.

8 Allow to boil gently, stirring occasionally with a wooden spoon.

9 Cook for approximately 12–15 minutes. Drain well in a colander.

10 Return to a clean pan containing the rest of the butter or oil (optional).

11 Correct the seasoning.

12 Serve with the sauce in centre of the spaghetti.

13 Serve grated cheese separately.

There are many variations on Bolognaise sauce, e.g. substitute lean beef with pork or lamb mince or use a combination of both; add 50 g each of chopped carrot and celery; add 100 g chopped pancetta or bacon.

Healthy Eating Tip

● Try using more pasta and extending the sauce with tomatoes.

PROFESSIONAL TIP

If serving large quantities of spaghetti or keeping it hot for service stir in a little olive oil to stop the pasta sticking together.

NOTE

Some traditional recipes for bolognaise sauce use finely diced fillet beef. This is now usually considered far too expensive an option but could be tried if you had some fillet beef off cuts and trimmings.

Alternative recipe for bolognaise sauce:

	4 portions	10 portions
Olive oil	1 tbsp	2½ tbsp
Good quality lean minced beef, pork or lamb	270 g	700 g
Onion, chopped	½	1
Mushrooms, sliced	100 g	250 g
Carrots peeled and cut as for paysanne	1	2
Red wine	75 ml	180 ml
Beef stock or meat stock, reduced	100 ml	250 ml
Tomato purée	1 tbsp	2 tbsp
Tabasco sauce (optional)	½ tsp	1 tsp
Salt and freshly ground black pepper, to taste		
Fresh parsley, chopped	2 tbsp	5 tbsp
Fresh chives, chopped, to garnish		

1 Heat the olive oil in a frying pan, over a medium heat.

2 Add the mince and the chopped onion and pan-fry for 4–6 minutes, stirring well, until the mince has browned and the onion has softened.

3 Add the mushrooms and carrots and cook for a further minute before adding the red wine, beef stock and tomato purée.

4 Add the Tabasco sauce and season to taste (optional).

5 Add the chopped parsley and cook for 2–4 minutes more to allow the wine and stock to reduce a little.

6 When mixing the pasta into the sauce, first drain the water thoroughly from the pasta then place into the bolognaise sauce.

7 Toss well, to evenly coat, then spoon into a serving bowl.

8 Garnish with the chopped chives, to serve (optional).

HEALTH AND SAFETY

Keep minced meats in the refrigerator until required.

VARIATION

In the second version use more finely chopped vegetables and less meat.

13 Cannelloni

Energy	Calories	Fat	Saturated fat	Carbohydrates	Sugar	Protein	Fibre	*
1823 kJ	435 kcal	20.3 g	5.6 g	44.2 g	4.3 g	21.6 g	4.1 g	

* 1 portion with beef

	4 portions	10 portions
Flour	200 g	500 g
Salt		
Olive oil	35 ml	150 ml
Water	105 ml	250 ml
Fresh egg pasta (recipe 8) can also be used		

Mise en place

1 Prepare the pasta dough.
2 Weigh the flour, add the salt and place in a mixing bowl.
3 Carefully measure the liquids.

Use the same ingredients as for ravioli dough (recipe 11).

Cooking

1 Roll out the paste as for ravioli.
2 Cut into squares approximately 6 cm × 6 cm.
3 Cook in gently boiling salted water for approximately 10 minutes. Refresh in cold water.

4 Drain well and lay out singly on the table. Pipe out the filling across each.
5 Roll up like sausage rolls but with only a small overlap on the pasta. Place in a greased earthenware dish with the pasta overlap underneath.
6 Add 250 ml demi-glace, jus-lié or tomato sauce.
7 Sprinkle with 25–50 g grated cheese.
8 Brown slowly under the salamander or in the oven, then serve.

A wide variety of fillings may be used, such as those given in recipe 11.

PROFESSIONAL TIP
Make sure that the dough is rolled thinly and evenly and there is a good proportion of filling to pasta

HEALTH AND SAFETY
Check fillings and sauces used for possible allergens in the ingredients (see Food Information for Consumers Regulation No. 1169/201).

Healthy Eating Tip
Use low fat and low salt fillings. Use low fat cheese or keep the cheese to a minimum.

ON A BUDGET
Using vegetable-based fillings can reduce the cost as well as providing a healthy option.

SERVING SUGGESTION
Serve in an individual dish and with a crisp mixed salad.

14 Macaroni cheese

Energy	Calories	Fat	Saturated fat	Carbohydrates	Sugar	Protein	Fibre
7596 kJ	1808 kcal	116.6 g	64.2 g	136.6 g	26.6 g	60.0 g	6.8 g

* For 4 portions

	4 portions	10 portions
Macaroni	400 g	1kg
Butter or oil, optional	25 g	60 g
Grated cheese	100 g	250 g
Thin béchamel sauce	500 ml	1.25 litres
Diluted English or continental mustard	¼ tsp	1 tsp
Salt, mill pepper		

Mise en place

1 Weigh, measure and prepare all ingredients.
2 Measure previously made béchamel.
3 Put on a saucepan of salted water to boil for macaroni.

Cooking

1 Plunge the macaroni into a saucepan containing plenty of boiling salted water.
2 Allow to boil gently and stir occasionally with a wooden spoon.
3 Cook for approximately 15 minutes and drain well in a colander.
4 Return to a clean pan containing the butter.
5 Mix with half the cheese and add the béchamel and mustard. Season.
6 Place in an earthenware dish and sprinkle with the remainder of the cheese.
7 Brown lightly under the salamander and serve.

Browning the macaroni well gives it a good flavour, texture and presentation.

Healthy Eating Tips
- Half the grated cheese could be replaced with a small amount of Parmesan (more flavour and less fat).
- Use semi-skimmed milk for the béchamel. No added salt is necessary.

VARIATION
Variations include the addition of cooked, sliced mushrooms, diced ham, sweetcorn, tomato or other suitable ingredients. Macaroni may also be prepared and served as for any of the spaghetti dishes.

PROFESSIONAL TIP
Macaroni cheese has become a popular restaurant item. It is an economical dish and can be suitable for vegetarians.

HEALTH AND SAFETY
Take care when draining the boiled macaroni to avoid burns from hot water and steam (Health and Safety at work regulations 1999).

FAULTS
Just as with spaghetti and other pasta it is important not to overcook the macaroni.

SERVING SUGGESTION
Finish and serve in individual dishes.

15 Tagliatelle carbonara

Energy	Calories	Fat	Saturated fat	Carbohydrates	Sugar	Protein	Fibre
2911 kJ	698 kcal	44.8 g	20.8 g	43.0 g	1.9 g	33.4 g	0.2 g

	4 portions	10 portions
Tagliatelle	400 g	1 kg
Olive oil	1 tbsp	2½ tbsp
Cloves of garlic, peeled and crushed	2	5
Smoked bacon, diced	200 g	500 g
Eggs	4	10
Double or single cream	4 tbsp	150 ml
Parmesan cheese	4 tbsp	150 ml

Mis en place

1 Weigh, measure and prepare all ingredients.

2 Bring a saucepan of salted water to the boil.

Cooking

1 Cook the tagliatelle in boiling salted water until al dente. Refresh and drain.

2 Heat the oil in a suitable pan. Fry the crushed and chopped garlic. Add the diced, smoked bacon.

3 Mix together the beaten eggs, cream and Parmesan. Season with black pepper.

4 Add the tagliatelle to the garlic and bacon. Add the eggs and cream, stirring until the eggs cook in the heat. Serve immediately.

PROFESSIONAL TIP

It is essential not to overheat the egg and cream mixture or the eggs will overcook and scramble.

HEALTH AND SAFETY

Although it is important not to overcook the eggs, it is still essential to cook them properly to avoid possibility of food poisoning (Food Safety Act 1990 and the Food Hygiene Regulations 2006).

Healthy Eating Tip

Try using low fat crème fraiche instead of cream to produce a lower fat version.

VARIATION

Make a vegetarian version by omitting the bacon and adding sauté mushrooms, courgettes or other vegetables.

FAULTS

Pasta should be cooked 'al dente' (to the bite) which means it is cooked and not gritty in the middle but still firm to bite. Overcooked pasta is too soft, lacking in texture, pale and loses shape.

SERVING SUGGESTION

Serve in individual bowls with the ingredients well distributed through the pasta.

16 Crisp polenta and roasted Mediterranean vegetables

Energy	Calories	Fat	Saturated fat	Carbohydrates	Sugar	Protein	Fibre
3267 kJ	790 kcal	71.4 g	18.9 g	28.6 g	14.0 g	10.1 g	6.6 g

	4 portions	10 portions
Polenta		
Water	200 ml	500 ml
Butter	30 g	75 g
Polenta flour	65 g	160 g
Parmesan, grated	25 g	60 g
Egg yolks	1	2
Crème fraiche	110 g	275 g
Seasoning		
Roasted vegetables		
Red peppers	2	5
Yellow peppers	2	5
Courgettes	2	5
Red onions	2	5
Vegetable oil	200 ml	500 ml
Seasoning		
Clove of garlic	1	3
Thyme, sprigs	2	5

Mise en place

1 Weigh, measure and prepare all ingredients.

Cooking

Polenta:

1 Bring the water and the butter to the boil.
2 Season the water well and whisk in the polenta flour.
3 Continue to whisk until very thick.
4 Remove from the heat and add the Parmesan, egg yolk and crème fraiche.
5 Whisk until all incorporated; check the seasoning.
6 Set in a lined tray.

7 Once set, cut using a round cutter or cut into squares.
8 Reserve until required.

Roasted vegetables:

1 Roughly chop the vegetables into large chunks. Ensure the seeds are removed from the peppers.
2 Toss the cut vegetables in the oil and season well.
3 Place the vegetables in an oven with the aromats for 30 minutes at 180 °C.
4 Remove from the oven and drain. Reserve until required.

To serve:

1 To serve the dish, shallow-fry the polenta in a non-stick pan until golden on both sides.
2 Warm the roasted vegetables and place them in the middle of the plate. Place the polenta on top.
3 Serve with rocket salad and balsamic dressing.

> **PROFESSIONAL TIP**
> Line the tray with cling film and silicone paper before pouring in the polenta – this will stop it from sticking to the tray when it sets.

> **NOTE**
> Aromats is a term used for fresh herbs.

> **HEALTH AND SAFETY**
> Keep the egg yolks, crème fraiche and cheese in the refrigerator until needed (Food Safety Act 1990 and the Food Hygiene Regulations 2006).

> **Healthy Eating Tip**
> Use low fat crème fraiche for a lower fat version.

> **VARIATION**
> Vary the types of roasted vegetables.

> **FAULTS**
> Do not overcook the roasted vegetables as they will lose colour, flavour and texture.

17 Couscous with chorizo sausage and chicken

Energy	Calories	Fat	Saturated fat	Carbohydrates	Sugar	Protein	Fibre
1590 kJ	380 kcal	13.1 g	4.3 g	34.1 g	1.7 g	33.1 g	0.3 g

	4 portions	10 portions
Couscous	250 g	625 g
Olive oil	1 tbsp	1½ tbsp
Garlic cloves, finely chopped	2	3
Chorizo sausage	150 g	400 g
Suprêmes of chicken, skinned	3	7
Sunblush tomatoes	75 g	200 g
Fresh parsley	¼ tsp	½ tsp

Mise en place

1 Weigh, measure and prepare all ingredients.

Cooking

1 Prepare the couscous in a suitable bowl and gently pour over 300 ml of boiling water (750 ml for 10 portions).

2 Stir well, cover and leave to stand for 5 minutes.

3 Heat the olive oil in a suitable frying pan, add the chopped garlic, then sauté for 1 minute.

4 Add the chorizo sausage (sliced 1 cm thick) and the chicken cut into fine strips. Cook for 5–6 minutes.

5 Add the couscous (if any water remains on the surface drain it off), sunblush tomatoes (skinned or diced) and parsley, mix thoroughly and heat for a further 2–3 minutes.

6 Drizzle with olive oil, serve as a warm salad.

7 Garnish with mixed leaves and flat parsley.

Healthy Eating Tips
- Use an unsaturated oil (sunflower or olive) to lightly oil the pan to sauté the garlic.
- Drain off any excess fat after cooking the sausage and chicken.
- Serve with mixed leaves.
- Chorizo and sunblush tomatoes can be salty so there is no need to add further salt.

PROFESSIONAL TIP
Replace the chorizo and chicken with roasted or grilled vegetables for a vegetarian version.

18 Scrambled eggs (basic recipe) (*oeufs brouillés*)

Energy	Calories	Fat	Saturated fat	Carbohydrates	Sugar	Protein	Fibre	*
1105 kJ	263 kcal	22.9 g	8.7 g	0.5 g	0.5 g	13.9 g	0.0 g	

* Using hard margarine instead of butter

	4 portions	10 portions
Eggs	6–8	15–20
Milk (optional)	2 tbsp	5 tbsp
Salt, pepper		
Butter or oil	50 g	125 g

Mise en place

1 Assemble the ingredients.
2 Break the eggs in a basin, add milk (if using), lightly season with salt and pepper and thoroughly mix with a whisk.

Cooking

1 Melt half the butter in a thick-bottomed pan, add the eggs and cook over a gentle heat, stirring continuously until the eggs are lightly cooked.
2 Remove from the heat, correct the seasoning and mix in the remaining butter. (A tablespoon of cream may also be added at this point.)
3 Serve in individual egg dishes or on a slice of freshly butter toast with the crust removed.

VARIATION
Scrambled eggs may be served with smoked salmon.

Healthy Eating Tips
● Try to keep the butter used in cooking to a minimum and serve with unbuttered toast.
● Garnish with a grilled tomato.

HEALTH AND SAFETY
Always wash hands thoroughly before and after handling eggs.

FAULTS
If scrambled eggs are cooked too quickly or for too long the protein will toughen, the eggs will discolour because of the iron and sulphur compounds being released, and syneresis (separation of water from the eggs) will occur. This means that they will be unpleasant to eat. The heat from the pan will continue to cook the eggs after it has been removed from the stove; therefore, the pan should be removed from the heat just before the eggs are cooked.

SERVING SUGGESTION
Serve with toast.

19 Soft-boiled eggs in the shell (*oeufs à la coque*)

Energy	Calories	Fat	Saturated fat	Carbohydrates	Sugar	Protein	Fibre	*
340KJ	81 kcal	6.0 g	1.9 g	0.0 g	0.0 g	6.8 g	0.0 g	

* Using 1 egg per portion

Cooking

Method 1 (soft):

1 Place the eggs in cold water and bring to the boil.

2 Simmer for 2–2½ minutes, then remove from the water.

3 Serve at once in an egg cup.

Method 2 (medium soft):

1 Plunge the eggs in boiling water, then reboil.

2 Simmer for 4–5 minutes.

3 Serve at once in an egg cup.

> **HEALTH AND SAFETY**
> Remember that raw egg can carry bacteria, so thorough hand washing is needed when handling eggs.

> **Healthy Eating Tip**
> Serve with wholemeal toast and minimum butter.

Mis en place

1 Two-thirds fill the saucepan with water to cover the eggs

2 Allow 1 or 2 eggs per portion.

20 Fried eggs (*oeufs frits*)

Energy	Calories	Fat	Saturated fat	Carbohydrates	Sugar	Protein	Fibre	*
536 kJ	128 kcal	31.0 g	9.8 g	0.0 g	0.0 g	7.6 g	0.0 g	

* Fried in sunflower oil

Frying eggs
http://bit.ly/2qbxhNZ

Mise en place

1 Allow 1 or 2 eggs per portion.

Cooking

1 Heat a little fat or oil in a frying pan. Add the eggs.

2 Cook gently until lightly set. Serve on a plate or flat dish.

21 Poached eggs with cheese sauce (*oeufs pochés mornay*)

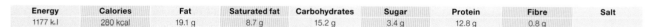

Energy	Calories	Fat	Saturated fat	Carbohydrates	Sugar	Protein	Fibre	Salt
1177 kJ	280 kcal	19.1 g	8.7 g	15.2 g	3.4 g	12.8 g	0.8 g	

Poached eggs Florentine (left), mornay (middle) and Washington (right)

	4 portions	10 portions
Eggs	4	10
Short paste tartlets	4	10
or		
Buttered toast cut into rounds	4	10
Mornay sauce (see page 100)	250 ml	625 ml

Cooking

1 Carefully break the eggs one by one into a bowl then slide into a pan of vinegar water (approximately 15 per cent acidulation) and make sure the water is just below boiling point.

2 Cook until lightly set, for approximately 3–3½ minutes.

3 Remove carefully with a perforated spoon into a bowl of ice water.

4 Trim the white if necessary.

5 Reheat, when required, by placing into hot salted water for approximately ½–1 minute.

6 Remove carefully from the water using a perforated spoon. Drain on a cloth.

7 Place toast or tartlets in an earthenware dish (the slices of toast may be halved, cut in rounds with a cutter, crust removed).

8 Add the hot, well-drained eggs.

9 Completely cover with the sauce, sprinkle with grated Parmesan cheese, brown under the salamander and serve.

Healthy Eating Tip
Poached eggs are a healthy eating option if served without the sauce.

Florentine – poached eggs on a bed of leaf spinach and finished as for mornay.

Washington – on a bed of sweetcorn coated with sûpreme sauce (page 104) or cream.

22 Omelette (*omelette nature*)

Energy	Calories	Fat	Saturated fat	Carbohydrates	Sugar	Protein	Fibre
990 kJ	236 kcal	20.2 g	9.1 g	0.0 g	0.0 g	13.6 g	0.0 g

* Using 2 eggs per portion. Using 3 eggs per portion, 1 portion provides: 1330 kJ/317 kcal energy; 26.2 g fat; 11.0 g Saturated fat; 0.0 g Carbohydrates; 0.0 g sugar; 20.3 g protein; 0.0 g fibre.

Mise en place

1 Break the eggs into a basin, season lightly with salt and pepper.
2 Beat well with a fork, or whisk until the yolks and whites are thoroughly combined and no streaks of white can be seen.

Cooking

1 Heat the omelette pan; wipe thoroughly clean with a dry cloth.
2 Add the butter; heat until foaming but not brown.
3 Add the eggs and cook quickly, moving the mixture continuously with a fork until lightly set; remove from the heat.
4 Half fold the mixture over at right angles to the handle.
5 Tap the bottom of the pan to bring up the edge of the omelette.

	1 portion
Eggs	2–3
Salt, pepper	
Butter or oil	10 g

Making an omelette
http://bit.ly/2oPEiQq

Healthy Eating Tip
Use salt sparingly and serve with plenty of starchy carbohydrate and vegetables or salad.

Whisk the eggs until the yolks and whites are combined.

Move the mixture continuously while it cooks.

Carefully tip the omelette out of the pan.

6 With care, tilt the pan completely over so as to allow the omelette to fall into the centre of the dish or plate.

7 Neaten the shape if necessary and serve immediately.

VARIATION

Variations to omelettes can easily be made by adding the ingredient that the guest or dish may require. For example:

● fine herbs (chopped parsley, chervil and chives)
● mushroom (cooked, sliced, wild or cultivated)
● cheese (25 g grated cheese added before folding)
● tomato (incision made down centre of cooked omelette, filled with hot tomato concassé; served with tomato sauce)
● bacon (grill and then julienne into small strips and fold in at the end).

HEALTH AND SAFETY

Wash hands carefully after handling raw eggs.

FAULTS

Take care not to overcook. Remove from the heat while still soft and slightly liquid. Cooking will continue for a little while after the pan is removed from the heat. There should be no colouring on the omelette.

SERVING SUGGESTION

Served folded into a neat cigar shape.

23 Spanish omelette

Energy	Calories	Fat	Saturated fat	Carbohydrates	Sugar	Protein	Fibre
1726 kJ	416 kcal	30.9 g	11.2 g	10.9 g	8.2 g	24.6 g	2.5 g

Cooking

1 Make up an omelette following recipe 28, steps 1–6, but including the tomato, onion, red pepper and parsley with the eggs.

2 Sharply tap the pan on the stove to loosen the omelette and toss it over as for a pancake.

3 This omelette is cooked and served flat. Many other flat omelettes can be served with a variety of ingredients.

	1 portion
Eggs	2–3
Salt, pepper	
Butter, or oil	10 g
Tomato concassé	50 g
Onions, cooked	100 g
Red pepper, diced	
Parsley, chopped	

When the butter is foaming, add the egg mixture.

The omelette is tipped out flat.

Healthy Eating Tip
Use less butter and make in a non-stick pan sprayed with oil.

SERVING SUGGESTION
Serve flat on individual plates.

24 Scotch eggs

Energy	Calories	Fat	Saturated fat	Carbohydrates	Sugar	Protein	Fibre
2906 kJ	692 kcal	30.9 g	8.5 g	80.2 g	4.7 g	28.2 g	2 g

Cooking

1 Place the eggs, still in their shells, in a pan of water.

2 Place over a high heat and bring to the boil, then reduce the heat to simmer for approximately 9 minutes.

3 Drain and refresh the eggs under cold running water, then peel.

4 Mix the sausage meat with the thyme, parsley and spring onion in a bowl, season well with salt and freshly ground black pepper.

5 Divide the sausage meat mixture into four and flatten each out on a clean surface into ovals about 12 cm long and 8 cm at the widest point.

6 Roll the boiled egg in the seasoned flour.

7 Place each egg onto a sausage meat oval, then wrap the sausage meat around the egg, making sure the coating is smooth and completely covers the egg.

8 Dip each meat-coated egg in the beaten egg, covering the entire surface area.

9 Roll in the breadcrumbs to coat completely.

	4 portions	10 portions
Eggs	4	10
Pork sausage meat	275 g	700 g
Fresh thyme leaves	1 tsp	2 tsp
Fresh parsley, chopped	1 tsp	2 tsp
Spring onion, very finely chopped	1	3
Plain flour, seasoned	125 g	300 g
Egg, beaten	1	2
Breadcrumbs	250 g	625 g
Salt and freshly ground black pepper		
Vegetable oil for deep-frying		

Flatten out an oval of the sausage meat mixture.

Flour the egg and wrap the meat around it until it is completely covered.

Dip in flour, then beaten egg.

Roll the egg in the breadcrumbs.

The egg should be completely coated in breadcrumbs, ready for frying.

10 Heat the oil in a thermostatically controlled deep fryer, to 180 °C.

11 Carefully place each Scotch egg into the hot oil and deep-fry for 6–8 minutes, until golden and crisp and the sausage meat is completely cooked.

12 Carefully remove from the oil with a frying basket or spider and drain on kitchen paper.

To serve, cut the egg in half and season slightly with rock salt. The Scotch eggs can be served hot, warm or cold.

HEALTH AND SAFETY

Take care when using a deep fat fryer. Do not let the oil overheat. Set the temperature correctly.

At the deep-frying stage the coated eggs must be fried for long enough to cook the sausage meat.

FAULTS

If hard-boiled eggs are overcooked it will result in a black line around the yolk which is visible when the eggs are cut. The ring is caused by a chemical reaction between sulphur from the egg white and iron from the egg yolk.

VARIATION

For a vegetarian version of the traditional pork Scotch egg, follow the same method as above, replacing the sausage meat with 350 g of dry mashed potato.

A fish version can be made. Follow the same method as above, using 300 g fish mousse (this works best using salmon) instead of sausage meat.

Scotch Eggs were first developed by Fortnum and Mason in London in the eighteenth century as a picnic or travelling snack.

25 Eggs in cocotte (*oeufs en cocotte*)

Energy	Calories	Fat	Saturated fat	Carbohydrates	Sugar	Protein	Fibre
534KJ	127 kcal	11.2 g	5.2 g	0.0 g	0.0 g	6.8 g	0.0 g

	4 portions	10 portions
Butter	25 g	60 g
Eggs	4	10
Salt, pepper		

Mise en place

1 Assemble the eggs, butter and cooking dishes.
2 Place 1 cm depth of water in a sauté pan with a lid

Cooking

1 Butter the appropriate number of egg cocottes.
2 Break an egg carefully into each and season.
3 Place the cocottes in a sauté pan containing 1 cm water.

4 Cover with a tight-fitting lid, place on a fierce heat so that the water boils rapidly.
5 Cook for 2–3 minutes until the eggs are lightly set, then serve.

VARIATION

Half a minute before the cooking is completed, add 1 tsp of cream to each egg and complete the cooking.

When cooked, add 1 tsp of jus-lié to each egg.

Place diced cooked chicken, mixed with cream, in the bottom of the cocottes; break the eggs on top of the chicken and cook.

As above, using tomato concassé in place of chicken.

Water can settle on top of the eggs as it drips down from the lid. Either very gently pour this off or use absorbent kitchen paper to remove it.

HEALTH AND SAFETY
Take care removing the pots from the boiling liquid.

Healthy Eating Tip
Without the additions described, this is a very healthy way to cook eggs.

SERVING SUGGESTION
Place the cooking dish on a small plate to serve.

26 Gnocchi parisienne

Energy	Calories	Fat	Saturated fat	Carbohydrates	Sugar	Protein	Fibre
1385 kJ	334 kcal	24.5 g	12.6 g	18.4 g	3.4 g	10.8 g	0.8 g

	4 portions	10 portions
Water	125 ml	300 ml
Margarine or butter	50 g	125 g
Salt		
Flour, white or wholemeal	60 g	150 g
Eggs	2	5
Cheese, grated	50 g	125 g
Béchamel (thin) (see page 100)	250 ml	625 ml
Salt, pepper, to season		

Mise en place

1 Weigh, measure and prepare all ingredients.

Cooking

1 Boil the water, margarine or butter, and salt in a saucepan. Remove from the heat.

2 Mix in the flour with a kitchen spoon. Return to a gentle heat.

3 Stir continuously until the mixture leaves the sides of the pan.

4 Cool slightly. Gradually add the eggs, beating well. Add half the cheese.

5 Place in a piping bag with ½ cm plain tube.

6 Pipe out in 1-cm lengths into a shallow pan of gently simmering salted water. Do not allow to boil.

7 Cook for approximately 10 minutes. Drain well in a colander.

8 Combine carefully with the béchamel. Correct the seasoning.

9 Pour into an earthenware dish.

10 Sprinkle with the remainder of the cheese.

11 Brown lightly under the salamander and serve.

NOTE
Gnocchi may be used to garnish goulash or navarin in place of potatoes.

PROFESSIONAL TIP
Mix the eggs into the paste carefully and slowly, but make sure they are well mixed in. If there is too much egg, the mixture will be slack.

Healthy Eating Tip
● No added salt is necessary because of the presence of the cheese.

Pipe the gnocchi paste into the simmering water; a string tied across the pan can be used to form the lengths of paste.

Allow the gnocchi to simmer.

Remove the gnocchi from the water and drain them well.

HEALTH AND SAFETY

Take care not to burn arms and face when working over a pan of boiling water (Health and Safety at Work Act 1999).

SERVING SUGGESTION

Serve with a mixed salad.

27 Potato gnocchi (gnocchi piemontese)

Energy	Calories	Fat	Saturated fat	Carbohydrates	Sugar	Protein	Fibre
045 kJ	248 kcal	9.7 g	4.7 g	35.2 g	2.1 g	7.2 g	2.1 g

* Using white flour.

	4 portions	10 portions
Potatoes	300 g	1 kg
Flour, white or wholemeal	100 g	250 g
Egg and egg yolk, beaten	1	2
Butter	25 g	60 g
Salt and pepper		
Nutmeg, grated		
Tomato sauce (page 114)	250 ml	625 ml
Grated cheese, to serve		

Mise en place

1 Weigh, measure and prepare all ingredients.

Cooking

1 Bake or boil the potatoes in their jackets.
2 Remove the skins and mash with a fork or pass through a sieve.
3 Mix with the flour, egg, butter and seasoning while hot.
4 Mould into balls the size of walnuts.
5 Dust well with flour and flatten slightly with a fork.
6 Poach in gently boiling water until they rise to the surface. Drain carefully.
7 Dress in a buttered earthenware dish, cover with tomato or any other pasta sauce.
8 Sprinkle with grated cheese, brown lightly under the salamander and serve.

Combine the mashed potato with the other ingredients while hot.

Mould the mixture into balls, then flatten with a fork.

Drop the gnocchi into gently boiling water and poach them.

Healthy Eating Tip
- No added salt is necessary.

PROFESSIONAL TIP
Oven baked potatoes will be drier and easier to handle and mould than boiled potatoes.

28 Gnocchi romaine (semolina)

Energy	Calories	Fat	Saturated fat	Carbohydrates	Sugar	Protein	Fibre
1066 kJ	254 kcal	12.5 g	6.9 g	27.5 g	7.0 g	9.8 g	0.8 g

* Using semi-skimmed milk

Cooking

1 Boil the milk in a thick-bottomed pan.

2 Sprinkle in the semolina, stirring continuously. Stir to the boil.

3 Season and simmer until cooked (approximately 5–10 minutes). Remove from heat.

4 Mix in the egg yolk, cheese and butter.

5 Pour into a buttered tray 1 cm deep.

6 When cold, cut into rounds with a 5 cm round cutter.

7 Place the offcuts in a buttered earthenware dish.

8 Neatly arrange the rounds on top.

9 Sprinkle with melted butter and cheese.

10 Lightly brown in the oven or under a salamander.

	4 portions	10 portions
Milk	500 ml	1.5 litre
Semolina	100 g	250 g
Salt, pepper		
Grated nutmeg		
Egg yolk	1	3
Cheese, grated	25 g	60 g
Butter or margarine	25 g	60 g
Tomato sauce (see page 114)	250 ml	625 ml

NOTE

Serve with a thread of tomato sauce round the gnocchi.

PROFESSIONAL TIP

Make sure the mixture has chilled until it is completely set, before cutting it into shape as this will make it much easier to handle.

Mise en place

1 Weigh, measure and prepare all ingredients including the sauce.

Healthy Eating Tip

● No added salt is necessary because of the presence of the cheese.

Stir the semolina into the milk.

Pour the mixture into a greased tray and leave to cool.

Cut the gnocchi into shape.

Sprinkle grated cheese over the dish.

NOTE
Make sure that the basic mixture is firm enough to cut cleanly with a pastry cutter.

VARIATION
The cut shapes of the gnocchi can be varied.

309 Desserts and puddings

Learning outcomes

In this unit you will be able to:

1 Understand how to produce desserts and puddings, including:
   - knowing examples of desserts and puddings, the differences between desserts and puddings and their suitability of different types of service
   - understanding the preparation and processing techniques used to produce desserts and puddings
   - understanding how techniques are used to produce desserts and puddings and knowing the types of desserts and puddings these techniques are used to produce
   - knowing differences between different types of commodities and understanding the implications of those differences for the production of desserts and pudding
   - understanding the correct storage procedures to use throughout production and on completion of the final products.

2 Produce desserts and puddings, including:
   - knowing the different specialist pieces of equipment used in a patisserie and understanding how to safely operate them in a professional competent manner to produce hot, cold and frozen desserts
   - being able to apply techniques to produce desserts and puddings
   - understanding the appropriate sauces and fillings to use with desserts and puddings and being able to produce fillings and sauces in the presentation of desserts and puddings
   - being able to produce decorative items to use in the presentation of desserts
   - being able to measure and evaluate against quality standards throughout preparation and cooking
   - being able to evaluate products against recipe specification and production standards and recognise any faults.

Recipes included in this chapter

No.	Recipe
Fillings	
1	Chantilly cream
2	Buttercream
3	Pastry cream *(crème pâtissière)*
3	Crème chiboust
3	Crème diplomat
3	Crème mousseline
4	Boiled buttercream
5	Ganache
6	Italian meringue
7	Swiss meringue
8	Frangipane (almond cream)
9	Apple purée
Sauces and glazes	
10	Apricot glaze
11	Fruit coulis
12	Fresh egg custard sauce *(sauce à la anglaise)*
13	Custard sauce
14	Chocolate sauce *(sauce au chocolat)*
15	Stock syrup
16	Caramel sauce
17	Butterscotch sauce

	Hot desserts and puddings
18	Steamed sponge pudding
19	Sticky toffee pudding
20	Eve's pudding with gooseberries
21	Chocolate fondant
22	Orange pancakes *(crêpes Suzette)*
23	Baked Alaska *(omelette soufflé surprise)*
24	Apple fritters *(beignets aux pommes)*
25	Choux paste fritters *(beignets souffleés)*
26	Griottines clafoutis
27	Bread and butter pudding
28	Rice pudding
29	Apple crumble tartlets
30	Apple charlotte
31	Bramley apple spotted dick
32	Jam roly-poly
33	Vanilla soufflé
34	Soufflé pudding *(pouding soufflé)*
	Cold desserts and puddings
35	Fresh fruit salad
36	Poached fruits or fruit compote *(compote de fruits)*
37	Fruit mousse
38	Chocolate mousse

39	Bavarois: basic recipe and range of flavours
40	Lime soufflé frappé
41	Vanilla panna cotta served on a fruit compote
42	Lime and mascarpone cheesecake
43	Trifle
44	Tiramisu torte
45	Crème brûlée (Burned, caramelised or browned cream)
46	Cream caramel (crème caramel)
47	Meringue
48	Vacherin with strawberries and cream (vacherin aux fraises)

	Iced desserts
49	Vanilla ice cream
50	Lemon curd ice cream
51	Apple sorbet
52	Chocolate sorbet
53	Peach melba (pêche Melba)
54	Pear belle Hélène (poire belle Hélène)
55	Raspberry parfait

Understand how to produce desserts and puddings

1.1 Examples of desserts and puddings

In British tradition, pudding is often defined as the dessert or sweet course of a meal. Puddings are often made from a base of eggs, milk and flour and combined with other ingredients before being baked or steamed and served either hot or cold. However, there are no set rules as some puddings, such as rice pudding, are made from pudding rice, sugar, vanilla, milk and cream. Traditional sweet puddings include bread and butter pudding and steamed sponge puddings, which are both served hot, whereas other sweet puddings, such as diplomat pudding, are served cold. Additionally, there are some puddings, such as Yorkshire pudding which are not served as dessert at all but as part of the main course, in this example with Roast Beef. Pease pudding is another example of a dish made from split lentils that is an accompaniment traditionally served with boiled joints of ham or bacon. Then there are savoury puddings in their own right, such as the traditional steak and kidney pudding, which is a pudding steamed in a case of suet paste.

However, the terms 'dessert' and 'pudding' are often used to refer to a sweet dish that is served at the end of a meal. The term dessert is singular in its meaning whereas pudding has more connotations and links to traditional British culture.

Although desserts or puddings can be served as singular, plated items or multiple items served by silver service or from a gueridon trolley, it is perhaps more traditional to serve them as multiple portions and present them to customers in this way.

Egg custard-based desserts

Egg custard mixture provides the chef with a versatile basic set of ingredients for a wide range of sweets. Often the mixture is referred to as **crème renversée**. Egg custard uses eggs, sugar, milk (full cream, semi-skimmed or skimmed) and cream (which is often added to egg custard desserts to enrich them and to improve the mouth-feel of the final product). For more information on these ingredients, see below.

Some examples of sweets produced using this mixture:
- crème caramel
- bread and butter pudding
- diplomat pudding
- cabinet pudding
- queen of puddings
- baked egg custard.

Crème caramel

Savoury egg custard is used to make:
- quiches
- tartlets
- flans.

When a starch such as flour is added to the ingredients for an egg custard mix, it changes the characteristic of the end product.

Basic egg custard sets by coagulation of the egg protein. Egg white coagulates at approximately 60°C and egg yolk at 70°C. Whites and yolks mixed together will coagulate at 66°C. If the egg protein is overheated or overcooked, it will shrink and water will be lost from the mixture, causing undesirable bubbles in the custard. This loss of water is called **syneresis**.

KEY TERMS

Crème renversée: egg custard mixture made using eggs, milk and cream (sweet egg custard mixes will also contain sugar).

Syneresis: loss of water from a mixture.

Traditional custard made from custard powder

Custard powder is used to make custard sauce. It is made from vanilla-flavoured cornflour with yellow colouring added, and is a substitute for eggs. Sweetness is adjusted by adding sugar before mixing with milk and heating. The fat content can be reduced by making it with semi-skimmed milk rather than full fat milk.

PROFESSIONAL TIP

Points to remember when making egg custard-based desserts;
- Always work in a clean and tidy way, complying with food hygiene regulations.
- Prevent cross-contamination by not allowing any potentially harmful substances to come into contact with the mixture.
- Always heat the egg yolk or egg mixes beyond 70°C or use pasteurised egg yolks or eggs.
- Follow the recipe carefully.
- Ensure that all heating and cooling temperatures are followed.
- Always store the end product carefully at the right temperature.
- Check all weighing scales.
- Check all raw materials for correct use-by dates.
- Always wash your hands when handling fresh eggs or dairy products and other pastry ingredients.
- Never use cream to decorate a product that is still warm.
- Always remember to follow the Food Safety and Hygiene (England) Regulations 2013.
- Check the temperature of refrigerators and freezers to ensure that they comply with the current regulations.

Ice cream

Traditional ice cream is made from a basic egg custard sauce. The sauce is cooled and mixed with fresh cream. It is then frozen by churning in an ice-cream machine where the water content forms ice crystals.

Ice cream should be removed from the freezer a few minutes before serving. Long-term storage should be between −18°C and −20°C.

The traditional method of making ice cream uses only egg yolks and sugar and the traditional anglaise base. The more modern approach to making ice cream uses **stabilisers** and different sugars, as well as egg whites.

KEY TERM

Stabiliser: a substance added to foods to help to preserve its structure.

Food Standards (Ice Cream) Regulations 1959 and 1963

The Food Standards (Ice Cream) Regulations 1959 and 1963 require ice cream to be pasteurised by heating to any of:
- 65°C for 30 minutes
- 71°C for 10 minutes
- 80°C for 15 seconds
- 149°C for 2 seconds (sterilised).

After heat treatment, the mixture is reduced to 7.1°C within one and a half hours and kept at this temperature until the freezing process begins. Ice cream needs this treatment in order to kill harmful bacteria. Freezing without the correct heat treatment does not kill bacteria – it allows them to remain dormant. The storage temperature for ice cream should not exceed −20°C.

Any ice cream sold must comply with the following standards.
- It must contain not less than 5 per cent fat and not less than 2.5 per cent milk protein (not necessary in natural proportions).
- It must conform to the Dairy Product Regulations 1995.

For further information contact the Ice Cream Alliance (see **www.ice-cream.org**).

Main components of ice cream

- Sucrose (common sugar) not only sweetens ice cream, but also gives it body. An ice cream that contains only sucrose (not recommended) has a higher freezing point.
- The optimum sugar percentage of ice cream is between 15 and 20 per cent.

- Ice cream that contains dextrose (another type of sugar) has a lower freezing point and better taste and texture. The quantity of dextrose used should be between 6 and 25 per cent of the substituted sucrose (by weight).
- As much as 50 per cent of the sucrose can be substituted with other sweeteners, but the recommended amount is 25 per cent.
- Glucose (another type of sugar) improves smoothness and prevents the crystallisation of sucrose. The quantity of glucose used should be between 25 and 30 per cent of the sucrose by weight.
- Atomised glucose (glucose powder) is more water absorbent, so helps to reduce the formation of ice crystals.
- Inverted sugar is a paste or liquid obtained from heating sucrose with water and an acid (e.g. lemon juice). Using inverted sugar in ice cream lowers the freezing point. Inverted sugar also improves the texture of ice cream and delays crystallisation. The quantity of inverted sugar used should be a maximum of 33 per cent of the sucrose by weight. It is very efficient at sweetening and gives the ice cream a low freezing point.
- Honey has similar properties to inverted sugar.
- The purpose of cream in ice cream is to improve creaminess and taste.
- Egg yolks act as stabilisers for ice cream due to the lecithin they contain; they help to prevent the fats and water in the ice cream from separating. Egg yolks improve the texture and viscosity of ice cream.
- The purpose of stabilisers (e.g. gum Arabic, gelatine, pectin) is to prevent crystal formation by absorbing the water contained in ice cream and making a stable gel. The quantity of stabilisers in ice cream should be between 3 grams and 5 grams per kilogram of mix, with a maximum of 10 grams. Stabilisers promote air absorption, making products lighter to eat and also less costly to produce, as air makes the product go further.

Ice cream-making process

1 Weighing – ingredients should be weighed precisely in order to ensure the best results and regularity and consistency.
2 Pasteurisation – this is a vital stage in making ice cream. Its primary function is to minimise bacterial contamination by heating the mixture of ingredients to 85 °C, then quickly cooling it to 4 °C.
3 Homogenisation – high pressure is applied to cause the explosion of fats, which makes ice cream more homogenous, creamier, smoother and much lighter. It is not usually done for homemade ice cream.
4 Ripening – this basic but optional stage refines flavour, further develops aromas and improves texture. This occurs during a rest period (4–24 hours), which gives the stabilisers and proteins time to act, improving the overall structure of the ice cream. This has the same effect on a crème anglaise, which is much better the day after it is made than it is on the same day.
5 Churning – here, the mixture is frozen while at the same time air is incorporated. The ice cream is removed from the machine at about –10 °C.

> **PROFESSIONAL TIP**
>
> What you need to know about ice cream:
> - Hygienic conditions are essential while making ice cream – personal hygiene and high levels of cleanliness in the equipment and the kitchen environment must be maintained.
> - An excess of stabilisers in ice cream will make it sticky.
> - Stabilisers should always be mixed with sugar before adding, to avoid lumps.
> - Stabilisers should be added at 45 °C, which is when they begin to act.
> - Cold stabilisers have no effect on the mixture, so the temperature must be raised to 85 °C.
> - Ice cream should be allowed to 'ripen' for 4–24 hours. This helps to improve its properties.
> - Ice cream should be cooled quickly to 4 °C, because micro-organisms reproduce rapidly between 20 °C and 55 °C.

ACTIVITY

1 Create your own dessert using ice cream, meringue and fresh fruit using a plate presentation.
2 Cost your dish and calculate a 70 per cent gross profit.

Sorbets

Sorbets belong to the ice cream family; they are a mixture of water, sucrose, atomised glucose, stabiliser, fruit juice, fruit pulp and, sometimes, liqueurs.

PROFESSIONAL TIP

What you need to know about sorbet:

- Sorbet is generally more refreshing and easier to digest than ice cream.
- Fruit for sorbets must always be of a high quality and perfectly ripe.
- The percentage of fruit used in sorbet varies according to the type of fruit, its acidity and the quality desired.
- The percentage of sugar will depend on the type of fruit used.
- The minimum sugar content in sorbet is about 13 per cent.
- As far as ripening is concerned, the syrup should be left to rest for 4–24 hours and never mixed with the fruit because its acidity would damage the stabiliser.
- Stabiliser is added in the same way as for ice cream.
- Sorbets are not to be confused with granitas, which are semi-solid.

Stabilisers

Gelling substances, thickeners and emulsifiers are all stabilisers. These products are used regularly and each has its own specific function, but their main purpose is to retain water to make a gel. In ice-cream they are used to prevent ice crystal formation. They are also used to stabilise the emulsion, increase the viscosity of the mix and give a smoother product that is more resistant to melting. There are many stabilising substances, both natural and artificial, as described below.

- Edible gelatine – this is extracted from animals' bones (usually pork and veal) and, more recently, fish skin. Sold in sheets of two grams, it is easy to precisely control the amount used and to manipulate it. Gelatine sheets should be soaked in cold water to soften and then drained before use. Gelatine sheets melt at 40 °C and should be melted in a little of the liquid from the recipe before being added to the base preparation.
- Pectin – another commonly used gelling substance because of its great absorption capacity. It comes from citrus peel (orange, lemon, etc.), though all fruits contain some pectin in their peel. It is a good idea to mix pectin with sugar before adding it to the rest of the ingredients.
- Agar-agar – a gelatinous marine algae found in Asia. It is sold in whole or powdered form and has a great absorption capacity. It dissolves very easily and, in addition to gelling, adds elasticity and resists heat (this is classified as a non-reversible gel).

Other stabilisers:

- Carob gum – this comes from the seeds of the carob tree; it makes sorbets creamier and improves heat resistance.
- Guar gum and carrageen – like agar-agar, these are extracted from marine algae and are some of the many existing gelling substances available, but they are used less often.

Fruit-based desserts

Fruit is used as an ingredient in many desserts. It can be the main feature of a dessert, for example, peach melba or pear belle Hélène, or in conjunction with other dessert bases and ingredients. Fruit can also form the main component of batter-based desserts, such as fruit fritters or the Griottines (cherries) used in the clafoutis recipe, for example.

Fruit is used to contribute to sponge-based desserts, such as baked Alaska, as a platform to support fruit mousses and/or bavarois, or as a component in steamed-sponge puddings. Fruit is also regularly used in a variety of pastry-based desserts, such as fruit tarts and barquettes.

Other dessert mediums include meringue (vacherins), egg custards (crème renversée) and purées (to flavour ice-creams and sorbets). These are just a few examples of the versatility of fruit and how it can be used alongside other dessert mediums to produce an endless variety of desserts.

Meringue-based desserts

Meringue is a whipped mixture of sugar and egg whites which is used to aerate soufflés, mousses and cake mixtures. It is also used to make pie toppings, such as lemon meringue pie, and to cover desserts such as baked Alaska. Meringue mixtures can also be piped and baked until the surface is crisp to produce shells and other shapes, as in vacherins, for example. There are three types of meringue, used for different purposes.

- French meringue – this uncooked meringue is probably the most common. Caster sugar is gradually whisked into the egg whites once they have reached soft peaks until the mixture reaches firm peaks. This type of meringue is the least stable but the lightest, which makes it perfect for aerating desserts such as soufflés.
- Italian meringue – this is the most stable form of meringue. It is made with sugar syrup that has been heated to the soft-ball stage (121 °C). The hot syrup is gradually beaten into the egg whites after soft peaks have formed and then whipped to firm glossy peaks. Its stability and smooth texture make it ideal for parfaits and mousses, particularly if the product is not going to be cooked or baked beyond the addition of the meringue. An uncooked meringue, particularly if made with unpasteurised eggs, would not be suitable due to this reason.
- Swiss meringue – this produces a firmer and slightly denser result than the other types of meringue. Swiss meringue is produced by whisking caster sugar and egg whites together over a pot of simmering water. The early addition of the sugar prevents the egg whites from

increasing as much in volume as they do in the other meringues, but adds to its fine texture. Swiss meringue is often used in buttercreams or baked to produce meringue shells.

Italian meringue is used to cover baked Alaska

Mousses and bavarois

Dessert mousses are generally made from a base of vanilla, fruit or chocolate. The base is usually mixed with whipped egg whites (meringue) to aerate or whipped cream and usually served chilled with an appropriate accompanying sauce. A gelling agent such as gelatine or agar-agar can also be added during the mixing stage to help the mousse set once it is chilled. Mousses can be served in a mould or de-moulded.

A bavarois or 'Bavarian cream' is a classic dessert originating from the repertoire of chef, Marie-Antoine Carême. A bavarois is a mousse-like flavoured (for example, vanilla, chocolate, fruit-based), egg custard-based cream, set with gelatine or agar-agar. It is lightened with whipped cream and meringue just before setting and prior to being molded. A classic true Bavarian cream was filled into a fluted mould, chilled until firm, before being de-moulded and served with an appropriate sauce. Modern presentation of mousses and bavarois use various shaped moulds to enhance creative presentation.

Rice-based desserts

Rice desserts are one of the classics, although they are not seen as often in restaurants in current times. Rice desserts are usually produced using a plump, short-grained (pudding) rice from which the starch is released into the cooking liquor to produce a sauce-like bind. Rice desserts can be served hot or cold and can also be set to produce a combination

similar to that of a mousse. Often the rice pudding will be flavoured with vanilla, cinnamon or nutmeg, for example.

Batter-based desserts

Batters are used in the production of fritters. They are often used to coat fruits before frying, as in apple fritters, for example. Batters can also be baked to produce desserts such as clafoutis. Batters are general made from a flour base utilising a liquid such as water, milk or even beer, as well as a raising and lightening agent such as egg white, baking powder or yeast.

Soufflés

Soufflés form part of the world's classic dessert repertoire and there are many different types. The classic 'dessert' soufflé is baked and served in a ramekin. The base of dessert soufflés is usually flavoured with vanilla, fruit or chocolate and then aerated with whisked egg whites. This causes the mix to rise beyond the rim of the ramekin or dish to give that classic soufflé appearance and lightness. However, this type of soufflé is not the most stable and the dish needs to be consumed quickly after baking as the soufflé will start to collapse and shrink after a short period of time.

Other types of soufflé include the 'pudding' soufflé which is usually baked in dariole moulds in a bain-marie and served de-moulded on a plate with an appropriate sauce. There are also cold and iced soufflés, which are mousse- and parfait-based desserts, set in lined moulds to provide the appearance of a classic soufflé, but served chilled or frozen, as appropriate. Finally, there are omelette soufflés, which is a classic dessert soufflé mix, baked inside a folded omelette.

Sponge-based desserts

Sponges usually form a component of desserts: the base for a mousse, for example, or sometimes as a lining for a mould before it is filled. Some desserts, such as steamed puddings and baked puddings such as 'sticky toffee pudding', have sponge-like characteristics and mouth-feel but sponges in their true sense are rarely served as a dessert on their own.

Milk puddings

Milk is used as a major ingredient in many desserts and regularly forms a component of desserts. For example, milk is used to make egg custards and is a key commodity in the production of a classic rice pudding. Milk-based puddings are not as commonly seen today as much as they were in the past. Semolina pudding, where semolina is cooked in milk with the possible inclusion of cream alongside other flavours and sweeteners is an example of a milk pudding.

Cereal

As per rice and milk puddings, cereal-based puddings are not seen as regularly as they were in the past, although cereals may be used to add a dimension to other desserts, for example, oats used as part of a crumble mix.

Suet paste-based desserts

Suet is traditionally associated with the production of steamed desserts such as jam roly-poly and spotted dick. As suet is an animal fat, taken from the fat surrounding the kidneys of cattle and sheep, it is often considered unhealthy or neglected by customers.

However, vegetable suet can be used to replace the animal suet, thereby resolving this issue and allowing these traditional puddings to be made without the use of animal fats.

Crêpes

Crêpes are a traditional pancake dessert belonging to France. In their simplest form, crêpes are served with a sprinkling of sugar and a splash of fresh lemon juice. They can also be taken further to produce flamed (flambé) dishes such as the classic Crêpes Suzette. Crêpes are fine pancakes and should not be mistaken for Scotch pancakes or American pancakes, which are much thicker and spongier.

Trifles

Trifle in British cuisine is a dessert made using a layer of sponge fingers, sponge or cake soaked in sherry or another fortified wine. This is traditionally topped with fruit, sometimes suspended and set in a fruit jelly with a third layer of custard. Some recipes will finish this with an additional layer of whipped cream. These ingredients are usually arranged to produce three or four layers.

There are similar desserts produced in various parts of Europe and other areas of the world. In Italy, the dessert 'tiramisu' is often referred to as the Italian trifle, although the Italians also produce a similar dessert to the English classic under the name 'Zuppa Inglese', literally meaning 'English Soup'.

1.2 Techniques for the production of desserts and puddings

For details of the techniques involved in preparing different types of desserts and puddings, see the recipe section at the end of this chapter.

1.3 Commodities in desserts and puddings

The following information about ingredients and commodities is relevant to several units of your course: in addition to this unit, it will be relevant to Unit 310 Paste products; Unit 311 Biscuits, cakes and sponges; and Unit 312 Fermented dough products.

The principal building blocks of pastry dishes are flour, fat, sugar, raising agents, eggs and cream.

Flour

Flour is probably the most common commodity in daily use. It forms the foundation of bread, pastry and cakes and is also used in soups, sauces, batters and other foods. It is one of the most important ingredients in patisserie, if not *the* most important.

There are a great variety of high-quality flours made from cereals, nuts or legumes, such as chestnut flour and cornflour. They have been used in patisserie, baking, dessert cuisine and savoury cuisine in all countries throughout history. The most significant flour is wheat flour.

Flour is used in a variety of recipes

Wheat flour

Wheat flour is composed of starch, gluten, sugar, fats, water and minerals.

- Starch is the main component, but gluten is also a significant element. Gluten is elastic and **impermeable** and therefore makes wheat flour the most common flour used in bread making.
- The quantity of sugar in wheat is very small but it plays a very important role in **fermentation**.
- Wheat contains a maximum of only 16 per cent water, but its presence is important.

● The mineral matter (ash), which is found mainly in the husk of the wheat grain and not in the kernel, determines the purity and quality of the flour.

Production of flour

From the ear to the final product, flour, wheat goes through several distinct processes. These are carried out in modern industrial plants, where wheat is subjected to the various treatments and phases necessary for the production of different types of flour. These arrive in perfect condition at our workplaces and are made into preparations like sponge cakes, yeast dough, puff pastries, biscuits, pastries and much more.

White flour is made up almost entirely of the part of the wheat grain known as the **endosperm**, which contains starch and protein. In milling, the whole grain is broken up, the parts separated, sifted, blended and ground into flour. Some of the outer coating of bran is removed, as is the **wheatgerm**, which contains oil and is therefore likely to become rancid and spoil the flour. For this reason wholemeal flour should not be stored for more than 14 days.

Flour is often categorised by its strength. Wheat varieties are called 'soft' or 'weak' if they are low in a protein known as gluten. Gluten become sticky when moistened, which is why flour mixed with water is converted into a sticky dough. Flours are referred to as 'hard' or 'strong' if they have a high protein (gluten) content. The relative proportion of starch and gluten varies in different wheats. Strong flour, or bread flour, is high in gluten, with 12 to 14 per cent gluten content. The dough produced from strong flour has elastic properties that holds its shape and structure well once baked. Softer flours are comparatively lower in gluten, 8 to 10 per cent, and are more suited to the production of products with finer textures such as pastries, biscuits and cakes. Soft flour is usually divided into cake flour, which is the lowest in gluten, and pastry flour, which has slightly more gluten than cake flour, although many kitchens may use one type of soft flour for both.

See Unit 312 Fermented dough products for more on the strength of flour.

Types of flour

● White flour contains 72–85 per cent of the whole grain (the endosperm only).
● Wholemeal flour contains 100 per cent of the whole grain.
● Wheatmeal flour contains 85–95 per cent of the whole grain.
● High-ratio or patent flour contains 40 per cent of the whole grain.
● Self-raising flour is white flour with baking powder added to it.
● Semolina is granulated hard flour prepared from the central part of the wheat grain. White or wholemeal semolina is available.

Storing flour

Flour is a particularly delicate living material and it must be used and stored with special care. It must always be in the best condition, which is why storing large quantities is not recommended. It must be kept in a suitable environment: a clean, organised, disinfected and aerated storeroom. Warm and humid places must be avoided.

Sugar

Sugar (or sucrose) is extracted from sugar beet or sugar cane. The juice is crystallised by a complicated manufacturing process. It is then refined and sieved into several grades, such as granulated, caster or icing sugars.

Syrup and treacle are produced as a by-product during the production of refined sugar. Simple sugar syrups are produced by dissolving sugar with water and heating to invert the sugar. The density and viscosity required to produce different products, such as sorbets and ice-creams, is based on the ratio or percentage of sugar to water. The higher the percentage of sugar, the more dense and viscous the syrup will be. The density and sweetness of sugar syrups can be measured using equipment such as a **refractometer**, which measures the level of sweetness on the Brix (°Bx) scale and a Sacchrometer or **hydrometer**, which measures density using the Baumé scale. For example, a fruit sorbet mixture will typically measure around 17–18 °Baumé/30–31 °Brix.

Loaf or cube sugar is obtained by pressing the crystals together while they are slightly wet, drying them in blocks, and then cutting the blocks into squares. Fondant is a cooked mixture of sugar and glucose which, when heated, is coloured and flavoured. It is used for decorating cakes, buns, gateaux and petits fours. Fondant is generally bought ready-made.

When sucrose is broken down with water in a chemical process called **hydrolysis**, it separates into its two constituent parts: fructose and glucose. Sugar that has been treated in this way is called inverted sugar and, after sucrose, is one of the most commonly used sugars in the catering profession, thanks to its sweetening properties.

Inverted sugar comes in liquid and syrup forms:

- Liquid inverted sugar is a yellowish liquid with no less than 62 per cent dry matter. It contains more than 3 per cent inverted sugar, but less than 50 per cent. It is used mainly in the commercial food industry.
- Inverted sugar syrup is a white, sticky paste and has no particular odour. It has no less than 62 per cent dry matter and more than 50 per cent inverted sugar. It is the form in which inverted sugar is most commonly used. With equal proportions of dry matter and sucrose, its sweetening capacity is 25–30 per cent greater than sucrose.

Inverted sugar:

- improves the aroma of products
- improves the texture of doughs
- prevents the dehydration of frozen products
- reduces or stops crystallisation
- is essential in ice cream making because it greatly improves its quality and lowers its freezing point.

> **KEY TERMS**
>
> **Refractometer:** equipment used to measure sweetness on the Brix scale.
>
> **Hydrometer:** equipment used to measure density on the Baumé scale.
>
> **Hydrolysis:** a chemical reaction in which a compound such as sugar breaks down by reacting with water.

Glucose

Glucose is available in various forms:

- the characteristics of a viscous syrup, called crystal glucose
- its natural state, in fruit and honey
- a dehydrated white paste (used mainly in the commercial food industry, but also used in catering)
- 'dehydrated glucose' (atomised glucose) – a glucose syrup from which the water is evaporated; this is used in patisserie, but mainly in the commercial food industry.

Glucose syrup is a transparent, viscous paste, which:

- prevents the crystallisation of boiled sugars, jams and preserves
- delays the drying of a product
- adds plasticity and creaminess to ice cream and the fillings of chocolate bonbons
- prevents the crystallisation of ice cream.

Honey

Honey, a sweet syrup that bees make with the nectar extracted from flowers, is the oldest known sugar. A golden-brown thick paste, it is 30 per cent sweeter than sucrose.

Honey lowers the freezing point of ice cream. It can also be used in the same way as inverted sugar, but unlike inverted sugar, will give flavour to the preparation. Honey is unsuitable for preparations that require long storage, since it re-crystallises after time.

Isomalt

Isomalt sugar is a sweetener that is less well known in the patisserie world, but it has been used for some time. It has different properties from those of the sweeteners already mentioned. It is produced through the hydrolysis of sugar, followed by hydrogenation (the addition of hydrogen).

Produced through these industrial processes, this sugar has been used for many years in industry in confectionery and chewing gum production and is now earning a place in the culinary kitchen.

One of its most notable characteristics is that it can melt without the addition of water or another liquid. This is a very interesting property for making artistic decorations in caramel. Its appearance is like that of confectioners' sugar: a glossy powder. Its sweetening strength is half that of sucrose and it is much less soluble than sugar, which means that it melts less easily in the mouth.

Over the past few years, isomalt has been used as a replacement for normal sugar or sucrose when making sugar decorations, blown sugar, pulled sugar or spun sugar. Isomalt is not affected by air humidity, so sugar pieces will keep for longer.

Fats

Pastry goods may be made from various types of fat, either a single named fat or a combination. Examples of fats are butter, margarine, shortening and lard.

Butter and other fats

Butter brings smoothness, rich aromas and impeccable textures to patisserie products.

Butter is a key ingredient in patisserie products.

Butter is an **emulsion** – the perfect interaction of water and fat. It is composed of a minimum of 82 per cent fat, a maximum of 16 per cent water and 2 per cent dry extracts.

> **PROFESSIONAL TIP**
>
> What you need to know about butter:
> - Butter is the most complete fat in terms of its contribution to patisserie products.
> - It is a very delicate ingredient that can quickly spoil if a series of basic rules are not followed in its use.
> - It absorbs odours very easily, so should be kept well covered and should always be stored away from anything that produces strong odours.
> - When kept at 15 °C, butter is stable and retains all its properties: finesse, aroma and creaminess.
> - It should not be kept too long: it is always better to work with fresh butter.
> - Good butter has a stable texture, pleasing taste, fresh odour, homogenous (even) colour and, most importantly, it must melt perfectly in the mouth.
> - It softens preparations like cookies and petit fours, and keeps products like sponge cakes soft and moist.
> - Butter enhances flavour – as in brioches, for example.
> - The melting point of butter is between 30 °C and 35 °C.

Margarine

Margarine is often made from a blend of oils that have been hardened or **hydrogenated** (had hydrogen gas added). Margarine may contain up to 10 per cent butterfat.

Cake margarine is also a blend of hydrogenated oils, to which is added an emulsifying agent that helps combine water and fat. Cake margarine may contain up to 10 per cent butterfat. Pastry margarine is used for puff pastry. It is a hard plastic or waxy fat that is suitable for layering.

Shortening

Shortening is another name for fat used in pastry making. It is made from oils and is 100 per cent fat, such as hydrogenated lard; another type of shortening is rendered pork fat.

> **KEY TERMS**
>
> **Emulsion:** a mixture of two or more liquids that are normally immiscible (non-mixable or un-blendable).
>
> **Hydrogenated:** a substance that has had hydrogen gas added to it. This is used to harden oils in margarine.
>
> **Shortening:** fat used in pastry making; it is made from oils and is 100 per cent fat.

Eggs

The egg is one of the principal ingredients of the culinary world. Its great versatility and extraordinary properties as a thickener, emulsifier and stabiliser make its presence important in numerous creations in patisserie, including sauces, creams, sponge cakes, custards and ice creams. Although it is not often the main ingredient, it plays specific and determining roles in terms of texture, taste and aroma. The egg is fundamental in preparations such as brioche, crème anglaise, sponge cake and crème pâtissière. A good custard cannot be made without eggs, as they cause the required **coagulation** and give it the desired consistency and finesse. Eggs are also an important ingredient in ice cream, where the yolk acts as an emulsifier. The extent to which eggs are used (or not) makes an enormous difference to the quality of the product.

Eggs are a key ingredient in quiche.

Egg yolk is high in saturated fat. The yolk is a good source of protein and also contains vitamins and iron. The egg white is made up of protein (albumen) and water. The egg yolk also contains lecithin, which acts as an **emulsifier** in dishes such as mayonnaise – it helps to keep the ingredients mixed, so that the oils and water do not separate.

Hens' eggs are graded in four sizes: small, medium, large and very large. For the dessert, paste and cake recipes in this book, use medium-sized eggs (approximately 50 grams).

PROFESSIONAL TIP

What you need to know about eggs:
- Eggs act as a texture agent in, for example, patisseries and ice creams.
- They intensify the aroma of pastries like brioche.
- They enhance flavours.
- They give volume to whisked sponges and batters.
- They strengthen the structure of preparations such as sponge cakes.
- They act as a thickening agent, for example, in crème anglaise.
- They act as an emulsifier in preparations such as mayonnaise and ice cream.
- A fresh egg should have a small, shallow air pocket inside it.
- The yolk of a fresh egg should be bulbous, firm and bright.
- The fresher the egg, the more viscous the egg white.
- Eggs should be stored away from strong odours as, despite their shells, odours are easily absorbed.
- In a whole 60-gram egg, the yolk weighs about 20 grams, the white 30 grams and the shell 10 grams.
- Eggs contain protein and fat.

Egg whites

- To avoid the danger of salmonella, if the egg white is not going to be cooked or will not reach a temperature of 70 °C, use pasteurised egg whites. Egg white is available chilled, frozen or dried.
- Equipment must be thoroughly clean and free from any traces of fat, as this will prevent the whites from whipping; fat or grease prevents the albumen strands from bonding and trapping the air bubbles.
- Take care that there are no traces of yolk in the white, as yolk contains fat.

- A little acid (cream of tartar or lemon juice) strengthens the egg white, extends the foam and produces more meringue. The acid also has the effect of stabilising the meringue.
- If the foam is over-whipped, the albumen strands, which hold the water molecules with the sugar suspended on the outside of the bubble, are overstretched. The water and sugar make contact and the sugar dissolves, making the meringue heavy and wet. This can sometimes be rescued by further whisking until it foams up, but very often you will have to discard the mixture and start again. Beaten egg white forms a foam that is used for aerating sweets and many other desserts, including meringues (see page 464).

Milk

Milk is a basic and fundamental element of our diet throughout our lives. It is composed of water, sugar and fat (with a minimum fat content of 3.5 per cent). It is essential to numerous preparations, from cream, ice cream, yeast dough, mousse and custard to certain types of ganache, cookies, tuiles and muffins. A yeast dough will change considerably in texture, taste and colour if made with milk instead of water.

Milk has a slightly sweet taste and little odour. Two distinct processes are used to conserve it:
- Pasteurisation – the milk is heated to between 73 °C and 85 ° for a few seconds, then cooled quickly to 4 °C.
- Sterilisation (UHT) – the milk is heated to between 140 °C and 150 °C for two seconds, then cooled quickly.

Milk is **homogenised** to disperse the fat evenly, since the fat has a tendency to rise to the surface (see 'Cream', below).

PROFESSIONAL TIP

What you need to know about milk:
- Pasteurised milk has a better taste and aroma than UHT milk.
- Milk is useful for developing flavour in sauces and creams, due to its **lactic fermentation**.
- Milk is an agent of colour, texture and aroma in dough.
- Because of its lactic ferments, it helps in the maturation of dough and cream.
- There are other types of milk, such as sheep's milk, that can be interesting to use in desserts.
- Milk is much more fragile than cream. In recipes, adding it in certain proportions is advisable for a much more subtle and delicate final product.

KEY TERMS

Pasteurisation: a process where heat is applied to products such as milk and cream for a short period of time before being cooled quickly. This helps to kill harmful bacteria and extend shelf-life while maintaining flavour properties.

Sterilisation: a process where heat is applied but at much higher temperatures than pasteurisation. This increases

shelf-life further than pasteurisation but products treated this way tend to lose some of their flavour properties.

Homogenise: a process in the production of milk in which the fats are emulsified so that the cream does not separate.

Lactic fermentation: a biological process in which sugars (e.g. sucrose or lactose) are converted into cellular energy.

Cream

Cream is used in many recipes because of its high fat content and great versatility. Cream is the concentrated milk fat that is skimmed off the top of the milk when it has been left to sit. A film forms on the surface because of the difference in density between fat and liquid. This process is speeded up mechanically in large industries by heating and using **centrifuges**.

Cream should contain at least 18 per cent butter fat. Cream for whipping must contain more than 30 per cent butter fat. Commercially frozen cream is available, usually in 2 kilogram and 10 kilogram slabs. Types, packaging, storage and uses of cream are listed in Table 9.1.

Table 9.1 Types of cream

Type of cream	Legal minimum fat (%)	Process and packaging	Storage	Characteristics and uses
Half cream	12	Homogenised and may be pasteurised or ultra-heat treated.	2–3 days	Does not whip; used for pouring; suitable for low fat diets.
Cream or single cream	18	Homogenised and pasteurised by heating to about 79.5 °C for 15 seconds and then cooled to 4.5 °C. Automatically filled into bottles and cartons after processing. Sealed with foil caps. Availability of bulk quantities according to local suppliers.	2–3 days in summer; 3–4 days in winter under refrigeration	A pouring cream suitable for coffee, cereals, soup or fruit A valuable addition to cooked dishes Makes delicious sauces. Does not whip.
Whipping cream	35	Not homogenised, but pasteurised and packaged as above.	2–3 days in summer; 3–4 days in winter under refrigeration	The ideal whipping cream. Suitable for piping, cake and dessert decoration, ice-cream, cake and pastry fillings.
Double cream	48	Slightly homogenised and pasteurised and packaged as above.	2–3 days in summer; 3–4 days in winter under refrigeration	A rich pouring cream which will also whip. The cream will float on coffee or soup.
Double cream 'thick'	48	Heavily homogenised, then pasteurised and packaged. Usually only available in domestic quantities.	2–3 days in summer; 3–4 days in winter under refrigeration	A rich spoonable cream that will not whip.
Clotted cream	55	Heated to 82 °C and cooled for about 4½ hours. The cream crust is then skimmed off. Usually packed in cartons by hand. Availability of bulk quantities according to local suppliers.	2–3 days in summer; 3–4 days in winter under refrigeration	A very thick cream with its own special flavour and colour. Delicious with scones, fruit and fruit pies.
Ultra-heat treated (UHT) cream	12, 18, 35	Half (12%), single (18%) or whipping cream (35%) is homogenised and heated to 132 °C for one second and cooled immediately. Aseptically packed in polythene and foil-lined containers. Available in larger packs for catering purposes.	6 weeks if unopened Needs no refrigeration. Usually date stamped.	A pouring cream.

Whipping and double cream may be whipped to make them lighter and to increase volume. Cream will whip more easily if it is kept at refrigeration temperature. All cream products must be kept in the refrigerator for health and safety reasons. They should be handled with care and, because they will absorb odour, they should never be stored near onions or other strong-smelling foods.

As with milk, there are two main methods for conserving cream:

● Pasteurisation – the cream is heated to between 85 °C and 90 °C for a few seconds and then cooled quickly; this cream retains all its flavour properties.
● Sterilisation (UHT) – this consists of heating the cream to between 140 °C and 150 °C for two seconds; cream treated this way loses some of its flavour properties, but it keeps for longer. Always use pasteurised cream when possible, for example, in the restaurant when desserts are made for immediate consumption.

KEY TERMS

Centrifuge: a piece of laboratory equipment, driven by a motor, which spins liquid samples at high speed. Centrifuges work by the sedimentation principle to separate substances of greater and lesser density.

PROFESSIONAL TIP

What you need to know about cream:
● Cream whips with the addition of air, thanks to its fat content. This retains air bubbles formed during beating.
● Cream adds texture.
● Once cream is boiled and mixed or infused with other ingredients to add flavour, it will whip again if first left to cool completely.
● To whip cream well, it must be cold (around 4 °C).
● Cream can be infused with other flavours when it is hot or cold. If cold, it requires sufficient infusion time to absorb the flavours.

Rice

Rice can be used to produce sweet puddings. For example, it is combined with milk, eggs and nutmeg to make English rice pudding. Usually, a short-grained, plump rice is used to produce rice pudding. This is because the starch levels and consistency of short-grain rice make it ideal for puddings. However, different varieties of rice including long-grained rices such as basmati, Patna and Carolina can be used in sweet dishes. In Asia, rice puddings are often flavoured with cardamom, cinnamon, saffron, rosewater, pistachios, almonds or fruit.

Different cultivation techniques, as well as cross-breeding, have resulted in thousands of varieties of rice, including sticky rices, wild rices and fragrant rices. They are generally categorised as long-, medium- or short-grain. Long-grain rices such as basmati and Patna are thin, dainty and pointed. Medium-grain and short-grain rices are plumper, starchier and more absorbent. Examples of medium-grain rices are risotto and paella rices such as Carnaroli and the shorter-grained Arborio. Short-grain rices include pudding rice and sushi rice.

Fruit

The natural flavourings, sweetness and fresh colours provided by fruits makes them perfect for inclusion in desserts and puddings. Many fruits, particularly soft fruits and berries, are quite perishable and therefore should be used in their prime. It is important to use fruit seasonally, at the time that they are at their very best in terms of flavour, ripeness and appearance. The storage of fresh fruit is also very important to maximise its quality and shelf-life.

Hard fruits, such as apples, should be left in boxes and kept in a cool store. Soft fruits, such as raspberries and strawberries, should be left in their punnets or baskets in a cold room. Stone fruits, such as apricots and plums, are best placed in trays so that any damaged fruit can be seen and discarded. Peaches and citrus fruits are left in their delivery trays or boxes. Bananas should not be stored in chilled conditions because their skins will turn black.

Quality and purchasing

Fresh fruit should be:
● whole and fresh looking (for maximum flavour the fruit must be ripe but not overripe)
● firm, according to type and variety
● clean and free from traces of pesticides and fungicides
● free from external moisture
● free from any unpleasant foreign smell or taste
● free from pests or disease
● sufficiently mature; it must be capable of being handled and travelling without being damaged
● free from any defects characteristic of the variety in terms of their shape, size and colour
● free of bruising and any other damage.

Soft fruits deteriorate quickly, especially if they are not sound. Take care to see that they are not damaged or overripe when purchased. Soft fruits should look fresh; there should be no signs of wilting, shrinking or mould. The colour of certain soft fruits is an indication of their ripeness (for example, strawberries or dessert gooseberries).

Flavourings, colourings and essences

Flavourings (see list)

Flavourings, colourings and essences are highly concentrated and should be used sparingly, a little at a time until the desired impact, flavour or colour is achieved. More can always be added but once added, it cannot be taken away. They can come as liquids, gels and pastes but are added carefully in the same way.

Flavourings, colourings and essences are available in different grades and quality. Some are synthetic, while others are natural. Generally, the more natural a product is, the more expensive it will be to purchase. The photo above shows some examples, which are described in more detail below.

1 Essential oils are pressed from plants or fruit and then purified so that they impart a very clean and sharp flavour to food. They are ideal for flavouring ganaches, chocolate and desserts.

2 Flavour drops are used to add intense flavour to ganaches, ice creams, desserts and fondant.

3 Alcohol concentrates are natural concentrates made in exactly the same way as the original spirit or liqueur. They have an enhanced flavour and higher alcohol strength, making them more economical to use.

4 Fruit pastes are highly concentrated to give a strong flavour. The normal dosage rate is 20–30 per cent of the product. Fruit pastes are used to flavour cream-based mousses, gâteaux, chocolate centres and ice creams.

5 Vanilla pods are dark brown, plump and moist pods with a sweet flavour and aroma. Vanilla powder can be used to add a visual effect to products. Vanilla is used to flavour crème anglaise, crème renversée, crème patissière, bavarois and so on.

6 Griottines are made using a type of morello cherry, the Oblachinska cherry, found only in the Balkans. The

cherries are macerated twice in Kirsch or Cointreau and their shape and flavour is preserved. Framboisines are a similar product, also macerated in liqueur.

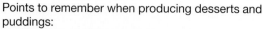

> **PROFESSIONAL TIP**
>
> Points to remember when producing desserts and puddings:
> - Check all weighing scales for accuracy.
> - Follow the recipe carefully.
> - Check all storage temperatures are correct.
> - Fat is better to work with if it is 'plastic' (i.e. at room temperature). This will make it easier to cream.
> - Always cream the fat and sugar well, before adding the liquid.
> - Always work in a clean, tidy and organised way; clean all equipment after use.
> - Handle all equipment carefully to avoid cross-contamination.
> - Always store ingredients correctly: eggs should be stored in a refrigerator; flour in a bin with a tightfitting lid; sugar and other dry ingredients in air-tight storage containers.
> - Ensure all cooked products are cooled before finishing.
> - Understand how to use fresh cream; remember that it is easily over-whipped.
> - Always plan your time carefully.
> - Use silicone paper for baking in preference to greaseproof.
> - Keep all small moulds clean and dry to prevent rusting.

Healthy eating and desserts and puddings

Desserts and puddings remain popular with the consumer, but there is now a demand for products with reduced fat and sugar content, as many people are keen to eat healthily. Chefs will continue to respond to this demand by modifying recipes to reduce the fat and sugar content; they may also use alternative ingredients, such as low-calorie sweeteners where possible and unsaturated fats. Although salt is an essential part of our diet, too much of it can be unhealthy (see page 31) and this should be taken into consideration.

Allergens found in desserts and puddings

Although it is essential to clearly list all potential allergens used to make products, the allergens that are most likely to be used in the production of hot and cold desserts include:

- gluten – flours and any products made from wheat, rye, barley and oats
- nuts – such as hazelnuts, walnuts, Brazil nuts and almonds. These can form part of desserts or be used to make components, such as praline

- peanuts
- eggs – used in many ways, for example, in sponge bases, mousses, meringues and sauces
- lactose – in milk and milk products such as yoghurt and cheese
- sesame seeds – regularly used in Greek and Eastern European desserts such as Baklava
- sulphites – commonly found in wine which may form part of a jelly or sauce as a component of a dessert.

Legislation requires restaurants to identify to the customer any of the listed allergens which the food may contain. Allergens in addition to the list above are soybeans, milk, celery, mustards, lupin (found in some flours and bakery products), fish and seafood (including molluscs).

1.4 Storage

- Store all goods according to the Food Safety and Hygiene Regulations 2013 and the Food Safety Temperature Control Regulation 1995.
- Always store the end product carefully at the right temperature.

- Desserts should always be stored at the appropriate temperature (refrigerated or frozen) and clearly date labelled.
- All dessert products should be wrapped and date labelled before storage.
- All food items should be stored in appropriate positions in the fridge or freezer, away from any potential contaminants.
- Dessert items, like all other food stocks, should form part of a functioning stock control system.
- Take special care when using cream and ensure that products containing cream are stored under refrigerated conditions.
- Always make sure that storage containers are kept clean and returned ready for re-use. On their return they should be hygienically washed and stored.
- Ice cream should be removed from the freezer a few minutes before serving. Long-term storage should be at between −18 °C and −20 °C.

Produce desserts and puddings

2.1 Specialist equipment

A selection of specialist equipment used in patisserie work (see list)

A great deal of specialist equipment is used in the production of desserts and puddings. The photo shows a selection of specialist equipment, and items that have particular uses in patisserie work:

1 Silicone moulds: used to set desserts in moulded shapes, and for baking.

2 Torte rings: used to make charlottes, mousses, parfaits and truffe au chocolate.
3 Specialist moulds: used to create shaped mousse-based desserts.
4 Thermoformed PVC moulds: used to mould buttercream products, bavarois and mousses. These moulds leave a design on the surface of the dessert.
5 Tartlet and barquette moulds: used to produce shaped pastry cases.
6 Flan ring: used to produce baked pastry cases for flans.
7 Conical strainer and passoir: used to strain fruit coulis and sauces.
8 Stepped palette knife: used to evenly spread out sponges such as roulades, jaconde sponge and dacquoise sponge.
9 Mousse frames: used in large-scale catering to set mousses, bavarois, buttercream products, gateau opera and other products.
10 Marzipan roller: used to give a pattern on the surface of marzipan and pastillage.
11 Digital scales: electronic scales used for weighing pastry products.

12 Metal ruler: used as a straight edge for cutting sponges, and for measuring the portion size of cut-out desserts when using a mousse frame.

13 Pastry brushes: used to apply eggwash to pastry products; to clean the sides of the pan when boiling sugar; and to apply glazes to fruit tartlets.

14 Wood grainer: used to apply a wood grain effect when using tempered **couverture** or chocolate cigarette paste.

15 Roller: used to apply a thin layer of chocolate to the base of cut pralines, or to roll tempered chocolate onto the flat surface of a centrepiece.

16 Sprayer: used to spray coloured cocoa butter, or chocolate spray made from equal parts of melted cocoa butter and couverture, when embellishing chocolate centrepieces.

17 Nitrogen flask: used to produce and apply espumas and foams.

18 Induction hob: ideal for sugar boiling because it cooks quickly and with no naked flames.

19 Non-stick mat: Used in place of lining baking trays with oil or using greaseproof or silicon paper. These mats are robust and re-useable, requiring no pre-lining before baking. These mats are often referred to as 'Silpat mats'.

Larger items used in patisserie work include:

- convection oven – a multi-function oven that can be used for steaming, convection or a combination of both
- microwave oven – used for tempering chocolate, softening fondant and heating liquids for pastry dishes (e.g. heating fruit coulis before adding gelatine)
- combination oven – a computerised, highly controllable and programmable oven that can be used as a convection oven, a steamer or a combination of the two, to very precise degrees of heat and humidity. For example, the bread-making process from proving to baking and even steam injection is fully controllable in a combination oven through a single programmed process
- food processor – used for a multitude of tasks, such as breaking down biscuits into crumbs
- stem blender – used to make purees and to blend sauces
- sous-vide water bath – used for cooking fruit and making fruit syrups
- pastry break – used to roll out pastry evenly
- pastry deck oven – a sectioned deck oven designed for patisserie work. It allows the chef to control top and bottom heat within the baking chamber, which gives a more uniform baked product
- dehydrator cabinet – used to dry pastry products such as meringues and fruit slices
- blast fridges and freezers – used to quickly chill or freeze pastry products

- mixing machines – used for numerous pasty functions such as mixing, whipping, **creaming** and beating. It has three attachments: whisk, dough hook and paddle beater.

KEY TERMS

Couverture: chocolate which has a very high percentage of cocoa butter; used in professional kitchens and patisserie.

Creaming: refers to the initial mixing of sugar and cream together using a wooden spoon or electric mixer until a smooth mixture is formed. This is often used in the production of sweet/sugar pastry.

A sous vide water bath

A pastry break

A deck oven

A dehydrator cabinet

The photo to the right shows some of the smaller equipment used in the production of ice cream and other frozen desserts:

1 Refractometer: measures the amount of sugar in a syrup. It is calibrated using the Brix scale, which gives a percentage reading. Used for water ices, sorbets, savarins and poaching syrups.

2 Saccharometer: measures the amount of sugar in a syrup. It is calibrated using the Baumé scale of the density of sugar solutions (Pesse syrup).

3 Ice cream scoops: available in various sizes. Used to dispense scoops of ice cream and sorbets.

4 Metal rings: used to set parfait glacé and mould ice cream.

Other specialist items used in the production of ice cream include:

● Tabletop ice cream machine: used to make traditional ice cream and water ices

● Upright commercial ice cream machine: designed to make traditional ice cream and water ices in large quantities

● Paco jet: designed to make ice cream and water ices. The base for the ice cream or water ice is frozen first, in special containers. The machine 'shaves' the surface of the frozen mixture, creating a smooth iced confection.

Small equipment for ice cream production (see list)

An upright ice cream machine

A Paco jet

2.2 Producing desserts and puddings

For details of the techniques involved in producing different types of desserts and puddings, see the recipe section at the end of this chapter.

2.3 Presenting desserts and puddings

Sauces, creams and coulis

Desserts are often complemented by the addition of a sauce. Some of the most commonly used sauces are as follows:

● Crème anglaise – the classic dessert sauce made from egg yolks and sugar, mixed with vanilla-infused milk and cooked gently to produce a delicious custard-style sauce. Crème anglaise can be served hot or cold, depending on the type of dessert.

● Powdered-based custard – this is a convenience powder that already contains a starch element to provide viscosity

(thickening), flavouring (vanilla, egg) and colouring (egg, yellow/orange). Powdered-based custard is mixed with heated milk with the option of sweetening further with sugar. The starch within the powder will thicken as it is heated and mixed with the liquid (milk).

● Chantilly cream (crème Chantilly) – this is a widely used and versatile cream. Chantilly cream is whipped so that it is spreadable and/or can be piped. It is made by whipping fresh cream which is usually sweetened with sugar and flavoured with vanilla.

● Fruit coulis – fruit coulis are strained purées of fruit that are sometimes sweetened with sugar or sugar syrup. The consistency of the desired coulis should be taken into consideration and adjusted to meet the requirements of the sauce. This can be achieved through the density of the syrup or through adjustment by reduction, the use of a starch or a stabilising agent such as ultratex.

● Cooked fruit sauces – depending on the ripeness of the fruit in question, it may be necessary to cook the fruit before it is blended or passed to make into a sauce. Additional sugar may be required to balance sweetness with any tart or bitter elements of the chosen fruit.

- Chocolate – chocolate sauce is usually made by melting chocolate and combining it with a liquid such as cream. Depending on the type of sauce required, additional ingredients such as butter (for richness) or sugar (for sweetness) may be added to the sauce, alongside any potential additional flavouring.
- Flavoured syrups – simple syrups are made by inverting (dissolving) sugar with water. The higher the percentage of sugar, the more dense and viscous the syrup will become. Additional flavours can be incorporated from fruits such as lemon or orange or spices such as vanilla, star anise or cinnamon.

Finishing and decorating techniques

Piping

- Piping can be used to enhance desserts – a swirl of cream or buttercream, for example, or a fine chocolate motif.
- Piping fresh cream is a skill and like all other skills it takes practice to become proficient. The finished item should look attractive, simple, clean and tidy, with neat piping.
- All piping bags should be sterilised after each use, as these may be a source of contamination; alternatively use a disposable piping bag.
- Make sure that all the equipment you need for piping is hygienically cleaned before and after use to avoid cross-contamination.

> **HEALTH AND SAFETY**
> Sterilise piping bags after each use.

Filling

Some dessert products are finished by being filled, such as profiteroles and crêpes. Other desserts might include a chocolate cup filled with a mousse or cream.

Saucing

The primary use of sauces is to complement the dessert and to make the eating experience more pleasurable and digestible for the customer. The presentation of sauces can also add to the look of the dessert. The use of sauce bottles can help apply sauces in artistic ways – swirls, drags, lines and dots, for example. Consistency and viscosity are important considerations for the stability of the sauce.

Glazing

Some desserts are finished by being glazed or gratinated, for example, the sugar coating on a crème brûlée or a light glaze to finish bread and butter pudding. This is usually achieved by lightly coating the surface of the dessert with sugar (icing, caster, Demerara, etc.) and then caramelising it with a blow-torch or under a hot grill.

Dusting

Dusting is the simple process of lightly sieving a delicate coating of products such as icing sugar or neige décor or cocoa powder to create a light dust-like finish. A fine sieve is best to create this affect.

Additions

Desserts can often have additional ingredients added to enhance the eating quality of the dish. Current menus often use terms such as 'textures of' commodities such as fruits. In this case the chef may add a variety of ways in which the fruit can be presented, for example, dried, puréed, as a jelly or as a crisp. Other alternatives include caramelised or sugared nuts, chopped and powdered nuts and textured nut products such as praline (a brittle textured product made from caramelised almonds or hazelnuts).

Motifs or run-outs

A motif or run-out refers to chocolate that has been tempered and piped finely into a design or shape, usually onto greaseproof or silicone paper. Once the chocolate has set, it can be lifted from the paper and used to garnish a dessert. Motifs vary from simple shapes to quite elaborate pieces and may even be three dimensional.

Cigarettes/pencils

Cigarettes/pencils are produced by spreading tempered couverture or chocolate finely over a flat surface such as granite. Once the chocolate is almost set, use a scraper at an angle in short, sharp motions to force the chocolate into fine rolls, approximately the length of a cigarette. Chocolate cigarettes/pencils are often used to provide dimension to desserts by being placed gently, perhaps at an angle, into a quenelle of cream, sorbet or ice-cream, for example.

Moulding

The use of moulds creates a neat, uniform finish to desserts. They also produce portions of consistent sizes. Moulds are made in many shapes and sizes, offering chefs the opportunity to create interesting presentations. Moulds can also be used to set items such as chocolate rings and tears before they are filled.

Chocolate transfer sheets

Chocolate transfer sheets are produced by lining sheets of acetate with designs printed using coloured cocoa butter. When tempered couverture or chocolate is spread over the sheet and left to set, the design will be transferred to the chocolate. As well as various designs, it is now possible to have signatures and logos professionally printed in this method, giving chefs a simple way of producing high-quality personalised chocolate products.

Portioning

Many desserts are produced in multiple portions, a raspberry délice or a chocolate tart, for example. It is important that when dividing multiple portion desserts this is done equally to produce portions of the same size. It is also important to use clean, hygienic equipment at all times. This will obviously help to maintain food safety but will also help to produce cleanly cut and neat portions to enhance their presentation when served.

2.4 Evaluation

A pastry chef or patisseur has an opportunity to delight and impress guests and customers with the luxurious and exquisite dishes and products that can be produced.

In terms of measuring the quality of desserts and puddings against recipe specifications and production standards, the following points should be considered as part of an overall evaluation:

Measurement, shape and dimension

The following should be considered:
- Is each dessert or portion equal in in size?
- Is each portion the right size?
- Is the dessert shaped in the way it was intended?
- Are the dimensions of the dessert suitable for the type of service being provided?
- Is the portion size right in terms of the stage of the meal in which it is served?

Consistency in presentation is a vital aspect for a pastry chef. Each dessert should be consistently served to the dish specifications so that customers are provided with a high-quality and reliable product.

A dessert served after a main course as part of a 'main course and dessert' offer may be considerably different to a dessert served as part of an eight-course tasting menu, that may even have more than one dessert. Therefore, it is important to consider portion size in the context of a combined meal. However, when producing meals for à la carte menus (customers selecting dishes individually), it remains important that the portion size is carefully considered in the context of the menu and the dessert is sized accordingly.

Tasting and consuming desserts and puddings

The following should be considered:
- Taste – are the tastes and flavours as they were planned? Does the dessert taste fresh and well balanced? Have the ingredients been mixed and balanced appropriately?
- Texture – is the texture as desired? Should the texture be light and delicate, as in a soufflé, or more robust and textured, as per a sticky toffee pudding?
- Consistency – is the consistency correct, both in terms of the physical consistency (as in the smooth, delicate and velvety consistency of a crème caramel), but also in the general consistency of one product's production compared with the next?
- Aroma – desserts and puddings tend to be luxurious items which are often consumed as treats. Aromas should be comforting and welcoming, lightly fragrant and reflective of the ingredients and processes that have led to their production.
- Colour – desserts and puddings should reflect the natural colour of the ingredients they are made with. Many desserts are fruit based and therefore should boast the vibrancy and seasonality of the fruit(s) in question. Contrasts in colour should be complimentary: a chocolate dessert with a crème anglaise provides a good example.
- Harmony – this refers to the way that the various aspects of the dessert or pudding work together to complement each other to produce a harmonious result. Classic texture components such as a crisp item served with a smooth, creamy item or a sweet flavour offset by something sharp can provide excellent contrasts in dishes, adding to their overall harmony.
- Presentation – Pastry chefs or patissuers have an excellent opportunity to impress their guests and customers in the presentation of their dishes and products. Creativity and food styling are almost endless for the pastry chef. Clean, sharp, creative and consistent presentation is an area that every pastry chef should think about and plan for carefully. Dishes should also be served at their planned temperature, which in the case of desserts, can be iced or frozen, hot or cold. It is vital that this is achieved, not only in terms of food safety, but also to satisfy the enjoyment and expectations of the customer.

TEST YOURSELF

1 Name **three** types of setting agent.
2 Describe what would happen to a crème caramel if it was cooked at 180°C.
3 Name the **three** main types of meringues and describe their production and use.
4 Describe the conditions that promote the foam of egg whites when making meringues.
5 Name **two** desserts that are finished with a glaze.
6 Name **five** fruits suitable for stewing.
7 Describe the differences between the production of a 'pudding soufflé' and a 'fruit soufflé'.
8 Name **four** types of sugar and give an example of their use in a dessert.
9 Describe why salt is sometimes used in the production of desserts.
10 Describe **three** ways in which chocolate can be used as a decorating medium.
11 Name the **two** pieces of equipment used to measure the density and sweetness of sugar syrups.
12 Describe the quality points to look for in the following desserts:
   a) ice cream
   b) crème brulée
   c) strawberry mousse.
13 Name **four** desserts produced from milk, eggs and sugar.
14 At what temperature should an individual vanilla soufflé be baked, and for how long?

1 Chantilly cream

Energy	Calories	Fat	Saturated fat	Carbohydrates	Sugar	Protein	Fibre	Sodium
9530 kJ	2301 kcal	202.0 g	126.0 g	118.0 g	118.0 g	103.0 g	0.0 g	130.0 g

	500 ml
Whipping cream	500 ml
Caster sugar	100 g
Vanilla essence or fresh vanilla pod	A few drops to taste or seeds from 1 vanilla pod

Preparing the dish

1 Place all ingredients in a bowl. Whisk over ice until the mixture forms soft peaks. If using a mechanical mixer, stand and watch until the mixture is ready – do not leave it unattended as the mix will over-whip quickly, curdling the cream.

2 Cover and place in the fridge immediately.

PROFESSIONAL TIP

Always keep cream very cold during the whipping process. This will help to achieve a smooth whipped finish and help to prevent the cream from over-whipping and curdling. It is best practice to use a spotlessly clean stainless steel bowl and whisk to whip cream.

HEALTH AND SAFETY

Cream is a milk product so customers should be made aware of its inclusion in recipes. It is also a high-risk product from a food safety perspective and should be kept cold between 1 °C and 5 °C.

FAULTS

Cream can over-whip very quickly so it is important to pay close attention during the whipping process to prevent this from happening.

2 Buttercream

Energy	Calories	Fat	Saturated fat	Carbohydrates	Sugar	Protein	Fibre	Sodium
8635 kJ	2077 kcal	164.0 g	104.0 g	151.0 g	149.0 g	151.0 g	0.0 g	1,478.0 g

	350 g
Icing sugar	150 g
Butter	200 g

Preparing the dish

1 Sieve the icing sugar.

2 Cream the butter and icing sugar until light and creamy.

3 Flavour and colour as required.

PROFESSIONAL TIP

Add colourings and flavours gradually in small amounts, literally a drop at a time. Flavourings and colourings are very strong so need to be used carefully. It is always possible to add more colour or flavour to the cream but it is not possible to remove it once it has been added.

VARIATION

- **Rum buttercream**: add rum to flavour and blend in.
- **Chocolate buttercream**: add melted chocolate, sweetened or unsweetened according to taste.

ON A BUDGET

It is possible to use margarine to produce a similar product but the consistency will be thinner with less 'buttery' flavour.

3 Pastry cream (*crème pâtissière*)

Energy	Calories	Fat	Saturated fat	Carbohydrates	Sugar	Protein	Fibre	Sodium
5736kJ	1357kcal	42.0g	18.1g	217.0g	154.0g	128.0g	3.1g	293.0g

* Using whole milk

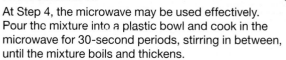

Left to right: pastry cream, crème diplomat and crème chiboust

	Approximately 750 ml
Milk	500 ml
Vanilla pod	1
Egg yolks	4
Caster sugar	125 g
Soft flour	75 g
Custard powder	10 g

Cooking

1 Heat the milk with the cut vanilla pod and leave to infuse.

2 Beat the sugar and egg yolks together until creamy white. Add the flour and custard powder.

3 Strain the hot milk, gradually blending it into the egg mixture.

4 Strain into a clean pan and bring back to the boil, stirring constantly.

5 When the mixture has boiled and thickened, pour into a plastic bowl, sprinkle with caster sugar and cover with cling film.

6 Chill over ice and refrigerate as soon as possible. Ideally, blast chill.

7 When required, knock back on a mixing machine with a little kirsch.

PROFESSIONAL TIP

At Step 4, the microwave may be used effectively. Pour the mixture into a plastic bowl and cook in the microwave for 30-second periods, stirring in between, until the mixture boils and thickens.

VARIATION

The recipes below are based on using the recipe for crème pâtissière, with additional ingredients.

● **Crème mousseline**: beat in 100g of soft butter (a pomade). The butter content is usually about 20 per cent of the volume but this can be raised to 50 per cent depending on its intended use.

● **Crème diplomat**: when the pastry cream is chilled, fold in an equal quantity of whipped double cream.

● **Crème chiboust**: when the pastry cream mixture has cooled slightly, fold in an equal quantity of Italian meringue (recipe 6).

● Additional flavourings can also be added to crème pâtissière, crème diplomat or crème chiboust.

FAULTS

Make sure that the cream is properly cooked through to thicken the starches in the flour and custard powder. If this does not take place, the final product will lack the viscosity (thickness) required.

4 Boiled buttercream

Energy	Calories	Fat	Saturated fat	Carbohydrates	Sugar	Protein	Fibre	Sodium
19445 kJ	4673 kcal	338.0 g	211.0 g	402.0 g	402.0 g	402.0 g	0.0 g	207.0 g

	750 ml
Eggs	2
Icing sugar	50 g
Granulated sugar or cube sugar	300 g
Water	100 g
Glucose	50 g
Unsalted butter	400 g

Mise en place

1 The use of a good-quality mixing machine would be very useful to produce this recipe.

Cooking

1 Beat the eggs and icing sugar until at ribbon stage (sponge).

2 Boil the granulated or cube sugar with water and glucose to 118 °C.

3 Gradually add the sugar at 118 °C to the eggs and icing sugar at ribbon stage, whisk continuously and allow to cool to 26 °C.

4 Gradually add the unsalted butter while continuing to whisk until a smooth cream is obtained.

VARIATION

Buttercream may be flavoured with numerous flavours and combinations of flavours, for example:

- chocolate and rum
- whisky and orange
- strawberry and vanilla
- lemon and lime
- apricot and passionfruit
- brandy and praline
- coffee and hazelnut.

PROFESSIONAL TIP

When boiling sugar, start off with a low heat until all the sugar has dissolved. The heat can then be raised to cook the sugar to the required temperature. Use a clean sugar thermometer to monitor the temperature of the sugar throughout the cooking process.

HEALTH AND SAFETY

The sugar becomes extremely hot during this recipe so ensure that arms are well covered and that good quality, thick, dry oven cloths (or gloves) are used when handling the sugar to produce this recipe.

FAULTS

Always use spotlessly clean utensils when boiling sugar. Contamination to the sugar from fat, for example, will cause the sugar to crystallise, making it unusable for this recipe.

Whisk the eggs.

Add the boiling sugar and water.

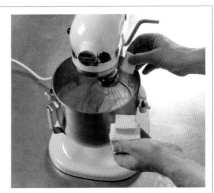

Add the butter.

5 Ganache

Energy	Calories	Fat	Saturated fat	Carbohydrates	Sugar	Protein	Fibre	Sodium
16374 kJ	3947 kcal	329.0 g	203.0 g	217.0 g	214.0 g	213.0 g	8.7 g	94.0 g

* Using 37.5% spirit and unsalted butter

NOTE

The two versions have different textures. Version 1 is ideal for truffles; version 2 for tortes or fillings.

PROFESSIONAL TIP

Pour the cream on to the couverture just off boiling point. This will help to prevent the chocolate from scalding.

	750 g
Version 1 (for decoration)	
Double cream	300 ml
Couverture, cut into small pieces	350 g
Unsalted butter	85 g
Spirit or liqueur	20 ml
Version 2 (for a filling)	1 kg
Double cream	300 ml
Vanilla pod	½
Couverture, cut into small pieces	600 g
Unsalted butter	120 g

Cooking

1 Boil the cream (and the vanilla for Version 2) in a heavy saucepan.

2 Pour the cream over the couverture.

3 Whisk with a fine whisk until the chocolate has melted.

4   Whisk in the butter (and the liqueur for version 1).

5   Stir over ice until the mixture has the required consistency.

HEALTH AND SAFETY
Once the cream is added to the couverture, whisk carefully and slowly to begin with to prevent the cream from leaving the bowl and scalding the arms.

ON A BUDGET
For best results, it is necessary to use a good-quality couverture in this recipe.

FAULTS
Insufficient whisking may lead to poor emulsion of the cream and chocolate. If some of the chocolate pieces were too big, it may not have melted completely by the time the cream had reduced in temperature, leaving small bits of chocolate in the ganache. This would leave a grainy eating texture.

6 Italian meringue

Energy	Calories	Fat	Saturated fat	Carbohydrates	Sugar	Protein	Fibre
3606 kJ	845 kcal	0.0 g	0.0 g	200.0 g	200.0 g	200.0 g	0.0 g

	250 g	625 g
Granulated or cube sugar	200 g	500 g
Water	60 ml	140 g
Cream of tartar	Pinch	Large pinch
Egg whites	4	10

Cooking

1 Boil the sugar, water and cream of tartar to hard-ball stage of 121 °C. (To ensure the sugar is not heated beyond this point, it is advisable to remove from the heat at 115 °C. The sugar will continue to rise in temperature and this will provide a little time to ensure the egg whites are whipped to the correct point.)

2 While the sugar is cooking, beat the egg whites to full peak and, while stiff, beating slowly, pour on the boiling sugar.

3 Use as required.

PROFESSIONAL TIP

When boiling sugar, start off with a low heat until all the sugar has dissolved. The heat can then be raised to cook the sugar to the required temperature. Use a clean sugar thermometer to monitor the temperature of the sugar throughout the cooking process

HEALTH AND SAFETY

The sugar becomes extremely hot during this recipe so ensure that arms are well covered and that good quality, thick, dry oven cloths (or gloves) are used when handling the sugar to produce this recipe.

FAULTS

Always use spotlessly clean utensils when boiling sugar. Contamination of the sugar from fat, for example, will cause the sugar to crystallise, making it unusable for this recipe.

Boil the sugar.

Combine with the beaten egg whites.

The mixture will stand up in stiff peaks when it is ready.

7   Swiss meringue

Energy	Calories	Fat	Saturated fat	Carbohydrates	Sugar	Protein	Fibre	Sodium
360 kJ	87 kcal	0.0 g	0.0 g	19.5 g	19.5 g	19.5 g	0.0 g	33.4 g

	10 portions	15 portions
Egg whites, pasteurised	190 ml	300 ml
Caster sugar	230 ml	340 ml

Cooking

1   Whisk the pasteurised egg whites and sugar over a bain-marie of simmering water until a light and aerated meringue is achieved.

8   Frangipane (almond cream)

Energy	Calories	Fat	Saturated fat	Carbohydrates	Sugar	Protein	Fibre	Sodium
991 kJ	238 kcal	18.3 g	7.3 g	14.2 g	13.0 g	12.7 g	1.7 g	22.5 g

* Using unsalted butter

Preparing the dish

1   Cream the butter and sugar until aerated.

2   Gradually beat in the eggs.

3   Mix in the almonds and flour (mix lightly).

4   Use as required.

> **PROFESSIONAL TIP**
> The eggs should be added slowly and gradually to the creamed butter and sugar. If too much egg is added at one time, it will split, leaving a curdled mixture. If the mixture curdles, it is possible to remedy this by mixing in a little of the flour at this point.

	8 portions
Butter	100 g
Caster sugar	100 g
Eggs	2
Ground almonds	100 g
Flour	10 g

> **VARIATION**
> ● Try adding lemon zest or vanilla seeds to the recipe.

Cut the butter into small pieces and add to the sugar

Cream the butter and sugar together

Beat in the eggs before adding to the flour

9 Apple purée (*marmalade de pommes*)

Energy	Calories	Fat	Saturated fat	Carbohydrates	Sugar	Protein	Fibre
1782 kJ	419 kcal	9.4 g	5.4 g	86.0 g	85.0 g	51.0 g	6.8 g

PROFESSIONAL TIP

Peel the apples as close to the cooking time as possible to prevent them from discolouring through aeration. If a large quantity of apples is required, it is recommended that they are kept in acidulated water (e.g. water with lemon juice) until required. It is important to dry the apples before cooking to drain off as much water as possible if they have been stored in this way.

Healthy Eating Tip
- Using sweeter apples will require less sugar or perhaps no additional sugar at all.

	400 g	1 kg
Cooking apples	400 g	1 kg
Butter or margarine	10 g	25 g
Sugar	50 g	125 g

Cooking

1 Peel, core and slice the apples.

2 Place the butter or margarine in a thick-bottomed pan; heat until melted.

3 Add the apples and sugar, cover with a lid and cook gently until soft.

4 Drain off any excess liquid and pass through a sieve or liquidise.

10 Apricot glaze

Energy	Calories	Fat	Saturated fat	Carbohydrates	Sugar	Protein	Fibre	Sodium
1073 kJ	251 kcal	0.0 g	0.0 g	66.0 g	59.0 g	21.4 g	0.0 g	0.1 g

	150 ml
Apricot jam	100 g
Stock syrup (Recipe 15) or water	50 ml

Cooking

1 Boil the apricot jam with a little syrup or water.

2 Pass through a strainer. The glaze should be used hot.

> **PROFESSIONAL TIP**
> A flan jelly (commercial pectin glaze) may be used as an alternative to apricot glaze. This is usually a clear glaze to which food colour may be added.

11 Fruit coulis

Energy	Calories	Fat	Saturated fat	Carbohydrates	Sugar	Protein	Fibre	Sodium
12763 kJ	3011 kcal	6.3 g	0.5 g	715.0 g	714.0 g	638.0 g	94.0 g	285.0 g

* Using passion fruit puree

3 Pour the soft-ball sugar into the warm fruit purée while whisking vigorously. Add the lemon juice. Bring back to the boil.

4 Store ready for use when required.

> **PROFESSIONAL TIP**
> The reason the soft-ball stage needs to be achieved when the sugar is mixed with the purée is that this stabilises the fruit and prevents separation once the coulis is presented on the plate.
>
> Adding lemon juice brings out the flavour of the fruit.

	1.4 litres
Fruit purée	1 litre
Caster sugar	500 g
Lemon juice	10 g

Cooking

1 Warm the purée.

2 Boil the sugar with a little water to soft-ball stage (121 °C).

> **VARIATION**
> With ripe and seasonal fruits, such as strawberries, it is possible to produce a good coulis by liquidising raw strawberries with icing sugar and a drop of lemon juice

12 Fresh egg custard sauce (sauce à l'anglaise)

Energy	Calories	Fat	Saturated fat	Carbohydrates	Sugar	Protein	Fibre	*
1583 kJ	377 kcal	20.7 g	8.8 g	35.3 g	35.2 g	24.2 g	0.0 g	

* Using whole milk

	300 ml	700 ml
Egg yolks, pasteurised	40 ml	100 ml
Caster or unrefined sugar	25 g	60g
Vanilla extract or vanilla pod (seeds)	2–3 drops per ½ pod	5–7 drops per 1 pod
Milk, whole or skimmed, boiled	250 ml	625 ml

Cooking

1 Mix the yolks, sugar and vanilla in a bowl.

2 Whisk in the boiled milk and return to a thick-bottomed pan.

3 Place on a low heat and stir with a wooden spoon until it coats the back of the spoon. Do not allow the mix to boil or the egg will scramble. A probe can be used to ensure the temperature does not go any higher than 85 °C.

4 Strain through a fine sieve into a clean bowl. Set on ice to seize the cooking process and to chill rapidly.

> **VARIATION**
> - Other flavours may be used in place of vanilla, for example, coffee, curacao, chocolate, Cointreau, rum, Tia Maria, brandy, whisky, star anise, cardamom seeds, kirsch, orange flower water.

13 Custard sauce

Energy	Calories	Fat	Saturated fat	Carbohydrates	Sugar	Protein	Fibre	*
1245 kJ	296 kcal	9.6 g	6.0 g	47.2 g	38.0 g	8.3 g	0.3 g	

* Using whole milk

	500 ml
Custard powder	30 g
Milk, whole or semi-skimmed	500 ml
Caster or unrefined sugar	35 g

Cooking

1 Mix the sugar and custard powder and dilute with a little of the milk.

2 Boil the remainder of the milk.

3 Pour a little of the boiled milk on to the diluted custard powder mix.

4 Return to the saucepan.

5 Bring to the boil slowly, stirring consistently to prevent the custard from catching.

> **Healthy Eating Tip**
> - The sugar can be reduced or removed completely. However, the final product will lose sweetness.

> **VARIATION**
> This recipe can be prepared in the microwave. Simply mix the sugar and custard powder and dilute with a little of the milk. Add the remaining milk at this point and then cook in 30-second batches in the microwave, mixing well between each cooking period.

14 Chocolate sauce (*sauce au chocolat*)

Energy	Calories	Fat	Saturated fat	Carbohydrates	Sugar	Protein	Fibre
1287 kJ	310 kcal	26.0 g	16.0 g	17.5 g	17.5 g	2.7 g	0.2 g

	300 ml	750 ml
Method 1		
Double cream	150 ml	375 ml
Butter	25 g	60 g
Milk or plain couverture callets	180 g	420 g
Method 2		
Caster sugar	40 g	100 g
Water	120 ml	300 ml
Dark chocolate couverture (75 per cent cocoa solids)	160 g	400 g
Unsalted butter	25 g	65 g
Single cream	80 ml	200 ml

Cooking

Method 1:

1 Place the cream and butter in a saucepan and gently bring to a simmer.

2 Add the chocolate and stir well until the chocolate has melted and the sauce is smooth.

Method 2:

1 Dissolve the sugar in the water over a low heat.

2 Remove from the heat. Stir in the chocolate and butter.

3 When everything has melted, stir in the cream and gently bring to the boil.

ON A BUDGET

Any chocolate can be used to make a chocolate sauce. However, the flavour and quality of the sauce will be directly linked to the quality of the chocolate used.

15 Stock syrup

Energy	Calories	Fat	Saturated fat	Carbohydrates	Sugar	Protein	Fibre
4982 kJ	1172 kcal	0.0 g	0.0 g	300.0 g	300.0 g	300.0 g	0.0 g

	750 ml	1.5 litres
Water	500 ml	1.25 litres
Granulated sugar	250 g	625 g
Glucose	50 g	125 g

Cooking

1 Boil the water, sugar and glucose together.

2 Strain and cool.

PROFESSIONAL TIP

The glucose helps to prevent crystallising.

16 Caramel sauce

Energy	Calories	Fat	Saturated fat	Carbohydrates	Sugar	Protein	Fibre
883 kJ	213 kcal	19.9 g	12.1 g	8.1 g	8.1 g	1.0 g	0.0 g

	750 ml
Caster sugar	100 g
Water	80 ml
Double cream	500 ml
Egg yolks, lightly beaten (optional)	2

Mise en place

1 Have a clean pastry brush in a measuring jug of iced water ready. This can be used to brush the sides of the pan to remove sugar crystals when boiling sugar. This will help to prevent crystallisation.

Cooking

1 In a large saucepan, dissolve the sugar with the water over a low heat and bring to boiling point.

2 Wash down the inside of the pan with a pastry brush dipped in cold water to prevent crystals from forming.

3 Cook until the sugar turns to a deep amber colour. Immediately turn off the heat and whisk in the cream.

4 Set the pan back over a high heat and stir the sauce with the whisk. Let it bubble for 2 minutes, then turn off the heat.

5 You can now strain the sauce and use it when cooled, or, for a richer, smoother sauce, pour a little caramel onto the egg yolks, then return the mixture to the pan and heat to 80 °C, taking care that it does not boil.

6 Pass the sauce through a conical strainer and keep in a cool place, stirring occasional to prevent a skin from forming.

HEALTH AND SAFETY

Working with sugar at such high temperatures can cause severe scalds if care is not taken. Always use thick, dry oven cloths (or gloves) when handling pans at such high temperatures. Stand back when adding the cream to the cooked sugar. The variance between the high temperature of the sugar and low temperature of the cream will cause some fierce bubbling which could spit beyond the rim of the boiling pan.

SERVING SUGGESTION

Excellent partner to ice creams, sponge-based desserts and desserts featuring bananas or apples.

17 Butterscotch sauce

Energy	Calories	Fat	Saturated fat	Carbohydrates	Sugar	Protein	Fibre
883 kJ	213 kcal	19.9 g	12.1 g	8.1 g	8.1 g	1.0 g	0.0 g

	300 ml	750 ml
Double cream	250 ml	625 ml
Butter	62 g	155 g
Demerara sugar	100 g	250 g

Cooking

1 Boil the cream, then whisk in the butter and sugar.

2 Simmer for 3 minutes.

18 Steamed sponge pudding

Energy	Calories	Fat	Saturated fat	Carbohydrates	Sugar	Protein	Fibre	Sodium
1748 kJ	418 kcal	23.8 g	14.0 g	46.0 g	26.8 g	25.7 g	1.0 g	229.0 g

* Using unsalted butter and 30 ml semi-skimmed milk

	10 individual puddings
Butter	250 g
Caster sugar	250 g
Self-raising flour	250 g
Baking powder	15 g
Eggs	6
Milk	20–40 g

Mise en place

1 Line the moulds with a little butter and sugar. Have disks of silicone paper and foil ready for covering. Some chefs tie the paper and foil covering with string to secure. If so, ensure that string is in place and cut to size.

Cooking

1 Cream the butter and sugar until lighter in colour and an increase in volume of approximately 30 per cent is achieved.

2 Sieve the flour and baking powder twice to ensure even dispersion.

3 Beat the eggs and gradually add to the butter and sugar, beating in each addition before adding the next.

4 Finally, fold in the dry ingredients.

5 Add enough milk to achieve a dropping consistency.

6 Fill buttered dariole moulds three-quarters full.

7 Place a disc of silicone paper on top and cover with foil; create a pleat to allow for expansion and crimp the edges around the lip of the mould to seal.

8 Steam for 45–50 minutes.

VARIATION

- **Vanilla**: scrape the seeds from a vanilla pod into the mixture at the creaming stage. Use the pod to flavour the accompanying crème anglaise.
- **Chocolate**: replace 50 g of the flour in the basic recipe with an equal amount of cocoa powder and add 150 g ground almonds, a pinch of sea salt, a splash of coffee essence and two more eggs. Add extra butter to the base of the mould and cover with muscovado sugar before filling. Serve with hot chocolate sauce.
- **Lemon or orange**: add the grated zest of a large lemon or an orange to the mixture at the creaming stage. Serve with lemon or orange sauce as appropriate.
- **Sultana**: add 100 g of washed sultanas and a few drops of vanilla compound. Serve with crème anglaise. The sultanas may be replaced by any dried fruit.
- **Ginger**: add 10 g of ground ginger (sieved in with the dry ingredients) and 50 g of finely diced stem ginger folded in at the end. Serve with crème anglaise.
- **Golden syrup**: after buttering the mould pour a generous layer of warmed golden syrup into the base before adding the sponge mixture. Serve with crème anglaise sweetened with golden syrup.
- **Chocolate and fig/date**: add six diced figs or dates to the chocolate recipe above.

NOTE

A good-quality steamed sponge pudding will be light, moist and slightly spongy in texture. As steaming at 100 °C imparts no colour, it should be as light in colour as the ingredients allow.

19 Sticky toffee pudding

Energy	Calories	Fat	Saturated fat	Carbohydrates	Sugar	Protein	Fibre
4103 kJ	980 kcal	60.4 g	36.7 g	106.7 g	78.9 g	9.1 g	1.8 g

	10 puddings
Dates	375 g
Water	625 g
Butter	125 g
Caster sugar	375 g
Eggs	5
Soft flour	375 g
Baking powder	10 g
Vanilla compound	1 tsp
Sticky toffee sauce	
Granulated sugar	600 g
Unsalted butter	300 g
Double cream	450 g
Brandy	30 ml

Cooking

1 Remove stones and chop the dates, then place in the water and simmer for about 5 minutes until soft. Set aside to cool.

2 Butter and sugar individual dariole moulds.

3 Cream the butter and sugar until light in colour and aerated.

4 Gradually add the beaten eggs, beating continuously.

5 Sieve the flour and baking powder twice and fold in.

6 Finally add the dates and water and vanilla compound.

7 Fill the moulds three-quarters full and bake at 180°C for 30–35 minutes.

Sticky toffee sauce:

1 Line the moulds with a little butter and sugar.

2 Make a dry caramel by carefully melting the sugar until a deep golden colour is achieved.

3 Cut the butter into small cubes add to the cream and heat.

4 Gradually add the hot cream and butter to the caramel a little at a time.

5 Finally stir in the brandy.

SERVING SUGGESTION

To serve, coat the pudding with the sauce and serve with vanilla or milk ice cream.

PROFESSIONAL TIP

Sticky toffee pudding can be steamed instead of baked. If steaming, remember to cover with a disc of silicone paper and seal with a pleated square of foil, as for steamed puddings (recipe 18).

20 Eve's pudding with gooseberries

Energy	Calories	Fat	Saturated fat	Carbohydrates	Sugar	Protein	Fibre
1234 kJ	294 kcal	14.7 g	8.5 g	37.9 g	20 g	4.9 g	2.6 g

	10 portions
Gooseberries, washed, topped and tailed	500 g
Caster sugar	150 g plus a little for the fruit
Butter	150 g
Vanilla essence	
Eggs, beaten	150 g
Self-raising flour	240 g
Milk	60 g

Cooking

1 Arrange the gooseberries in the bottom of a buttered dish or individual ramekins. Sprinkle with caster sugar.

2 Cream the sugar, butter and vanilla essence together until white.

3 Gradually add the eggs to the butter mixture.

4 Sieve the flour twice, then fold it in to the eggs and butter. Adjust the consistency with milk.

5 Spread the mixture over the fruit, to a thickness of about 2 cm. This will form the sponge.

6 Bake for 30–40 minutes in a pre-heated oven at 180 °C.

7 Rest for 5 minutes before removing from the dish. Brush with boiling apricot glaze.

SERVING SUGGESTION
Serve with custard or crème anglaise.

21 Chocolate fondant

Energy	Calories	Fat	Saturated fat	Carbohydrates	Sugar	Protein	Fibre
2830 kJ	675 kcal	46.8 g	29.6 g	55.9 g	40.9 g	11.0 g	0.6 g

	10 portions
Unsalted butter	260 g
Dark couverture	260 g
Eggs, pasteurised	120 g
Egg yolks, pasteurised	40 g
Caster sugar	150 g
Instant coffee	5 g
Plain flour	110 g
Baking powder	5 g
Cocoa powder	75 g
Salt	Pinch

Mise en place

1 Have all dry ingredients sieved together twice. This will ensure that they are evenly mixed and well incorporated.

2 Chop the chocolate into small, evenly sized pieces and cut the butter into cubes. This will allow them to melt evenly when melting together.

Cooking

1 Melt the butter and couverture together.

2 Warm the eggs, egg yolks, sugar and coffee and whisk to the ribbon stage.

3 Sieve all the dry ingredients twice.

4 Fold the chocolate and butter into the eggs.

5 Fold in the dry ingredients.

6 Pipe into individual stainless steel rings lined with silicone paper and placed on a silicone paper-lined baking sheet.

7 Bake at 190°C for 5 minutes.

8 Carefully slide off the rings and serve with vanilla ice cream.

PROFESSIONAL TIP

These fondants can be kept in the refrigerator and cooked to order. If they are chilled, then extend the cooking time by 2 minutes.

Chocolate fondant should have a liquid centre with a rich, buttery, chocolate taste. Because of the liquid centre, they are very delicate; if piped inside a ring they are much easier and quicker to serve, rather than trying to turn them out of a mould.

Precise timing is essential or the centre of the fondant will not be liquid.

Like most recipes, the quality of the finished product relies on the quality of the ingredients. Always use good quality chocolate (couverture) which contains a high percentage of cocoa butter and solids.

VARIATION

- Try adding salted caramel to the centre by making and freezing it in ice cube trays.
- Serve with malt ice cream (just add malt powder instead of vanilla and mix in some crushed chocolates) or replace the cream with crème fraiche to give a less rich ice cream.
- Prepare fondants in moulds lined with melted butter and roasted sesame seeds.

Melt the chocolate and butter in small pieces.

Fold the melted chocolate into the egg mixture.

Add the dry ingredients.

To make a contrasting centre, add white chocolate pieces on a base of the chocolate mixture.

Pipe in more of the chocolate mixture until the mould is full.

22 Orange pancakes (*crêpes Suzette*)

Energy	Calories	Fat	Saturated fat	Carbohydrates	Sugar	Protein	Fibre
1258 kJ	299 kcal	11.3 g	6.5 g	42.3 g	25.7 g	6.4 g	1.7 g

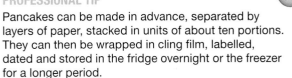

For the crêpes	
Eggs	2
Salt	Pinch
Caster sugar	112 g
Milk	575 ml
Strong flour	225 g
Melted butter	30 g
For the Suzette mixture	
Unsalted butter	60 g
Zest and juice of orange	1
Caster sugar	60 g
Grand Marnier	30 ml
Cognac	30 ml
Orange, peeled and divided into segments	2

Cooking

1 Preparing and cooking the pancakes

2 Sieve the flour and salt into a bowl.

3 Make a well add the eggs and half the milk whisk to a thick batter and gradually add the rest of the milk and the melted butter.

4 Pass through a conical strainer.

5 Heat the pancake pan and wipe thoroughly with kitchen paper, add a little oil and heat until it starts to smoke.

6 Add some of the batter, swirling the pan until covered in a thin layer. Cook for a few seconds until lightly coloured.

7 Turn or 'toss' and cook on the other side.

8 Turn out onto a plate and sprinkle with sugar.

9 Repeat the process until all the mixture has been used, stacking the pancakes one on top of the other.

For the Suzette:

1 Melt the butter in a shallow pan, add the zest, sugar, orange juice and Grand Marnier. Boil rapidly to reduce until thickened.

2 One by one add the pancakes to the pan, fold into four, lift out and keep hot on the plate.

3 Add the cognac and flambé.

4 Pour over the crêpes and serve.

5 Some orange segments may be added at the flambé stage.

> **PROFESSIONAL TIP**
>
> Pancakes can be made in advance, separated by layers of paper, stacked in units of about ten portions. They can then be wrapped in cling film, labelled, dated and stored in the fridge overnight or the freezer for a longer period.
>
> This is not a dish suitable for large numbers of people – up to four servings is manageable.

> **NOTE**
>
> Crêpe pans are heavy, cast iron, flat-bottomed, low-lipped pans specifically for making crêpes and should not be used for any other purpose, or they will start to stick. It is necessary to 'temper' the pans occasionally, by filling with salt and heating to a high temperature, so any moisture is drawn out of the metal. The salt is tipped out and the pan wiped with kitchen paper after which it is ready to use.
>
> Traditionally this dish was finished and flambéed at the table, which provided the customers with visual entertainment and allowed them to enjoy the aromas provided by the oranges and alcohol.

> **VARIATION**
>
> Other fruits can be used with the appropriate spirit or liqueur, such as apples and calvados, or pears with poire William. These fruits would need to be cooked in the liquor before the dish is flambéed.

23 Baked Alaska (*omelette soufflé surprise*)

Energy	Calories	Fat	Saturated fat	Carbohydrates	Sugar	Protein	Fibre
290 kJ	521 kcals	16.4 g	7.3 g	91.3 g	81.2 g	7.7 g	0.6 g

	10 portions
Vanilla ice cream or parfait	10 × 5-cm diameter rings
Roulade sponge (see page 556)	1 sheet
Stock syrup flavoured with rum or kirsch	50 ml
Italian meringue	500 g

Mise en place

1 The production of this dessert is heavily reliant on mise en place. If made in advance and stored in a freezer, its finishing is fairly straightforward because the external meringue is heated (either in the oven or using a blow-torch) and then served immediately.

Cooking

1 Sit the ice cream or parfait on a base of sponge.

2 Cut more sponge to fit and completely cover (as in the photo).

3 Brush all over with the syrup.

4 Set on squares of silicone paper, coat with the meringue and decorate by piping on a design with a small plain tube.

5 Dust with icing sugar and place in a very hot oven at 230 °C for 2–3 minutes until the meringue is coloured.

> **SERVING SUGGESTION**
> Serve immediately with crème anglaise or a fruit coulis.

Brush the sponge with syrup.

Pipe meringue to cover.

Pipe swirls of meringue to decorate.

NOTE
This dessert does not clearly fit into a category and it is debatable as to whether it should be classified as a hot dessert as the filling will remain frozen. However, as it is baked and coloured in a very hot oven, this is probably the most appropriate place for it.

HEALTH AND SAFETY
Under no circumstances should this dessert be re-frozen once it has been removed from the freezer and baked. Ice cream is highly susceptible to contamination by bacteria which can cause food poisoning.

PROFESSIONAL TIP

Baked Alaska is best made in advance and held in the freezer, then flashed through the oven just before serving. It is now common practice to colour the meringue with a blowtorch, but the meringue will have a much better texture and more even colouring if it is finished in the oven.

If making individual baked Alaskas, as in the photographs, take care not to upset the balance between filling and meringue – when scaled down it is easy to pipe on too much meringue.

VARIATION

- Classic variations are **omelette soufflé milady**, which contains poached sliced peaches with vanilla or raspberry ice cream, and **omelette soufflé milord**, which contains poached sliced pear with vanilla ice cream.

24 Apple fritters (*beignets aux pommes*)

Energy	Calories	Fat	Saturated fat	Carbohydrates	Sugar	Protein	Fibre
1034 kJ	246 kcal	10.2 g	1.9 g	38.9 g	25.0 g	2.1 g	3.0 g

* Fried in peanut oil

Left to right: apple, fig and banana fritters

	4 portions	10 portions
Cooking apples	400 g	1 kg
Flour, as needed		
Frying batter	150 g	375 g
Apricot sauce	125 ml	300 ml

Mise en place

1 Make sure the oil is pre-heated to the required temperature.

2 Have a spider ready to remove the fritters and a tray lined with absorbent paper to drain.

3 Have the tray or bowl with flavoured sugar in place to coat the fried fritters.

Cooking

1 Peel and core the apples and cut into ½-cm rings.

2 Pass through flour, shake off the surplus.

3 Dip into the frying batter (see page 357)

4 Lift out with the fingers, into fairly hot deep fat: 185 °C.

5 Cook for about 5 minutes on each side.

6 Drain well on kitchen paper, dust with icing sugar and glaze under the salamander.

7 Serve with hot apricot sauce.

HEALTH AND SAFETY

Take extreme care when placing the fritters into the hot oil. This should be done very calmly to avoid splashing. The fritters should be placed into the oil carefully and never thrown in. A spider is ideal to drain off the oil when removing the fritters.

25 Choux paste fritters (*beignets soufflés*)

Energy	Calories	Fat	Saturated fat	Carbohydrates	Sugar	Protein	Fibre
344 kJ	82 kcal	3.9 g	1.3 g	11.5 g	8.8 g	0.9 g	0.2 g

Mise en place

1 Follow the instructions for Recipe 24 Apple fritters.

Cooking

1 Using two spoons of the same size, shape the paste into quenelles.

2 Place on to strips of lightly greased, greaseproof paper.

3 Lower the pieces into moderately hot deep fat at 170 °C. Allow to cook gently for 10–15 minutes.

4 Drain well and roll in the caster sugar. The sugar can be flavoured at this point (ground cinnamon is

5 Serve with a sauceboat of hot fruit-based sauce.

	8 portions
Choux paste	125 ml
Icing sugar or caster sugar, to serve	
Apricot sauce, to serve	125 ml

HEALTH AND SAFETY

Take extreme care when placing the fritters into the hot oil. This should be done very calmly to avoid splashing. The fritters should be placed into the oil carefully and never thrown in. A spider is ideal to drain off the oil when removing the fritters.

Shape the fritter with two spoons.

Lower the fritters into hot oil on a strip of greaseproof paper.

Lift them out with a spider.

26 Griottines clafoutis

Energy	Calories	Fat	Saturated fat	Carbohydrates	Sugar	Protein	Fibre
790 kJ	187 kcal	4.3 g	1.5 g	28.8 g	21.4 g	6.7 g	1.3 g

	4 portions	10 portions
Eggs	2	5
Caster sugar	40 g	100 g
Flour	40 g	100 g
Milk	175 ml	440 ml
Kirsch	1 tsp	1 tbsp
Griottine cherries	28	70
Neige décor or icing sugar to dust		

Mise en place

1 Brush the sur le plat dishes with soft butter.
2 Ensure the oven is preheated.

Cooking

1 Whisk the eggs and sugar together.
2 Add the flour and whisk until smooth.
3 Add the milk and kirsch; mix well and pass through a conical strainer.
4 Brush 4 (or 10) sur le plat dishes with soft butter.
5 Add seven cherries to each dish and pour over the batter.

6 Bake at 200 °C for 12–15 minutes until the batter has risen and set.
7 Serve warm, dusted with neige décor or icing sugar, and a kirsch sabayon.

> **VARIATION**
> - Griottines work particularly well in this dish but could be substituted for any other hard or fleshy fruit.
> - Substitute 25 per cent of the flour with ground almonds or hazelnuts.
> - To make a richer mixture, substitute up to half the milk with cream.

> **PROFESSIONAL TIP**
> - Good-quality clafoutis should be golden brown and will rise around the edges but stay flatter in the centre. Check to make sure they are set; if not, bake for a little longer.
> - Like most batters, this one benefits from resting before cooking and can be made the day before.
> - Clafoutis can also be made in Yorkshire pudding tins and served unmoulded.

> **NOTE**
> This dish originates from the Limousin region of France and is traditionally made with black cherries.

> **ON A BUDGET**
> Griottines cherries can be quite expensive but many other fruits could be used. Cherries themselves are traditional, but the recipe could be made using fruits such as apricots, plums or blueberries.

27 Bread and butter pudding

Energy	Calories	Fat	Saturated fat	Carbohydrates	Sugar	Protein	Fibre
1463 kJ	350 kcal	23.0 g	14.4 g	34.0 g	27.0 g	3.7 g	0.3 g

	10 portions
Washed sultanas	100 g
Thin slices of white bread	Approximately 5
Melted butter	200 g
Custard	
Vanilla pod	1
Milk	300 ml
Cream	300 ml
Eggs	5
Caster sugar	100 g
Nutmeg	
Apricot jam	100 g

Cooking

1 Butter an earthenware or other suitable dish and sprinkle with the sultanas.

2 Cut the crusts off the bread, dip in melted butter on both sides and cut in half diagonally.

3 Arrange overlapping bread slices neatly in the dish.

4 Sprinkle with more sultanas and cover with another layer of bread.

5 To make the custard, split the vanilla pod, add to the milk and cream and slowly bring to the boil.

6 Whisk the eggs and sugar together and add the boiling liquid, leave to infuse for 5 minutes before passing through a conical strainer.

7 Pour the custard over the bread and grate on some fresh nutmeg.

8 Place dish in a bain-marie and put into a moderate oven at 160 °C; pour hot water into the bain-marie until it comes half way up the dish.

9 Bake for around 45 minutes until the custard is just set.

10 Once removed from the oven sprinkle with sugar and place under the salamander to crisp up and colour the top.

11 Finally, brush with boiled apricot glaze and serve with pouring cream or crème fraiche.

PROFESSIONAL TIP

Add the custard in two or three lots, allowing it to soak in before adding the next. This will prevent the bread floating to the top.

VARIATION

- This pudding can be made in individual dishes or baked in a tray and cut and plated.
- Try using alternatives to bread such as fruit loaf, brioche, baguette slices or panettone.
- Soak the sultanas in rum the day before.
- Try adding a layer of caramelised apple slices.
- A chocolate version can be made by adding couverture (good-quality chocolate) to the custard.

Healthy Eating Tips

- There are several ways to make this dessert healthier. Reduce the sugar content and add more dried fruit, apricots or cranberries.
- This dessert can also be made using milk only, semi-skimmed or skimmed.
- You can also dip the bread in the butter on just one side to reduce the fat content.

ACTIVITY

1 What is the purpose of a bain-marie?

2 Why is steaming the preferred method when cooking puddings made with suet?

3 Describe the correct procedure when scaling up (increasing) or scaling down (decreasing) a recipe.

4 List the differences between a soufflé and a pudding soufflé.

28 Rice pudding

Energy	Calories	Fat	Saturated fat	Carbohydrates	Sugar	Protein	Fibre
644 kJ	153 kcal	2.7 g	1.7 g	28.6 g	18.7 g	5.2 g	0.0 g

6 Ladle a quarter of the boiling milk and rice onto the liaison, mix well and return all to the pan, carefully cook out until the mixture thickens, before removing from the heat (it must not be allowed to boil).

7 Place into suitable individual or large (usually china) dishes.

8 Grate with nutmeg and glaze under the salamander.

SERVING SUGGESTION

Serve with a warm seasonal fruit compote.

VARIATION

- Serve with the apple compote (see page 490).
- Place some good-quality jam in the base of the dish for an extra dimension and flavour.
- Place in a serving dish before piping meringue on top and baking in a hot oven until coloured.

	10 portions
Milk	650 ml
Vanilla pod	1
Short grain rice	60 g
Liaison	
Pasteurised egg	60 g
Caster sugar	60 g
Butter (diced)	30 g

HEALTH AND SAFETY

As this recipe requires some ingredients that are not boiled, ensure all work surfaces and equipment are kept scrupulously clean.

It is recommended that pasteurised eggs are used.

Rice pudding can be held over service at a temperature of no less than 75°C for 2 hours.

Any leftover rice pudding must be cooled to below 5°C within 20 minutes, labelled and stored in a refrigerator.

Cooking

1 Rinse a heavy pan with cold water and add the milk.

2 Split the vanilla pod, scrape out the seeds and add along with the pod.

3 Slowly bring to the boil.

4 Wash the rice and sprinkle into the boiling milk, stir, cover with a lid and allow to simmer until the rice is tender.

5 In a bowl whisk the eggs and sugar and drop in the butter.

PROFESSIONAL TIP

Rinsing out the saucepan with cold water and adding the milk to the wet pan will help prevent the milk from catching on the bottom.

ON A BUDGET

Leftover rice pudding can be used as an alternative to crème pâtissière or frangipane as a filling for a baked flan.

29 Apple crumble tartlets

Energy	Calories	Fat	Saturated fat	Carbohydrates	Sugar	Protein	Fibre
2200 kJ	541 kcal	36 g	19.3 g	51 g	20 g	7.3 g	2.6 g

Cooking

1 Line individual tartlet moulds with the sweet paste.

2 Peel, core and finely slice the apples and divide between the tartlets.

3 Whisk together the soured cream, sugar, flour, egg and a few drops of vanilla, and pass through a conical strainer.

4 Pour over the apples and bake at 190°C for 10 minutes.

5 Combine the dry crumble ingredients and mix with the melted butter.

6 Divide the crumble mixture between the tartlets and bake for a further 10 minutes.

7 Allow to cool slightly before unmoulding. Dust with icing sugar and serve with *sauce à l'anglaise*.

> **VARIATION**
> This dish could be made with pears or plums instead of apples.

	10 tarts
Sweet paste	500 g
Dessert apples	5
Filling	
Soured cream	500 ml
Caster sugar	70 g
Plain flour	75 g
Egg	1
Vanilla extract	
Crumble	
Plain flour	80 g
Walnuts, chopped	60 g
Brown sugar	65 g
Ground cinnamon	Pinch
Salt	Pinch
Unsalted butter, melted	65 g
Icing sugar, to garnish	

30 Apple charlotte

Energy	Calories	Fat	Saturated fat	Carbohydrates	Sugar	Protein	Fibre
2163 kJ	515 kcal	22.3 g	9.3 g	74.5 g	23.4 g	9.4 g	6.1 g

	8 individual or 2 small charlotte moulds
Dessert apples (Cox's)	1 kg
Butter	30 g
Caster sugar	100 g
Lemon zest	1
Breadcrumbs	
Large thin sliced bread loaf	1
Clarified butter - melted	100 g

Mise en place

1 Butter moulds with softened butter (melted butter will run down to the base of the mould).

2 Cut the crusts efficiently from the slices of bread.

Cooking

1 Peel, core and cut the apples into thick slices.

2 Melt the butter in a pan, add the sugar and finely grated lemon zest.

3 Add the apples and simmer until barely cooked, stir in some breadcrumbs to absorb any liquid.

4 Cut out circles of bread for the top and base of the moulds.

5 Cut the crusts and the rest of the bread into fingers 2–3 cm wide depending on whether you are making individual or larger charlottes.

6 Butter the moulds, dip half the circles in the clarified butter and place them butter side down in the base of each mould.

7 Next dip the fingers and line around the outside of the moulds, slightly overlapping.

8 Fill the centre with the apple filling, pressing in carefully.

9 Dip the rest of the circles in the butter and place on top, press firmly.

10 Bake at 230°C for 30–40 minutes until the bread is coloured and crisp.

11 Allow to cool slightly before unmoulding, serve with hot apricot sauce or crème anglaise.

FAULTS

A common fault is that the charlotte collapses when unmoulded (the larger versions are more prone to this). The main reasons for this are:

- the filling is too wet. To avoid this ensure that cooking apples are never used, and do not overcook
- the bread is not baked crisp enough to withstand the pressure of supporting the filling.

If making individual charlottes, make sure the ratio of bread to filling is not compromised by using bread that is cut too thick or overlapping too much.

VARIATION

- Replace the apples with pears or use a mixture of both.

PROFESSIONAL TIP

As this dessert will most likely be plated and sent from the kitchen, it will usually be made in individual portions. Larger versions (such as that in the photograph) would be suitable for family or silver service, but could not be cut and made to look presentable if served from the kitchen.

31 Bramley apple spotted dick

Energy	Calories	Fat	Saturated fat	Carbohydrates	Sugar	Protein	Fibre
1456 kJ	346 kcal	14.5 g	8.2 g	53 g	23.8 g	4.6 g	2.4 g

	10 puddings
Soft flour	350 g
Salt	pinch
Baking powder	20 g
Suet	150 g
Light brown sugar	100 g
Currants	150 g
Lemon zest	1
Bramley apples	2 medium sized
Milk	250 ml

Mise en place

1 Butter moulds with softened butter (melted butter will run down to the base of the mould).

Cooking

1 Sieve the flour, salt and baking powder into a bowl.

2 Stir in the suet, sugar, currants and grated lemon zest.

3 Peel and dice the apple into small cubes and add to the ingredients above.

4 Stir in the milk to form a sticky dough.

5 Divide the mixture between buttered dariole moulds (or similar).

6 Cover and seal the tops with foil and steam for 1½ hours.

SERVING SUGGESTION

Serve with crème anglaise or custard sauce.

PROFESSIONAL TIP

Vegetarian suet can be used in this recipe.

Alternatively, this recipe can be cooked in the oven in a bain-marie.

VARIATION

Serve with an apple and vanilla compote:

Vanilla pod	1
Sugar	300 g
Water	75 ml
Apples	4–5
Sultanas	50 g
White wine	50 ml

1 Split the vanilla pod and bring to the boil with the sugar and water.

2 Prepare the apples as for the spotted dick above.

3 Add the apples, sultanas and wine to the boiling syrup, remove from the heat and leave to stand before serving.

32 Jam roly-poly

Energy	Calories	Fat	Saturated fat	Carbohydrates	Sugar	Protein	Fibre
2425 kJ	575 kcal	21 g	11.8 g	91 g	47 g	7.4 g	5.3 g

* Using extra 10g flour for dusting and semi-skimmed milk. Based on 5 portions.

	4–6 portions
Strawberry jam	
Strawberries, hulled and quartered	350 g
Caster sugar	150 g
Lemon, zest and juice	½
Milk	250 ml
Roly poly	
Self-raising flour	275 g, plus extra for dusting
Caster sugar	40 g
Suet	110 g
Lemon, zest only	1

Cooking

Strawberry jam:

1 Heat a small frying pan until hot, then add the strawberries and cook for one minute.

2 Add the sugar and lemon zest and juice and cook for 3–4 minutes, until the sugar has dissolved and the mixture has thickened slightly.

3 Transfer half the mixture to a blender, blend to a purée, then return it to the pan with the remaining cooked strawberries, mix well and cook for a further 3–5 minutes, until thick.

4 Remove from the heat and leave to cool.

Roly-poly:

1 Put the flour, sugar, suet and lemon zest in a bowl and mix well. Make a well in the centre, then gradually mix in enough water to make a soft dough.

2 Knead lightly until smooth.

3 Turn onto a floured surface and roll out to an oblong about 1 cm thick.

4 Spread the jam onto the pastry leaving a 2.5 cm border. Roll up from the long side into a pinwheel. Place the roly-poly in a loaf tin and cover with kitchen foil, sealing tightly.

5 Put a plate upside-down in the base of a large saucepan, place the loaf tin on top and fill the pan two-thirds full of water (do not allow the water to reach the top of the tin).

6 Bring to the boil. Cover, reduce the heat and simmer for one hour.

7 Remove the pudding from the tin and place on a baking tray. Bake at 200 °C for 15–30 minutes, or until browned.

Alternatively, the roly-poly may be steamed in a steamer and then finished in the oven as in the recipe.

SERVING SUGGESTION
Serve the roly-poly with hot, fresh custard.

Healthy Eating Tip
● Vegetable suet can be used to produce this recipe.

33 Vanilla soufflé

Energy	Calories	Fat	Saturated fat	Carbohydrates	Sugar	Protein	Fibre
932 kJ	223 kcal	13.7 g	6.7 g	19.5 g	13.9 g	6.6 g	0.2 g

10 individual soufflés	
Base mixture	
Milk	500 ml
Vanilla pods	2
Butter	75 g
Strong flour	60 g
Egg yolks	10
Caster sugar	50 g
Additional ingredients for every 400 g of base mixture	
Egg whites	150 g
Caster sugar	60 g
Cornflour	12 g
Lemon juice	2–3 drops

Cooking

1 Rinse a heavy saucepan with cold water and add the milk. Split the vanilla pod, scrape out the seeds, and add both to the milk. Put on the heat to boil.

2 Melt the butter in another heavy pan. Add the flour and cook out to form a white roux. Gradually add the boiling milk, mixing in each addition before adding the next.

3 When all the milk has been added, allow to simmer for a few minutes.

4 Whisk the egg yolks and sugar. Add to the mixture in the saucepan and keep stirring over the heat until the mixture starts to bubble around the edges. This forms the panada.

5 Pour on to a clean tray and cover with cling film to prevent a skin forming. Allow to cool. (This can be kept in the fridge until needed, as soufflés must be cooked to order.)

6 Take 400 g of the base and beat in a clean bowl until smooth.

7 Whisk the egg whites, sugar, cornflour and lemon juice to form firm peaks.

8 Add one-third of the whites to the base and mix in, then very carefully fold in the remaining whites.

9 Carefully fill prepared individual china ramekins (see below). Level the top and run your thumb around the edge, moving the mixture away from the lip of the mould.

10 Space well apart on a solid baking sheet (if they are close together they will not rise evenly and will bake stuck together).

11 Place immediately in the oven at 215 °C for 12–14 minutes.

12 The soufflés should rise out of the moulds by around 5–6 cm and have a flat top with no cracks.

13 Dust with icing sugar and serve immediately with fruit coulis and/or ice cream (chocolate or vanilla).

> **PROFESSIONAL TIP**
>
> Preparation of a soufflé mould is very important. The rule is to butter the sides twice and the bottom once. Always use soft, not melted, butter, so that it stays where you put it. Give the moulds one coat all over to start with, then place in the fridge to set before giving a second coat only to the sides. (Giving the base two coats results in a puddle of butter in the bottom.)
>
> After giving the sides a second coat of butter, the mould is usually coated with sugar.

Prepare the mould.

Thumb the edge.

Fold in the remaining meringue.

Knock one-third of the meringue into the panada.

FAULTS

Soufflé does not rise:
- under- or over-beaten whites
- wrong proportion of whites to base
- mixture left to stand before cooking
- moulds not buttered correctly.

Soufflé does not rise evenly:
- moulds not prepared correctly (mixture has stuck to the mould on one side)
- uneven heat in the oven.

Soufflé rises but drops back:
- too much egg white used.

Soufflé has a cracked top:
- too much egg white used
- egg white is overbeaten.

PROFESSIONAL TIP

Work as methodically as possible and make sure that all bowls and equipment are as clean as possible.

Be organised with your timings. Make sure you have the moulds fully prepared and that the oven is up to heat well before the egg whites are whisked.

Before whisking the egg whites, scald the bowl to remove any trace of fat.

When the egg whites are whisked with the sugar, lemon juice and cornflour, they should be firm but creamy. Over-whisking will ruin them.

Do not use more egg white than is needed for the recipe.

A soufflé should have flavour as well as an impressive 'rise'. Too much egg white can dilute the flavour.

VARIATION

- This vanilla recipe is basic and can easily be adapted. For example, a **chocolate soufflé** can be made by adding melted chocolate and cocoa powder to the base. This will firm up the base mixture, so the whites mixture will need to be increased by 25–30 per cent to compensate.
- Soufflés can be made in many different flavours and combinations. Recipes can vary considerably. For example, **fruit soufflés** can be made using a sabayon or a boiled sugar base.
- As an alternative to coating the moulds with sugar, if compatible, dust them with cocoa powder, grated chocolate or try adding some cinnamon to the sugar.

34 Soufflé pudding (*pouding soufflé*)

Energy	Calories	Fat	Saturated fat	Carbohydrates	Sugar	Protein	Fibre
510 kJ	122 kcal	7.6 g	3.2 g	5.9 g	4.8 g	0.2 g	0.0 g

* Using white flour and hard margarine

8 Whisk the egg whites and sugar to a creamy but firm consistency, and fold into the base mixture.

9 Three-quarters fill buttered and sugared dariole moulds and place in a tray of boiling water (bain-marie) on top of the stove. Allow to gently simmer. When the mixture has risen to the top of the mould, place the whole tray in the oven at 210°C and bake for 15–20 minutes.

10 Unmould and serve with an appropriate sauce.

> **NOTE**
> A pudding soufflé is the only type of soufflé that is traditionally unmoulded. All other types of soufflé are served in the mould in which they are cooked or prepared.
>
> By leaving them to stand in the bain-marie after cooking, pudding soufflés can be held for up to 15–20 minutes before serving.

	10 puddings
Milk	375 ml
Butter	50 g
Flour	50 g
Caster sugar	50 g
Eggs	6

Cooking

1 Slowly bring the milk to boil in a sauteuse or heavy-bottomed pan.

2 Cream together the butter and flour (*beurre manié*).

3 Add small pieces of the *beurre manié* to the boiling milk, one at a time, whisking in each addition before adding the next.

4 Bring to the boil and simmer for a couple of minutes before removing from the heat.

5 Separate the eggs, put the whites to one side and whisk the yolks into the hot mixture.

6 Cover with cling film and set aside to cool.

7 At this stage add any flavouring (see variations below).

> **VARIATION**
> - **Chocolate:** Add to the milk 60 g grated couverture.
> - **Coffee:** Add to the milk 15 g instant coffee or coffee essence
> - **Lemon, orange** or both (**St Clements**): Add to the milk the grated zest of lemon or orange or both.
> - **À l'indienne:** Add 60 g finely diced stem ginger to the mixture after adding the egg yolks. Serve with sauce à l'anglaise after infusing the milk with grated fresh ginger.
> - **Grand Marnier:** Add a small dice of sponge fingers which have been sprinkled with Grand Marnier. Fold in just before the egg whites.
> - **Saxon:** Flavour with vanilla.
> - **Sans souci:** Add diced cooked apples and currants.

35 Fresh fruit salad

Energy	Calories	Fat	Saturated fat	Carbohydrates	Sugar	Protein	Fibre
493 kJ	117 kcal	0.0 g	0.0 g	30.3 g	29.5 g	0.9 g	3.0 g

3 Peel and cut the orange into segments.

4 Peel, quarter and core the apple and pear, then cut each quarter into two or three slices, place in the bowl with the syrup and mix in the orange segments.

5 Stone the cherries but leave them whole.

6 Cut the grapes in half, peel if required, and remove any pips. Add the cherries and grapes to the fruit and syrup mix.

7 Mix the fruit salad carefully and place in a glass bowl in the refrigerator to chill.

8 Just before serving, peel and slice the banana and mix in.

	4 portions	10 portions
Stock syrup		
Caster sugar	50 g	125 g
Water	125 ml	310 ml
Lemon, juice of	½	1¼
Fruit		
Orange	1	2½
Dessert apple	1	2½
Dessert pear	1	2½
Cherries	50 g	125 g
Grapes	50 g	125 g
Banana	1	2½

VARIATION

- Any of the following fruits may be used: dessert apples, pears, pineapple, oranges, grapes, melon, strawberries, peaches, raspberries, apricots, bananas, cherries, kiwi fruit, plums, mangoes, paw paws and lychees. Allow about 150 g unprepared fruit per portion. All fruit must be ripe.
- Kirsch or an orange liqueur could be added to the syrup.

Healthy Eating Tip

- Fruit juice (such as apple, orange, grape or passion fruit) can be used instead of syrup.

Cooking

1 For the syrup, boil the sugar with the water and place in a bowl.

2 Allow to cool, then add the lemon juice.

36 Poached fruits or fruit compote (*compote de fruits*)

Energy	Calories	Fat	Saturated fat	Carbohydrates	Sugar	Protein	Fibre
531 kJ	126 kcal	0.0 g	0.0 g	33.5 g	33.5 g	0.2 g	2.2 g

* Using pears

Poached rhubarb and pear

	4 portions	10 portions
Stock syrup (see page 469)	250 ml	625 ml
Fruit	400 g	1 kg
Sugar	100 g	250 g
Lemon, juice of	½	1

Cooking

Apples, pears:

1 Boil the water and sugar.

2 Quarter the fruit, remove the core and peel.

3 Place in a shallow pan in sugar syrup.

4 Add a few drops of lemon juice.

5 Cover with greaseproof paper.

6 Allow to simmer slowly, preferably in the oven, cool and serve.

Soft fruits (raspberries, strawberries):

1 Pick and wash the fruit. Place in a glass bowl.

2 Pour on the hot syrup. Allow to cool and serve.

Stone fruits (plums, damsons, greengages, cherries):

1 Wash the fruit, barely cover with sugar syrup and cover with greaseproof paper or a lid.

2 Cook gently in a moderate oven until tender.

Rhubarb:

1 Trim off the stalk and leaf and wash.

2 Cut into 5-cm lengths and cook as above, adding extra sugar if necessary. A little ground ginger may also be added.

Gooseberries, blackcurrants, redcurrants:

1 Top and tail the gooseberries, wash and cook as for stone fruit, adding extra sugar if necessary.

2 The currants should be carefully removed from the stalks, washed and cooked as for stone fruits.

Dried fruits (prunes, apricots, apples, pears):

1 Dried fruits should be washed and soaked in cold water overnight.

2 Gently cook in the liquor with sufficient sugar to taste.

> **VARIATION**
> - A piece of cinnamon stick and a few slices of lemon may be added to the prunes or pears, one or two cloves to the dried or fresh apples.
> - Any compote may be flavoured with lavender and/or mint.

> **Healthy Eating Tips**
> - Use fruit juice instead of stock syrup.
> - If dried fruits are used, no added sugar is needed.

37 Fruit mousse

Energy	Calories	Fat	Saturated fat	Carbohydrates	Sugar	Protein	Fibre
950 kJ	227 kcal	12.4 g	7.0 g	26.0 g	26.0 g	4.0 g	2.0 g

	10 portions
Egg yolks	4
Sugar	50 g
Fruit purée	250 g
Gelatine	4 leaves
Lemon juice	
Lightly whipped cream	250 g
Italian meringue	
Sugar	112 g
Egg whites	2
Cream of tartar	pinch
Glaze topping	
Stock syrup	150 ml
Fruit purée	150 ml
Gelatine	3 leaves, soaked in cold water

Cooking

1 Mix the egg yolks and sugar together and slowly add the boiled fruit purée which has been flavoured with a squeeze of lemon juice.

2 Return to the stove and cook to 80°C until slightly thickened. Do not boil.

3 Add the previously softened gelatine to the warm purée and mix until fully dissolved. Chill down.

4 Prepare the Italian meringue by placing the sugar in a pan and saturating in water.

5 Boil the sugar to 115°C, then whisk the egg whites with a pinch of cream of tartar.

6 Once the egg whites are at full peak, gradually add the boiled sugar, which now should have reached the temperature of 121°C. Whisk until cool.

7 Once the purée is cold, but not set, fold in the Italian meringue and whipped cream.

8 Place into piping bag and pipe into the desired ring mould, normally lined with a suitable sponge, such as a jaconde.

9 Level the surface using a palette knife and refrigerate.

10 Once set, glace the surface, refrigerate.

11 To remove from the mould, warm the outside of the mould with a blow torch and remove the ring mould.

Glaze:

1 Warm the syrup. Add the gelatine and stir until dissolved, then add the desired fruit purée.

2 Apply to the surface of the chilled mousse whilst in a liquid state, but not hot.

> **VARIATION**
> ● Practically any fruit puree could be used here, including raspberry, strawberry, mango, etc. In this example, passion fruit has been used with an orange coulis to accompany the mousse.

38 Chocolate mousse

Energy	Calories	Fat	Saturated fat	Carbohydrates	Sugar	Protein	Fibre
1760 kJ	419 kcal	37.0 g	22.0 g	13.7 g	12.8 g	5.1 g	2.6 g

	8 portions	16 portions
Stock syrup at 30° Baumé (equal quantities of sugar and water will give 30° Baume)	125 ml	250 ml
Pasteurised egg yolks	80 ml	160 ml
Bitter couverture, melted	250 g	500 g
Gelatine	2 leaves	4 leaves
Whipping cream, whipped	500 ml	1 litre

Mise en place

1 Line individual moulds with acetate, if using.

2 If using a base of sponge (as in the picture), have these cut and in place before making the mousse.

Cooking

1 Boil the syrup.

2 Place the yolks into the bowl of a food mixer. Pour over the boiling syrup and whisk until thick. Remove from the mixer (to make a *pâte à bombe*)

3 Add all the melted couverture at once and fold it in quickly.

4 Drain the gelatine, melt it in the microwave and fold it into the chocolate sabayon mixture.

5 Add all the whipped cream at once and fold it in carefully.

6 Place the mixture into prepared moulds. Refrigerate or freeze immediately.

FAULTS

Possible causes of a heavy texture in chocolate mousse include:
- the *pâte à bombe* is under-aerated
- the cream is insufficiently whipped
- the mix has been over-worked when folding in the cream and Italian meringue.

PROFESSIONAL TIP

Specialist plastic moulds may be used, as shown in the photo. The smaller individual portions have been set in metal rings lined with acetate. It is also possible to place the mix directly into the mould. If this is done, the exterior of the mould will need to be heated slightly, using a blow-torch for example, in order to enable the ring to be removed.

SERVING SUGGESTION
- Serve with a complementary sauce such as crème anglaise.

39 Bavarois: basic recipe and a range of flavours

Energy	Calories	Fat	Saturated fat	Carbohydrates	Sugar	Protein	Fibre
970 kJ	231 kcal	18.2 g	10.9 g	11.8 g	11.8 g	5.8 g	0.0 g

Cream-based bavarois

Cream-based bavarois:

	6–8 portions
Gelatine	10 g
Eggs, pasteurised, separated	2
Caster sugar	50 g
Milk, whole, semi-skimmed or skimmed	250 ml
Whipping or double cream or non-dairy cream	125 ml

Mise en place

1 If using leaf gelatine, soak in cold water.

Cooking

1 Cream the yolks and sugar in a bowl until almost white.

2 Whisk in the milk, which has been brought to the boil, mix well.

3 Clean the milk saucepan, which should be a thick-based one, and return the mixture to it.

4 Return to a low heat and stir continuously with a wooden spoon until the mixture coats the back of the spoon. The mixture must not boil.

5 Remove from the heat; add the gelatine and stir until dissolved.

6 Pass through a fine strainer into a clean bowl, leave in a cool place, sstirring occasionally until almost at setting point.

7 Allow to set in the refrigerator.

8 Shake and turn out on to a flat dish or plates.

> **VARIATION**
> - **Raspberry or strawberry bavarois**: when the custard is almost cool, add 200 g of picked, washed and sieved raspberries or strawberries. Decorate with whole fruit and whipped cream.
> - **Chocolate bavarois**: dissolve 50 g chocolate couverture in the milk. Decorate with whipped cream and grated chocolate.
> - **Coffee bavarois**: proceed as for a basic bavarois, with the addition of coffee essence to taste.
> - **Orange bavarois**: add grated zest and juice of 2 oranges and 1 or 2 drops orange colour to the mixture, and increase the gelatine by 2 leaves. Decorate with blanched, fine julienne of orange zest, orange segments and whipped cream.
> - **Lemon or lime bavarois**: as orange bavarois, using lemons or limes in place of oranges.
> - **Vanilla bavarois**: add a vanilla pod or a few drops of vanilla essence to the milk. Decorate with vanilla-flavoured sweetened cream (*crème Chantilly*).

NOTE

Bavarois may be decorated with sweetened, flavoured whipped cream (crème Chantilly). It is advisable to use pasteurised egg yolks and whites.

VARIATION

- **Charlotte royale**: Line a charlotte mould with thin slices of jam-filled Swiss roll, then fill with cream-based bavarois. Remove from the mould and finish with apricot glaze.
- **Charlotte Russe**: Line a charlotte mould with fingers of *biscuit a la cuillière*, then fill with cream-based bavarois and remove from the mould.

Fruit-based bavarois:

	8 portions
Fruit purée	300 g
Lemon juice	1
Caster sugar	75 g
Gelatine	5 leaves
Lightly whipped cream	400 g

Mise en place

1 If using leaf gelatine, soak in cold water.

Cooking

1 Heat the fruit purée, but do not boil. Add the lemon juice.

2 Add the previously soaked gelatine and add to the warm fruit purée.

3 Set the fruit purée over ice following the same guidelines as the cream-based bavarois.

4 Quickly fold in the lightly whipped cream.

5 Pour into mould.

6 Finish as for a cream-based bavarois.

VARIATION

- **Chartreuse aux fruits**: Line a charlotte mould with a jelly appropriate to the fruit. When the jelly is set, line the inside and base of the mould with thinly sliced fruit, dipping it first into liquid jelly. Fill the mould with fruit bavarois and allow to set. Remove from the mould by quickly dipping the mould in hot water to soften the jelly coating, being careful not to fully melt the jelly.

40 Lime soufflé frappé

Energy	Calories	Fat	Saturated fat	Carbohydrates	Sugar	Protein	Fibre
2734 kJ	655 kcal	40.8 g	20.7 g	66.9 g	61.2 g	9.2 g	0.8 g

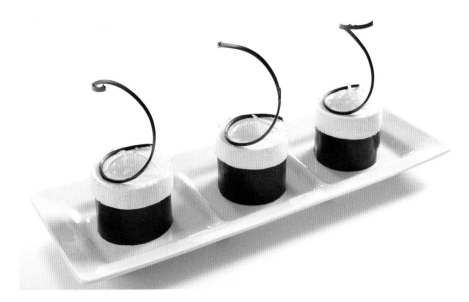

	10 portions	15 portions
Couverture	150g	200g
Sponge, thin slices, cut into rounds	10	15
Lime syrup	100ml	150ml
Swiss meringue		
Egg whites, pasteurised	190 ml	300 ml
Caster sugar	230 ml	340 ml
Sabayon		
Whipping cream	600 ml	900 ml
Lime zest, finely grated and blanched, and juice	8	12
Egg yolks	10	15
Caster sugar	170 g	250 g
Leaf gelatine, soaked in iced water	9½	14
To decorate		
Confit of lime segments		
Moulded chocolate		

Mise en place

1 Use individual stainless steel ring moulds. Cut a strip of acetate, 8 cm wide, to fit inside each ring. Cut a 6-cm strip to fit inside the first, spread it with tempered couverture and place inside the first strip, in the mould.

2 Ensure tempered chocolate/couverture is available.

Cooking

1 Place a round of sponge in the base of each mould and moisten with lime syrup.

2 Make a Swiss meringue by whisking the pasteurised egg whites and sugar over a bain-marie of simmering water until a light and aerated meringue is achieved.

3 Whisk the cream until it is three-quarters whipped, then chill.

4 Whisk together the egg yolks, sugar and blanched lime zest. Boil the juice and pour it over the mixture to make the sabayon. Whisk over a bain-marie until it reaches 75 °C, then continue whisking away from the heat until it is cold.

5 Drain and melt the gelatine. Fold it into the sabayon.

6 Fold in the Swiss meringue, and then the chilled whipped cream.

7 Fill the prepared moulds. Level the tops and chill until set.

8 To serve, carefully remove the mould, peel away the acetate, plate and decorate.

495

41 Vanilla panna cotta served on a fruit compote

Energy	Calories	Fat	Saturated fat	Carbohydrates	Sugar	Protein	Fibre
1565kJ	378 kcal	34.0 g	21.1 g	16.1 g	16.1 g	2.9 g	1.5 g

	6 portions
Milk	125 ml
Double cream	375 ml
Aniseeds	2
Vanilla pod	½
Gelatine (soaked)	2 leaves
Caster sugar	50 g
Fruit compote	
Apricot purée	75 g
Vanilla pod	½
Peach	1
Kiwi fruit	1
Strawberries	75 g
Blueberries	75 g
Raspberries	50 g

Cooking

1 Prepare the fruit compote by boiling the apricot purée and infusing with vanilla pod. Remove pod, allow purée to cool.

2 Finely dice the peach and the kiwi and quarter the strawberries. Mix, then add blueberries and raspberries.

3 Bind the fruit with the apricot purée. A little stock syrup (recipe 15) may be required to keep the fruit free flowing.

4 For the panna cotta, boil the milk and cream, add aniseeds, infuse with the vanilla pod, remove after infusion.

5 Heat again and add the soaked gelatine and caster sugar. Strain through a fine strainer.

6 Place in a bowl set over ice and stir until it thickens slightly; this will allow the vanilla seeds to suspend throughout the mix instead of sinking to the bottom.

7 Fill individual dariole moulds.

8 Place the fruit compote with individual fruit plates, turn out the panna cotta, place on top of the compote, finish with a tuile biscuit.

FAULTS

If too much gelatine is used, the panna cotta will set too firmly. A panna cotta should be just set and have a slight wobble.

SERVING SUGGESTION

● A biscuit, such as a tuille, will provide a contrast in texture. Think of the smooth creaminess of the panna cotta and the short, crispness of the biscuit.

42 Lime and mascarpone cheesecake

Energy	Calories	Fat	Saturated fat	Carbohydrates	Sugar	Protein	Fibre
2064 kJ	633 kcal	54.0 g	32.3 g	31.5 g	24.3 g	7.1 g	0.0 g

	1 cheesecake
Base	
Packet of ginger biscuits	200 g approx
Butter, melted	200 g
Cheesecake	
Egg yolks, pasteurised	125 g
Caster sugar	75 g
Cream cheese	250 g
Mascarpone	250 g
Gelatine, softened in cold water	15 g
Limes, juice and grated zest of	2
Semi-whipped cream	275 ml
White chocolate, melted	225 g

Cooking

1 Blitz the biscuits in a food processor. Mix in the melted butter. Line the cake ring with this mixture and chill until required.

2 Make a sabayon by whisking the egg yolks and sugar together over a pan of simmering water.

3 Stir the cream cheese and mascarpone into the sabayon until soft.

4 Meanwhile, warm the gelatine in the lime juice, and pass through a fine chinois. Also whip the cream.

5 Pour the gelatine and melted white chocolate into the cheese mixture.

6 Remove from the food mixer and fold in the whipped cream with a spatula. Finally, whisk in the lime zest.

7 Pour over the prepared base. Chill for 4 hours.

> **PROFESSIONAL TIP**
> Use a microplane or micrograter for the lime zest.

> **FAULTS**
> Adding the whipped cream while the mix is too warm will melt the cream, turning it back to a liquid from its whipped form. This will lose the aeration and cause result in a dense finish.
>
> Over-whipping the cream will also increase the density and a loss of aeration.
>
> If the mix gets too cold after the gelatine has been added and starts to set, it will become very difficult to fold in the cream and the mix may become overworked and split.

43 Trifle

Energy	Calories	Fat	Saturated fat	Carbohydrates	Sugar	Protein	Fibre
2280 kJ	543 kcal	29.1 g	17.1 g	66.2 g	51.3 g	8.2 g	1.9 g

	6–8 portions
Sponge (made with 3 eggs)	1
Jam	25 g
Tinned fruit (pears, peaches, pineapple)	1
Sherry (optional)	
Custard	
Custard powder	35 g
Milk, whole or skimmed	375 ml
Caster sugar	50 g
Cream (¾ whipped) or non-dairy cream	125 ml
Whipped sweetened cream or non-dairy cream	250 ml
Angelica	25 g
Glacé cherries	25 g

Cooking

1 Cut the sponge in half, sideways, and spread with jam.

2 Place in a glass bowl or individual dishes and soak with fruit syrup drained from the tinned fruit; a few drops of sherry may be added.

3 Cut the fruit into small pieces and add to the sponge.

4 Dilute the custard powder in a basin with some of the milk and add the sugar.

5 Boil the remainder of the milk, pour a little on the custard powder, mix well, return to the saucepan and over a low heat and stir to the boil. Allow to cool, stirring occasionally to prevent a skin forming; fold in the three-quarters whipped cream.

6 Pour on to the sponge. Leave to cool.

7 Decorate with the whipped cream, angelica and cherries.

> **VARIATION**
> - Other flavourings or liqueurs may be used in place of sherry (such as whisky, rum, brandy, Tia Maria).
> - For raspberry or strawberry trifle use fully ripe fresh fruit in place of tinned and decorate with fresh fruit in place of angelica and glacé cherries.
> - A fresh egg custard may be used with fresh egg yolks (see page 468).

44 Tiramisu torte

Energy	Calories	Fat	Saturated fat	Carbohydrates	Sugar	Protein	Fibre	Sodium
17755 kJ	4267 kcal	306 g	189 g	317 g	232 g	52 g	6.2 g	1985 g

	2 tortes
Biscuit or sponge bases	4
Egg yolks, pasteurised	60 g
Sugar	150 g
Gelatine	3 leaves
Mascarpone cheese	600 g
Double cream	200 g
Coffee syrup	100 ml
Rum	40 ml
Cocoa powder	

Cooking

1 Cut the biscuit or sponge bases into shape: cut two to the size of the flan ring, and two to the same shape, but slightly smaller.

2 Mix the egg yolks and sugar. Cook over a bain-marie to 75 °C, to form a sabayon.

3 Soak the gelatine in iced water. Drain and add it to the sabayon.

4 Beat the cheese well. Add the sabayon.

5 Lightly whip the cream and fold it into the mixture.

6 Place a large biscuit or sponge base into each flan ring, on a board. Soak the base with a mixture of coffee syrup and rum.

7 Half-fill each ring with the cheese mixture.

8 Place the smaller circles of biscuit or sponge on top of the filling. Again, soak with syrup and rum.

9 Fill the rest of the ring with the cheese mixture, to a level top.

10 Chill in the fridge overnight.

11 Dust with cocoa powder and decorate with chocolate pieces.

> **PROFESSIONAL TIP**
> Use an accurate thermometer, such as a thermopen, to accurately measure the temperature of the sabayon.

> **VARIATION**
> ● This recipe could be made into individual portions using stainless steel rings lined with acetate.
> ● To make this dessert with a looser consistency, and perhaps more like a traditional tiramisu, line a pudding bowl with sponge fingers and soak them (using a brush) with a mix of strong coffee and coffee liqueur (e.g. Tia Maria or Kahlua). Make the mix as per the recipe, excluding the gelatine, and pure on top of the sponge fingers. Chill for a few hours and dust with cocoa powder before serving.

45 *Crème brûlée* (Burned, caramelised or browned cream)

Energy	Calories	Fat	Saturated fat	Carbohydrates	Sugar	Protein	Fibre
1154 kJ	278 kcal	21.9 g	12.1 g	14.8 g	14.8 g	6.2 g	0.0 g

	4 portions	10 portions
Milk	125 ml	300 ml
Double cream	125 ml	300 ml
Natural vanilla essence or pod	3–4 drops	7–10 drops
Eggs	2	5
Egg yolk	1	2–3
Caster sugar	25 g	60 g
Demerara sugar		

Cooking

1 Warm the milk, cream and vanilla essence in a pan.

2 Mix the eggs, egg yolk and caster sugar in a basin and add the warm milk. Stir well and pass through a fine strainer.

3 Pour the cream into individual dishes and place them into a tray half-filled with warm water.

4 Place in the oven at approximately 160 °C for about 30–40 minutes, until set.

5 Sprinkle the tops with Demerara sugar and glaze under the salamander or by blowtorch to a golden brown.

6 Clean the dishes and serve.

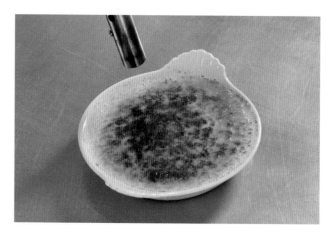

Use a blowtorch carefully to glaze the top

VARIATION
- Sliced strawberries, raspberries or other fruits (e.g. peaches, apricots) may be placed in the bottom of the dish before adding the cream mixture, or placed on top after the creams are caramelised.

SERVING SUGGESTION
Serve with a biscuit, such as a tuille or shortbread finger, to provide a contrast to the creamy texture the dessert.

46 Cream caramel (*crème caramel*)

Energy	Calories	Fat	Saturated fat	Carbohydrates	Sugar	Protein	Fibre
868 kJ	207 kcal	7.2 g	3.3 g	30.2 g	30.2 g	7.3 g	0 g

* Using whole milk

3 Prepare the cream by warming the milk and whisking on to the beaten eggs, sugar and essence (or vanilla pod).

4 Strain and pour into the prepared moulds.

5 Place in a roasting tin half full of water.

6 Cook in a moderate oven at 150–160 °C for 30–40 minutes.

7 When thoroughly cold, loosen the edges of the cream caramel with the fingers, shake firmly to loosen and turn out on to a flat dish or plates.

8 Pour any caramel remaining in the mould around the creams.

	4–6 portions	10–12 portions
Caramel		
Sugar, granulated or cube	100 g	200 g
Water	125 ml	250 ml
Cream		
Milk, whole or skimmed	0.5 litres	1 litre
Eggs	4	8
Sugar, caster or unrefined	50 g	100 g
Vanilla essence or a vanilla pod	3–4 drops	6–8 drops

SERVING SUGGESTION

Cream caramels may be served with whipped cream or a fruit sauce such as passion fruit, and accompanied by a sweet biscuit (e.g. shortbread, palmiers).

PROFESSIONAL TIP

Adding a squeeze of lemon juice to the caramel will invert the sugar, thus preventing re-crystallisation.

Cooking

1 Prepare the caramel by placing three-quarters of the water in a thick-based pan, adding the sugar and allowing to boil gently, without shaking or stirring the pan.

2 When the sugar has cooked to a golden-brown caramel colour, add the remaining quarter of the water, reboil until the sugar and water mix, then pour into the bottom of dariole moulds.

FAULTS

This caramel was baked at too high a temperature. The custard has curdled.

47 Meringue

Energy	Calories	Fat	Saturated fat	Carbohydrates	Sugar	Protein	Fibre
913 kJ	214 kcal	0.1 g	0 g	52 g	52 g	4 g	0 g

	4 portions	10 portions
Lemon juice or cream of tartar		
Egg whites, pasteurised	4	10
Caster sugar	200 g	500 g

Mise en place

1 Ensure all tools are scrupulously clean and free from fat or grease.

Cooking

1 Whip the egg whites stiffly with a squeeze of lemon juice or cream of tartar.

2 Sprinkle on the sugar and carefully mix in.

3 Place in a piping bag with a large plain tube and pipe onto silicone paper on a baking sheet.

4 Bake in the slowest oven possible or in a hot plate (110 °C). The aim is to dry out the meringues without any colour whatsoever.

NOTE

The reason egg whites increase in volume when whipped is because they contain so much protein (11 per cent). The protein forms tiny filaments, which stretch on beating, incorporate air in minute bubbles then set to form a fairly stable puffed-up structure expanding to seven times its bulk. To gain maximum efficiency when whipping egg whites, the following points should be observed.

Because of possible weakness in the egg-white protein, it is advisable to strengthen it by adding a pinch of cream of tartar and a pinch of dried egg-white powder. If all dried egg-white powder is used no additions are necessary.

Other points to note:
● Eggs should be fresh.
● When separating yolks from whites, no speck of egg yolk must be allowed to remain in the white; egg yolk contains fat, the presence of which can prevent the white being correctly whipped.
● The bowl and whisk must be scrupulously clean, dry and free from any grease.
● When egg whites are whipped, the addition of a little sugar (15 g to 4 egg whites) will assist efficient beating and reduce the chances of over-beating.

PROFESSIONAL TIP

A small amount of an acid such as cream of tartare or a squeeze of lemon juice will help to strengthen the protein structure of the foam when egg whites are whipped.

FAULTS

Over-whipped eggs whites will collapse, so it important to gauge progress through the whipping process.

48 Vacherin with strawberries and cream (*vacherin aux fraises*)

Energy	Calories	Fat	Saturated fat	Carbohydrates	Sugar	Protein	Fibre
1436 kJ	341 kcal	12.6 g	7.9 g	56.3 g	56.3 g	3.9 g	0.6 g

	4 portions	10 portions
Egg whites	4	10
Caster sugar	200 g	500 g
Cream (whipped and sweetened) or non-dairy cream	125 ml	300 ml
Strawberries, picked and washed)	100–300 g	250–750 g

Cooking

1 Stiffly whip the egg whites. (Refer to the notes in recipe 47 for more guidance.)

2 Carefully fold in the sugar.

3 Place the mixture into a piping bag with a 1 cm plain tube.

4 Pipe on to silicone paper on a baking sheet.

5 Start from the centre and pipe round in a circular fashion to form a base of 16 cm then pipe around the edge 2–3 cm high.

6 Bake in a cool oven at 100 °C until the meringue case is completely dry. Do not allow to colour.

7 Allow the meringue case to cool then remove from the paper.

8 Spread a thin layer of cream on the base. Add the strawberries.

9 Decorate with the remainder of the cream.

NOTE

A vacherin is a round meringue shell piped into a suitable shape so that the centre may be filled with sufficient fruit (such as strawberries, stoned cherries, peaches and apricots) and whipped cream to form a rich sweet. The vacherin may be prepared in one-, two- or four-portion sizes, or larger.

VARIATION

- Melba sauce may be used to coat the strawberries before decorating with cream.
- Raspberries can be used instead of strawberries.

Healthy Eating Tip

- Try 'diluting' the fat in the cream with some low fat fromage frais.

49 Vanilla ice cream (*glace vanille*)

Energy	Calories	Fat	Saturated fat	Carbohydrates	Sugar	Protein	Fibre
616 kJ	147 kcal	8.1 g	4.2 g	15.8 g	15.8 g	3.5 g	0.0 g

* Using whole milk and single cream

	8–10 portions
Egg yolks	4
Caster or unrefined sugar	100 g
Milk, whole or skimmed	375 ml
Vanilla pod or essence	
Cream or non-dairy cream	125 ml

Cooking

1 Whisk the yolks and sugar in a bowl until almost white.
2 Boil the milk with the vanilla pod or essence in a thick-based pan.
3 Whisk on to the eggs and sugar; mix well.
4 Return to the cleaned saucepan, place on a low heat.
5 Stir continuously with a wooden spoon until the mixture coats the back of the spoon.
6 Pass through a fine strainer into a bowl.
7 Freeze in an ice cream machine, gradually adding the cream.

Whisk boiling milk into the egg yolks and sugar.

Return the mixture to the hot pan used for the milk.

Test the consistency on the back of a spoon.

Pass through a fine strainer into a cold pot.

The mixture will cool down; if it was left in the hot pan it would continue to cook.

Gradually add cream to the mixture in the ice cream machine.

HEALTH AND SAFETY
If frozen ice cream melts or thaws, it should not be re-frozen. This is largely due to the wide range of temperatures that is required from heating through to freezing when producing ice cream. This makes ice-cream a high-risk product and therefore it needs to be produced and handled very precisely. It is also made using milk, cream and eggs, which are all high-risk products in their own right.

50 Lemon curd ice cream

Energy	Calories	Fat	Saturated fat	Carbohydrates	Sugar	Protein	Fibre
774 kJ	185 kcal	8.9g	5.2g	24.8g	16.7g	2.8g	0.0g

	6–8 portions
Lemon curd	250 g
Crème fraiche	125 g
Greek yoghurt	250 g

Preparing the dish

1 Mix all ingredients together.
2 Churn in the ice cream machine.

51 Apple sorbet

Energy	Calories	Fat	Saturated fat	Carbohydrates	Sugar	Protein	Fibre
673 kJ	158 kcal	0.1 g	0.0 g	41.5 g	38.7 g	0.4 g	2.4 g

Cooking

1 Cut the apples into 1-cm pieces and place into lemon juice.
2 Bring the water, sugar and glucose to the boil, then allow to cool.
3 Pour the water over the apples. Freeze overnight. Blitz in a food processor.
4 Pass through a conical strainer, then churn in an ice cream machine.

PROFESSIONAL TIP
For best results, after freezing, process in a Pacojet.

VARIATION
- **Fruits of the forest sorbet**: use a mixture of forest fruits instead of apples.

	8–10 portions
Granny Smith apples, washed and cored	4
Lemon, juice of	1
Water	400 ml
Sugar	200 g
Glucose	50 g

52 Chocolate sorbet

Energy	Calories	Fat	Saturated fat	Carbohydrates	Sugar	Protein	Fibre
544 kJ	129 kcal	3.0 g	1.8 g	25.4 g	24.9 g	1.6 g	6.3 g

Cooking

1 Combine the water, milk, sugar, stabiliser and cocoa powder. Bring to the boil slowly. Simmer for 5 minutes.

2 Add the couverture and allow to cool.

3 Pass and churn.

SERVING SUGGESTION
This sorbet would be a very good accompaniment with a slice of chocolate tart.

	8 portions
Water	400 ml
Skimmed milk	100 ml
Sugar	150 g
Ice-cream stabiliser	40 g
Cocoa powder	30 g
Dark couverture	60 g

53 Peach Melba (*pêche Melba*)

Energy	Calories	Fat	Saturated fat	Carbohydrates	Sugar	Protein	Fibre
607 kJ	145 kcal	2.6g	1.3g	30.5g	30.2g	1.6g	1.3g

	4 portions	10 portions
Peaches	2	5
Vanilla ice cream	125 ml	300 ml
Melba sauce or raspberry coulis	125 ml	300 ml

Cooking

1 Poach the peaches. Allow to cool, then peel, halve and remove the stones.

2 Dress the fruit on a ball of the ice cream in an ice cream coupe or in a tuile basket.

3 Finish with the sauce.

54 Pear belle Hélène (*poire belle Hélène*)

Energy	Calories	Fat	Saturated fat	Carbohydrates	Sugar	Protein	Fibre
673 kJ	158 kcal	0.1 g	0.0 g	41.5 g	38.7 g	0.4 g	2.4 g

Preparing the dish

1 Serve a poached pear on a ball of vanilla ice cream in a coupe.

2 Decorate with whipped cream. Serve with a sauce boat of hot chocolate sauce.

55 Raspberry parfait

Energy	Calories	Fat	Saturated fat	Carbohydrates	Sugar	Protein	Fibre
981 kJ	234 kcal	10.7 g	5.6 g	31.3 g	30.2 g	4.7 g	0.5 g

	6–8 portions
Italian meringue	
Caster sugar	150 g
Glucose	20 g
Water	80 ml
Egg whites	200 g
Additional ingredients	
Egg yolks, pasteurised	80 g
Caster sugar	60 g
Gelatine, soaked	1½ leaves
Raspberry liqueur	10 ml
Lemon juice	10 ml
Raspberry purée	120 g
Whipped cream	150 ml
Sponge	

Cooking

1 Make up the Italian meringue (see page 464).

2 Combine the egg yolks and caster sugar in a stainless steel bowl. Whisk over a bain-marie to make a sabayon.

3 Drain the gelatine and dissolve it in the liqueur and lemon juice.

4 Fold the gelatine mixture into the sabayon, then fold in the raspberry purée.

5 Fold in the Italian meringue, then fold in the whipped cream.

6 Place into prepared moulds lined with sponge and freeze.

7 Once set, remove from the moulds.

SERVING SUGGESTION

In this presentation, the parfait has been lined with sponge only on the base. It is served with a raspberry coulis and fresh and dried raspberry garnishes.

PROFESSIONAL TIP

Remove from the mould by quickly heating the outside of the ring using a blow torch.

310 Paste products

In this unit you will be able to:

1 Understand how to produce paste products, including:
   - understanding different types of pastes and their uses
   - understanding the preparation and processing techniques used in the production of pastes and knowing the types of pastes these techniques are used to produce
   - knowing differences between types of commodities and understanding the implications of those differences for the production of paste products
   - understanding the correct storage procedures to use throughout production and on completion of the final products.

2 Produce paste products, including:
   - understanding how to use and operate equipment
   - being able to apply techniques to produce paste products
   - being able to finish paste products
   - being able to measure and evaluate against quality standards throughout preparation and cooking
   - being able to evaluate products against dish requirements and production standards and recognise any faults.

The key ingredients for pastry work, such as flour, eggs and sugar, are described in Unit 309 Desserts and puddings. Make sure you read and understand this section.

Recipes in this chapter

No.	Recipe
	Pastes
1	Short paste (*pâte à foncer*)
2	Sweet (or sugar) paste (*pâte à sucre*)
3	Choux paste (*pâte à choux*)
4	Puff paste
5	Suet paste
	Pastry goods
	Short paste
6	Flan case
7	Quiche Lorraine (cheese and ham savoury flan)
8	Treacle tart
	Sweet paste
9	Mince pies
10	Egg custard tart
11	Lemon tarte (*tarte au citron*)
12	Bakewell tart
13	Baked chocolate tarte
14	Fruit tart, tartlets and barquettes

No.	Recipe
15	Pear and almond tart
16	French apple flax (*flan aux pommes*)
17	Lemon meringue pie
	Choux paste
18	Chocolate éclairs (*éclairs au chocolat*)
19	Profiteroles and chocolate sauce (*profiteroles au chocolat*)
20	Gâteau Paris-Brest
	Puff paste
21	Cheese straws (*paillettes au fromage*)
22	Sausage rolls
23	Eccles cakes
24	Fruit slice (*bande aux fruits*)
25	Pear jalousie
26	Puff pastry slice (*mille-feuilles*)
27	Gâteau Pithiviers
28	Palmiers
	Suet paste
29	Spotted dick

Understand how to produce paste products

1.1 Examples of paste

Short paste

The shortness of a paste refers to the crisp, light and sometimes crumbly texture of the finished paste. The term 'shortening' describes the effect of the fat when rubbed gently into the flour, breaking down the gluten strands in the flour and producing the short texture qualities. Short pastry is typically used as a lining for savoury and sweet pies, tarts and flans.

Sweet paste

Sweet paste is a short paste that has been sweetened with the addition of sugar and often enriched by the addition of egg. The type of sugar used is normally caster sugar or icing sugar. This helps to achieve a fine, smooth paste. Sweet paste is commonly used to line sweet tarts, tartlets and flans.

Choux paste

Choux paste is made by melting butter in water, then binding this mixture by cooking to a paste with flour, before beating in eggs to produce a fairly thick but slack paste of a 'dropping consistency'. It is then piped into the desired shapes and baked. During the baking process the moisture from the water content helps to produce an air bubble around which the paste bakes to form a light batter-like product.

Choux paste is used to produce products such as profiteroles, éclairs, gâteaux Paris-Brest and the famous French dessert, Croquembouche. It can also be used to produce savoury products – small choux buns filled with a savoury mousse, for example.

Puff paste

Puff paste is a laminated paste. The term '**lamination**' refers to the layers that are produced when making puff paste. To make puff paste, a dough is produced using a strong flour as the dough needs to be elastic and robust enough to incorporate layers of butter without splitting or oozing. The butter is added to the paste in a layer which is multiplied hundreds of times through the process known as '**turning**'.

Each turn multiplies the layers of paste and butter until the desired amount is reached, usually four 'double' or 'book' turns. When the paste is baked, the layers of fat produce steam, resulting in a rising between the layers of dough. This is what causes the rising of puff paste and the development of hundreds of fine, delicate layers in the finished, baked product. Puff paste is used to make sweet and savoury products such as turnovers, pastry cases (bouchées, vol-au-vents), pies, palmiers, fruit bands and mille-feuilles.

Suet paste

Suet is the raw fat of beef or mutton, especially the hard fat found around the loins and kidneys, although vegetable-based alternatives are also available. To make suet paste, the suet is coarsely grated (though often it is purchased ready grated) and mixed with self-raising flour (or plain flour and baking powder) and water to produce a soft, slightly sticky dough. Salt and pepper, as well as other seasonings such as herbs, can be added, particularly for savoury puddings and dumplings.

Suet produces a soft-textured pastry and is used to make dumplings and traditional puddings. It is also integral to the production of haggis, mincemeat, Christmas pudding and a rendered fat called tallow.

Suet paste is used across a variety of boiled, steamed or baked savoury and sweet puddings, such as steak and kidney pudding, spotted dick and jam roly-poly.

Hot water paste

Hot water paste is used to produce a range of savoury pies such as pork pies and game pies as well as terrines and pâtés. It is traditionally associated with the artisan production of what is referred to as hand-raised pies.

Hot water pastry is made by melting fat in heated water before mixing with the flour. The result is a hot, slightly sticky paste that can be used for hand-raising: shaping by hand, sometimes using shaping tools, dishes or bowls as an inner mould. As the crust cools, its shape is retained and it is filled and covered with a crust before baking. Hand-raised hot water paste does not produce a neat and uniform finish associated with other types of paste. This, however, is a sign of a traditional hand-raised pie.

Filo paste

Filo paste is a very thin paste used for making pastry products in Middle Eastern and Greek cuisines, such as baklava. Filo-based pastry products are made by layering many sheets of filo brushed with olive oil or butter; the pastry is then baked. Filo dough is made with flour, water and a small amount of oil (traditionally olive).

Hand-made filo takes time and skill, requiring progressive rolling and stretching to a single thin and very large sheet. A large table or surface area and a long roller are used, with continuous flouring between layers to prevent sticking and/ or tearing. Once the paste is rolled, it is important to keep it moist (for example, with a clean, slightly dampened cloth) to prevent the paste from drying out and cracking.

It is quite rare that filo paste would be produced from scratch in a modern kitchen. Manufactured filo paste is very reliable, economic to use and often of high quality. This type of paste is usually purchased and utilised in this way.

1.2 Techniques for the production of paste products

Adding fat to flour

Fats act as a **shortening agent**. The fat has the effect of shortening the gluten strands, which are easily broken when eaten, making the texture of the product more crumbly. The development of gluten in puff pastry is very important as long strands are needed to trap the expanding gases, and this is what makes the paste rise.

Fat can be added to flour by:
- rubbing in by hand: short pastry
- rubbing in by machine: short pastry
- creaming method by machine or by hand: sweet pastry
- lamination: puff pastry
- boiling: choux pastry.

KEY TERMS

Lamination: the process of alternating layers of dough and butter when making puff pastry, croissants or Danish pastries.

Turning: describes the process of producing layers in laminated pastry. Each time the paste is rolled and folded, it is referred to as a turn.

Shortening agent: a fat used to help shorten the development of gluten strands when making pastry. This helps to make the texture of the product more crumbly.

Glazes

A glaze is sometimes used to give a product a smooth, shiny surface. Examples of glazes used for pastry dishes are as follows:
- A hot clear gel produced from a pectin source obtainable commercially for finishing flans and tartlets; always use while still hot. A cold gel is exactly the same except that it is used cold. The gel keeps a sheen on the goods and keeps out all oxygen, which might otherwise cause discoloration.

- Apricot glaze, produced from apricot jam, which acts in the same way as a gel.
- Eggwash, applied prior to baking, produces a rich glaze during the cooking process.
- Icing sugar dusted on the surface of the product caramelises in the oven, under the grill or heated using a blow torch.
- Fondant gives a rich sugar glaze, which may be flavoured and/or coloured.
- Water icing gives a transparent glaze, which may also be flavoured and/or coloured.

Key processes and techniques when producing pastes

- **Rubbing in** – a technique where flour is rubbed into a fat to make products such as short pastry and crumbles. Using the fingertips, flour and butter are rubbed gently together until the mixture resembles fine breadcrumbs.
- Creaming – the initial combining of sugar and cream together using a wooden spoon or electric mixer until a smooth mixture is formed. This is often used in the production of sweet or sugar pastry.
- Folding – an example is folding puff pastry to create its layers.
- Rolling – roll the pastry on a lightly floured surface; turn the pastry to prevent it sticking. Keep the rolling pin lightly floured and free from the pastry. Always roll with care, treating the pastry lightly – never apply too much pressure.
- Resting or relaxing – keeping pastry covered with a damp cloth, cling film or plastic to prevent a skin forming on the surface. Relaxing allows the pastry to lose some of its resistance to rolling.
- Beating – an example of beating is provided in the production of choux paste. Once the butter and water have boiled, flour is added and this mix is beaten into a paste. Once cooled, beaten eggs are gradually beaten into the mix until a paste of 'dropping consistency' is achieved.
- Boiling or melting – this method is unique to the production of choux paste, where the butter is initially melted in boiling water before being made into a paste with the addition of flour and then eggs.

KEY TERM

Rubbing in: a technique where flour is rubbed into a fat to make products such as short pastry and crumbles. Using the fingertips, flour and butter are rubbed gently together until the mixture resembles fine breadcrumbs.

Additional techniques used in the production of pastes

- Kneading – using your hands to work dough or puff pastry in the first stage of making.
- Blending – mixing all the ingredients carefully by weight.
- Cutting – always cut with a sharp, clean, damp knife:
  - When using cutters, always flour them before use by dipping in flour. This will give a sharp, neat cut.
  - When using a lattice cutter, use only on firm pastry; if the pastry is too soft, you will have difficulty lifting the lattice.
  - Always apply even pressure when using a rolling pin.
- Shaping – this refers to producing flans, tartlets, barquettes and other such goods with pastry. Shaping also refers to crimping with the back of a small knife using the thumb technique.
- Docking – this is piercing raw pastry with small holes to prevent it from rising during baking, as when cooking tartlets blind.

1.3 Commodities for paste products

The commodities you will need to know about for paste products are flour, sugar and salt – see Unit 309 Desserts and puddings for more information on these.

1.4 Storage

- Store all goods according to the Food Safety and Hygiene Regulations 2013/Food Safety Temperature Control Regulation 1995 and General Food Regulations (2004).
- Always make sure that storage containers are kept clean and returned ready for re-use. On their return they should be hygienically washed and stored.
- Freshly made, raw paste should be wrapped tightly in secure film or placed in an air-tight, sealed bag. It should then be clearly labelled and dated before storing in a refrigerator or freezer.
- Finished paste products can be refrigerated to maintain food safety. However, pastry does not tend to maintain its quality in refrigerated conditions. The moist atmosphere leads to pastes softening, losing their crisp and short properties. Any additional ingredients also have to be

considered. Creams can lose their viscosity and can retract from the pastry lining and prepared fruits can weep, losing their structure.
- Some cooked pastry products are suitable for freezing. For example, unfilled, blind-baked pastry cases freeze well for use at a later stage. Other completed products need to be analysed as to their suitability for freezing, based on the additional ingredients used and their suitability.
- Humidity refers to the moisture in the atmosphere or air. Cooked pastry items will soften in moist conditions, so it is important that they are kept in dry, cool conditions. Storing pastry in refrigerators should be minimised due to the humid conditions.
- Although storage of cooked pastry items in refrigerators should be minimised, it may be necessary due to the other ingredients that form part of the pastry item, such as cream.
- As with other food items, pastry products that are stored should be date labelled and stored in appropriate, hygienic conditions. Pastry items should be positioned separately from other foods with strong aromas.
- Pastry items for long-term storage, by freezing for example, should be rotated to ensure that older products are utilised before newer ones.

Allergies

Although it is essential to clearly list all potential allergens when making paste products, the allergens that are most like to be used in their production include:

- gluten – flours and any products made from wheat, rye, barley and oats
- nuts – such as ground hazelnuts and almonds. These can be added to flavour pastes such as sablé
- eggs – used in the production of sweet and choux paste.

Beyond the basic preparation of pastes, attention is also required with regard to the additional ingredients that are used to complete pastry products. Tarts are often filled with creams, produced with milk and/or cream (lactose). Other fillings may include nuts, such as frangipane, so it is vitally important to assess any of the other potential allergens that are incorporated into pastry products as well as the paste itself.

Produce paste products

2.1 Specialist equipment

In addition to the products described in Unit 309 Desserts and puddings, the items described below are used regularly in the production of pastry dishes.

Flour dredger

A metal or plastic hand-held container with a perforated top for sifting or dredging flour, icing sugar or cocoa onto food or a surface. Flour is shaken or dredged onto a board or surface before rolling out pastry, for example.

Double boiler

A double boiler is basically two pans, one being placed on top of the other. One is slightly larger than the other. The large pan resembles a regular saucepan with a smaller, shallower pan nestled inside. A double boiler is used for cooking delicate ingredients that have a tendency to seize or separate over direct heat, such as tempering chocolate or whisking an egg-based sabayon. They can also be use when steaming puddings.

2.2 Producing paste products

See the recipe section at the end of the unit for details of how to produce paste products.

2.3 Finishing paste products

It is essential that all products are finished according to the recipe requirements. Finishing and presentation is often a key stage in the process, as failure at this point can affect sales. The way goods are presented is an important part of the sales technique. Each product of the same type must be of the same shape, size, colour and finish. The decoration should be attractive, delicate and in keeping with the product range. All piping should be neat, clean and tidy.

Fillings, glazes, cream and icings

Many different fillings are used in pastry products, including include crème pâtissière, frangipane and fresh fruit. Cream and butter cream, preserves and jam can also be used.

Finishing and decorating techniques

Some methods of finishing and presentation are as follows:
- Dusting – sprinkling icing sugar on a product using a fine sugar dredger or sieve.
- Piping – using fresh cream, chocolate or fondant.
- Filling – with fruit, cream, pastry cream, etc. Avoid overfilling as this can give the product a clumsy appearance.
- Icing – some paste products, such as a Bakewell tart, are glazed using an icing, such as water icing in this example.
- Glazing – a glaze helps to add a shine to some pastry products and brightens foods such as the fruit on a fruit tart. Tart glazes can be purchased commercially but can be made by heating some slightly slaked jam. This is then strained, if necessary, and brushed onto products while the glaze is still hot. Apricot jam is regularly used for this purpose due to its versatility.

2.4 Evaluation

When evaluating pastry dishes, the following should be considered:
- Portion size – is the portion size accurate in terms of the planned portion? Customers will expect value for money and consistency in the size of one portion compared to another. If the portion size is too small, customers may feel disappointed and complain. If a portion size is too big, this may lead to wastage and it will certainly contribute to unnecessary additional costs as well as making it more challenging to meet profit targets.
- Colour – the colour of baked pastry will reflect how it has been baked. Ideally, most pastries will have a light, golden-brown colour when baked correctly. However, it is important that the colours of additional ingredients including fillings, fruits, glazes, sauces and any other additional ingredients are considered and well balanced to enhance the presentation of the product.
- Taste – are the tastes and flavours as they were planned? Does the pastry item taste fresh and well balanced? Have the ingredients been mixed and balanced appropriately?
- Texture – is the texture as desired? Is the texture light and flakey, such as in puff pastry when making a mille-feuille, or short and crisp, as when making sweet pastry for a fruit tartlet?

- Consistency – is the consistency correct, both in terms of the physical consistency (as in the light, aired texture of a choux pastry), but also in the general consistency of one product's production compared to the next?
- Harmony – this refers to the way that the various aspects of pastry items come together to complement each other to produce a harmonious result. Classic texture components such as a crisp item served with a smooth, creamy item or a sweet flavour offset by something sharp can provide excellent contrasts in dishes, adding to their overall harmony.
- Presentation – pastry chefs/patissuers have an excellent opportunity to impress their guests and customers in the presentation of their dishes and products. Creativity and food styling are almost endless for the pastry chef. Clean, sharp, creative and consistent presentation is an

area that every pastry chef should think about and plan for very carefully. Pastry items should deliver the various textures they provide, combined with fillings, fruits and other ingredients, to ensure enjoyment and to meet the expectations of the customer.
- Temperature – this is important when producing pastry dishes. Pastries can be served cold, such as tarts with cream, or hot, such as an apple pie. It is important that foods are served at the correct temperature to maximise the customer's enjoyment of the food and also to promote good food safety practices.
- Aroma – pastry items are often enjoyed as luxurious items or consumed as treats. Aromas should be comforting and welcoming. They may be lightly fragrant and should reflect the ingredients and processes that have led to the production of the particular pastry item.

TEST YOURSELF

1 What is the ratio of fat to flour for:
   a) short pastry
   b) puff pastry
   c) sugar pastry?

2 How is the fat added to the flour in the production of choux pastry?

3 What type of fat is required for the production of suet paste?

4 What is meant by the term 'lamination'?

5 What is the filling for a classical gâteau Pithiviers?

6 Provide **five** examples of products that can be produced using puff pastry.

7 Name **one** pastry product, eaten as a dessert, which would be unsuitable for a vegetarian customer.

8 Describe **three** fillings that can be used in the production of sweet tarts.

9 Other than éclairs and profiteroles, name **three** products that are made using choux paste.

10 What quality points indicate a well-produced lemon tart?

11 Describe the finishing stages when producing mille-feuilles.

12 Describe **three** considerations when refrigerating a freshly baked strawberry tart, filled with crème patissière.

1   Short paste (*pâte à foncer*)

Short pastry is used in fruit pies, Cornish pasties and so on.

Energy	Cals	Fat	Saturated fat	Carbohydrates	Sugar	Protein	Fibre	Sodium
7152 kJ	1708 kcal	95.0 g	32.3 g	185.0 g	2.4 g	0.0 g	10.0 g	1,273.0 g

* Using margarine

From left to right: short paste, rough puff paste (Recipe 4) and sweet paste (Recipe 2)

	400 g	850 g
Flour (soft)	250 g	500 g
Salt	Pinch	Large pinch
Butter or block/cake margarine	125 g	250 g
Water	40–50 ml	80–100 ml

Mise en place

1   Cutting the butter or fat into small pieces can help with the rubbing in process.

Preparing the dish

1   Sieve the flour and salt.

2   Rub in the fat to achieve a sandy texture.

3   Make a well in the centre.

4   Add sufficient water to make a fairly firm paste.

5   Handle as little and as lightly as possible. Refrigerate until firm before rolling.

PROFESSIONAL TIP

The amount of water used varies according to:
● the type of flour (a very fine soft flour is more absorbent)
● the degree of heat (for example, prolonged contact with hot hands, and warm weather conditions).

Different fats have different shortening properties. For example, paste made with a high ratio of butter to other fat will be harder to handle.

VARIATION

● **Wholemeal short pastry**: use wholemeal flour in place of half to three-quarters of the white flour.
● Short pastry for sweet dishes such as baked jam roll may be made with self-raising flour.
● Lard can be used in place of some or all of the fat (the butter or cake margarine). Lard has excellent shortening properties and would lend itself, in terms of flavour, to savoury products, particularly meat-based ones. However, many people view lard as an unhealthy product as it is very high in saturated fat. It is also unsuitable for anyone following a vegan or vegetarian diet as it is an animal product.

FAULTS

Possible reasons for faults in short pastry are detailed below.

Hard:
● too much water
● too little fat
● fat rubbed in insufficiently
● too much handling and rolling
● over-baking.

Soft–crumbly:
● too little water
● too much fat.

Blistered:
● too little water
● water added unevenly
● fat not rubbed in evenly.

Soggy:
● too much water
● too cool an oven
● baked for insufficient time.

Shrunken:
● too much handling and rolling
● pastry stretched while handling.

From left to right: correct, blistered and shrunken short paste

515

2 Sugar (or sweet) paste (*pâte à sucre*)

Sugar pastry is used for products such as flans, fruit tarts and tartlets.

Energy	Cals	Fat	Saturated fat	Carbohydrates	Sugar	Protein	Fibre	Sodium
7515 kJ	1794 kcal	99.0 g	33.4 g	198.0 g	52.0 g	51.0 g	8.0 g	1,352.0 g

* Using margarine

	400 g	1 kg
Sugar	50 g	125 g
Butter or block/cake margarine	125 g	300 g
Eggs	1	2–3
Flour (soft)	200 g	500 g
Salt	Pinch	Large pinch

Mise en place

1 Cutting the butter or fat into small pieces can help with the rubbing in process.

Preparing the dish

Method 1 – sweet lining paste (rubbing in):

1 Sieve the flour and salt. Lightly rub in the margarine or butter to achieve a sandy texture.

2 Mix the sugar and egg until dissolved.

3 Make a well in the centre of the flour. Add the sugar and beaten egg.

4 Gradually incorporate the flour and margarine or butter and lightly mix to a smooth paste. Allow to rest before using.

Method 2 – traditional French sugar paste (creaming):

1 Taking care not to over-soften, cream the butter and sugar.

2 Add the beaten egg gradually and mix for a few seconds.

3 Gradually incorporate the sieved flour and salt. Mix lightly until smooth.

4 Allow to rest in a cool place before using.

> **PROFESSIONAL TIP**
>
> The higher the percentage of butter, the shorter and richer the paste will become. However, as the butter will soften and melt during handling, the paste will become softer and more difficult to work with. Therefore chilling and light, quick handling are required when using a sweet paste with a high butter content.
>
> This also applies to the working environment. For example, in a particularly warm kitchen, it will be more difficult to work with a paste of this structure than in a cooler kitchen.
>
> The butter in this recipe could be reduced from 125 g to 100 g to make handling easier.

Measure out the sugar and cut the butter into small chunks.

Cream the butter and sugar together.

Add the beaten egg in stages, thoroughly mixing each time.

Incorporate the flour and salt.

Press into a tray and leave to chill.

The paste will need to be rolled out before use in any recipe.

> **FAULTS**
>
> Overworking sweet paste will make it crumbly and difficult to use. It is also likely that the paste will shrink or retract during the baking process as the gluten will be developed

3 Choux paste (*pâte à choux*)

Choux paste is used to make products such as éclairs, profiteroles and gâteau Paris-Brest.

Energy	Cals	Fat	Saturated fat	Carbohydrates	Sugar	Protein	Fibre	Sodium
6377 kJ	1527 kcal	97.0 g	31.8 g	111.0 g	3.5 g	2.0 g	5.0 g	1,481.0 g

* Using margarine and 5 medium eggs

	750 g	1.5 kg
Water	250 ml	500 ml
Sugar	Pinch	Large pinch
Salt	Pinch	Large pinch
Butter or block/cake margarine	100 g	200 g
Flour (strong)	150 g	300 g
Eggs	4–5	8–10

Cooking

1 Beat the eggs in a measuring jug.
2 Bring the water, sugar, salt and fat to the boil in a saucepan. Remove from the heat.
3 Add the sieved flour and mix in with a wooden spoon.
4 Return to a moderate heat and stir continuously until the mixture leaves the sides of the pan (this is known as a panada).
5 Remove from the heat and allow to cool.
6 Gradually add the beaten eggs, beating well. Do not add all the eggs at once – check the consistency as you go. The mixture should just flow back when moved in one direction (it may not take all the egg).

Cut the butter into cubes and then melt them in the water.

Add the flour.

When the panada is ready, it will start to come away from the sides.

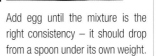
Add egg until the mixture is the right consistency – it should drop from a spoon under its own weight.

Pipe the paste into the shape required – these rings can be used for Paris-Brest (recipe 20).

A selection of shapes in raw choux paste.

> ### FAULTS
>
> Greasy and heavy paste:
> * the basic mixture was over-cooked.
>
> Soft paste, not aerated:
> * flour insufficiently cooked
> * eggs insufficiently beaten in the mixture
> * oven too cool
> * under-baked.
>
> Split or separated mixture:
> * egg added too quickly.
>
> When baking choux paste items, it is important not to have fluctuations in the baking temperature. Opening of oven doors during baking can cause choux paste products not to rise.
>
>
>
> The choux buns on the left are light and well risen; those on the right are poorly aerated.

> ### VARIATION
> * 50 per cent, 70 per cent or 100 per cent wholemeal flour may be used to make choux paste.

> ### PROFESSIONAL TIP
> If piping into circles, these can be pre-marked on the baking sheet to provide a guide. This will help to provide accuracy and consistency

4 Puff paste

Energy	Cals	Fat	Saturated fat	Carbohydrates	Sugar	Protein	Fibre	Sodium
23687 kJ	5692 kcal	416.0 g	142.0 g	408.0 g	7.2 g	0.0 g	18.5 g	8,660.0 g

* Using margarine and lemon juice

	Approximately 1.5 kg
Flour (strong)	560 g
Salt	12 g
Butter or pastry margarine	60 g
Water, ice-cold	325 ml
Butter or pastry margarine	500 g
Lemon juice or ascorbic or tartaric acid	a few drops

Mise en place

1 Ensure that the butter used for laminating is the same temperature and consistency as the paste. If they are the same, they will roll well together when folding during the lamination process.

Preparing the dish

1 Sieve the flour and salt. Rub in the 60 g of the butter or pastry margarine.

2 Make a well in the centre.

3 Add the water and lemon juice or acid (to make the gluten more elastic) and knead well into a smooth dough in the shape of a ball.

4 Relax the dough in a cool place for 30 minutes.

5 Cut a cross halfway through the dough and pull out the corners to form a star shape.

6 Roll out the points of the star square, leaving the centre thick.

7 Knead the remaining butter or pastry margarine to the same texture as the dough. This is most important – if the fat is too soft it will melt and ooze out, if too hard it will break through the paste when being rolled.

8 Place the butter or margarine on the centre square, which is four times thicker than the flaps.

9 Fold over the flaps.

10 Roll out to 30 × 15 cm, cover with a cloth or plastic and rest for 5–10 minutes in a cool place.

11 Roll out to 60 × 20 cm, fold both the ends to the centre, fold in half again to form a square. This is one double turn.

12 Allow to rest in a cool place for 20 minutes.

13 Half-turn the paste to the right or the left.

14 Give one more double turn; allow to rest for 20 minutes.

15 Give two more double turns, allowing to rest between each.

16 Allow to rest before using.

Rub one-quarter of the butter into the flour.

Mix in the water and lemon juice.

Knead into a smooth dough.

Roll the dough out into a cross shape.

Knead the remaining butter in a plastic bag, then place it on the centre of the dough.

Fold over each flap.

The folded dough forms a parcel.

Roll out in a rectangle, then fold the ends to the middle.

When resting the turned and folded dough, leave a finger mark to show the number of turns completed.

PROFESSIONAL TIP

Care must be taken when rolling out the paste to keep the ends and sides square.

The lightness of the puff pastry is mainly due to the air that is trapped when folding the pastry during preparation.

The addition of lemon juice (acid) helps to strengthen the gluten in the flour, thus helping to make a stronger dough so that there is less likelihood of the fat oozing out; 3 g (7.5 g for 10 portions) ascorbic or tartaric acid may be used in place of lemon juice.

The rise is caused by the fat separating layers of paste and air during rolling. When heat is applied by the oven, steam is produced causing the layers to rise and give the characteristic flaky formation.

Do not use excess pressure when using the rolling pin. Let the rolling pin do the work. Too much pressure is likely to indent the paste and cause the fat to ooze from the paste

FAULTS

The pastry on the left is unevenly laminated. Possible reasons for this are:
- paste was not folded equally
- paste was rolled too thinly
- re-used scraps of paste were used instead of making up a virgin paste.

5 Suet paste

Suet paste is used for steamed fruit puddings, steamed jam rolls, steamed meat puddings and dumplings.

Energy	Cals	Fat	Saturated fat	Carbohydrates	Sugar	Protein	Fibre	Sodium
6473 kJ	1546 kcal	89.0 g	51.0 g	162.0 g	1.3 g	0.2 g	8.5 g	1,577.0 g

PROFESSIONAL TIP

Self-raising flour already contains baking powder so this element could be reduced in the recipe if using self-raising flour.

Vegetarian suet is also available to enable products to be meat-free.

FAULTS

Heavy and soggy paste:
- cooking temperature may have been too low.

Tough paste:
- handled too much or over-cooked.

VARIATION

Vegetable suet can be used in place of beef suet.

	400 g	1 kg
Flour (soft) or self-raising flour	200 g	500 g
Baking powder	10 g	25 g
Salt	Pinch	Large pinch
Prepared beef or vegetarian suet	100 g	250 g
Water	125 ml	300 ml

Preparing the dish

1 Sieve the flour, baking powder and salt.
2 Mix in the suet. Make a well. Add the water.
3 Mix lightly to a fairly stiff paste.

6 Flan case

Energy	Cals	Fat	Saturated fat	Carbohydrates	Sugar	Protein	Fibre

Video: Lining a flan
http://bit.ly/2oMJV1b

Mise en place

1 Allow 25 g flour per portion and prepare sugar pastry as per recipe 1.

Preparing the dish

1 Grease the flan ring and baking sheet.
2 Roll out the pastry 2 cm larger than the flan ring. The pastry may be rolled between greaseproof or silicone paper.
3 Place the flan ring on the baking sheet.
4 Carefully place the pastry on the flan ring, by rolling it loosely over the rolling pin, picking up and unrolling it over the flan ring.
5 Press the pastry into shape without stretching it, being careful to exclude any air.
6 Allow a ½ cm ridge of pastry on top of the flan ring.
7 Cut off the surplus paste by rolling the rolling pin firmly across the top of the flan ring.
8 Mould the edge with thumb and forefinger. Decorate (a) with pastry tweezers or (b) with thumbs and forefingers, squeezing the pastry neatly to form a corrugated pattern.

> **VARIATION**
> - Some chefs prefer to have a straight cut at the top of the flan case instead of a crimpled edge. This can be achieved by gently pressing and raising the paste above the top of the ring. Using a small, sharp knife, the excess paste can be cut away using the lip of the ring as a guide.

> **FAULTS**
> If the paste is overworked during preparation and/or rolling, it is likely that it will become difficult to handle and may shrink/retract during the baking process.

7 Quiche Lorraine (cheese and ham savoury flan)

Energy	Cals	Fat	Saturated fat	Carbohydrates	Sugar	Protein	Fibre
2955 kJ	704 kcal	48.4 g	22.6 g	38.1 g	6.5 g	31.6 g	1.8 g

	4 portions	10 portions
Short paste	100 g	250 g
Ham, chopped	75 g	150 g
Cheese, grated	50 g	125g
Egg	1	2
Milk	125 ml	300 ml
Cayenne	1-2g	3g
Sea salt (e.g. Maldon)	2g	5g

Mise en place

1 Pre-heat the oven and ensure all equipment is prepared and in place (ring, baking tray, ladles). Ensure ham and cheese are chopped and grated.

Cooking

1 Lightly grease an appropriately sized flan ring or barquette, or tartlet moulds if making individual portions. Line thinly with pastry.

2 Prick the bottom of the paste two or three times with a fork to dock.

3 Cook in a hot oven at 200 °C for 3–4 minutes or until the pastry is lightly set. Reduce the oven temperature to 160 °C.

4 Remove from the oven; press the pastry down if it has tended to rise.

5 Add the chopped ham and grated cheese.

6 Mix the egg, milk, salt and cayenne thoroughly. Strain over the ham and cheese.

7 Return to the oven at 160°C and bake gently for approximately 20 minutes or until nicely browned and the egg custard mix has set.

VARIATION

- The filling can be varied by using lightly fried lardons of bacon (in place of the ham), chopped cooked onions and chopped parsley.
- A variety of savoury flans can be made by using imagination and experimenting with different combinations (for example, stilton and onion; salmon and dill; sliced sausage and tomato).

FAULTS

If the egg is undercooked, it will not set and it will be runny when cut. If the egg overheats, it will scramble and appear split. Therefore, it is vital to pay close attention to the cooking of egg custard-based products.

---

8 Treacle tart

Energy	Calories	Fat	Saturated fat	Carbohydrates	Sugar	Protein	Fibre
1100 kJ	262 kcal	10.7 g	5.8 g	41.1 g	20.3 g	2.8 g	0.8 g

Mise en place

1 Lightly grease an appropriately sized flan ring, or barquette or tartlet moulds if making individual portions.

Cooking

1 Line with pastry.

2 Warm the treacle, water and lemon juice; add the crumbs.

3 Place into the pastry ring and bake at 170 °C for about 20 minutes.

VARIATION

- This tart can also be made in a shallow flan ring. Any pastry debris can be rolled and cut into ½-cm strips and used to decorate the top of the tart before baking.
- Try sprinkling with vanilla salt as a garnish.

---

	4 portions	10 portions
Short paste	125 g	300 g
Treacle	100 g	250 g
Water	1 tbsp	2½ tbsp
Lemon juice	3–4 drops	8–10 drops
Fresh white bread or cake crumbs	15 g	50 g

9 Mince pies

Energy	Calories	Fat	Saturated fat	Carbohydrates	Sugar	Protein	Fibre
2009 kJ	479 kcal	23.4 g	0.0 g	66.6 g	32.0 g	4.6 g	2.8 g

	12 small pies
Sweet paste	200 g
Mincemeat (see below)	200 g
Egg wash	1 egg
Icing sugar	

Mise en place

1 Ensure you have cutters (fluted) ready as well as two pastry brushes – one in water for sealing the edges of the paste and the other in egg-wash for brushing the surface of the mince pies before baking. You will also need a shaker with icing sugar or a small sieve to dust once the pies are baked.

Cooking

1 Roll out the pastry 3 mm thick.
2 Cut half the pastry into fluted rounds 6 cm in diameter.
3 Place on a greased, dampened baking sheet.
4 Moisten the edges. Place a little mincemeat in the centre of each.
5 Cut the remainder of the pastry into fluted rounds, 8 cm in diameter.
6 Cover the mincemeat with pastry and seal the edges. Brush with eggwash.

7 Bake at 210 °C for approximately 20 minutes.
8 Sprinkle with icing sugar and serve warm.

> **SERVING SUGGESTION**
> Accompany with a suitable sauce, such as custard, brandy sauce or brandy cream.
>
> Serve warm with brandy butter.

Mincemeat:

Suet, chopped	100 g
Mixed peel, chopped	100 g
Currants	100 g
Sultanas	100 g
Raisins	100 g
Apples, chopped	100 g
Barbados sugar	100 g
Mixed spice	5 g
Lemon, grated zest and juice of	1
Orange, grated zest and juice of	1
Rum	60 ml
Brandy	60 ml

Preparing the dish

1 Mix the ingredients together.
2 Seal in jars and use as required.

> **VARIATION**
> • Short or puff pastry may also be used. Various toppings can also be added, such as crumble mixture or flaked almonds and an apricot glaze.

> **Healthy Eating Tip**
> • Vegetable suet can be used in place of beef suet.

10 Egg custard tart

Energy	Calories	Fat	Saturated fat	Carbohydrates	Sugar	Protein	Fibre
1998 kJ	482 kcal	39 g	22 g	27 g	15.8 g	6.4 g	0.6 g

	8 portions
Sweet paste	250 g
Egg yolks	9
Caster sugar	75 g
Whipping cream, gently warmed and infused with 2 sticks of cinnamon	500 ml
Nutmeg, freshly grated	

Mise en place

1 Roll out the pastry on a lightly floured surface, to 2 mm thickness. Use it to line a 20 cm flan ring, placed on a baking sheet.

Cooking

1 Line the pastry with food-safe cling film or greaseproof paper and fill with baking beans. Bake blind in a preheated oven at 190 °C for about 10 minutes or until the pastry is turning golden brown. Remove the paper and beans, and allow to cool. Turn the oven down to 130 °C.

2 To make the custard filling, whisk together the egg yolks and sugar. Add the cream and mix well.

3 Pass the mixture through a fine sieve into a saucepan. Heat to 37 °C.

4 Fill the pastry case with the custard to ½ cm below the top. Place it carefully into the middle of the oven and bake for 30–40 minutes or until the custard appears to be set but not too firm.

5 Remove from the oven and cover liberally with grated nutmeg. Allow to cool to room temperature.

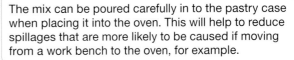

PROFESSIONAL TIP

The mix can be poured carefully in to the pastry case when placing it into the oven. This will help to reduce spillages that are more likely to be caused if moving from a work bench to the oven, for example.

11 Lemon tart (*tarte au citron*)

Energy	Calories	Fat	Saturated fat	Carbohydrates	Sugar	Protein	Fibre
1907 kJ	456 kcal	25 g	13.4 g	53 g	44 g	8.5 g	

	8 portions
Sweet paste	200 g
Lemons	Juice of 3, zest from 4
Eggs	8
Caster sugar	300 g
Double cream	250 ml

Cooking

1 Prepare 200 g of sweet paste, adding the zest of one lemon to the sugar.

2 Line a 20 cm flan ring with the paste.

3 Bake blind at 190 °C for approximately 15 minutes.

4 Prepare the filling: mix the eggs and sugar together until smooth, add the cream, lemon juice and zest. Whisk well.

5 Seal the pastry, so that the filling will not leak out. Pour the filling into the flan case and bake for 30–40 minutes at 150 °C until just set. (Take care when almost cooked as overcooking will cause the filling to rise and possibly crack.)

6 Remove from the oven and allow to cool.

7 Dust with icing sugar and glaze under the grill or with a blowtorch. Portion and serve.

NOTE

The mixture will fill one 16 × 4 cm or two 16 × 2 cm flan rings. If using two flan rings, double the amount of pastry and reduce the baking time when the filling is added.

PROFESSIONAL TIP

If possible, make the filling one day in advance. The flavour will develop as the mixture matures.

VARIATION

- Limes may be used in place of lemons. If so, use the zest and juice of 5 limes or use a mixture of lemons and limes.

SERVING SUGGESTION

Serve with a sharp, sweet accompaniment such as raspberry sorbet, coulis and/or fresh raspberries.

12 Bakewell tart

Energy	Calories	Fat	Saturated fat	Carbohydrates	Sugar	Protein	Fibre
1308 kJ	313 kcal	18.5 g	8.6 g	34 g	24 g	5 g	tbc

	8 portions
Sugar paste	200 g
Raspberry jam	50 g
Egg wash	1 egg
Apricot glaze	50 g
Icing sugar	35 g
Frangipane (almond cream)	250g

Mise en place

1 Have the apricot glaze and water icing ready, with clean brushes for each.

Cooking

1 Line a 20 cm flan ring using three-quarters of the paste, 2 mm thick.

2 Pierce the bottom with a fork.

3 Spread with jam and the frangipane.

4 Roll the remaining paste, cut into neat 0.5-cm strips and arrange neatly criss-crossed (lattice) on the frangipane; trim off surplus paste. Brush with egg wash.

5 Bake in a moderately hot oven at 200–210 °C for 30–40 minutes. Brush with hot apricot glaze.

6 When cooled brush over with very thin water icing.

> **VARIATION**
> - In place of using lattice strips of pastry, lightly cover the surface with water icing. Feather the icing using a small amount coloured (usually red or brown) icing by piping thin lines of the coloured icing across the white icing. Cut through (feather) these lines intermittently using the tip of a sharp small knife, a cocktail stick or similar. An example of feathering in this way can be seen in the recipe for mille-feuilles (recipe 26).

13 Baked chocolate tart

Energy	Calories	Fat	Saturated fat	Carbohydrates	Sugar	Protein	Fibre
2414 kcal	580 kcal	42.4 g	24.8 g	44.9 g	35 g	7.4 g	1.8 g

	8 portions
Sweet paste	200 g
Filling	
Eggs	3
Egg yolks	3
Caster sugar	60 g
Butter	200 g
Chocolate pistoles (55% cocoa, unsweetened)	300 g

Cooking

1 Roll out the sweet paste and line a 20 cm flan ring. Bake the flan case blind.

2 For the filling, whisk the eggs, yolks and sugar together to make a sabayon.

3 Bring the butter to the boil, remove and mix in the chocolate pistoles until they are all melted.

4 Once the sabayon is light and fluffy, fold in the chocolate and butter mixture, mixing very carefully so as not to beat out the air.

5 Pour into the cooked flan case and place in a deck oven at 150°C until the edge crusts (approximately 5 minutes). Chill to set.

6 Once set, remove from fridge and then serve at room temperature.

Add chocolate pistoles to the melted butter.

Fold in the chocolate.

Pour the mixture into the flan case.

SERVING SUGGESTION

Serve a slice of the chocolate tart with a quenelle of chocolate sorbet. Alternatively, a sourer sorbet such as a yoghurt-based or raspberry sorbet would complement this tart.

FAULTS

If the chocolate and butter mix is folded in too harshly, this will result in a loss of aeration in the sabayon, resulting in a densely textured, low volume product.

NOTE

Pistoles or pellets are one form in which chocolate is sold. They are very versatile and easy to use for melting purposes due to their uniform size.

14 Fruit tart, tartlets and barquettes

Energy	Calories	Fat	Saturated fat	Carbohydrates	Sugar	Protein	Fibre
1907 kJ	454 kcal	18.7 g	10.7 g	68 g	39 g	6.8 g	3.6 g

	4 portions
Sweet paste	250g
Fruit (e.g. strawberries, raspberries, grapes, blueberries)	500g
Pastry cream	
Glaze	5 tbsp

Mise en place

1 Do not prepare fruit, particularly soft and ripe fruits, too far in advance as they are likely to lose shape, structure and possibly weep.

2 Have plenty of glaze prepared and a clean brush to apply it.

Cooking

Fruit tart:

1 Line a flan ring with paste and cook blind at 190°C. Allow to cool.

2 Pick and wash the fruit, then drain well. Wash and slice or segment any larger fruit being used.

3 Pipe pastry cream into the flan case, filling it to the rim. Dress the fruit neatly over the top.

4 Coat with the glaze. Use a glaze suitable for the fruit chosen, for example, with a strawberry tart, use a red glaze.

Tartlets:

1 Roll out pastry 3 mm thick.

2 Cut out rounds with a fluted cutter and place them neatly in greased tartlet moulds. If soft fruit (such as strawberries or raspberries) is being used, the pastry should be cooked blind first.

3 After baking and filling (or filling and baking) with pastry cream, dress neatly with fruit and glaze the top.

PROFESSIONAL TIP

Brush the inside of the pastry case with melted couverture before filling. This forms a barrier between the pastry and the moisture in the filling.

FAULTS

Although this strawberry tart may appear to be fine at first glance, the husks of the strawberries are visible. It would be better to present the strawberries with their tops pointing upwards or sliced and overlapping.

There is also quite a wide gap between the rows of strawberries, showing the crème pâtissière underneath. This should be avoided.

The second photo shows the importance of ensuring that fillings are prepared and/or cooked properly. In this case, the crème pâtissière has not been cooked sufficiently or prepared accurately as the filling is not structured sufficiently to support the fruit once the tart has been cut.

15 Pear and almond tart

Energy	Calories	Fat	Saturated fat	Carbohydrates	Sugar	Protein	Fibre
1500 kJ	359 kcal	23 g	10.3 g	34 g	24 g	5.7 g	3.6 g

	8 portions
Sweet paste	200 g
Apricot jam	25 g
Almond cream	350 g
Poached pears	4
Apricot glaze	
Flaked almonds	
Icing sugar	

Mise en place

1 Ensure the pears are dried well to avoid bleeding or thinning of the frangipane.

2 Have a piping bag with a fairly wide open nozzle for piping the frangipane.

3 Prepare enough apricot glaze to cover the tart and a clean brush ready for glazing.

Cooking

1 Line a buttered 20 cm flan ring with sweet paste. Trim and dock.

2 Using the back of a spoon, spread a little apricot jam over the base.

3 Pipe in almond cream until the flan case is two-thirds full.

4 Dry the poached pears. Cut them in half and remove the cores and string.

5 Score across the pears and arrange on top of the flan.

6 Bake in the oven at 200 °C for 25–30 minutes.

7 Allow to cool, then brush with apricot glaze.

8 Sprinkle flaked almonds around the edge and dust with icing sugar.

16 French apple flan (*flan aux pommes*)

Energy	Calories	Fat	Saturated fat	Carbohydrates	Sugar	Protein	Fibre
1428 kJ	340 kcal	13.8 g	5.8 g	53.8 g	36 g	3.5 g	2.9 g

	4 portions	10 portions
Sweet paste	100 g	250 g
Pastry cream (*crème patissiere*) (see page 460)	250 ml	625 ml
Cooking apples	400 g	1 kg
Sugar	50 g	125 g
Apricot glaze	2 tbsp	6 tbsp

Mise en place

1 Do not slice the apples too far in advance of them being required. This will cause discolouration.

2 Have a piping bag ready with a fairly wide, plain nozzle for piping the pastry cream (*crème patisserie*) into the base of the lined flan ring.

Cooking

1 Line a flan ring with sugar paste. Pierce the bottom several times with a fork.

2 Pipe a layer of pastry cream into the bottom of the flan.

3 Peel, quarter and wash the selected apple.

4 Cut into neat thin slices and lay carefully on the pastry cream, overlapping each slice. Ensure that each slice points to the centre of the flan then no difficulty should be encountered in joining up the pattern neatly.

5 Sprinkle a little sugar on the apple slices and bake the flan at 200–220 °C for 30–40 minutes.

6 When the flan is almost cooked, remove the flan ring carefully, return to the oven to complete the cooking. Mask with hot apricot glaze or flan jelly.

Pipe the filling neatly into the flan case.

Slice the apple very thinly for decoration.

Arrange the apple slices on top of the flan.

SERVING SUGGESTION

Apples and vanilla complement each other very well. On this basis, try serving a slice with a creamy crème anglaise.

17 Lemon meringue pie

Energy	Calories	Fat	Saturated fat	Carbohydrates	Sugar	Protein	Fibre
12,138 kJ	2895 kcal	147.7 g	84.8 g	379.4 g	305.0	36.4 g	4.2 g

Cooking

1 Place the sugar into a bowl and grate the zest of lemon into it, rubbing together.

2 Strain the lemon juice into a non-reactive pan. Add the eggs, egg yolks, butter and zested sugar. Whisk to combine.

3 Place over a medium heat and whisk continuously for 3–5 minutes, until the mixture begins to thicken.

4 At the first sign of boiling, remove from the heat. Strain into a bowl and cool before filling the pastry cases.

5 Make the meringue (see page 464). Pipe it on top of the filled pie.

6 Colour in a hot oven at 220 °C.

> **PROFESSIONAL TIP**
> Use a stainless steel pan to avoid any reactions when making the lemon curd.

	2 × 20 cm flan rings (16 portions)
Sweet paste flan cases (pre-baked)	2
Granulated sugar	450 g
Lemons, grated zest	2
Fresh lemon juice	240 ml
Eggs, large	8
Large egg yolks	2
Unsalted butter, cut into small pieces	350 g
Meringue	
Egg whites	6
Caster sugar	600 g

> **VARIATION**
> - An Italian meringue is more stable than a traditional French meringue. Also, as the sugar is boiled before being mixed with the foamed egg white when making an Italian meringue, the egg white is pasteurised making it safe to consume from this point. Therefore, if Italian meringue is used in this recipe, once piped onto the tart, it can be browned using a blow torch rather than being baked in the oven.

Mise en place

1 Have a clean bowl and chinois (conical strainer) ready to strain the lemon curd once it has boiled and thickened.

18 Chocolate éclairs (*éclairs au chocolat*)

Energy	Calories	Fat	Saturated fat	Carbohydrates	Sugar	Protein	Fibre
516 kJ	123 kcal	9.5 g	5.7 g	8.8 g	7.3 g	1.1 g	0.1 g

	12 portions
Choux paste	200 ml
Whipped cream/Chantilly cream	250 ml
Fondant	100 g
Chocolate couverture	25 g

Mise en place

1 The baking sheet can be pre-marked with guide-lines to assist with accurate piping. This will help to ensure consistently sized and shaped products once they are baked.

Cooking

1 Place the choux paste into a piping bag with a 1 cm plain tube.

2 Pipe into 8-cm lengths onto a lightly greased, dampened baking sheet.

3 Bake at 200–220 °C for about 30 minutes.

4 Allow to cool. Slit down one side, with a sharp knife.

5 Fill with Chantilly cream (or whipped cream) using a piping bag and small tube. The continental fashion is to fill with pastry cream.

6 Warm the fondant, add the finely cut chocolate, allow to melt slowly, adjusting the consistency with a little sugar and water syrup if necessary. Do not overheat or the fondant will lose its shine.

7 Glaze the éclairs by dipping them in the fondant; remove the surplus with the finger. Allow to set.

NOTES
Traditionally, chocolate éclairs were filled with chocolate pastry cream.

VARIATION
- For coffee éclairs (*éclairs au café*) add a few drops of coffee extract to the fondant instead of chocolate; coffee éclairs may also be filled with pastry cream (see page 460) flavoured with coffee.

Pierce the éclair.

Pipe in the filling.

Dip the éclair in fondant; wipe the edges to give a neat finish.

FAULTS

Try not to open the oven doors while the eclairs are baking. The rapid drop in temperature can make the eclairs lose aeration and therefore the light open texture required.

19 Profiteroles and chocolate sauce (*profiteroles au chocolat*)

Energy	Calories	Fat	Saturated fat	Carbohydrates	Sugar	Protein	Fibre
919 kJ	219 kcal	16.2 g	9.7 g	16.4 g	12.8 g	2.9 g	0.2 g

Cooking

1 Spoon the choux paste into a piping bag with a plain nozzle (approx. 1.5 cm diameter).

2 Pipe walnut-sized balls of paste onto the greased baking sheet, spaced well apart. Level the peaked tops with the tip of a wet finger.

3 Bake for 18–20 minutes at 200 °C, until well risen and golden brown. Remove from the oven, transfer to a wire rack and allow to cool completely.

4 Make a hole in each and fill with Chantilly cream.

5 Dredge with icing sugar and serve with a sauceboat of cold chocolate sauce or coat the profiteroles with the sauce.

	10 portions
Choux paste	200 ml
Chocolate sauce (see page 469)	250 ml
Chantilly cream	250 ml
Icing sugar, to serve	

VARIATION

- Alternatively, coffee sauce may be served and the profiteroles filled with non-dairy cream. Profiteroles may also be filled with chocolate-, coffee- or rum-flavoured pastry cream.

Mise en place

1 Have a small bowl of iced-water to hand to dab the spikes of the piped profiteroles before baking.

2 Once baked, the hole for the filling can be made using the end of a clean metal piping nozzle.

SERVING SUGGESTION

Profiteroles are usually served as a dessert, whereas other choux pastries, such as eclairs, are served as pastries on occasions such as afternoon tea or as an accompaniment or treat with a coffee.

20 Gâteau Paris-Brest

Energy	Calories	Fat	Saturated fat	Carbohydrates	Sugar	Protein	Fibre
2089 kJ	501 kcal	32.1 g	10.8 g	47.5 g	39.8 g	8.2 g	0.4 g

	8 portions
Choux paste	200ml
Crème diplomat or pastry cream (see page 460)	800ml
For the praline	
Flaked almonds, hazelnuts and pecans (any combination)	375 g
Granulated sugar	500 g

Cooking

Praline:

1 Place the nuts on a baking sheet and toast until evenly coloured. Sprinkle with flaked almonds before baking.

2 Place the sugar in a large, heavy, stainless steel saucepan. Set the pan over a low heat and allow the sugar to caramelise. Do not over-stir, but do not allow the sugar to burn.

3 When the sugar is lightly caramelised and reaches a temperature of 170 °C, remove from the heat and stir in the nuts.

4 Immediately deposit the mixture on a silpat mat. Place another mat over the top and roll as thinly as possible.

5 Allow to cool completely. Break up and store in an airtight container.

Paris-Brest:

1 Pipe choux paste (recipe 3) into rings and bake. A light sprinkle of sliced almonds can be place on top at this stage.

2 Slice each ring in half. Fill with a mixture of crème diplomat or pastry cream and praline. Dust lightly with icing sugar.

NOTES

The elements for a gâteau Paris-Brest are made separately. The choux rings themselves, the crème diplomat and the praline. It is then a case of neatly combing these products together to complete the gâteau.

PROFESSIONAL TIP

Mark out the circles on the baking parchment before piping the choux paste. This will help with accuracy of size and shape as well as consistency.

SERVING SUGGESTION

Gâteau Paris-Brest can be served as a dessert or as a pastry at an afternoon tea.

21 Cheese straws (*paillettes au fromage*)

Energy	Calories	Fat	Saturated fat	Carbohydrates	Sugar	Protein	Fibre
2562 kJ	610 kcal	48.1 g	24.1 g	28.7 g	0.6 g	17.4 g	1.4 g

	8–10 portions	16–20 portions
Puff paste or rough puff paste	100 g	250 g
Cheese, grated	50 g	125 g
Cayenne pepper		

Cooking

1 Roll out the pastry to 60 × 15 cm, 3 mm thick.
2 Sprinkle with the cheese and cayenne pepper.
3 Roll out lightly to embed the cheese.
4 Cut the paste into thin strips by length.
5 Twist each strip to form rolls in the strip.
6 Place on a silicone mat.
7 Bake in a hot oven at 230–250 °C for 10 minutes or until a golden brown. Cut into lengths as required.

VARIATION
● Parmesan cheese is a good cheese to use for cheese straws. It grates very finely and packs a good depth of flavour.

SERVING SUGGESTION
Cheese straws can be served as a canapé or with cheese at the end of a meal.

22 Sausage rolls

Energy	Calories	Fat	Saturated fat	Carbohydrates	Sugar	Protein	Fibre	Sodium
708 kJ	170 kcal	13.2 g	4.7 g	7.6 g	1.0 g	0.4 g	1.1 g	258.0g

	12 × 8-cm rolls
Puff pastry	200 g
Sausage meat	400 g
Egg wash	

Cooking

1 Roll out the pastry 3-mm thick into a strip 10 cm wide.

2 Shape the sausage meat into a roll 2 cm in diameter.

3 Place on the pastry. Moisten the edges of the pastry.

4 Fold over and seal. Cut into 8-cm lengths.

5 Mark the edge with the back of a knife. Brush with eggwash.

6 Place on to a greased, dampened baking sheet.

7 Bake at 220 °C for approximately 20 minutes.

PROFESSIONAL TIP

Place the sausage meat into a piping bag with a wide plain nozzle. This can then be piped on to the rolled pastry in an even tubular shape (long sausage).

VARIATION

● Vegetable or cheese-based fillings could be used for customers who do not to eat meat.

23 Eccles cakes

Energy	Calories	Fat	Saturated fat	Carbohydrates	Sugar	Protein	Fibre
691 kJ	164 kcal	8.6 g	3.7 g	22.1 g	17.3 g	1.1 g	1.4 g

	12 cakes
Puff pastry or rough puff pastry	300 g
Egg white, to brush	
Caster sugar, to coat	
Filling	
Butter	50 g
Raisins	50 g
Demerara sugar	50 g
Currants	200
Mixed spice (optional)	pinch

Cooking

1 Roll out the pastry 2 mm thick.

2 Cut into rounds 10–12 cm diameter. Damp the edges.

3 Mix together all the ingredients for the filling and place 1 tbsp of the mixture in the centre of each round.

4 Fold the edges over to the centre and completely seal in the mixture.

5 Brush the top with egg white and dip into caster sugar.

6 Place on a greased baking sheet.

7 Cut two or three incisions with a knife so as to show the filling.

8 Bake at 220 °C for 15–20 minutes.

24 Fruit slice (*bande aux fruits*)

Energy	Calories	Fat	Saturated fat	Carbohydrates	Sugar	Protein	Fibre
847 kJ	202 kcal	10.3 g	3.5 g	26 g	14.3 g	3.2 g	1.6 g

	8–10 portions
Puff pastry	250 g
Fruit (see note)	400 g
Pastry cream	250 ml (approximately)
Apricot glaze	2 tbsp

Cooking

1 Roll out the pastry 2 mm thick in a strip 12 cm wide.

2 Place on a greased, dampened baking sheet.

3 Moisten two edges with eggwash; lay two 1.5 cm-wide strips along each edge.

4 Seal firmly and mark at the sides using the back of a knife. Dock the bottom of the slice.

5 Depending on the fruit used, either put the fruit (such as apple) on the slice and cook together, or cook the slice 'blind' and afterwards place the pastry cream and fruit (e.g. soft fresh fruits) on the pastry. Glaze and serve as for fruit flans.

NOTES
Fruit slices may be prepared from any fruit suitable for flans or tarts.

VARIATION
Alternative methods:
- short or sweet pastry for the base and puff pastry for the two side strips
- sweet pastry in a slice mould.

SERVING SUGGESTION
When cutting or slicing a bande aux fruits, make sure that a sharp, serrated knife is used and that the slice incorporates a portion of all the fruits being used. Bandes are sliced across the width of the bande rather than the length.

25 Pear jalousie

Energy	Calories	Fat	Saturated fat	Carbohydrates	Sugar	Protein	Fibre
1178 kJ	282 kcal	17.8 g	5.1 g	27.2 g	17.5 g	3.8 g	0.8 g

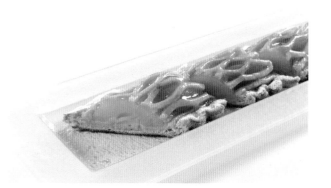

	8–10 portions
Puff pastry	200 g
Frangipane	200 g
Pears, poached or tinned (cored and cut in half lengthways)	5

Mise en place

1 Prepare the frangipane, poach the pears (if this is required), prepare glaze and egg wash.

Cooking

1 Roll out two-thirds of the pastry 3 mm thick into a strip 25 × 10 cm and place on a greased, dampened baking sheet.

2 Pierce with a docker. Moisten the edges.

3 Pipe on the frangipane, leaving 2 cm free all the way round. Place the pears on top.

4 Roll out the remaining one-third of the pastry to the same size. Chill before cutting.

5 Cut the dough with a trellis cutter to make a lattice.

6 Carefully open out this strip and neatly place onto the first strip.

7 Trim off any excess. Neaten and decorate the edge. Brush with egg wash.

8 Bake at 220 °C for 25–30 minutes.

9 Glaze with apricot glaze. Dust the edges with icing sugar and return to a very hot oven to glaze.

PROFESSIONAL TIP

Use a lattice cutter when the paste is very cold. The blades of the cutter will penetrate more cleanly when the paste is cold rather than at an ambient temperature.

VARIATION

● Other fruits could be used, for example, peaches, plums and apricots.

SERVING SUGGESTION

Serve simply with a quenelle of crème Chantilly or crème anglaise.

26 Mille-feuilles (puff pastry slices)

Energy	Calories	Fat	Saturated fat	Carbohydrates	Sugar	Protein	Fibre
1158 kJ	369 kcal	10.9 g	1.3 g	67.7 g	52.3 g	4.9 g	0.1 g

	10 portions
Puff pastry trimmings	600 g
Pastry cream	400 ml
Apricot glaze	
Fondant	350 g
Chocolate	100 g

Mise en place

Have all the components for the mille-feuille ready to put together. It is important that the temperatures and textures are correct to ensure that the assembly is consistent with the specifications for the product. This is particularly relevant for the feathering of the fondant.

Cooking

1 Roll out the pastry 2-mm thick into an even-sided square.

2 Roll up carefully on a rolling pin and unroll onto a greased, dampened baking sheet.

3 Dock well.

4 Bake in a hot oven at 220 °C for 15–20 minutes; turn over after 10 minutes. Allow to cool.

5 Using a large knife, cut into three even-sized rectangles.

6 Keeping the best strip for the top, pipe the pastry cream on one strip.

7 Place the second strip on top and pipe with pastry cream.

8 Place the last strip on top, flat side up. Press gently. Brush with boiling apricot glaze to form a key.

Decorate by feather icing:

1 Warm the fondant to 37 °C (warm to the touch) and correct the consistency with sugar syrup if necessary.

2 Pour the fondant over the mille-feuilles in an even coat.

3 Immediately pipe the melted chocolate over the fondant in strips, 0.5 cm apart.

4 With the back of a small knife, wiping after each stroke, mark down the slice at 2-cm intervals.

5 Quickly turn the slice around and repeat in the same direction with strokes in between the previous ones.

6 Allow to set and trim the edges neatly.

7 Cut into even portions with a sharp, thin-bladed knife; dip into hot water and wipe clean after each cut.

For a traditional finish, crush the pastry trimmings and use them to coat the sides.

Pipe cream between layers of pastry.

Ice the top with fondant.

Decorate with chocolate.

VARIATION
- Whipped fresh cream may be used as an alternative to pastry cream.
- The pastry cream or whipped cream may also be flavoured with a liqueur if so desired, such as curaçao, Grand Marnier or Cointreau.

27 Gâteau pithiviers

Energy	Calories	Fat	Saturated fat	Carbohydrates	Sugar	Protein	Fibre
1452.8 kJ	349.2 kcal	25.8 g	9.4 g	25.6 g	10.4 g	5.3 g	1.7 g

	2 × 22cm gâteaux
Puff pastry	1 kg
Pastry cream	60 g
Frangipane	600 g
Egg wash (yolks only)	
Granulated sugar	

Cooking

1 Divide the paste into four equal pieces. Roll out each piece in a circle with a 22 cm diameter and 4 mm thick.

2 Rest in the fridge between sheets of cling film, preferably overnight.

3 Lightly butter two baking trays and splash with water. Lay one circle of paste onto each tray and dock them.

4 Mark a 16 cm diameter circle in the centre of each.

5 Beat the pastry cream, if desired, and mix it with the frangipane.

6 Using a plain nozzle, pipe the cream over the inner circles, making them slightly domed.

7 Egg wash the outer edges of the paste. Lay one of the remaining pieces over the top of each one, smooth over and press down hard.

8 Mark the edges with a round cutter. Cut out a scallop pattern with a knife or use a cut piping nozzle as shown in the photo sequence.

9 Egg wash twice, leaving time for the egg-wash to set in between each time. Mark the top of both with a spiral pattern.

10 Bake at 220 °C for 10 minutes. Remove from the oven and sprinkle with granulated sugar. Turn the oven down to 190 °C and bake for a further 20–25 minutes.

11 Glaze under a salamander.

Adding the filling to the rolled base.

Trimming the edge.

Marking the top.

PROFESSIONAL TIP
Resting puff pastry will help to hold the desired sped and prevent retraction.

28 Palmiers

Energy	Calories	Fat	Saturated fat	Carbohydrates	Sugar	Protein	Fibre
174 kJ	42 kcal	2.5 g	0.8 g	4.3 g	2.1 g	2.0 g	0.1 g

* Based on 25 portions per recipe

	20–30 (depending on thickness and size)
Puff pastry (see page 518)	200 g
Caster sugar	50 g
Egg wash	50 ml

Cooking

1 Roll out puff pastry 3 mm thick, into a square.

2 Sprinkle liberally with caster sugar on both sides and roll into the pastry.

3 Fold into three from each end so as to meet in the middle; brush with egg wash and fold in two.

4 Cut into strips approximately 2 cm thick; dip one side in caster sugar.

5 Place on a greased baking sheet, sugared side down, leaving a space of at least 2 cm between each.

6 Bake in a very hot oven for about 10 minutes.

7 Turn with a palette knife, cook on the other side until brown and the sugar is caramelised.

> **NOTE**
> Palmiers are usually made from leftover or off-cuts of puff pastry. As a more biscuit-like property is sought, previously rolled pastry can be utilised because the rise required is not as much as in products such as bouchées and vol-au-vents.

> **VARIATION**
> Puff pastry trimmings are suitable for this recipe. Palmiers may be made in a wide variety of sizes. Two joined together with a little whipped cream may be served as a pastry or small ones for petits fours. They may be sandwiched together with soft fruit, whipped cream and/or ice cream and served as a sweet.

29 Spotted dick

Energy	Calories	Fat	Saturated fat	Carbohydrates	Sugar	Protein	Fibre
2552 kJ	607 kcal	28.5 g	16.2 g	81.0 g	32.9 g	15.5 g	4.0 g

* Based on 5 portions per recipe; using semi-skimmed milk

	4–6 portions
Plain flour	300 g
Baking powder	10 g
Shredded suet	150 g
Caster sugar	75 g
Currants	110 g
Lemon, zest only	1
Milk	200 ml
Butter, for greasing	

Mise en place

1 Line the pudding basins before starting the mix and have the greaseproof paper or baking parchment, cloth and string cut and ready to use.

Cooking

1 Place the flour, baking powder, shredded suet, caster sugar, currants and lemon zest into a bowl and mix to combine.

2 Add the milk and stir to make a soft dough.

3 Grease a pudding basin with butter and spoon the mixture into the basin. Cover with a piece of folded greaseproof paper.

4 Tie around the edge with string to secure the paper and place a damp piece of towel or muslin over the top. Tie once more with string to secure the towel.

5 Place the basin into a large lidded saucepan and fill the pan two-thirds of the way up with water.

6 Cover with the lid, bring to a boil and simmer for one hour (a steamer could be used if available)

7 To serve, slice a wedge of spotted dick for each person and serve with fresh, hot custard.

SERVING SUGGESTION

Serve with hot, fresh crème anglaise or traditional custard.

Biscuits, cakes and sponges

Learning outcomes

In this unit you will be able to:

1 Understand how biscuits, cakes and sponges are produced, including:
   - knowing examples of, and the differences between, biscuits, cakes and sponges, and their suitability for different types of service
   - understanding the preparation and processing techniques used to produce biscuits, cakes and sponges and the types of biscuits, cakes and sponges that are produced with these techniques
   - knowing the difference between types of commodities and understand the implications of these in the production of biscuits, cakes and sponges
   - understanding the correct storage procedures to use throughout production and on completion of the final products.

2 Produce biscuits, cakes and sponges, including:
   - understanding how to use and operate specialist equipment to produce biscuits, cakes and sponges
   - being able to apply techniques to produce biscuits, cakes and sponges
   - understanding the appropriate fillings to use with different types of cakes and sponges
   - being able to apply finishing techniques to biscuits, cakes and sponges
   - being able to measure and evaluate against quality standards throughout preparation and cooking
   - being able to evaluate products against dish requirements and production standards and recognise any faults.

The key ingredients for biscuits, cakes and sponges, such as flour, eggs, sugar and raising agents are described in Unit 309. Make sure you read and understand this section.

Recipes in this chapter

No	Recipe
	Sponges
1	Genoese sponge (*génoise*)
2	Chocolate genoise sponge (*génoise au chocolat*)
3	Fresh cream and strawberry gâteau
4	Chocolate gâteau
5	Coffee gâteau
6	Roulade sponge
7	Swiss roll
	Cakes
8	Rich fruit cake
9	Banana bread
10	Lemon drizzle cake
11	Victoria sandwich

No	Recipe
12	Scones
13	Cupcakes
	Biscuits
15	Sponge fingers (*biscuits à la cuillère*)
16	Shortbread biscuits
17	Cats' tongues (*langues de chat*)
18	Piped biscuits (*sablés à la poche*)
19	Viennese biscuits
20	Chocolate pecan brownie
21	Tuiles
22	Brandy snaps
23	Madeleines
24	Cigarette or pencil paste cornets

Understand how biscuits, cakes and sponges are produced

Biscuits, cakes and sponges come in a variety of forms. They are produced in different ways to achieve a range of qualities and characteristics.

1.1 Examples of biscuits, cakes and sponges

Biscuits

Some biscuits are short, in that they are crisp and crumbly, such as a shortbread, whereas other biscuits vary in texture from sponge-like to open-textured cookies. Depending on the type of biscuit being produced, the method of production differs to produce a variety of desired textures, such as the dense sponge–biscuit-cross texture of chocolate brownies or the biscuit discs used in the production of dobos torte. This is described below in further detail, with examples of biscuits suited to each method provided.

Cakes

Various methods are used to make cakes, as described below. However, the sugar batter/creaming method is the most popular and practical method of making a cake. Cakes are produced in many different shapes and sizes. They vary from single portion-sized cakes, as in cupcakes, to large, multi-portioned products such as wedding cakes, intended to serve hundreds of guests. The production of cakes can vary slightly to incorporate flavourings and additional ingredients such as mashed bananas in banana cake, red wine for red wine cake, lemon syrup for lemon drizzle cake and dried fruits and spices when making a rich fruit cake.

However, cakes share similar quality points in that they should be aerated, slightly domed with a consistent, moist crumb. They should also deliver good clean flavours and a consistent spread of any additional ingredients, such as fruit, throughout the mix.

Batters and whisked sponges

Sponges are generally a lot lighter and more aerated than cakes. Their production normally follows the whisking of warmed eggs and sugar until they reach what is referred to as the 'ribbon stage' or a 'sabayon', before folding in sieved, soft flour and melted butter, if this is required to enrich the sponge. In the production of a chocolate sponge, a percentage of the flour is replaced with cocoa powder, which is sieved together with the flour to ensure an even distribution. Sponges are very versatile. They are used as the primary component in the production of gâteaux, where a Genoese sponge is often used, for example. Alternatively, when multiple layers are a requirement, sponges are sliced into discs or baked in thin sheets. The classic gâteau 'Opera' provides a good example, or when making a roulade or Swiss roll.

Batters and sponges allow us to make a large assortment of desserts and cakes. Basically, they are a mix of eggs, sugar, flour and the air incorporated when these are beaten. Certain other raw materials can be combined – for example, almonds, hazelnuts, walnuts, chocolate, butter, fruit, ginger, anise, coffee and vanilla.

Service of biscuits, cakes and sponges

The service of biscuits, cakes and sponges can vary according to the occasion. This could range from a single portion of a cake or gâteau served at the counter in a patisserie or artisan bakery, to a whole cake at a celebration such as a birthday.

Although not desserts, biscuits, cakes and sponges often form parts of desserts. For example, sponges may be used to line mousses, sweet terrines and to surround the ice cream when making baked Alaska. Biscuits such as tuilles and brandy snaps are often used as carrying bases for mousses and other fillings.

1.2 Techniques used to produce biscuits, cakes and sponges

Preparation methods – biscuits and cookies

There are a wide variety of preparation methods used to produce biscuits and cookies, as described below.

Rubbing in

This is probably the best-known method and is used to produce some of the most famous types of biscuits, such as shortbread. The method is exactly the same as that for producing short pastry.

- Rub the fat into the flour, by hand or by machine, adding the liquid and the sugar and mixing in the flour to produce a smooth biscuit paste.
- Do not overwork the paste otherwise it will not combine and as a consequence you will not be able to roll it out.

Foaming (by whisking)

This is where a foam is produced from egg whites, egg yolks or both. Sponge fingers are an example of a two-foam mixture. Meringue is an example of a single-foam mixture using egg whites. Great care must be taken not to over-mix the product.

Sugar batter method (by creaming)

Fat and sugar are mixed together to produce a light and fluffy cream. Beaten egg is added gradually. The dry ingredients are then carefully folded in. Cats' tongues and sablé biscuits use this method.

Depositing

Depositing refers to the process of transferring mixtures and batters in to the tins or trays that they are to be baked in. In large productions, depositing machines can be used to ensure even distribution. In manufacturing, this is often an automated process.

Piping

When making biscuits such as cats' tongues, sablé biscuits or sponge fingers (*biscuits à la cuillère*), once the biscuit mix is made, it is piped into the desired shape before baking. With the three examples provided, the texture of the biscuit mix is quite different. The sponge finger shape is a very light mix which is piped into finger length, fairly wide strips. The cats' tongues produce a denser mix, which is piped into small, fairly flat lengths that are slightly thicker at the ends. The sablé biscuit mix is again denser and usually piped into various shapes of a more substantial size using a fluted piping nozzle.

Rolling

Some biscuits are rolled into shape following the production of the biscuit mix. This applies to biscuits that are cut into shape after being rolled into the desired width. Examples include ginger snaps and shortbread.

Flour batter method

Half the sieved flour is creamed with the fat. The eggs and sugar are beaten together before they are added to the fat and flour mixture. Finally, the remainder of the flour is folded in, together with any other dry ingredients. Cookies use the flour batter method.

Blending method

In several biscuit recipes, the method requires the chef only to blend all the ingredients together to produce a smooth paste. Almond biscuits (using basic almond commercial mixture) use the blending method.

Preparation methods – cakes

There are three basic methods of making cake mixtures, also known as cake batters. The working temperature of cake batter should be 21 °C.

Sugar batter or creaming method

For this method, the fat (cake margarine, butter or shortening) is blended in a machine with caster sugar. This is the basic or principal stage; usually the other ingredients are then added in the order shown in the steps below.

Step by step: sugar batter method

1　Soften the butter.

2　Cream the sugar and butter thoroughly to ensure a good, light crumb.

3　Slowly add the egg, a little at a time.

4　Once all the egg is incorporated, add the flour.

5　Fold in the flour.

Flour batter method

For this method the eggs and sugar are whisked to a half sponge; this is the basic or principal stage, which aims to foam the two ingredients together until half the maximum volume is achieved. Other ingredients are added as shown in the diagram.

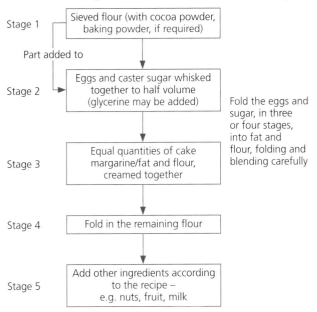

A type of product called a **humectant**, which helps the product to stay moist, may be added (e.g. glycerine or honey); if so, add this at stage 2.

Blending method

This is used for **high-ratio cake** mixtures. It uses high-ratio flour specially produced so that it will absorb more liquid. It also uses a high-ratio fat, made from oil, to which a quantity of emulsifying agent has been added, enabling the fat to take up a greater quantity of liquid.

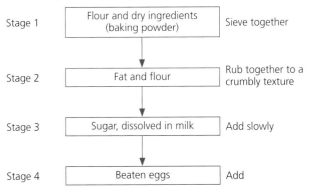

> **KEY TERMS**
>
> **Humectant:** a hygroscopic (helps to retain water) substance used to keep products moist. Honey is an example used in this way.
>
> **High-ratio cakes:** prepared with a relatively high proportion of sugar and eggs compared to flour.

All-in-one method

The all-in-one method is a more user-friendly modern variation. This is perhaps one example of where a method is suited to making a specific type of cake, and other methods would not be suitable.

The all-in-one method is often used for cakes which contain oil or little or no fat – a tea loaf or banana cake, for example. For this reason, the all-in-one is the most relevant method after the sugar batter or creaming method.

High-ratio cakes contain more liquid and sugar, resulting in a fine stable crumb, extended shelf life, good eating and excellent freezing qualities. The principal or basic stage is the mixing of the fat and flour to a crumbling texture. It is essential that each stage of the batter is blended into the next to produce a smooth batter, free from lumps.

When using mixing machines, it is important to remember to:
- blend on a slow speed
- beat on a medium speed, using a paddle attachment.

When blending, always clear the mix from the bottom of the bowl to ensure that any first- or second-stage batter does not remain in the bowl.

Melting method

Some traditional cakes, such as Yorkshire parkin or ginger cake, are produced by the melting method. In this case, the fats and sugars (for example, butter, sugar, treacle, syrup) are melted slowly before being mixed (folded) with the dry ingredients (such as self-raising flour, spices, oatmeal). Liquid ingredients such as eggs and milk will then be mixed in before being placed into a lined baking tin and baked.

Aeration

Aeration of biscuits (if desired), cakes and sponges is created in two ways predominantly. The first is referred to as 'mechanical aeration'. This is where air is created through the energy of whisking to produce foams, usually from whole eggs, egg whites or egg yolks (depending on the recipe), with sugar, that capture air bubbles which are then stabilised during the baking process. Care must be taken when adding additional ingredients to foams as excess folding and movement will damage the delicate foam and reduce the amount of air captured in the mix. Many sponges are created using this method such as Genoese and Swiss roll. The whisking of the warmed egg (whole) or yolks until the volume increases by double, leaving a ribbon effect, is known as a 'sabayon'. In the case of sponge fingers (biscuits à la cuillère), the eggs are split. The yolks are whisked to a sabayon, as described above, while the whites are whisked to the meringue stage. The two sets of eggs are then folded together with a mixture of soft flour and cornflour to produce a pipeable batter. Separating the eggs in this way is sometimes referred to as the 'separated egg method'.

The second method of aeration is known as 'chemical aeration'. In this case, a mix will have a **raising agent** as part of the recipe. The most commonly known and used raising agent is baking powder (see below). Products such as Victoria sandwich and scones contain the ingredient baking powder to create the aeration desired. Note that self-raising flour contains raising agents equivalent to baking powder, which is why it is often used in recipes requiring a chemical raising agent.

> **KEY TERM**
>
> **Raising agent:** a substance added to a cake or bread mixture while produces gases that give lightness to the product. Baking powder is an example of a raising agent.

Baking powder

Baking powder may be made from one-part sodium bicarbonate to two parts cream of tartar. In commercial baking the powdered cream of tartar may be replaced by another acid product, such as acidulated calcium phosphate.

When used under the right conditions, with the addition of liquid and heat, baking powder produces carbon dioxide gas. As the acid has a delayed action, only a small amount of gas is given off when the liquid is added, and the majority is released when the mixture is heated. Therefore, when cakes are mixed they do not lose the properties of the baking powder if they are not cooked right away.

Possible reasons for faults in cakes

Uneven texture:
- fat insufficiently rubbed in
- too little liquid
- too much liquid.

Close texture:
- too much fat
- hands too hot when rubbing in
- fat to flour ratio incorrect.

Dry:
- too little liquid
- oven too hot.

Poor shape:
- too much liquid
- oven too cool
- too much baking powder.

Fruit sunk:
- fruit wet
- too much liquid
- oven too cool.

Cracked:
- too little liquid
- too much baking powder.

Preparing batters and mixed sponges

Sponge mixtures are produced from a foam of eggs and sugar. The eggs may be whole eggs or separated. Examples of sponge products are gâteaux, sponge fingers and sponge cakes. The egg white traps the air bubbles. When eggs and sugar are whisked together, they thicken until maximum volume is reached, then flour is carefully folded in by a method known as **cutting in**. This is the most difficult operation, as the flour must not be allowed to sink to the bottom of the bowl, otherwise it becomes lumpy and difficult to clear. However, the mixture must not be stirred as this will disturb the aeration and cause the air to escape, resulting in a much heavier sponge.

If butter, margarine or oil is added, it is important that this is added at about 36 °C, otherwise overheating will cause the fat or oil to act on the flour and create lumps, which are difficult and often impossible to get rid of.

Stabilisers are often added to sponges to prevent them from collapsing. The most common are ethyl methylcellulose and glycerol monostearate; these are added to the eggs and sugar at the beginning of the mixing.

> **KEY TERMS**
>
> **Cutting in:** the method used to describe the way in which flour is carefully folded into sponge batters.
>
> **Stabiliser:** a product added to sponges to help prevent them collapsing.

> **PROFESSIONAL TIP**
>
> What you need to know about sponge cakes:
> - Never add flour or ground dry ingredients to a batter until the end because they prevent the air absorption in the first beating stage.
> - When making sponge cakes, always sift the dry ingredients (flour, cocoa powder, ground nuts, etc.) to avoid clumping.
> - Mix in the flour as quickly and delicately as possible, because a rough addition of dry ingredients acts like a weight on the primary batter and can remove part of the air already absorbed.

- Flours used in sponge cakes are low in gluten content. In certain sponge cakes, a portion of the flour can be left out and substituted with cornstarch. This yields a softer and more aerated batter.
- The eggs used in sponge cake batters should be fresh and at room temperature so that they take in air faster.
- Adding separately beaten egg whites produces a lighter and fluffier sponge cake.
- Once sponge cake batters are beaten and poured into moulds or baking trays, they should be baked as soon as possible, otherwise, the batter loses volume.

Methods of making sponges

- Foaming method – whisking eggs and sugar together to ribbon stage; folding in or cutting flour.
- Melting method – as with foaming, but adding melted butter, margarine or oil to the mixture. The fat content enriches the sponge, improves the flavour, texture and crumb structure, and will extend shelf life.
- Boiling method – sponges made by this method have a stable crumb texture that is easier to handle and crumbles less when cut than the standard basic sponge containing fat (known as Genoese sponge). This method will produce a sponge that is suitable for dipping in fondant.
- Blending method – this is used for high-ratio sponges, which follow the same principles as high-ratio cakes. As with cakes, high-ratio goods produce a fine, stable crumb, an even texture, excellent shelf life and good freezing qualities.
- Creaming method – this is the traditional method and is still used today for Victoria sandwiches and light fruit cakes. The fat and sugar are creamed together, then beaten egg is added and, finally, the sieved flour is added with the other dry ingredients as desired. Despite its title, a Victoria sponge sandwich really falls into the category of a cake rather than a sponge.
- Separate yolk and white method – this method is used for sponge fingers (recipe 15). Sponge fingers fall into the category of biscuits.

Preparing and lining moulds

All equipment should be prepared and ready for use before any mixing begins. Line moulds evenly, ensuring a light, even distribution. The fat used to line a mould should be soft rather than melted. A melted fat will tend to run to the bottom of the mould and collect in a puddle where the base meets the side.

For extra security and protection from sticking, collars and circles of greaseproof or silicone paper are often used to line moulds. This allows the product to be slipped from the mould in the paper, which is then gently removed once the product has cooled slightly. Many modern moulds have improved non-stick properties, which help to prevent sticking and the damage caused to cakes and sponges when they stick to their moulds.

Baking – temperature and humidity control

Because they become too dry while baking (due to the oven temperature producing a dry atmosphere), some cakes require the injection of steam. Combination ovens are ideally suited for this purpose. The steam delays the formation of the crust until the cake batter has become fully aerated and the proteins have set. Alternatively, add a tray of water to the oven while baking. If the oven is too hot, the cake crust will form early and the cake batter will rise into a peak.

Possible reasons for faults in sponges

Close texture:
- under-beating
- too much flour
- oven too cool or too hot

'Holey' texture:
- flour insufficiently folded in
- tin unevenly filled.

Cracked crust:
- oven too hot.

Sunken:
- oven too hot
- tin removed during cooking.

White spots on surface:
- insufficient beating.

Possible reasons for faults in Genoese sponges

Close texture:
- eggs and sugar overheated
- eggs and sugar under-beaten
- too much flour
- flour insufficiently folded in
- oven too hot.

Sunken:
- too much sugar
- oven too hot
- tin removed during cooking.

Heavy:
- butter too hot
- butter insufficiently mixed in
- flour over-mixed.

Points to remember when producing biscuits, cakes and sponges:

- Check all ingredients carefully.
- Make sure scales are accurate; weigh all ingredients carefully.
- Check ovens are at the right temperature and that the shelves are in the correct position.
- Check that all work surfaces and equipment are clean.
- Check that all other equipment required, such as cooling wires, is within easy reach.
- Always sieve flour to remove lumps and any foreign material.
- Make sure that eggs and fats are at room temperature.
- Check dried fruits carefully; wash, drain and dry if necessary.
- Always follow the recipe carefully.
- Always scrape down the sides of the mixing bowl when creaming mixtures.
- Always seek help if you are unsure or lack understanding.
- Try to fill the oven space when baking by planning production carefully; this saves time, labour and money.
- Never guess quantities. Time and temperature are important factors; they too should not be guessed.
- The shape and size of goods will determine the cooking time and temperature: the wider the cake, the longer and more slowly it will need to cook.
- Where cakes contain a high proportion of sugar in the recipe, this will caramelise the surface quickly before the centre is cooked. Therefore, cover the cake with sheets of silicone or dampened greaseproof paper and continue to cook.
- When cake tops are sprinkled with almonds or sugar, the baking temperature needs to be lowered slightly to prevent over-colouring of the cake crust.
- When glycerine, glucose and invert sugar, honey or treacle is added to cake mixtures, the oven temperature should be lowered as these colour at a lower temperature than sugar.
- Always work in a clean and hygienic way; remember the hygiene and safety rules, in particular the Food Safety Act 1990, and the Food Safety and Hygiene Regulations 2013.
- All cakes and sponges benefit from being allowed to cool in their tins as this makes handling easier. If sponges need to be cooled quickly, place a wire rack over the top of the tin and invert, then remove the lining paper and cool on a wire rack.

Convenience cake, biscuit and sponge mixes

There is now a vast range of prepared mixes and frozen goods available on the market. Premixes enable the caterer to calculate costs more effectively, reduce labour costs (with less demand for highly skilled labour) and limit the range of stock items to be held.

Every year, more and more convenience products are introduced to the market by food manufacturers. The caterer should be encouraged to investigate these products, and to experiment in order to assess their quality, value and contribution to the business.

1.3 Using commodities in biscuits, cakes and sponges

See Unit 309 for details of the commodities used in biscuits, cakes and sponges.

1.4 Storage

- Store all goods according to the Food Safety and Hygiene Regulations 2013, Food Safety Temperature Control Regulation 1995 and General Food Regulations (2004).
- Handle all equipment carefully to avoid cross-contamination.
- Always make sure that storage containers are kept clean and returned ready for re-use. On their return they should be hygienically washed and stored.
- Freshly baked cakes and sponges should be wrapped tightly in secure film or placed in an air tight, sealed bag. They should then be clearly labelled and dated before storing in a refrigerator or freezer.
- Finished cakes and sponges can be refrigerated to maintain food safety. However, attention to quality must be observed in refrigerated conditions. The moist atmosphere leads to sponges losing moisture and hardening in texture. Any additional ingredients also have to be considered. Creams can lose their viscosity and piped cream can retract, losing its fresh appearance. Prepared fruits, such as sliced strawberries, will weep, losing their structure and staining other elements of the product, such as the sponge or cream.
- Some baked products are suitable for freezing. Sponge bases, for example, freeze well for use at a later stage. Other completed products need to be analysed as to their suitability for freezing, based on the additional ingredients used and their suitability and composition.
- Biscuits are best stored in airtight containers. If left in open air for too long, biscuits will lose their crispness and short qualities.
- Humidity refers to the moisture in the atmosphere or air. Baked biscuits, cakes and sponges will soften in moist conditions, so it is important that they are kept in dry, cool conditions. Storing these items in refrigerators should be minimised due to the humid conditions.

- As with other food items, biscuits, cakes and sponges that are stored for longer-term storage should be date labelled and stored in appropriate, hygienic conditions. These items should also be positioned separately from other foods with strong aromas.

- Biscuits, cakes and sponges for long-term storage, by freezing for example, should be rotated to ensure that older products are utilised before newer ones.

Produce biscuits, cakes and sponges

2.1 Specialist equipment

Stencils

To produce fine biscuits such as tuilles and cornets, a stencil is often used. Stencils can be made in various shapes and sizes. They can be purchased from commercial manufacturers, but are often home-made by cutting shapes into plastic lids from finished ice cream tubs, for example. Creating shapes and possibilities is endless and the added advantage is that the biscuit is malleable for a few seconds once it is removed from the oven and baking tray. This allows the biscuit to be shaped even further by bending, rolling or shaping using a mould. Once cooled (literally after a few seconds in the air), the biscuit will become crisp but will have taken on the desired shape and size from the stencil that was used.

- To apply the biscuit mix across the stencil, the stencil is placed on top of a silicone mat on a baking tray.
- A small amount of biscuit mix is taken and spread finely across the stencil using a palate knife, smoothing away any ridges to make an even layer.
- Once the stencil is lifted, the biscuit paste will be left behind. It will carry the shape of the stencil and be ready to bake.

Such biscuits are usually baked in batches and will remain malleable until they are lifted from the baking tray. Timing is essential as the biscuits will need to cool slightly before working with them, as they will be molten and loose when they come out of the oven. However, if left too long, they will harden and will snap when trying to shape.

Other specialist equipment that you may use when preparing biscuits, cakes and sponges includes:
- induction hobs
- micro scales
- deck oven
- combination oven
- mixing machine
- food processor

- non-stick mats (often referred to as Silpat) mats.

You can find more details of these in Unit 309 Desserts and puddings.

2.2 Producing biscuits, cakes and sponges

For details of how to produce different types of biscuits, cakes and sponges, see the recipe section at the end of this unit.

2.3 Fillings and inserts

Cakes, sponges and biscuits may be filled or sandwiched together with a variety of different types of filling, including:
- creams – buttercream (plain, flavoured and/or coloured), pastry cream or crème patisserie (flavoured and/or coloured), whipped cream (*crème Chantilly*) or clotted cream
- fruit – fresh fruit purée, jams, fruit pastries, fruit mousses, preserves and fruit gels
- pastes and spreads – chocolate, praline, nuts and curds.

2.4 Finishing biscuits, cakes and sponges

Soaking

To add additional flavour and moisture to a sponge or cake, it can be soaked carefully with a liquid flavouring. For example, when making a Black Forest gâteau, the cut sponge is soaked by gently brushing with Kirsch to enhance the cherry flavour of the sponge. This could be neat Kirsch or a Kirsch-flavoured sugar syrup. Similarly, a rich fruit cake, such as Christmas cake, is soaked by injecting or seeding the cake with alcohol such as brandy. This can be done many times over months before the cake is actually consumed. This type of cake develops flavour over time and the alcohol not only adds flavour but also helps to preserve the cake.

Spreading and coating

This involves covering the top and sides of smaller cakes and gâteaux with any of the following:

- fresh whipped cream
- fondant
- chocolate
- royal icing
- buttercream
- water icing
- meringue (ordinary, Italian or Swiss)
- commercial preparations.

Piping

Piping is a skill that takes practice. There are many different (plain or fluted) types and sizes of piping tube available. The following may be used for piping:

- royal icing
- meringue
- chocolate
- boiled sugar
- fondant
- fresh cream.

Dusting, dredging and sprinkling

These techniques are used to give the product a final design or glaze during cooking, using sugar:

- dusting – a light dusting, giving an even finish
- dredging – heavier dusting with sugar
- sprinkling – a very light sprinkle of sugar.

The sugar used may be icing, caster or granulated white, Demerara, Barbados or dark brown sugar. The product may be returned to the oven for glazing or glazed under the salamander.

Feathering

A feathered finish can enhance the presentation of cakes and sponges. To create this finish, an icing is piped in straight lines, spiralling circles or dots on top of a base of another icing or topping before it has set. Once this piping is finished, a fine object such as a cocktail stick is used to cut through the lines to create the feathered finish desired. Although not a cake, probably the most well-known product that is traditionally feathered as part of its finish is the mille-feuille (page 537). The same principle applies to the feathering of cakes.

Portioning

As many cakes and sponges are multi-portioned items, they have to be cut or sliced into portions. Portions should be calculated before making cuts. Cake dividers can be used to help with this process. To use them, they are placed lightly on top of the cake or gâteau. This leaves indented marks on the top of the gâteau or sponge as an accurate indicator showing where cuts can be made. When cutting or slicing finished cakes, sponges or gâteaux, it is important to use a clean knife between each cut. Cleaning the knife between each cut will help to produce clean, neat and accurate cuts and greatly help with presentation.

Other decorative media

Remember that decorating is an art form and there is a range of equipment and materials available for this purpose. Some examples of decorative media are as follows:

- glacé and crystallised fruits – cherries, lemons, oranges, pineapple, figs
- crystallised flowers – ros petals, violets, mimosa, lilac
- crystallised stems – angelica
- nuts – almonds (nibbed, flaked), coconut (fresh slices, desiccated), hazelnuts, brazil nuts, pistachio
- chocolate – rolls, vermicelli, flakes, piping chocolate, chips.

Biscuit pastes

Piped biscuits can be used for decoration. For example:

- cats' tongues (recipe 17)
- piped sable pasté (recipe 18)
- almond biscuits.
- tuilles.

2.5 Evaluation

When producing biscuits, cakes and sponges, it is essential that the following quality standards are achieved to maximise the enjoyment and pleasure for customers consuming the products and to ensure customer satisfaction in all categories.

- Colour – the product is presented in its most natural colour or is baked to reflect the colour expected in the particular item concerned. For example, a baked shortbread biscuit should have a very light golden appearance, whereas a sponge or cake crust would be slightly darker. Coatings and icings should ideally reflect natural colours rather than garish, bright, unnatural colours.
- Flavours and taste – products should be pleasant to taste, providing clean, natural flavours that are enjoyable to consume.
- Aroma – many baked products provide that comforting aroma that customers appreciate and desire. Biscuits, cakes and sponges should have a delicate and pleasant aroma and one that is reflective of the any other ingredients contained within the product itself – for example, fruits, nuts, flavours, spices (vanilla, cinnamon).

- Texture – biscuits, cakes and sponges vary in texture, from light, airy sponges to sweet, crisp biscuits and cakes that can be full of rich fruit. It is part of the enjoyment of eating to consume foods of varying textures. It is, however, very important that the texture of the finished product meets the intended specification and not an undesired texture created by mistakes in the preparation and/or baking of the product.
- Consistency – it is vital that products are consistent in achieving a high quality output in production. Customers will expect a high standard product on every occasion and will be dissatisfied to be served an inferior product on another occasion. High-quality, consistent standards are essential.
- Presentation – the production of biscuits, cakes and sponges provides a great opportunity for creative and imaginative presentation, demonstrating the skill and techniques of the chef(s) responsible for the creation of the product. Again, this must be consistent and practical.
- Harmony – products should be harmonious across the points described above. All of the quality points mentioned should complement one another, creating a very high quality product.
- Temperature – most biscuits, cakes and sponges are served at an ambient temperature as refrigeration would damage the texture in many cases, especially if stored for a long period of time. It is, however, important to consider the additional ingredients used in the final product and recognise that any high risk commodities, such as cream, should be carefully considered in both storage and service. Such products are best served immediately to prevent the need for storage as this will reduce the eating qualities and presentation of the products concerned.

Allergies

Although it is essential to clearly list all potential allergens when making cakes, sponges and biscuits, the allergens that are most like to be used in their production include:

- gluten – flours and any products made from wheat, rye, barley and oats
- nuts – such as ground hazelnuts and almonds; these can also be added to flavour pastes such as sable
- eggs – widely used in the production of cakes, sponges and biscuits
- lactose – found in milk and milk products (used in some cake recipes)
- sesame seeds – may be incorporated into biscuits – biscuits served with cheese is a common example.

Also refer to Unit 301 Legal and social responsibilities in the professional kitchen for the 14 listed allergens required by EU Regulation FIR 1169/201. Beyond the basic preparation of the cake, sponge or biscuit base, attention is also required with regard to the additional ingredients that are used to complete this range of products. Cakes and gâteaux are often filled with creams, produced with milk and/or cream (lactose). Other fillings and decorations may include nuts, such as frangipane or sugared hazelnuts, so it is vitally important to assess any of the other potential allergens that are incorporated into products as well as the base product itself.

TEST YOURSELF

1 How much flour is required to produce a four-egg Genoese sponge?
2 Describe what is meant by the 'creaming' method.
3 Describe the production of biscuit à la cuillère.
4 What is the ratio of fat to flour for shortbread biscuits?
5 Describe tuiles biscuits and what they are used for.
6 Describe the preparation and baking of the following:
   a) Madeleines
   b) Sablé à la poche.
7 List the ingredients and method for a traditional Victoria sandwich.
8 List the various shapes that can be produced from a brandy snap mixture.
9 Describe three faults that are common when producing cakes and sponges, giving the reasons why the fault may have occurred.
10 Describe the quality points to look for in a fresh cream fruit gâteau.

1 Genoese sponge (*génoise*)

Energy	Calories	Fat	Saturated fat	Carbohydrates	Sugar	Protein	Fibre *
5978 kJ	1423 kcal	65.8 g	25.6 g	182.8 g	106.6 g	36.5 g	3.6 g

* Using hard margarine (4 portions)

Mise en place

1 Ensure that the cake tins are lined and ready to use.

Cooking

1 Whisk the eggs and sugar with a balloon whisk in a bowl over a pan of hot water.

2 Continue until the mixture is light and creamy and has doubled in bulk.

3 Remove from the heat and whisk until cold and thick (ribbon stage). Fold in the flour very gently.

4 Take a small amount of the mixture and combine it with the melted butter. Then return this to the rest of the mixture and fold through.

5 Place in a greased, floured Genoese mould.

6 Bake in a moderately hot oven, at 200–220 °C, for about 30 minutes.

	Single sponge	Double sponge
Eggs	4	10
Caster sugar	100 g	250 g
Flour (soft)	100 g	250 g
Butter, margarine or oil	50 g	125 g

Ingredients for Genoese sponge and boiling water ready for use.

Add the sugar to the eggs.

Whisk them together over boiling water.

Carry on whisking as the mixture warms up.

When the mixture is ready, it will form ribbons and you can draw a figure eight with it.

Fold in the flour.

Add part of the flour mixture to the butter.

Place the mixture into greased cake tins.

After baking, turn the sponges out to cool on a rack.

2 Chocolate Genoese sponge (*génoise au chocolat*)

Energy	Calories	Fat	Saturated fat	Carbohydrates	Sugar	Protein	Fibre *
1428 kJ	340 kcal	13.9 g	6.6 g	47 g	30 g	9.5 g	1.6 g

* Using hard margarine (4 portions)

Video: Making a Genoese
http://bit.ly/2ppqi30

	2 x 16 cm sponges
Eggs	8
Caster sugar	225 g
Flour	175 g
Cocoa powder	50 g
Butter, melted	65 g

Mise en place

1 As for Genoese but ensure that the flour and cocoa powder are sieved together and are well incorporated.

Cooking

1 Whisk the eggs and sugar together to form a sabayon.

2 Slowly fold in the flour and cocoa powder.

3 Take a small amount of the mixture and combine it with the melted butter. Then return this to the rest of the mixture and fold through.

4 Place in a lined mould and bake at 180°C for 15–20 minutes.

> **FAULTS**
> If the flour is folded in too harshly or the butter is not conditioned by mixing with a little of the sponge mix (too hot), the mix will lose aeration and will not rise as it should during the baking process.

3 Fresh cream and strawberry gâteau

Energy	Calories	Fat	Saturated fat	Carbohydrates	Sugar	Protein	Fibre
1975 kJ	473 kcal	28.4 g	11.1 g	530 g	39.7 g	4.5 g	0.9 g

	8 portions
Genoese sponge made with vanilla	1
Stock syrup (see page 469)	100 ml
Raspberry jam	50 ml
Whipping or double cream	500 ml
Icing sugar	75 g
Strawberries, sliced	1 punnet

Mise en place

1 Slice the strawberries, lightly whip the cream and chill.

2 Have a serrated knife ready to slice the sponge and a comb-scraper and board ready for assembly.

Preparing the dish

1 Carefully slice the sponge cake into three equal discs. Brush each with syrup.

2 Slowly whip the cream with the icing sugar to achieve the correct consistency.

3 Place the first piece of sponge on a cake board. Soak with syrup. Spread with a layer of jam, then a layer of cream. Scatter sliced strawberries on top.

4 Place the next piece of sponge on top. Repeat the layers of syrup, cream and strawberries. Top with additional cream.

5 Place the final piece of sponge on top.

6 Coat the top and sides with cream. Chill.

7 Comb scrape the sides of the gâteau. Pipe 12 rosettes on top.

> **VARIATION**
> ● Other seasonal fruits could be used for this gâteau. Examples include raspberries, tangerine and blueberries.

> **FAULTS**
> Do not over whip the cream. Cream will continue to thicken as it is worked with, so this needs to be considered during the production of the gâteau.

4 Chocolate gâteau

Energy	Calories	Fat	Saturated fat	Carbohydrates	Sugar	Protein	Fibre *
20113 kJ	4789 kcal	260.9 g	148.7 g	606.0 g	533.2 g	41.6 g	4.8 g

* Using hard margarine and butter (4 portions)

	Single gâteau	Double gâteau
Chocolate genoise sponge (recipe 2)		
Eggs	4	10
Chocolate vermicelli or flakes	50 g	125 g
Stock syrup (see page 469) as required		
Buttercream		
Unsalted butter	200 g	500 g
Icing sugar	150 g	375 g
Block chocolate (melted in a basin in a bain-marie)	50 g	125 g

Preparing the dish

1 Cut the Genoese into three slices crosswise.

2 Prepare the buttercream and mix in the melted chocolate.

3 Lightly moisten each slice of Genoese with stock syrup, which may be flavoured with kirsch, rum, etc.

4 Lightly spread each slice of Genoese with buttercream and sandwich together.

5 Lightly coat the sides with buttercream, then cover with chocolate vermicelli or flakes.

6 Neatly smooth the top using a little more buttercream if necessary.

> **NOTE**
> Chocolate glaze may be purchased as a commercial product.

5 Coffee gâteau

Energy	Calories	Fat	Saturated fat	Carbohydrates	Sugar	Protein	Fibre
7881 kJ	1873 kcal	75 g	41 g	303 g	282 g	12 g	1.8 g

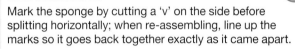

9 Starting in the centre and moving outwards, pour over the fondant to completely cover. Draw a palette knife across the top to remove the excess.

10 Add some melted chocolate to some of the fondant, adjust the consistency and squeeze through muslin. Decorate the gâteau by piping on a fine line design.

11 Finish the sides with squares of chocolate and the top with crystallised violets.

	1 × 16 cm gâteau
Plain Genoese sponge (recipe 1)	1 × 16 cm
Stock syrup flavoured with rum	50 ml
Coffee buttercream	750 g
Coffee marzipan	100 g
Fondant	500 g
Crystallised violets	
Chocolate squares	

Preparing the dish

1 Carefully split the sponge into three and line up the three pieces

2 Place the sponge base on a cake card and moisten with rum syrup.

3 Pipe on an even layer of buttercream, no thicker than that of the sponge.

4 Place on the next layer of sponge, moisten with the syrup and repeat to give three layers of sponge and two of buttercream. Moisten the top with syrup.

5 Put in the fridge for 1–2 hours to firm up.

6 Work some coffee essence into the marzipan, roll out to 2 mm thick and lay over the gâteau, working the sides to prevent any creases.

7 Warm the fondant to blood heat, flavour with coffee essence and the adjust consistency with syrup.

8 Place the gâteau on a wire rack with a tray underneath to catch the fondant.

PROFESSIONAL TIP

Mark the sponge by cutting a 'v' on the side before splitting horizontally; when re-assembling, line up the marks so it goes back together exactly as it came apart.

Turn the sponge upside down before splitting so the base becomes the top; this is the flattest surface and will give the best finish.

It is best practice to use a Genoese that was made the day before – fresh sponges do not cut well and are susceptible to falling apart.

Fondant should never be heated above 30 °C, as the shine will be lost.

NOTE

A good-quality coffee gâteau should have a moist sponge and a good balance between sponge and filling (as a guide, the thickness of the sponge and the depth of the buttercream should be equal). The coffee flavour should not be in question, and the decoration should reflect and complement the coffee theme. (It is sometimes easy to get carried away, so it is good to remember when decorating, 'less is definitely more'.)

VARIATION

- Instead of enrobing with fondant, the top and sides can be covered with buttercream, the sides can be either comb-scraped or masked with toasted nibbed or flaked almonds or grated chocolate. The top can be piped with buttercream and/or decorated with coffee marzipan cut-out shapes.
- To add another texture, place a disc of meringue or dacquoise on the bottom layer. Dacquoise is an Italian meringue with the addition of toasted ground hazelnuts, spread or piped onto a silicone mat and baked at 180 °C for 15–20 minutes. Cut out the desired shape half way through cooking.

6 Roulade sponge

Energy	Calories	Fat	Saturated fat	Carbohydrates	Sugar	Protein	Fibre	Sodium
4779 kJ	1129 kcal	23.2 g	6.6 g	199.0 g	137.0 g	133.0 g	3.4 g	313.0 g

You should be able to bend a roulade sponge.

	2 sheets	4 sheets
Eggs	8	16
Egg yolk	2	4
Caster sugar	260 g	520 g
Soft flour	170 g	340 g

Cooking

1 Make sure the mixing bowl is clean, dry and free from grease.

2 Line the baking sheets with silicone paper cut to fit.

3 Set the oven at 230 °C.

4 Place the eggs and sugar in a mixing bowl, and stir over hot water until warm.

5 Whisk to the 'ribbon' stage and sieve the flour onto greaseproof paper.

6 Carefully fold in the flour.

7 Divide equally between the baking sheets and spread evenly with a drop blade palette knife. Place immediately in the oven for between 5 and 7 minutes.

8 As soon as the sponge is cooked, turn it out onto sugared paper, place a damp, clean cloth over it and lay the hot baking sheet back on top, then leave to cool. (This will help keep the sponge moist and flexible as it cools.)

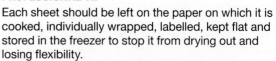

PROFESSIONAL TIP

Each sheet should be left on the paper on which it is cooked, individually wrapped, labelled, kept flat and stored in the freezer to stop it from drying out and losing flexibility.

FAULTS

There are two reasons why a roulade sponge might become hard and crisp, instead of being pliable:
- baked at too low a temperature for too long
- mixture spread too thin.

VARIATION

- A chocolate version can be made by substituting 30–40 per cent of the flour for cocoa powder. For a coffee sponge, add coffee extract to the eggs after whisking.
- A roulade sponge may also be made using a 'split-egg' method, separating the eggs.

SERVING SUGGESTION

Due to its flexibility, roulade sponge is ideal for lining moulds as well as to make a roulade dessert (a sponge roll usually filled with a cream and fruit filling).

7 Swiss roll

Energy	Calories	Fat	Saturated fat	Carbohydrates	Sugar	Protein	Fibre
4445 kJ	1058 kcal	25.3 g	8.0 g	182.7 g	106.5 g	36.5 g	3.6 g

	4 portions	10 portions
Eggs	4	10
Caster sugar	100 g	250 g
Self-raising flour	100 g	250 g
Jam, as required		

Mise en place

1 Ensure the Swiss roll tin is greased and lined with greased greaseproof or silicone paper.

2 Have a sheet of parchment paper sprinkled with additional caster sugar in place to turn the Swiss Roll onto once it has been baked.

Cooking

1 Whisk the eggs and sugar with a balloon whisk in a bowl over a pan of hot water.

2 Continue until the mixture is light, creamy and double in bulk.

3 Remove from the heat and whisk until cold and thick (ribbon stage).

4 Fold in the flour very gently.

5 Grease a Swiss roll tin and line with greased greaseproof or silicone paper.

6 Pour in the mixture and bake at 220 °C for about 6 minutes.

7 Turn out on to a sheet of paper sprinkled with additional caster sugar.

8 Remove the paper from the Swiss roll, spread with warm jam.

9 Immediately roll up as tight as possible and leave to cool completely.

8 Rich fruit cake

Energy	Calories	Fat	Saturated fat	Carbohydrates	Sugar	Protein	Fibre
2367 kJ	563 kcal	24.8 g	11.3 g	81.9 g	69.4 g	8.8 g	2.8 g

	16 cm diameter, 8 cm deep	21 cm diameter, 8 cm deep	26 cm diameter, 8 cm deep
Butter	150 g	200 g	300 g
Soft brown sugar	150 g	200 g	300 g
Eggs	4	6	8
Black treacle	2 tsp	3 tsp	1 tbsp
Soft flour	125 g	175 g	275 g
Salt	6 g	8 g	10 g
Nutmeg	3 g	4 g	5 g
Mixed spice	3 g	4 g	5 g

Ground cinnamon	3 g	4 g	5 g
Ground almonds	75 g	100 g	125 g
Currants	150 g	200 g	300 g
Sultanas	150 g	200 g	300 g
Raisins	125 g	150 g	225 g
Mixed peel	75 g	100 g	125 g
Glacé cherries	75 g	100 g	125 g
Grated zest of lemon	½	¾	1
Oven temperatures			
	150 °C	140 °C	130 °C
Approximate cooking times			
	2 hours	3 hours	4 ½ hours

Mise en place

1 There are a lot of ingredients in this cake. Weigh everything out accurately before starting this recipe.

Cooking

1 Cream the butter and sugar until soft and light.

2 Break up the eggs and beat in gradually.

3 Add the black treacle.

4 Sieve all the dry ingredients together and fold in.

5 Finally fold in the dried fruit and lemon zest.

6 Deposit into buttered cake tins lined with silicone paper.

7 Level the mix and make a well in the centre.

8 Check the cakes during baking, turn to make sure they are being cooked evenly.

9 Test to see if cooked with a metal skewer or needle.

10 Allow to cool completely before wrapping and storing in an airtight container.

NOTE

In the UK, fruit cakes are traditionally used as a base for celebration cakes such as Simnel and Christmas cakes and for weddings.

If using square cake tins increase the quantities by a quarter.

This cake can be made less rich by cutting down on the fruit and spices – these are all 'carried' ingredients and will not affect the cake as long as the basic ingredients (butter, sugar, eggs, flour) are not tampered with.

PROFESSIONAL TIP

This is a dense mixture. To prevent the outside becoming overcooked insulate the cake tins by standing on newspaper and tying several layers of newspaper around the sides.

The dried fruit can be pre-soaked in brandy or rum the day before, or, as the cake matures it can be given a 'drink' every so often.

Take a spoon and make a well in the centre of the mixture, this will help stop the cake doming as it bakes.

Unless filling the oven with cakes it is advisable to place a tray of water in the oven when baking. This will create steam and allow the cake to expand before the crust sets.

To test, insert a needle in the centre; when the cake is cooked the needle should come out clean and hot.

After cooling wrap in paper or foil and place in an airtight container and leave to mature for 3–4 weeks before covering with marzipan and decorating with icing.

FAULTS

Cake A is domed ('cauliflower top'). This occurs if:
● the flour used is too strong
● the oven is too hot and/or too dry,
● there is not enough fat
● the ingredients are over-mixed after adding flour.

Cake B has a sunken top ('M' fault). This is caused by adding too much baking powder and/or too much sugar.

Cake C has a sunken top and sides ('X' fault). This occurs if the mixture is too wet.

Cake D has a low volume. This occurs if:
● too much fat is added
● the mixture is too dry
● there is not enough aeration.

A good-quality fruit cake will:
● have straight sides, a flat top and good height
● have an even distribution of fruit
● be moist and dark.

9 Banana bread

Energy	Calories	Fat	Saturated fat	Carbohydrates	Sugar	Protein	Fibre
1421 kJ	338 kcal	13.6 g	2.4 g	52.7 g	36.3 g	4.4 g	2.0 g

Cooking

1 Beat the bananas, oil and butter together at a medium speed with the paddle attachment in a food mixer.

2 Add sugar and eggs and mix until smooth.

3 Add the dry ingredients and mix well.

4 Place mixture into three well-greased or silicon-lined tins (7.5 cm x 17.5 cm long x 10 cm wide).

5 Bake in a medium fan oven for 35 minutes and then check if cooked by inserting a skewer into the cake mixture. If the skewer comes out clean, the banana bread is done.

6 Allow to cool.

	3 cakes
Ripe bananas	460 g
Vegetable oil	110 ml
Melted butter	140 g
Caster sugar	460 g
Eggs	4
Soft flour	460 g
Baking powder	20 g
Salt	½ tsp

10 Lemon drizzle cake

Energy	Calories	Fat	Saturated fat	Carbohydrates	Sugar	Protein	Fibre
1437 kJ	342 kcal	14.6 g	8.6 g	51.7 g	33.5 g	4 g	0.9 g

	2 x 16 cm cakes
Butter	250 g
Caster sugar	400 g
Grated zest of lemons	3
Soft flour	380 g
Baking powder	10 g
Eggs	4
Vanilla extract	½ tsp
Milk	25 ml
Syrup	
Lemons, juice of	3
Caster sugar	100 g

Mise en place

1 Make the syrup and have it ready to pour onto the cake.

Cooking

1 Cream the butter, sugar and zest until soft and light.

2 Sieve the flour and baking powder twice.

3 Mix together the eggs and vanilla extract.

4 Beat the eggs into the butter and sugar mixture.

5 Fold in the flour.

6 Add milk to achieve a dropping consistency.

7 Deposit into buttered and floured cake tins.

8 Bake at 165 °C for approximately 45 minutes.

9 Boil the lemon juice and sugar.

10 When the cake is cooked, stab with a skewer and pour over the syrup.

11 Leave to cool in the tin.

> **SERVING SUGGESTION**
> This cake can be finished with lemon icing and decorated with strips of crystallized lemon peel.

11 Victoria sandwich

Energy	Calories	Fat	Saturated fat	Carbohydrates	Sugar	Protein	Fibre
6866 kJ	1635 kcal	94.3 g	39.3 g	184.7 g	106.6 g	23.3 g	3.6 g

3 Mix together the eggs and vanilla extract.

4 Beat the eggs gradually into the butter and sugar mixture.

5 Fold in the flour.

6 Deposit into buttered and floured cake tins and level.

7 Bake at 180 °C for approximately 15–20 minutes.

8 Turn out onto a wire rack to cool.

9 Spread the bottom half with softened jam.

10 Place on the top sponge and dust with icing sugar.

> **VARIATION**
> - In addition to jam the sponge can be filled with either butter icing or Chantilly cream.
> - The official Women's Institute version specifies the cake is filled with jam only and dusted with caster not icing sugar.

> **NOTE**
> This is a classic afternoon tea cake named after Queen Victoria. Although traditionally made in two halves, a slimmer version can be made by using a single sponge and splitting it.

	2 x 18 cm cakes
Butter	250 g
Caster sugar	250 g
Soft flour	250 g
Baking powder	10 g
Eggs	5
Vanilla extract	½ tsp
Jam to fill	

Cooking

1 Cream the butter, sugar until soft and light.

2 Sieve the flour and baking powder twice.

> **FAULTS**
> Adding the egg to the creamed butter and sugar will curdle the mixture. Add the egg slowly, mixing well to incorporate the egg and emulsify into the mix.

Beat the sugar and butter together.

Place the mixture into buttered cake tins.

Flatten the top before baking.

12 Scones

Energy	Calories	Fat	Saturated fat	Carbohydrates	Sugar	Protein	Fibre
678 kJ	162 kcal	5.8 g	2.5 g	26.3 g	7.5 g	2.7 g	1.0 g

4 Add the liquid to the dry ingredients and cut in with a plastic scraper, mix lightly and do not overwork. Wrap in cling film and chill for 1 hour.

5 Set the oven at 180 °C and line a baking sheet with silicone paper.

6 Roll out 2-cm thick on a floured surface, cut out with a plain or fluted cutter.

7 Brush with milk or eggwash and bake at 180 °C for approximately 15–20 minutes.

8 After 15 minutes, one scone could be tested by pulling apart to see if they are cooked through.

9 Allow to cool and dust with icing sugar before serving.

Scones are traditionally served at afternoon tea with jam and butter or clotted cream, and are best served on the day they are made.

	16 scones
Plain flour	450 g
Baking powder	25 g
Pinch of salt	
Butter	225 g
Caster sugar	170 g
Sour cream	300 ml

Cooking

1 Sieve the flour, baking powder and salt.

2 Cut the butter into small pieces and rub into the flour to achieve a sandy texture.

3 Dissolve the sugar in the cream.

> **VARIATION**
> ● **Fruit scones:** Add 50 g sultanas to the basic mix, or try adding dried cranberries or apricots as alternatives.

> **PROFESSIONAL TIP**
> After cutting out the scones turn upside down on the baking sheet, this will help them rise with straight sides.

13 Cupcakes

Energy	Calories	Fat	Saturated fat	Carbohydrates	Sugar	Protein	Fibre
947 kJ	225 kcal	11.6 g	4.8 g	28.8 g	13.4 g	3.3 g	0.7 g

* Using hard margarine

	20 portions
Flour (soft) or self-raising	200 g
Baking powder (if using plain flour)	1 level tsp
Salt (optional)	pinch
Margarine or butter	125 g
Caster sugar	125 g
Eggs	2–3

Cooking

Method 1 – rubbing in:

1 Sieve the flour, baking powder and salt (if using).

2 Rub in the butter or margarine to achieve a sandy texture. Add the sugar.

3 Gradually add the well-beaten eggs and mix as lightly as possible until combined.

Method 2 – creaming:

1 Cream the margarine and sugar in a bowl until soft and fluffy.

2 Slowly add the well-beaten eggs, mixing continuously and beating really well between each addition.

3 Lightly mix in the sieved flour, baking powder and salt (if using).

NOTE

In both cases, the consistency should be a light dropping one and, if necessary, it may be adjusted with the addition of a few drops of milk.

This is a great base for a cupcake.

VARIATION

- **Cherry cakes**: add 50 g glacé cherries cut in quarters and 3–4 drops vanilla essence to the basic mixture (method 2) and divide into 8–12 lightly greased cake tins or paper cases. Bake in a hot oven at 220 °C for 15–20 minutes.
- **Coconut cakes**: in place of 50 g flour, use 50 g desiccated coconut and 3–4 drops vanilla essence to the basic mixture (method 2) and cook as for cherry cakes.
- **Raspberry buns**: divide basic mixture (method 1) into 8 pieces. Roll into balls, flatten slightly, dip tops into milk then caster sugar. Place on a greased baking sheet, make a hole in the centre of each and add a little raspberry jam. Bake in a hot oven at 200 °C for 15–20 minutes.
- **Queen cakes**: to the basic mixture (method 2) add 100 g washed and dried mixed fruit and cook as for cherry cakes.

14 Sponge fingers (*biscuits à la cuillère*)

Energy	Calories	Fat	Saturated fat	Carbohydrates	Sugar	Protein	Fibre
372 kJ	88 kcal	2 g	0.6 g	16 g	8.8 g	1.3 g	0.2 g

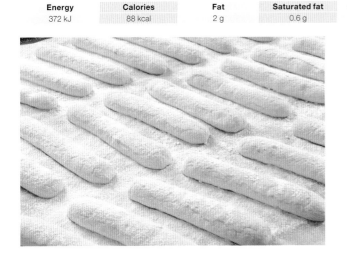

	Approximately 60 × 8cm fingers
Egg yolks	180 g
Caster sugar	125 g
Vanilla essence	Few drops
Soft flour	125 g
Cornflour	125 g
Egg whites	270 g
Caster sugar	125 g

Mise en place

1 Prepare a baking sheet by lining with silicone paper cut to fit.

2 Have ready a piping bag fitted with a medium plain tube.

3 Scald two mixing bowls to ensure they are clean and free of grease.

Cooking

1 Whisk the yolks, sugar and vanilla over a bain-marie until warm, then continue whisking off the heat until a thick, sabayon-like consistency is reached.

2 Sieve the flours onto paper.

3 In a second mixing bowl, whisk the whites with the sugar to a soft meringue.

4 Add the whisked yolks to the meringue and start folding in. Add the flour in 2 or 3 portions, working quickly but taking care not to overwork the mixture.

5 Using a plain piping tube, immediately pipe onto the prepared baking sheet in neat rows.

6 Dust evenly with icing sugar and immediately place in the oven for approximately 25 minutes.

7 When cooked, slide the paper (and biscuits) onto a cooling rack.

8 When cool, remove from the paper and store in an airtight container at room temperature, or leave on the pap er and store in a dry cabinet.

NOTE

The literal translation of this is 'spoon biscuits', which comes from a time when the mixture would have been shaped between two spoons instead of being piped. They are traditionally used to line the mould for a Charlotte Russe, although the 'spooned' version would not lend itself to that.

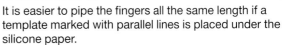

PROFESSIONAL TIP

It is easier to pipe the fingers all the same length if a template marked with parallel lines is placed under the silicone paper.

A common problem with this recipe is over-mixing and/or not working fast enough or being disorganised, which results in biscuits that collapse.

Sponge fingers should be pale in colour, very light in texture, be dusted with icing sugar and have a rounded shape. They should also be identical in length and width.

VARIATION

- For a chocolate version, instead of 125 g each of soft flour and cornflour, use 120 g soft flour, 60 g cornflour and 70 g cocoa powder.
- **Othellos**: use the above recipe to make small, domed sponges. Hollow them out and fill with crème mousseline (see page 460). Sandwich pairs together and coat with coloured fondant.

FAULTS

Be careful not to mix too much when folding in the flour. This will lose aeration and create a dense texture once baked.

15 Shortbread biscuits

Energy	Calories	Fat	Saturated fat	Carbohydrates	Sugar	Protein	Fibre
507 kJ	121 kcal	7.0 g	4.4 g	14.1 g	4.6 g	1.2 g	0.5 g

* Using butter

Method 1:

	12 portions
Flour (soft)	150 g
Salt	pinch
Butter or margarine	100 g
Caster sugar	50 g

Cooking

1 Sift the flour and salt.

2 Mix in the butter or margarine and sugar with the flour.

3 Combine all the ingredients to a smooth paste.

4 Roll carefully on a floured table or board to the shape of a rectangle or round, 0.5 cm thick. Place on a lightly greased baking sheet.

5 Mark into the desired size and shape. Prick with a fork.

6 Bake in a moderate oven at 180–200 °C for 15–20 minutes.

Method 2:

	12 portions
Flour (soft), white or wholemeal	100 g
Rice flour	100 g
Butter or margarine	100 g
Caster or unrefined sugar	100 g
Egg, beaten	1

Cooking

1 Sieve the flour and rice flour into a basin.

2 Rub in the butter until the texture of fine breadcrumbs. Mix in the sugar.

3 Bind the mixture to a stiff paste using the beaten egg.

4 Roll out to 3 mm using caster sugar, prick well with a fork and cut into fancy shapes. Place the biscuits on a lightly greased baking sheet.

5 Bake in a moderate oven at 180–200 °C for 15 minutes or until golden brown.

6 Remove with a palette knife on to a cooling rack.

Method 3:

	12 portions
Butter or margarine	100 g
Icing sugar	100 g
Egg	1
Flour (soft)	150 g

Cooking

1 Cream the butter or margarine and sugar thoroughly.

2 Add the egg and mix in. Mix in the flour.

3 Pipe on to lightly greased and floured baking sheets using a large star tube.

4 Bake at 200–220 °C, for approximately 15 minutes.

FAULTS

Do not over-mix the paste as it will develop the gluten which may lead to retraction when rolling and shrinkage or products becoming misshapen after baking.

SERVING SUGGESTION

A simple finger of shortbread is an ideal accompaniment for a dessert such as crème brûlée.

16 Cats' tongues (*langues de chat*)

Energy	Calories	Fat	Saturated fat	Carbohydrates	Sugar	Protein	Fibre	Sodium
676 kJ	161 kcal	8.3 g	5.2 g	20.7 g	13.2 g	2.1 g	0.4 g	0.1 g

Pipe cats' tongues into their distinctive shape, thicker at the ends

	Approximately 40
Icing sugar	125 g
Butter	100 g
Vanilla essence	3–4 drops
Egg whites	3–4
Flour (soft)	100 g

Cooking

1 Lightly cream the sugar and butter, add the vanilla essence.

2 Add the egg whites one by one, continually mixing and being careful not to allow the mixture to curdle.

3 Gently fold in the sifted flour and mix lightly.

4 Pipe on to a lightly greased baking sheet using a 3 mm plain tube, 2½ cm apart.

5 Bake at 230–250 °C, for a few minutes.

6 The outside edges should be light brown and the centres yellow.

7 When cooked, remove on to a cooling rack using a palette knife.

PROFESSIONAL TIP

Mark lines on the baking sheet to provide a guide before piping. This can be on the underside of parchment or greaseproof paper so that the markings do not come into direct contact with the product itself.

FAULTS

If the consistency of the paste is too thick or thin, it may hold too tightly or spread out during the baking process.

SERVING SUGGESTION

Langue de chat are regularly served as petits fours.

17 Piped biscuits (*sablés à la poche*)

Energy	Calories	Fat	Saturated fat	Carbohydrates	Sugar	Protein	Fibre
993 kJ	237 kcal	15.2 g	8.2 g	23.3 g	8.4 g	3.4 g	0.8 g

Cooking

1 Cream the sugar and butter until light in colour and texture.

2 Add the egg gradually, beating continuously, add the vanilla essence or lemon zest.

3 Gently fold in the sifted flour and almonds, mix well until suitable for piping. If too stiff, add a little beaten egg.

4 Pipe on to a lightly greased and floured baking sheet using a medium-sized star tube (a variety of shapes can be used).

5 Some biscuits can be left plain, some decorated with half almonds or neatly cut pieces of angelica and glacé cherries.

6 Bake in a moderate oven at 190 °C for about 10 minutes.

7 When cooked, remove on to a cooling rack using a palette knife.

	20–30 biscuits
Caster or unrefined sugar	75 g
Butter or margarine	150 g
Egg	1
Vanilla essence or Grated zest of one lemon	3–4 drops
Soft flour, white or wholemeal	200 g
Ground almonds	35 g

18 Viennese biscuits

Energy	Calories	Fat	Saturated fat	Carbohydrates	Sugar	Protein	Fibre	Sodium	*
1258 kJ	301 kcal	17.9 g	11.3 g	33.2 g	19.0 g	18.9 g	0.7 g	4.3 g	

* Using unsalted butter and vanilla extract

	16 biscuits
Butter, softened	250 g
Icing sugar, plus extra for sieving to decorate	50 g
Plain flour	250 g
Cornflour	50 g
Vanilla pod or pure vanilla extract	1 pod or ½ tsp
Filling	
Soft butter	100 g
Icing sugar	200 g, plus ½ tsp for dusting
Vanilla pod or pure vanilla extract	1 pod or ½ tsp
Seedless raspberry jam	75 g

Mise en place

1 Have the jam ready and the buttercream pre-prepared at the correct consistency to pipe on to the baked and cooled biscuits before sandwiching together.

Cooking

1 Cream the butter and icing sugar.

2 Add the plain flour, cornflour and vanilla extract and blend until smooth (a food mixer could be used for stages 2 and 3).

3 Transfer the mixture into a piping bag fitted with a large star nozzle. Pipe into 6-cm rosettes, spacing well apart.

4 Bake in the centre of the oven for 13–15 minutes or until pale golden-brown and firm. Cool on the baking tray for a few minutes then transfer to a cooling rack.

5 For the filling, cream the butter with the sifted icing sugar. Add the vanilla extract and beat until very light and smooth. Spoon into a clean piping bag fitted with a large star nozzle.

6 Put the jam in a bowl and stir until smooth.

7 Spoon a little jam onto the flat side of half the biscuits and place jam-side up on the cooling rack. Pipe the buttercream icing onto the remaining biscuits and sandwich with the jam.

8 Serve with a little sifted icing sugar on the top surface.

> **SERVING SUGGESTION**
> Viennese biscuits are a perfect addition to an afternoon tea.

19 Chocolate pecan brownie

Energy	Calories	Fat	Saturated fat	Carbohydrates	Sugar	Protein	Fibre	Sodium *
1762 kJ	424 kcal	30.1 g	16.8 g	31.6 g	27.9 g	25.2 g	3.5 g	24.0 g

* Using unsalted butter

	24 slices
Butter	330 g
Dark chocolate	330 g
Soft flour	150 g
Cocoa powder	50 g
Eggs	6
Caster sugar	450 g
Vanilla extract	½ tsp
Roasted and chopped pecans	100 g
Topping	
Whipping cream	250 ml
Glucose	25 g
Dark chocolate	250 g

Cooking

1 Cut the butter into small pieces and melt with the chocolate over hot water.

2 Sieve the flour and cocoa powder onto paper.

3 Add the vanilla extract to the eggs and whisk with the sugar to the ribbon stage. This can be done over a bowl of hot water to assist the aeration process.

4 Fold the chocolate and butter into the eggs followed by the flour, cocoa and pecans.

5 Pour into a silicone-lined deep tray, smooth top and bake at 165 °C for approximately 30 minutes. The mixture should still be soft in the middle.

6 For the topping boil the cream and glucose, whisk in the chocolate.

7 Allow to cool before pouring over the brownie, spread evenly and at setting point use a comb scraper to produce wavy lines.

8 When set cut into even sized rectangles.

> **VARIATION**
> ● The mixture can be baked in individual dariole moulds and served as a dessert with cream or ice cream.

20 Tuiles

Energy	Calories	Fat	Saturated fat	Carbohydrates	Sugar	Protein	Fibre
417 kJ	100 kcal	5.5 g	3.4 g	12.1 g	7 g	1.1 g	0.3 g

Mise en place

Have the moulds ready and close to the work-station. You will need a palette knife to lift the biscuits from the tray.

Cooking

1 Mix all ingredients; allow to rest for 1 hour.

2 Spread to the required shape and size.

3 Bake at approximately 200–210 °C.

4 While hot, mould the biscuits to the required shape and leave to cool.

	15–20 portions
Butter	100 g
Icing sugar	100 g
Flour	100 g
Egg whites	2

PROFESSIONAL TIP

When the tuiles first come out of the oven, they will be too hot and fluid to mould easily. They will need a few seconds to acclimatise slightly and then they will be ready to mould. Try to avoid placing the baking tray directly onto a cool surface. This will cool the biscuits quicker and give less time to lift and mould the biscuits.

21 Brandy snaps

Energy	Calories	Fat	Saturated fat	Carbohydrates	Sugar	Protein	Fibre
2146 kJ	510 kcal	21 g	13.1 g	82 g	66 g	1.5 g	1 g

	Approximately 20
Strong flour	225 g
Ground ginger	10 g
Golden syrup	225 g
Butter	250 g
Caster sugar	450 g

Mise en place

1 Have the moulds ready and close to the work-station. Use a palette knife to lift the biscuits from the tray before shaping on the moulds.

Cooking

1 Combine the flour and ginger in a bowl on the scales. Make a well.

2 Pour in golden syrup until the correct weight is reached.

3 Cut the butter into small pieces. Add the butter and sugar.

4 Mix together at a slow speed.

5 Divide into 4 even pieces. Roll into sausage shapes, wrap each in cling film and chill, preferably overnight.

6 Slice each roll into rounds. Place on a baking tray, spaced well apart.

7 Flatten each round using a fork dipped in cold water, keeping a round shape.

8 Bake in a pre-heated oven at 200 °C until evenly coloured and bubbly.

9 Remove from oven. Allow to cool slightly, then lift off and shape over a dariole mould.

10 Stack the snaps, no more than 4 together, on a stainless steel tray and store.

PROFESSIONAL TIP

As with the tuile biscuits, the timing of the removal of the brandy snaps from the baking tray is very important. If the biscuits are lifted too early they will be soft and slightly liquid, and will form back into a paste. They will need a short time (20 to 30 seconds) to acclimatise slightly and then they will be ready to mould. Try to avoid placing the baking tray directly onto a cool surface. This will cool the biscuits quicker and give less time to lift and mould the biscuits.

SERVING SUGGESTION

Brandy snaps are the perfect accompaniment to a wide range of items, such as mousses, ice creams, creams and fresh fruit.

22 Madeleines

Energy	Calories	Fat	Saturated fat	Carbohydrates	Sugar	Protein	Fibre
392 kJ	94 kcal	5.2 g	3 g	10.8 g	5.8 g	0.7 g	0.2 g

	Makes 45
Caster sugar	125 g
Eggs	3
Vanilla pod, seeds from	1
Flour	150 g
Baking powder	1 tsp
Beurre noisette	125 g

Mise en place

1 Prepare the vanilla pod by splitting it lengthways in half and scraping out the seeds carefully with the tip of a small knife.

Cooking

1 Whisk the sugar, eggs and vanilla seeds to a hot sabayon.

2 Fold in the flour and the baking powder.

3 Fold in the *beurre noisette* and chill for up to 2 hours.

4 Pipe into well-buttered madeleine moulds and bake in a moderate oven.

5 Turn out and allow to cool.

PROFESSIONAL TIP

Beurre noisette translates from French to English as 'hazelnut butter'. It is made by heating slices or cubes of cold butter in a dry pan. The butter will melt and as it heats it will start to bubble and foam. It will also start to darken in colour. Once the butter reaches a light, golden brown (hazelnut) colour, remove from the heat and pour it into a clean bowl.

FAULTS

Take care not to overheat the eggs when making the sabayon as they will start to scramble as the proteins denature and start to coagulate.

23 Cigarette or pencil paste cornets

Energy	Calories	Fat	Saturated fat	Carbohydrates	Sugar	Protein	Fibre
444 kJ	106 kcal	5.6 g	3.6 g	14 g	8.8 g	0.3 g	0.2 g

	Approximately 30
Icing sugar	125 g
Butter, melted	100 g
Vanilla essence	3–4 drops
Egg whites	3–4
Soft flour	100g

Mise en place

1 Have the moulds ready and close to the work-station. You will need a palette knife to lift the biscuits from the tray before shaping on the moulds.

Cooking

1 Proceed as for steps 1–3 of recipe 17 (cats' tongues).

2 Using a plain tube, pipe out the mixture onto a lightly greased baking sheet into bulbs, spaced well apart. Place a template over each bulb and spread it with a palette knife.

3 Bake at 150 °C, until evenly coloured.

4 Remove the tray from the oven. Turn the cornets over but keep them on the hot tray.

5 Work quickly while the cornets are hot and twist them into a cornet shape using the point of a cream horn mould. (For a tight cornet shape it is best to set the pieces tightly inside the cream horn moulds and leave them until set.) If the cornets set hard before you have shaped them all, warm them in the oven until they become flexible.

The same paste may also be used for cigarettes russes, coupeaux and other shapes.

> **PROFESSIONAL TIP**
> As with the tuile and brandy snaps biscuits, the timing of the removal of the cornets from the baking tray is very important.

> **SERVING SUGGESTION**
> These biscuits can be used as garnishes, sometimes filled with a cream, for desserts or served as a petits fours.

312 Fermented products

Learning outcomes

In this unit you will be able to:

1 Understand how to produce fermented products, including:
   - knowing the types of fermented doughs these techniques are used to produce
   - understanding conditions required for yeast fermentation
   - understanding the function of salt, eggs, fats, flour and sugar on the fermentation process and production of fermented dough products
   - understanding the preparation and processing techniques used to produce fermented doughs and
   - understanding the difference between fresh yeast, dried yeast and fast action yeast in the production of fermented dough products
   - knowing the extraction rates of gluten in different types of flour and the importance of gluten to fermented doughs and the effect of working doughs to develop gluten.
   - understanding the correct storage procedures to use throughout production and on completion of final products.

2 Produce and finish fermented products, including:
   - understanding how to use and operate tools and equipment to produce fermented dough products
   - being able to use techniques in preparation for baking of fermented dough products
   - understanding and being able to use techniques to finish fermented dough products after baking
   - being able to measure and evaluate against quality standards throughout preparation and cooking
   - being able to evaluate products against dish requirements and production standards and recognise any faults.

Recipes included in this chapter

No.	Recipe
Bread loaves and rolls	
1	Wholemeal rolls
2	Seeded rolls
3	Parmesan rolls
4	Red onion and sage rolls
5	Sun-dried tomato bread
6	Wholemeal bread
7	Rye bread
8	Soda bread
9	Olive bread
Enriched doughs	
10	Bun dough and varieties of bun
11	Bath buns
12	Hot cross buns

No.	Recipe
13	Swiss buns
14	Doughnuts
Speciality doughs	
15	Bagels
16	Focaccia
17	Pizza
18	Cholla bread
Fermented batters	
19	Savarin dough and derivatives
20	Savarin with fruit
21	Blueberry baba
22	Marignans Chantilly
23	Pitta bread
24	Blinis

Additional recipes available online at www.hoddereducation.co.uk/practical-cookery-resources

Understand how to produce fermented products

Bread and dough products contain wheat flour and yeast. Bread and bread products form the basis of our diet and are staple products in our society. We eat bread and fermented dough products at breakfast, lunch and dinner, as sandwiches, bread rolls, croissants, French sticks and so on. Bread is also used as an ingredient for many other dishes, either as slices or as breadcrumbs.

Flour-based products provide us with variety, energy, vitamins and minerals. Wholemeal bread products also provide roughage, an essential part of a healthy diet.

1.1 Techniques used to produce fermented doughs

Kneading and proving

Dough consists of strong flour, water, salt and yeast, which are kneaded together to the required consistency at a suitable temperature. It is then allowed to **prove** (to rise and increase in size), when the yeast produces carbon dioxide and water, which aerates the dough.

When baked it produces a light digestible product with flavour and colour. Proving allows the dough to ferment; the second prove is essential for giving dough products the necessary volume and a good flavour.

Video: Principles of breadmaking

http://bit.ly/2qa8ve4

Knead the dough

Before and after proving: the same amount of dough is twice the size after it has been left to prove.

Salted dough is much more manageable than unsalted dough. Salt is usually added a few moments before the end of the kneading, since its function is to help expand the dough's volume.

> **HEALTH AND SAFETY**
> Always remember the health and safety rules when using machinery. When using machines such as planetary mixers, check that they are in safe working order.

Knocking back

Remember to **knock the dough back** (re-knead it) carefully once proved, as this will expel the gas and allow the yeast to be dispersed properly, coming back into direct contact with the dough.

Fermentation

For dough to become **leavened** bread (bread that has risen, rather than flat bread) it must go through a **fermentation** process. This is brought about by the action of yeast, a living micro-organism rich in protein and vitamin B. The yeast reacts with enzymes in the dough, which convert sugar into alcohol, producing the characteristic flavour of bread. The action also produces carbon dioxide, which makes the bread rise.

Yeast requires ideal conditions for growth. These are:
- warmth – a good temperature for dough production is 22–30°C
- moisture – the liquid should be added at approximately 37°C
- food – this is obtained from the starch in the flour
- time – this is needed to allow the yeast to grow.

Dried yeast has been dehydrated and must be creamed with a little water before use. It will keep for several months in its dry state. Some types of dried yeast can be used straight from the packet.

Yeast will not survive in a high concentration of sugar or salt, and its growth will slow down in a very rich dough with a high fat and egg content.

When mixing yeast in water or milk, make sure that the liquid is at the correct temperature (37 °C) and disperse the yeast in the liquid. (As a living organism cannot be dissolved, the word 'disperse' is used.)

Yeast should be removed from the refrigerator and used at room temperature.

Developing

Flavours and textures are developed in doughs during the fermentation process. Doughs that are developed over long periods of time and from natural yeast found in flour will gain more flavour than doughs developed very quickly using fast-acting yeasts and dough development improvers. This is one reason why artisan breads made from natural starters and developed slowly are more expensive to buy than breads produced on a large scale using single fermentation processes with dough-development ingredients and stabilisers.

> **KEY TERMS**
>
> **Prove:** process of allowing dough to rise and increase in size.
>
> **Knocking back:** re-kneading the dough once proved.
>
> **Leavened:** bread that has risen.
>
> **Fermentation:** this occurs when yeast and/or bacteria convert carbohydrates to carbon dioxide, causing gas bubbles to form. This has a leavening effect on products such as dough when making bread.

The process of dough fermentation is extraordinary, but, because it is so frequently used in the profession, it is often taken for granted, without much thought as to what is actually happening during the process. It is useful to know why doughs ferment and what the effects are on the end product. In order to understand why yeast dough rises, it is important to note that the main ingredients of natural **leavening** are water, air and, most importantly, sugar, which is transformed into carbon dioxide and alcohol, causing the leavening. The carbon dioxide forms bubbles inside the dough and makes it rise. This is a scientific process where the fermentation is a transformation undergone by organic matter (sugars).

> **KEY TERM**
>
> **Leavening:** the process where a substance, such as yeast or baking powder, is used to produce fermentation in a dough or batter.

Weighing, measuring and portion control

- Check that all ingredients are weighed carefully.
- Divide the dough with a dough divider, hard scraper or hydraulic cutting machine.
- Check the divided dough pieces for weight. When weighing, remember that doughs lose up to 12.5 per cent of their water during baking.

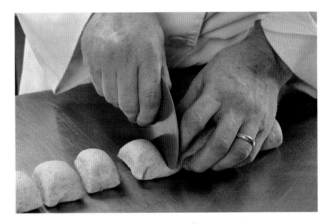

Dividing dough into evenly sized pieces

Scaling

The term for standardising bread into evenly distributed and pre-determined weights is 'scaling'. This is where a batch of dough is split into smaller amounts before being shaped into loaves or rolls. A roll could be scaled between 30 grams and 50 grams, for example, depending on the size of the roll being made. A loaf could be scaled from 250 grams to 500 grams, again depending on the size and shape of the bread being produced.

Shaping

Bread loaves and rolls can be shaped into many forms to enhance their appearance. The shape of bread also has an impact on the ways in which the bread is consumed. For example, a French stick (baguette) or an Italian focaccia or ciabatta, would not be used to make a sandwich in the same way that a tinned loaf could be.

Rolls can be enhanced by shaping into many styles. (See the recipes at the end of the unit for some of the various shapes and finishes that can be applied when producing a batch of bread rolls.)

Glazing

To add a glaze to fermented products such as breads or bread rolls, they can be brushed with egg wash prior to baking. In France, this is referred to as 'dorure' and is generally made up of a mixture of egg yolks, slaked with a little milk or cream. It is important to ensure a complete and even coverage of egg wash as gaps will become more apparent after baking. It is also important to brush as much of the surface area as possible as the products might rise further upon initial baking. Therefore, glazing of this nature should be completed as close to the baking time as possible.

Dusting

This is a very simple process in which a light dusting of flour is sieved onto the top surface of fermented products.

Scoring

Scoring refers to lights cuts that are made to the surface of the raw dough during the final proving stage and before baking. Any cuts or scoring will be emphasised further as the dough expands. Scoring enhances the appearance of fermented products and allows for differentiation of appearance.

Steam injection

Steam injection is an important aspect in baking. Steam helps to produce a crust in fermented products. Many commercial baking ovens have the option to inject steam during the baking process. The timing of the injection of the steam will impact the development of the crust and is usually done towards the end of the baking time.

De-moulding

De-moulding is the process in which fermented products are separated from the moulds in which they were baked. If goods are baked sufficiently, and the moulds are in good condition and lined properly, fermented products should come away from the moulds easily. Moulds can be tapped gently to release the product and then carefully tipped out, at an angle, onto a cooling rack. The product, such as a loaf, can then be placed onto its base to cool properly.

1.2 Fermented dough commodities

Types of flour

Strong flour

This is the essential ingredient when making fermented products. Strong flour is a white wheatflour which has been processed to remove the outer skin (bran), the husk and germ. It has a high gluten or protein content. The gluten provides increased water absorption and elasticity, which is essential for allowing the dough to expand by trapping the carbon dioxide gas produced by the yeast.

Legislation requires that four specific nutrients in specific quantities are added to wheat flour; they are: iron, vitamin B1, nicotinic acid (niacin) and calcium carbonate.

Wholemeal and wheatmeal flour

As the names would suggest, these flours contain the whole wheat grain. Nothing is added or taken away, and they are considered to be a healthier alternative to white flour. Because the nutrients mentioned above are naturally present, these flours do not require any additions.

Rye flour

Rye is a type of grass which grows in harsh climates and is associated with Northern and Eastern Europe, where there is a tradition of rye breads. Rye was looked upon as being inferior to wheat. It has a low gluten content and will add flavour and texture when used with strong flour.

Spelt

A grain related to wheat but with less gluten and a nutty flavour.

Rice cones and rice flour

Coarsely ground rice can be sprinkled on the baking sheet before placing dough products on the baking tray. It can also be sprinkled on top of the bread, which adds texture and crunch to the products. Rice flour is ground much finer and is used with gram flour to make the gluten-free products

Gram flour

This is flour made from chickpeas.

Flour extraction rate

This refers to the yield of flour obtained from wheat in the milling process. A 100 per cent extraction (or straight-run) is wholemeal flour containing all of the grain. Lower extraction rates are the whiter flours from which progressively more of the bran and germ (B vitamins and iron) are excluded, down to a figure of 72 per cent extraction, producing regular white flour. Patent flours are of lower extraction rate, 30–50 per cent, and so comprise mostly the endosperm of the grain. Patent flour is the purest and highest-quality commercial wheat flour available.

Wholemeal bread is baked with 100 per cent extraction flour, i.e. containing the whole of the cereal grain. White bread is made from 72 per cent extraction flour. Genuine brown bread is made with flour of extraction rate intermediate between that of white bread (72 per cent) and wholemeal (100 per cent). A loaf may not legally be described as brown unless it contains at least 0.6 per cent fibre on a dry weight basis, although some producers are known to dye white flour with brown colourings to make its appearance seem healthier.

Flour is often categorised by its strength. Wheat varieties are called 'soft' or 'weak' if they are low in a protein known as gluten. High-protein (gluten) flours and are referred to as 'hard' or 'strong' if they have high gluten content. Strong flour, or bread flour, is high in gluten, with 12–14 per cent gluten content. The dough produced from strong flour has elastic properties that holds its shape and structure well once baked. Softer flours are comparatively lower in gluten, 8–10 per cent, and are more suited to the production of products with finer textures such as pastries, biscuits and cakes. Soft flour is usually divided into cake flour, which is the lowest in gluten, and pastry flour, which has slightly more gluten than cake flour, although many kitchens may use one type of soft flour for both.

Yeast

Yeast is a living organism which, when fed (on sugar), watered and kept warm, will multiply and produce carbon dioxide gas and ethyl alcohol. Yeast is essential to lighten or leaven fermented products. It comes compressed in a block (fresh) or dried, sometimes with the addition of ascorbic acid (vitamin C) which is an improver and helps to speed up the fermentation process (fast-acting yeast). Most recipes will say to dissolve yeast in (usually warm) liquid, but more bakers are moving away from this and feel it is better to rub the yeast into the flour rather than 'drown' it in the liquid.

Dried yeast is concentrated so remember to use half the quantity if using it in place of fresh yeast.

Using too much yeast can affect flavour and the products will stale more quickly. The best results are achieved by using the minimum quantity of yeast and allowing it to prove and develop over a longer time.

Improvers

Available in powder form, these usually have a vitamin C or ascorbic acid base, which speeds up the action of the yeast. This eliminates the need for bulk proving or BFT (bulk fermentation time) – see the ADD method below. Fast-action dried yeast is a combination of dried yeast with an improver added.

Salt

In the last few years salt has had a bad press, mainly because of its overuse in processed foods, but in truth we cannot live without salt in our diet, as a lack of it can lead to dehydration. It is best to use sea or rock salt because, unlike table salt, where most of the mineral content is destroyed by the high temperatures used in its production, sea and rock salts are relatively natural products.

As salt plays such a huge role in fermented goods, it must be measured carefully. It helps to stabilise the fermentation and strengthen the gluten, improves the crust texture, colour and flavour and lengthens the shelf life of products. Salt must not come into direct contact with the yeast as it will slow, or at worst kill the yeast and stop the fermentation.

Cooking methods

Not all dough products are baked. Doughnuts, for example, are deep-fried. In Chinese cookery, dough buns are steamed and filled with fillings, pork being the most common (pork buns). Steam is injected into the oven during the baking process; the steam helps to keep the outside surface of loaves moist and supple so that the bread can 'spring' for as long as possible. Once the outside of the loaf begins to dry out it hardens, preventing further spring and the crust begins to form.

Dough products come in a variety of forms and styles. The variety of flour alongside the additional ingredients incorporated, and whether the dough is leavened or unleavened, gives dough products their own uniqueness. The majority of doughs have some sort of leavening agent, commonly yeast, or a form of starter from which the natural yeast in flour is developed slowly over a long period of time and replaced with fresh flour and water as the required amount of the starter is used to produce the bread in question.

Some dough products do not require yeast as a raising agent. Soda bread, for example, uses bicarbonate of soda, an alkali which reacts with the acidic components within doughs and batters to release carbon dioxide.

The carbon dioxide raises the product and helps with the development of texture. Other dough products, such as pitta bread and flatbreads, do not use a raising agent and are referred to as unleavened breads.

Categories of dough

Doughs are divided into different categories, which are dependent on the way they are produced and the ingredients they contain:

- Simple dough – contains flour, salt, yeast, water and sometimes fat.
- Enriched doughs – contain the same base as a simple dough, but are enriched with additional ingredients such as eggs, sugar, butter, milk in place of water and often include dried fruits and spices. Chelsea buns with dried fruit and spices and brioche, a rich yeast dough with a high fat and butter content, are good examples.
- Laminated doughs – simple doughs that are laminated with layers of fat, traditionally butter for its rich flavour. Croissants and Danish pastries are enriched with fat in layers referred to as lamination. This makes them softer to eat because the fat in the dough insulates the water molecules, keeping the moisture level higher during baking. The raising agent is a combination of the fermentation of yeast and the steam produced from the lamination of the butter. Danish pastries may be filled with, or contain, fruit, frangipane, apple, custard, cherries, crème pâtissière and many other ingredients.
- Batter – this type of dough is looser in structure than other types of dough, because of its wet consistency. Therefore, batters are shaped and baked using moulds. Rum babas and savarins provide good examples.

Speciality doughs

Speciality doughs from around the world include:

- blinis – a type of savoury pancake traditionally made from buckwheat flour
- naan bread – a leavened Indian bread traditionally cooked in a tandoor (oven)
- pitta bread – Middle Eastern and Greek unleavened bread
- chapatti – Indian unleavened bread made from a fine ground wholemeal flour, known as 'atta'

- pizza dough – a dough of Italian origin traditionally made using 00 flour. This is a fine flour produced from the central part of the wheat grain, which produces light, crisp doughs. Pizza dough is often enriched with the addition of olive oil to provide moisture and flavour.

> **KEY TERM**
>
> **Lamination:** the dough that is produced by building alternating layers of dough and butter when making puff pastry, croissants or Danish pastries

Bread

It is customary today for restaurants to offer a range of different flavoured breads. Internationally there is a wide variety available; different nations and regions have their own speciality breads. Bread plays an important part in many religious festivals, especially Christian and Jewish.

Bulk fermentation

The traditional breadmaking process is known as the **bulk fermentation** process. This was used by many bakers before the introduction of high-speed mixing and dough conditioners, which both eliminate the need for bulk fermentation time. However, this traditional method produces a fine flavour due to the fermentation and is evident in the final product.

Bulk fermentation time (BFT) is the term used to describe the length of time that the dough is allowed to ferment in bulk. BFT is measured from the end of the mixing method to the beginning of the scaling (weighing) process. The length of BFT can be from one to six hours and is related to the level of salt and yeast in the recipe, as well as the dough temperature.

It is important during the bulk fermentation process that ideal conditions are adhered to:

- the dough must be kept covered to prevent the surface of the dough developing a skin
- the appropriate temperature must be maintained to control the rate of fermentation.

> **KEY TERMS**
>
> **Bulk fermentation:** a proving/fermentation process in which a leavened dough will double in size.
>
> **Bulk fermentation time:** the time it takes for dough to double in size.

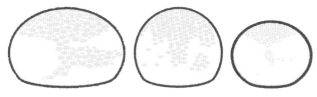

From left to right: overripe dough (proven too long), well-proven dough (good shape, structure and volume) and underripe dough.

1.3 Storage

Raw dough can be stored for use at a later time. For long-term storage, dough should be wrapped securely, labelled and placed in a freezer. For short-term storage, or to **retard** a dough, it can be placed into a refrigerator.

The dough should be covered to prevent a skin forming on the surface of the dough as well as following good food safety practices.

> **KEY TERM**
>
> **Retarding dough:** reducing the fermentation process of a dough by placing it in chilled conditions.

Storage of dough products

Crusty rolls and bread are affected by changes in storage conditions; they are softened by a damp environment and humid conditions, so should be stored in a dry environment to keep them crusty.

Always store dough products in suitable containers at room temperature or in a freezer for longer storage. Do not store in a refrigerator unless you want the bread to stale quickly for use as breadcrumbs. Staling will also occur quickly in products that contain a lot of fat and milk. Many commercial dough products contain anti-staling agents. As with all other products, ensure that fermented products are correctly wrapped and clearly labelled and dated before storing.

Positioning

It is always good practice to consider the positioning of products for storage. As well as the atmosphere in which they are stored, the type, size, weight and shape of the product needs to be considered. Heavy boxed products should, ideally, be stored on low level or safely accessible shelving.

Stock rotation

Stock rotation is an essential aspect of stock management. The rule of 'first in, first out' is a simple process where older stock is utilised before newer stock. This helps to ensure that stock does not go out of date before use and that it remains safe to use. For example, when buying ambient goods, such as tins, the process would involve bringing the current stock to the front of the shelf and placing the newer stock behind it. It is, however, still important to check that foods are in date and in a safe and hygienic condition to use.

Convenience dough products

There are many different types of convenience dough product on the market.

- Fresh and frozen pre-proved dough products: rolls; croissants; Danish pastries; French breads.
- Bake-off products that are finished and ready for baking. These can be bought either frozen or fresh, or in modified-atmosphere packaged forms (this method replaces most of the oxygen around the product to slow down spoilage). These products have to be kept refrigerated. They include garlic bread, rolls and Danish pastries.

> **HEALTH AND SAFETY**
>
> When using frozen dough products, always follow the manufacturer's instructions. Contamination can occur if doughs are defrosted incorrectly.

Produce and finish fermented products

2.1 Tools and equipment

Equipment that may be useful in the production of fermented products is described below.

- Loaf tin – usually rectangular shaped with raised, sometimes slightly angled, sides. They are produced in a range of sizes, normally in metal. There are a number of non-stick varieties and also those made from silicone.

The tin is designed to support the rising of the dough during the baking process resulting in the loaf shape once baked and cooled.

- French stick baking trays – these are usually made in long, half sleeves, with perforated bases. They are often joined into more than one sleeve (for example, four or five sleeves) to allow the baking of four or five sticks from one tray. The perforated bases allow direct heat penetration and creation of a crusty base as well as the lightly dimpled effect that is achieved on the finish of a French stick or baguette.

● Pastry scrapers – these are multi-functional, hand-held, items of 10–12 centimetres wide and 8–10 centimetres deep (although some may vary from this) They are designed to assist in the clean and efficient separation or removal of doughs, pastes and other mixes from one container to another. They can also be used to cut raw doughs and pastes into smaller sizes for batching purposes. Scrapers can be made from plastic or metal and can be curved or square, depending on the design and intended use.

Other pieces of equipment that may be useful which you can find details of in Unit 309 include:

● weighing scales
● mixer with dough hook
● proving cabinet
● baking sheets
● convection ovens
● deck ovens.

2.2 Producing fermented dough products

For details of the techniques involved in producing different types of fermented dough products, see the recipe section at the end of this chapter.

2.3 Additional ingredients and finishes

Depending of the type of product being made, there are a number of different finishes that can be applied to dough products. Certain breads, such as baguettes, are scored before they are baked which helps to produce an appealing visual finish to the baked product.

Fermented products, such as breads and bread rolls, can be manipulated into a variety of different shapes. The recipes for rolls provide examples of different shapes. It is also possible to plait bread by slitting it into three, or more, evenly weighted long strips of dough and plaiting in the same way as a hairdresser would when plaiting hair.

Many breads are sprinkled with seeds such as poppy, sesame and fennel before they are baked. Others may have herbs added, such as the red onion and sage rolls (recipe 4), or sprinkled with cheese as seen with Parmesan rolls (recipe 3). A common but perhaps the most simple of finishes applied to breads is a light dusting of the same flour used to produce the bread. This is sieved over the bread once it has cooled and applies to many of the examples shown in this chapter.

Bagels sprinkled with poppy seeds

Many dough products, including breads, bread rolls and laminated doughs such as croissant and pain-au-chocolat, are brushed with eggwash before they are baked. This produces a light, golden-brown glaze during the baking process. Other products are glazed after baking, such as Swiss buns with fondant and Danish pastries, which are often finished with a fruit glaze and water icing. Other enriched doughs such as Chelsea buns are soaked with bun syrup while still hot after baking and sprinkled with nibbed sugar.

2.4 Evaluation

Faults

If yeast dough has a close texture this may be because:

● it was insufficiently proved
● it was insufficiently kneaded
● it contains insufficient yeast
● the oven was too hot
● too much water was added
● too little water was added.

If dough has an uneven texture this may be because:

● it was insufficiently kneaded
● it was over-proved
● the oven was too cool.

If dough has a coarse texture this may be because:

● it was over-proved, uncovered
● it was insufficiently kneaded.
● too much water was added
● too much salt was added.

If dough is wrinkled this may be because:

● it was over-proved.

If dough is sour this may be because:

● the yeast was stale
● too much yeast was used.

If the crust is broken this may be because:
- the dough was under-proved at the second stage.

If there are white spots on crust this may be because:
- the dough was not covered before second proving.

Allergies

Gluten is the protein found in wheat, barley and rye, and to a lesser extent in yeast. An increasing number of people are intolerant to gluten, which results in damage to the lining of the small intestine. This is known as coeliac disease and people who are intolerant must avoid the consumption of wheat and flour-based products.

Although it is essential to clearly list all potential allergens when making fermented dough products, the allergens that are most likely to be used in their production include:

- gluten – flours and any products made from wheat, rye, barley and oats
- nuts – such as hazelnuts and nibbed almonds added to garnish Danish pastries, for example
- sesame seeds – often sprinkled on the surface of breads and rolls
- eggs – used in the production of enriched doughs
- lactose – for example, milk used in place of water in the production of dough; cheese used to glaze breads; yoghurt used in speciality doughs, such as naan.

Beyond the basic preparation of fermented doughs, attention is also required with regard to the additional ingredients that are used to complete the product. It is vitally import ant to assess any of the other potential allergens that are incorporated into fermented doughs, as well as the dough itself.

TEST YOURSELF

1 Give **two** examples, with descriptions, of dough products in the following categories:
   a) enriched doughs
   b) laminated doughs
   c) speciality doughs
2 List four products produced from a basic bun dough.
3 What is meant by fermentation?
4 What is the difference between strong and soft flour?
5 What is the difference between leavened and unleavened bread?
6 Why is temperature so important when making bread dough using yeast?
7 What is yeast and what is its use in the pastry kitchen?
8 What raising agent is used in the production of soda bread?
9 Name three seeds that are often sprinkled on bread rolls before baking.
10 What is the meaning of the term 'scaling'?

1 Wholemeal rolls

Energy	Calories	Fat	Saturated fat	Carbohydrates	Sugar	Protein	Fibre	Sodium
803 kJ	189 kcal	1.3 g	0.0 g	34.9 g	0.6 g	0.0 g	5.7 g	492.0 g

Strong stoneground wholemeal flour	450 g, plus a little extra
Packet fast action dried yeast	7 g
Fine salt	1 level dessertspoon
Warm water (approximately 37 °C)	350 ml

Cooking

1 Measure the flour into a large mixing bowl and sprinkle on the salt and dried yeast. Mix together thoroughly.

2 Make a well in the centre and pour in most of the warm water.

3 Mix the water into the flour gradually to form a dough. The exact amount of water needed will depend on the flour, so do not add all the water at once.

4 Mix until a smooth dough is formed, checking that it leaves the bowl clean. There should be no sign of flour or dough remaining on the sides of the bowl. The dough should be soft and pliable with a slight stickiness at this point.

5 Transfer the dough to a lightly floured flat surface and knead to develop the gluten.

6 Divide the dough into 8 equal pieces (cut in half, then quarters and then each quarter to make eight evenly-sized pieces). Quickly weigh to check accuracy.

7 Take each piece of the dough in turn and knead it briefly, bringing the edges into the centre and turning over. Shape each into a round rolling under the cup of the hand.

8 Place the rolls on a baking sheet and prove until doubled in size.

9 When the rolls have doubled in size, dust with a little flour and bake at 200 °C for 20–25 minutes. Check that the rolls are cooked through by analysing their colour and by tapping on the base of the roll, which should sound hollow.

10 Once baked sufficiently, cool on a wire rack.

> **PROFESSIONAL TIP**
> Shape rolls quickly on a clean, smooth surface. Any lumps from flour or small pieces of dough will leave an impression on the surface of the roll and ultimately impact negatively on the final presentation.

2 Seeded rolls

Energy	Calories	Fat	Saturated fat	Carbohydrates	Sugar	Protein	Fibre
633 kJ	150 kcal	3.4 g	0.5 g	26 g	1.2 g	5.3 g	1.4 g

	30 rolls
Strong flour	1 kg
Yeast	30 g
Water at 37°C	600 ml
Salt	20 g
Caster sugar	10 g
Milk powder	20 g
Sunflower oil	50 g
Egg wash	
Poppy seeds	
Sesame seeds	

Cooking

1 Sieve the flour onto paper.

2 Dissolve the yeast in half the water.

3 Dissolve the salt, sugar and milk powder in the other half.

4 Add both liquids and the oil to the flour at once and mix on speed 1 for 5 minutes or knead by hand for 10 minutes.

5 Cover with cling film and leave to prove for 1 hour at 26 °C.

6 'Knock back' the dough and scale into 50 g pieces.

7 Shape and place in staggered rows on a silicone paper-covered baking sheet.

8 Prove until the rolls almost double in size.

9 Egg wash carefully and sprinkle with seeds.

10 Bake immediately at 230 °C with steam for 10–12 minutes.

11 Break one open to test if cooked.

12 Allow to cool on a wire rack.

PROFESSIONAL TIP

Instead of weighing out each 50 g piece of dough, weigh out 100 g pieces and then halve them.

Placing bread rolls in staggered rows means they are less likely to 'prove' into each other. The spacing allows them to cook more evenly and more will fit on the baking sheet.

VARIATION

Try using other types of seed such as sunflower, linseed or pumpkin.

For a beer glaze, mix together 150 ml beer with 100 g rye flour and brush on before baking.

3 Parmesan rolls

Energy	Calories	Fat	Saturated fat	Carbohydrates	Sugar	Protein	Fibre
2,169 kJ	512 kcal	9.9 g	2.2 g	94.3 g	1.8 g	17.5 g	5.2 g

	30 rolls
Grated Parmesan cheese	200 g (approximately)

Cooking

1 Follow method for seeded rolls (recipe 2) up to Step 5.

2 Lightly flour work surface and roll the dough into a rectangle until 3 cm thick.

3 Make sure the dough is not stuck to the surface.

4 Brush with water and cover with Parmesan.

5 Using a large knife, cut into squares 6 × 6 cm.

6 Place on a silicone paper-covered baking sheet and leave to prove until almost double in size.

7 Bake at 230 °C for 10–12 minutes with steam.

8 Cool on a wire rack.

PROFESSIONAL TIP

When making bread that requires rolling out as opposed to being individually shaped, it is helpful to decrease the liquid content by 10 per cent so it will be easier to process.

To ensure the squares are all the same size, mark a grid using the back of the knife before cutting.

VARIATION

Other cheeses could be used such as Pecorino or Grana Padano.

4   Red onion and sage rolls

Energy	Calories	Fat	Saturated fat	Carbohydrates	Sugar	Protein	Fibre
2169 kJ	512 kcal	9.9 g	2.2 g	94.3 g	1.8 g	17.5 g	5.2 g

	Serves 8	Serves 10
Red onion	1/2	1
Dried sage	1/4tsp	1/2 tsp
Bread roll dough (see recipe 2 but omit the seeds)		
Oil	1 tbsp	2 tbsp

Mise en place

1   Chop the red onion and sage. Prepare the egg wash.

Cooking

1   Sweat the red onion, then leave to cool.

2   Add the sage to the onion.

3   Pin out bread dough in a rectangle. Spread the onion and sage mixture over seven-eighths of the dough. Egg wash the exposed edge.

4   Roll the dough as you would for a Swiss roll, and seal the edge.

5   Cut into 50 g slices.

6   Place the slices on a prepared baking sheet and eggwash. Bake in a pre-heated oven at 220 °C for approximately 10 minutes. Cool on a wire rack.

5   Sun-dried tomato bread

Energy	Calories	Fat	Saturated fat	Carbohydrates	Sugar	Protein	Fibre
10192 kJ	2,414 kcal	70.8 g	12.8 g	401.8 g	29.6 g	67.4 g	20.7 g

	2 × 450 g loaves
Sun-dried tomatoes, chopped	100 g
Water	300 ml
Bread flour	500 g
Salt	10 g
Skimmed milk powder	12.5 g
Shortening	12.5 g
Yeast (fresh)	20 g
Sugar	12.5 g

Mise en place

1   Soak the sun-dried tomatoes in boiling water for 30 minutes.

Cooking

1   Sieve the flour, salt and skimmed milk powder.

2   Add the shortening and rub through the dry ingredients.

3   Disperse the yeast into warm water, approximately 37 °C. Add and dissolve the sugar. Add to the above ingredients.

4   Mix until a smooth dough is formed. Check for any extremes in consistency and adjust as necessary until a smooth elastic dough is formed.

5   Cover the dough, keep warm and allow to prove.

6   After approximately 30–40 minutes, knock back the dough and mix in the chopped sun-dried tomatoes (well drained).

7 Mould and prove again for another 30 minutes (covered).

8 Divide the dough into two and mould round.

9 Rest for 10 minutes. Keep covered.

10 Re-mould into ball shape.

11 Place the dough pieces into 15-cm diameter hoops laid out on a baking tray. The hoops must be warm and lightly greased.

12 With the back of the hand flatten the dough pieces.

13 Prove at 38–40 °C in humid conditions, preferably in a prover.

14 Bake at 225 °C for 25–30 minutes.

15 After baking, remove the bread from the tins immediately and place on a cooling wire.

6 Wholemeal bread

Energy	Calories	Fat	Saturated fat	Carbohydrates	Sugar	Protein	Fibre
6839 kJ	1,628 kcal	32.5 g	16.7 g	302.3 g	62.6 g	50.5 g	40.0 g

	2 loaves
Unsalted butter or oil	60 g
Honey	3 tbsp
Water, lukewarm	500 ml
Fresh yeast, or Dried yeast	25 g 18 g
Salt	1 tbsp
Flour, unbleached strong white	125 g
Flour, stoneground wholemeal	625 g

Cooking

1 Melt the butter in a saucepan.

2 Mix together 1 tbsp of honey and 4 tbsp of the water in a bowl.

3 Disperse the yeast into the honey mixture.

4 In a basin, place the melted butter, remaining honey and water, the yeast mixture and salt.

5 Add the white flour and half the wholemeal flour. Mix well.

6 Add the remaining wholemeal flour gradually, mixing well between each addition.

7 The dough should pull away from the side of the bowl and form a ball. The resulting dough should be soft and slightly sticky.

8 Turn out onto a floured work surface. Sprinkle with white flour, knead well.

9 Brush a clean bowl with melted butter or oil. Place in the dough, cover with a damp cloth and allow to prove in a warm place. This will take approximately 1–1½ hours.

10 Knock back and further knead the dough. Cover again and rest for 10–15 minutes.

11 Divide the dough into two equal pieces.

12 Form each piece of dough into a cottage loaf or place in a suitable loaf tin.

13 Allow to prove in a warm place for approximately 45 minutes.

14 Place in a pre-heated oven, 220 °C and bake until well browned (approximately 40–45 minutes).

15 When baked, the bread should sound hollow and the sides should feel crisp when pressed.

16 Cool on a wire rack.

> **VARIATION**
> Alternatively, the bread may be divided into 50 g rolls, brushed with eggwash and baked at 200 °C for approximately 10 minutes.

> **Healthy Eating Tip**
> Only a little salt is necessary to 'control' the yeast. Many customers will prefer less salty bread.

7 Rye bread

Energy	Calories	Fat	Saturated fat	Carbohydrates	Sugar	Protein	Fibre
7174 kJ	1,690 kcal	24.7 g	2.9 g	331.8 g	12.5 g	46.7 g	46.2 g

	1 loaf
Fresh yeast (or dried yeast may be used)	15 g
Water	60 ml
Black treacle	1 tbsp
Vegetable oil	1 tbsp
Caraway seeds (optional)	15 g
Salt	15 g
Lager	250 ml
Rye flour	250 g
Unbleached bread flour	175 g
Polenta	
Egg wash	

Mise en place

1 Prepare a baking sheet by lightly sprinkling with polenta.

Cooking

1 Disperse the yeast in the warm water (at approximately 37°C)

2 In a basin mix the black treacle, oil, two-thirds of the caraway seeds (if required) and the salt. Add the lager. Add the yeast and mix in the sieved rye flour. Mix well.

3 Gradually add the bread flour. Continue to add the flour until the dough is formed and it is soft and slightly sticky.

4 Turn the dough onto a lightly floured surface and knead well.

5 Knead the dough until it is smooth and elastic.

6 Place the kneaded dough into a suitable bowl that has been brushed with oil.

7 Cover with a damp cloth and allow the dough to prove in a warm place until it is double in size. This will take about 1½–2 hours.

8 Turn the dough onto a lightly floured work surface, knock back the dough to original size. Cover and allow to rest for approximately 5–10 minutes.

9 Shape the dough into an oval approximately 25 cm long.

10 Place onto a baking sheet lightly sprinkled with polenta.

11 Allow the dough to prove in a warm place, preferably in a prover, until double in size (approximately 45 minutes to 1 hour).

12 Lightly brush the loaf with eggwash, sprinkle with the remaining caraway seeds (if required).

13 Using a small, sharp knife, make three diagonal slashes, approximately 5 mm deep into the top of the loaf.

14 Place in a pre-heated oven at 190 °C and bake for approximately 50–55 minutes.

15 When cooked, turn out. The bread should sound hollow when tapped and the sides should feel crisp.

16 Allow to cool.

Healthy Eating Tip

Only a little salt is necessary to 'control' the yeast. Many customers will prefer less salty bread.

8 Soda bread

Energy	Calories	Fat	Saturated fat	Carbohydrates	Sugar	Protein	Fibre
8341 kJ	1,970 kcal	39.8 g	20.9 g	357.8 g	18.5 g	67.9 g	20.9 g

Cooking

1 Sift the flours, salt and bicarbonate of soda into a bowl.
2 Make a well and add the buttermilk, warm water and melted butter.
3 Work the dough for about 5 minutes.
4 Mould into 2 round loaves and mark the top with a cross.
5 Bake at 200 °C for about 25 minutes. When the bread is ready, it should make a hollow sound when tapped.

	2 loaves
Flour, wholemeal	250 g
Flour (strong)	250 g
Bicarbonate of soda	1 tsp
Salt	1 tsp
Buttermilk	200 g
Water, warm	60 ml
Butter, melted	25 g

9 Olive bread

Energy	Calories	Fat	Saturated fat	Carbohydrates	Sugar	Protein	Fibre
5174 kJ	1,229 kcal	44.2 g	6.5 g	188.3 g	3.5 g	30.9 g	15.9 g

4 loaves	
Starter	
Yeast	40 g
Water at 37 °C	180 ml
Strong flour	225 g
Sugar	5 g
Dough	
Strong flour	855 g
Sugar	40 g
Salt	20 g
Water at 37 °C	450 ml
Olive oil	160 ml
Green olives, cut into quarters	100 g

Mise en place

1 Cut the green olives into quarters.

Cooking

1 For the starter, dissolve the yeast in the water, add the flour and sugar, mix well, cover and leave to ferment for 30 minutes.

2 For the dough, sieve the flour, sugar and salt into a mixing bowl, add the water followed by the starter and start mixing slowly.

3 Gradually add the oil and continue mixing to achieve a smooth dough.

4 Cover with cling film and prove for 1 hour or until double in size.

5 Knock back, add the olives and divide the dough into four.

6 Roll into long shapes and place on a baking sheet sprinkled with rice cones, return to the prover and leave until double in size.

7 Brush with olive oil and bake at 220 °C for 20–25 minutes.

8 When cooked, the bread should sound hollow when tapped on the base.

9 Leave to cool on a wire rack.

Make up the starter.

Starter ready for use after proving.

Start mixing in the ingredients for the main dough, tearing up the starter.

Continue mixing in the ingredients and working the dough.

Shape the dough.

Divide and roll into loaves.

10 Bun dough and varieties of bun

Energy	Calories	Fat	Saturated fat	Carbohydrates	Sugar	Protein	Fibre	Sodium
1060 kJ	250 kcal	6.0 g	3.3 g	41.0 g	10.8 g	9.4 g	1.4 g	65.0 g

	12 buns	24 buns
Strong flour	500 g	1 kg
Yeast	25 g	50 g
Milk (scalded – heated to 82 °C and cooled to 40 °C)	250 ml	500 ml
Butter	60 g	120 g
Eggs	2	4
Sugar	60 g	120 g

Mise en place

Prepare the milk by scalding, which means heating it to 82 °C. At this temperature, bacteria are killed, enzymes in the milk are destroyed and many of the proteins are denatured. After scalding, cool the milk to 40 °C for use.

Cooking

Sponge and dough:

1 Sieve the flour.

2 Dissolve the yeast in half the milk and add enough of the flour to make a thick batter, cover with cling film and place in the prover to ferment.

3 Rub the butter into the rest of the flour.

4 Beat the eggs and add the salt and sugar.

5 When the batter has fermented, add to the flour together with the liquid.

6 Mix slowly for 5 minutes to form a soft dough.

7 Place in a lightly oiled bowl, cover with cling film and prove for 1 hour at 26 °C.

8 Knock back the dough and knead on the table, rest for 10 minutes before processing.

Bun wash:

	250 ml
Milk	250 ml
Caster sugar	100 g

1 Bring both ingredients to the boil and brush over liberally as soon the buns are removed from the oven. The heat from the buns will set the glaze and prevent it from soaking in, giving a characteristic sticky coat.

Sift the flour.

Rub in the fat.

Make a well in the flour and pour in the beaten egg.

Pour in the liquid.

Fold the ingredients together.

Knead the dough.

Before and after proving: the same amount of dough is twice the size after it has been left to prove.

11 Bath buns

Energy	Calories	Fat	Saturated fat	Carbohydrates	Sugar	Protein	Fibre
827 kJ	196 kcal	6.5 g	3.6 g	31.9 g	13.4 g	4.5 g	1.4 g

	12–14 buns
Basic bun dough (recipe 10)	1 kg
Bun spice	20 ml
Sultanas	200 g
Sugar nibs	360 g
Egg yolks	8

Cooking

1 Mix the bun spice into the basic dough and knead.

2 Add the sultanas, two-thirds of the sugar nibs and all the egg yolks.

3 Using a plastic scraper, cut in the ingredients (it is usual for the ingredients not to be fully mixed in).

4 Scale into 60-g pieces.

5 Place on a paper-lined baking sheet in rough shapes.

6 Sprinkle liberally with the rest of the nibbed sugar.

7 Allow to prove until double in size.

8 Bake at 200 °C for 15–20 minutes.

9 Brush with bun wash as soon as they come out of the oven.

12 Hot cross buns

Energy	Calories	Fat	Saturated fat	Carbohydrates	Sugar	Protein	Fibre
744 kJ	177 kcal	6.7 g	3.6 g	26.8 g	8.3 g	4.6 g	1.2 g

	12–14 buns	24 buns
Basic bun dough (recipe 6)	1 kg	2 kg
Currants	75 g	150 g
Sultanas	75 g	150 g
Mixed spice	5 g	10 g
Crossing paste		
Strong flour	125 g	250 g
Water	250 ml	500 ml
Oil	25 ml	50 ml

Mise en place

1 Prepare a few piping bags from greaseproof or baking parchment. Mix together the ingredients for the crossing paste so this is immediately ready for use.

Cooking

1 Add the dried fruit and spice to the basic dough, mix well.

2 Scale into 60-g pieces and roll.

3 Place on a baking sheet lined with silicone paper in neat rows opposite each other and eggwash.

4 Mix together the ingredients for the crossing paste. Pipe it in continuous lines across the buns.

5 Allow to prove.

6 Bake at 220 °C for 15–20 minutes.

7 Brush with bun wash as soon as they come out of the oven.

> **VARIATION**
> - To make fruit buns, proceed as for hot cross buns without the crosses.

13 Swiss buns

Energy	Calories	Fat	Saturated fat	Carbohydrates	Sugar	Protein	Fibre
697 kJ	165 kcal	5.6 g	3.4 g	27.0 g	8.4 g	3.4 g	1.0 g

Cooking

1 Scale the dough into 60-g pieces.

2 Roll into balls then elongate to form oval shapes.

3 Place on a baking sheet lined with silicone paper, egg wash.

4 Allow to prove.

5 Bake at 220 °C for 15–20 minutes.

6 Allow to cool then dip each bun in lemon-flavoured fondant.

	12–14 buns
Basic bun dough (recipe 10)	1 kg
Fondant	500 g
Lemon oil	5 ml

> **PROFESSIONAL TIP**
> The fondant can be prepared by melting it in a bowl over steaming water (bain-marie). A little stock syrup can be used to thin the fondant to the right consistency for dipping. This is in addition to the lemon oil used in this recipe.

14 Doughnuts

Energy	Calories	Fat	Saturated fat	Carbohydrates	Sugar	Protein	Fibre
918 kJ	218 kcal	13.3 g	4.0 g	22.6 g	4.0 g	3.6 g	1.2 g

	12 doughnuts
Basic bun dough (recipe 10)	1 kg
Caster sugar	500 g
Raspberry jam	250 g

Mise en place

1 Have a tray of caster sugar ready alongside the jam for piping into the centre of the doughnuts.

Cooking

1 Scale the dough into 60-g pieces.

2 Roll into balls and make a hole in the dough using a rolling pin.

3 Prove on an oiled paper-lined tray.

4 When proved, carefully place in a deep fat fryer at 180 °C.

5 Turn over when coloured on one side and fully cook.

6 Drain well on absorbent paper.

7 Toss in caster sugar.

8 Make a small hole in one side and pipe in the jam.

> **VARIATION**
> The caster sugar can be mixed with ground cinnamon.

> **HEALTH AND SAFETY**
> As a fryer is not a regular piece of equipment found in a patisserie, a portable fryer is often used. Always make sure it is on a very secure surface in a suitable position. Never attempt to move it until it has completely cooled down. In addition, extreme care must be taken to avoid serious burns.
> - Only use a deep fat fryer after proper training.
> - Make sure the oil is clean and the fryer is filled to the correct level.
> - Pre-heat before using but never leave unattended.
> - Always carefully place the products into the fryer – never drop them in. Use a basket if appropriate.
> - Never place wet products into the fryer.

15 Bagels

Energy	Calories	Fat	Saturated fat	Carbohydrates	Sugar	Protein	Fibre
670 kJ	158 kcal	2.1 g	0.5 g	32.7 g	5.0 g	6.2 g	1.6 g

	10–12 bagels
Strong flour	450 g
Yeast	15 g
Warm water	150 ml
Salt	10 g
Caster sugar	25 g
Oil	45 ml
Egg yolk	20 g
Milk	150 ml
Poppy seeds	

Cooking

Method: ferment and dough

1 Sieve the flour, place in a mixing bowl.

2 Make a well and add the yeast which has been dissolved in the water.

3 Mix a little of the flour into the yeast to form a batter, sprinkle over some of the flour from the sides and leave to ferment.

4 Mix together the salt, sugar, oil, egg yolk and milk.

5 When the batter has fermented add the rest of the ingredients and mix to achieve a smooth dough.

6 Cover and prove for 1 hour (BFT).

7 Knock back and scale at 50-g pieces, shape into rolls and make a hole in the centre using a small rolling pin.

8 Place on a floured board and prove for 10 minutes.

9 Carefully drop into boiling water and simmer until they rise to the surface.

10 Lift out and place on a silicone-covered baking sheet, egg wash, sprinkle or dip in poppy seeds and bake at 210 °C for 30 minutes.

Mise en place

1 Use a wide pan when boiling the bagels. This will allow more bagels to be poached as a batch.

Use a rolling pin to make a hole in the centre of each bagel.

Poach the bagels in water.

Egg wash the bagels and sprinkle with seeds before baking.

16 Focaccia

Energy	Calories	Fat	Saturated fat	Carbohydrates	Sugar	Protein	Fibre
12915 kJ	3052 kcal	78.7 g	11.5 g	553.6 g	14.7 g	66.7 g	30.0 g

	1 loaf
Active dry yeast	2 packets
Sugar	1 tsp
Lukewarm water (about blood temperature)	230 ml
Extra virgin olive oil, plus extra to drizzle on the bread	70 g
Salt	1½ tsp
Flour, unbleached all-purpose	725 g
Coarse salt	
Picked rosemary	

Cooking

1 Dissolve the yeast and sugar in half of the lukewarm water in a bowl; let sit until foamy. In another bowl, add the remaining water, the olive oil, and the salt.

2 Pour in the yeast mixture.

3 Blend in the flour, a quarter at a time, until the dough comes together. Knead on a floured board for 10 minutes, adding flour as needed to make it smooth and elastic. Put the dough in an oiled bowl, turn to coat well, and cover with a towel.

4 Let rise in a warm draught-free place for 1 hour, until doubled in size.

5 Knock back the dough, knead it for a further 5 minutes, and gently roll it out in to a large disc or sheet to approximately 2 cm thick.

6 Let rise for 15 minutes, covered. Oil your fingers and make impressions with them in the dough, 3 cm apart. Let prove for 1 hour.

7 Preheat the oven to 210 °C. Drizzle the dough with olive oil and sprinkle with coarse salt and picked rosemary.

8 Bake for 15–20 minutes in a very hot oven at 200 °C, until golden brown. Sprinkle with additional oil if desired. Cut into squares and serve warm.

17 Pizza

	Energy	Calories	Fat	Saturated fat	Carbohydrates	Sugar	Protein	Fibre
	3956 kJ	941 kcal	46.3 g	13 g	114.4 g	20.1 g	23.6 g	8.4 g

* Using 100 per cent strong white flour

	2 × 18 cm
Flour, strong white	250 g
Pinch of salt (sea salt)	5 g
Olive oil	25 g
Yeast	5 g
Water or milk at 24 °C	150 g
Caster sugar	10 g
Onions, finely chopped	100 g
Cloves of garlic, crushed	2
Sunflower oil	60 ml
Plum tomatoes, canned	200 g
Tomato purée	100 g
Oregano	3 g
Basil	3 g
Sugar	10 g
Cornflour	10 g
Mozzarella cheese	10 g

Cooking

1 Sieve the flour and the salt. Rub in the margarine.

2 Disperse the yeast in the warm milk or water; add the caster sugar. Add this mixture to the flour.

3 Mix well, knead to a smooth dough, place in a basin covered with a damp cloth and allow to prove until doubled in size.

4 Knock back, divide into two and roll out into two 18-cm discs. Place on a lightly greased baking sheet.

5 Sweat the finely chopped onions and garlic in the oil until cooked.

6 Add the roughly chopped tomatoes, tomato purée, oregano, basil and sugar. Bring to the boil and simmer for 5 minutes.

7 Dilute the cornflour in a little water, stir into the tomato mixture and bring back to the boil.

8 Take the discs of pizza dough and spread 125 g of filling on each one.

9 Sprinkle with grated mozzarella cheese or lay the slices of cheese on top.

10 Bake in a moderately hot oven at 200 °C, for about 10 minutes.

The pizza dough may also be made into rectangles so that it can be sliced into fingers for buffet work.

> **NOTE**
> Pizza is a traditional dish originating from southern Italy. In simple terms it is a flat bread dough that can be topped with a wide variety of ingredients and baked quickly. The only rule is not to add wet ingredients, such as tomatoes, which are too juicy, otherwise the pizza will become soggy. Traditionally, pizzas are baked in a wood-fired brick oven, but they can be baked in any type of hot oven for 8–15 minutes depending on the ingredients. The recipe given here is a typical one.

> **VARIATION**
> Oregano is sprinkled on most pizzas before baking. This is a basic recipe and many variations exist. Some have the addition of olives, artichoke hearts, prawns, mortadella sausage, garlic sausage or anchovy fillets; other combinations include:
> - mozzarella cheese, anchovies, capers and garlic
> - mozzarella cheese, tomato and oregano
> - ham, mushrooms, egg and parmesan cheese
> - prawns, tuna, capers and garlic
> - ham, mushrooms and olives.

> **PROFESSIONAL TIP**
> The sauce can be left with a course texture or blended and passed for a finer texture.
>
> Pizzas baked on stone in traditional, wood-fired ovens will cook very quickly compared to regular ovens. The temperature inside a pizza oven can reach in excess of 400 °C. This produces excellent thin crust traditional pizzas that will cook in just a few minutes.

> **ON A BUDGET**
> To save time and expense, a good quality ready-made passata can be used in place of making the sauce.

18 Cholla bread

Energy	Calories	Fat	Saturated fat	Carbohydrates	Sugar	Protein	Fibre
4876 kJ	1154 kcal	30.1 g	16.1 g	198.1 g	13.1 g	35.0 g	10.3 g

	2 loaves
Butter or margarine	56 g
Flour (strong)	500 g
Caster sugar	18 g
Salt	1 tsp
Egg	1
Yeast	25 g

Cooking

1 Rub the butter or margarine into the sieved flour in a suitable basin.

2 Mix the sugar, salt and egg together.

3 Disperse the yeast in the water.

4 Add all these ingredients to the sieved flour and mix well to develop the dough. Cover with a damp cloth or plastic and allow to ferment for about 45 minutes.

5 Divide into 125–150 g strands and begin to plait as follows:

4–strand plait	5–strand plait
2 over 3	2 over 3
4 over 2	5 over 2
1 over 3	1 over 3

6 After moulding, place on a lightly greased baking sheet and eggwash lightly.

7 Prove in a little steam until double in size. Egg wash again lightly and decorate with maw seeds (poppy seeds).

8 Bake in a hot oven, at 220 °C for 25–30 minutes.

19 Savarin dough and derivatives

Energy	Calories	Fat	Saturated fat	Carbohydrates	Sugar	Protein	Fibre
896 kJ	212 kcal	4.6 g	2.5 g	42.8 g	33.2 g	2.6 g	0.8 g

35 items	
Basic dough	
Strong flour	450 g
Yeast	1 g
Water at 40 °C	125 ml
Eggs	5
Caster sugar	60 g
Salt	Pinch
Melted butter	150 g

Cooking

Ferment and dough:

1 Sieve the flour and place in a bowl. Make a well.

2 Make a ferment by dissolving the yeast in the water and pour into the well.

3 Gradually mix the flour into the liquid, forming a thin batter. Sprinkle over a little of the flour to cover, then leave to ferment.

4 Whisk the eggs, sugar and salt.

5 When the ferment has erupted through the flour, add the eggs and mix to a smooth batter. Cover and leave to prove until double in size.

6 Add the melted butter and beat in.

7 Pipe the batter into prepared (buttered and floured) moulds one-third full.

8 Prove until the mixture reaches the top of the mould and bake at 220 °C for 12–20 minutes, depending on the size of the mould.

9 Unmould and leave to cool.

1 Cream the yeast in milk to make a ferment.

2 Add the dissolved yeast to the flour and sprinkle a little flour over it.

3 The mixture after fermentation.

4 Add beaten eggs, sugar and salt.

5 The dough after proving.

6 Add the butter.

7 After proving, pipe into moulds.

8 After proving for the final time.

PROFESSIONAL TIP

Savarin and savarin-based products are never served without first soaking in a flavoured syrup. They are literally dry sponges and should be:
- golden brown in colour with an even surface
- smooth, with no cracks, breaks or tears
- evenly soaked without any hard or dry areas
- sealed by brushing with apricot glaze after soaking.

Cooked products can be stored in the fridge overnight, but if left for too long they will dry out and cracks will appear. They are best wrapped in cling film, labelled and stored in the freezer.

Savarin syrup:

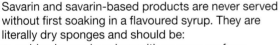

	3 litres
Oranges	2
Lemons	2
Water	2 litres
Sugar	1 kg
Bay leaf	2
Cloves	2
Cinnamon sticks	2

1 Peel the oranges and lemons and squeeze the juice.

2 Add all the ingredients into a large pan and bring to the boil, simmer for 2–3 minutes and pass through a conical strainer.

3 Allow to cool and measure the density (it should read 22 Baumé). Adjust if necessary. (More liquid will lower the density, more sugar will increase it.)

4 Re-boil then dip the savarins into the hot syrup until they swell slightly. Check they are properly soaked before carefully removing and placing onto a wire rack with a tray underneath to drain.

5 When cooled, brush with boiling apricot glaze.

20 Savarin with fruit

Energy	Calories	Fat	Saturated fat	Carbohydrates	Sugar	Protein	Fibre
1021 kJ	241 kcal	4.6 g	2.5 g	49 g	40 g	2.7 g	0.7 g

Preparing the dish

1 Use a savarin that is baked in a ring mould, either large or individual.

2 Before glazing, sprinkle with kirsch and fill the centre with prepared fruit. Serve with a crème anglaise or raspberry coulis.

21 Blueberry baba

Energy	Calories	Fat	Saturated fat	Carbohydrates	Sugar	Protein	Fibre
1415 kJ	337 kcal	14.4 g	8.7 g	52 g	42 g	3 g	0.9 g

Mise en place

1 Have the soaking syrup prepared in advance.

Preparing the dish

1 For a baba, currants are usually added to the basic savarin dough before baking.

2 As for marignans (recipe 22), split the glazed baba and fill with crème diplomate and blueberries or Chantilly cream with other seasonal fruits.

22 Marignans Chantilly

Energy	Calories	Fat	Saturated fat	Carbohydrates	Sugar	Protein	Fibre
1407 kJ	335 kcal	14.4 g	8.7 g	51 g	41 g	2.9 g	0.5 g

Preparing the dish

1 Split the glazed marignans and fill with Chantilly cream.

2 Once split and before filling, they can be sprinkled with rum or Grand Marnier.

> **PROFESSIONAL TIP**
>
> A marignan is baked in an individual boat-shaped mould or barquette.
>
> Savarin paste is notorious for sticking in the mould. Always butter moulds carefully and flour. After use do not wash the moulds but wipe clean with kitchen paper.

23 Pitta bread

Energy	Calories	Fat	Saturated fat	Carbohydrates	Sugar	Protein	Fibre	Sodium
781 kJ	184 kcal	3.8 g	0.6 g	30.7 g	0.3 g	0.0 g	1.4 g	331.0 g

	6 breads
Strong white flour	250 g, plus extra for dusting
Instant dried yeast	7 g
Nigella seeds or black onion seeds	20 g
Salt	1 tsp
Water	160 ml
Olive oil	2 tsp, plus extra for kneading

Cooking

1 Mix together the flour, yeast, nigella seeds and salt. Add 120 ml of the water and 1½ teaspoons of oil. Mix the ingredients together.

2 Gradually add the remaining water and oil until all the flour has come away from the sides and a soft dough is formed (not all the water may be required; the dough should be soft and not sticky.)

3 Pour a little oil onto the work top. Place the dough on top and knead for 5–10 minutes. The dough will be wet at first but will form a smooth dough once kneaded. Once a smooth dough is achieved, place into a clean, oiled bowl. Cover and leave to prove until doubled in size.

4 Preheat the oven to 250 °C and place a clean baking tray or baking stone on the middle shelf.

5 Once the dough has doubled in size, tip it onto a work surface dusted with flour. Knock the dough back by folding it inwards over and over again until all the air is knocked out. Split the dough into

6 equally-sized balls. Roll each ball into an oval shape 3–5 mm thick.

6 Bake directly on the pre-heated trays or stone for 5–10 minutes, or until the breads just start to colour. Remove them from the oven and cover with a clean, dry cloth until they are cool.

NOTE

Pitta breads are best eaten as fresh as possible on the same day they are baked. If they are not for immediate consumption, it is recommended that they are cooled quickly, wrapped securely, labelled and frozen until required.

24 Blinis

Energy	Calories	Fat	Saturated fat	Carbohydrates	Sugar	Protein	Fibre	Sodium
3460 kJ	827 kcal	31.2 g	8.5 g	86.0 g	3.8 g	0.0 g	12.0 g	489.0 g
431 kJ	103 kcal	3.9 g	1.1 g	10.7 g	0.5 g	0.0 g	1.5 g	61.0 g

* Using 15 ml of semi-skimmed milk
** For a portion of approx. 8 blinis

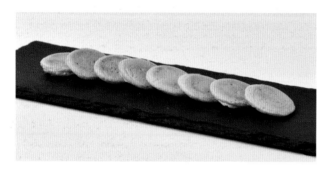

	Approx. 8 blinis
Eggs	6
Fresh yeast	15 g
Buckwheat flour	120 g
Milk	

Mise en place

1 Carefully separate the egg yolks and whites into two separate bowls ready for use.

Cooking

1 Disperse the yeast in a small amount of warm water.

2 Mix in the egg yolks and then the buckwheat flour.

3 Adjust the texture with a little milk until smooth.

4 Whisk the egg whites to a firm peak and then fold into the buckwheat base mix.

5 Allow to rest for 20 minutes and then fry in small rounds in a non-stick pan with a little oil.

6 Turn the blinis when the first side is golden brown (30–40 seconds) and continue to fry the other side to the same colour.

7 Allow to cool before serving.

PROFESSIONAL TIP

Whisk the egg whites in a scrupulously clean bowl. This will help to achieve a good peak and prevent the white from collapsing.

313 Chocolate products

Learning outcomes

In this unit you will be able to:

1 Understand how to work with chocolate, including:
   - understanding the meaning of hygroscopic when applied to chocolate
   - understanding the differences between chocolate compound and couverture
   - knowing the different uses of chocolate couverture
   - knowing different types of couverture
   - knowing cocoa butter contents and understanding the temperature ranges for tempering different types of couverture
   - understanding how to use flavourings, fillings, inclusions and coatings to enhance the end product to meet dish requirement
   - understanding how to produce inclusions and fillings to enhance the end product to meet dish requirement
   - understanding the quality points of correctly tempered couverture using any of the recommended tempering techniques
   - knowing the three recognised methods to temper chocolate couverture
   - knowing the most appropriate tools for each technique and understanding the most appropriate technique to meet dish requirements
   - understanding the importance of correct temperature sand movement when tempering chocolate couverture and how handling consistencies are achieved

   - understanding the common faults which may occur on tempered chocolate
   - knowing the correct storage for chocolate and finished chocolate work and the effect of poor storage.

2 Produce chocolate products, including:
   - understanding the most appropriate tools and equipment that used for working with chocolate, including size, type and materials
   - understanding the importance of correct emulsification techniques when mixing melted couverture to boiled liquids such as milk, cream, fruit purée, crème Anglaise
   - being able to apply appropriate techniques, tools and equipment to produce runouts to decorate desserts and petit fours
   - being able to apply appropriate techniques, tools and equipment to produce truffles to meet dish requirements
   - being able to apply appropriate techniques, tools and equipment to produce a moulded chocolate praline to meet dish requirements
   - being able to apply appropriate techniques, tools and equipment to produce moulded figures/eggs to meet dish requirements
   - being able to evaluate finished products.

Recipes included in this chapter

No.	Recipe
1	Chocolate run-outs
2	Chocolate spirals
3	Transfer sheet swirls
4	Chocolate cigarettes or pencils
5	Chocolate lollipops with infusions
6	Chocolate truffles
7	Chocolate and ginger tart with insert of mango jelly and praline

No.	Recipe
8	Convenience shells used for pralines
9	Cut pralines
10	Moulded pralines
11	Mandarin ganache
12	Green tea ganache
13	Caramel and orange ganache
14	Cremeux
15	Molded chocolate eggs

Understand how to work with chocolate

Chocolate is a luxurious product that is enjoyed by millions of people worldwide every day. There are many different grades, styles and flavours of chocolate and this unit provides an overview of the two most commonly and generic used types of chocolate – **couverture** and **compound chocolate** – as used in professional kitchens and patisseries.

Chocolate can be consumed as a base product, in a bar form for example. It can also be melted and processed in many ways to produce a huge range of chocolate products. This ranges from hand-made truffles and pralines to the flavouring and enriching properties that chocolate can provide. For example, chocolate can be used within a crème anglaise or crème pâtissière to provide flavour and richness as well as colour to these highly versatile and commonly used products within patisserie and confectionery. Chocolate can even be used within the savoury kitchen: its enriching profile is often utilised in sauces to accompany game dishes such as venison. Chocolate also harmonises extremely well with many aromatic commodities. Vanilla is the most obvious but chocolate is exceptional with some more unusual partnerships, including commodities such as chilli, green tea, wasabi and even Marmite!

> **KEY TERMS**
>
> **Couverture:** chocolate which has a very high percentage of cocoa butter; used in professional kitchens and patisserie.
>
> **Compound chocolate:** less expensive than couverture, compound chocolate contains cocoa powder instead of cocoa liquor and oil instead of cocoa butter.

1.1 Couverture and chocolate

Translated, the French word 'couverture' means 'covering' or 'coating' in English. Couverture has a very high percentage of cocoa butter (at least 30 per cent) and is used to flavour patisserie products such as ice creams, mousses, ganaches, soufflés and so on.

Types of couverture

Couverture comes in dark chocolate, milk chocolate, and white chocolate varieties. You can also purchase coloured and flavoured couverture. The purchasing unit is either in **callets** or a solid block. For **tempering** purposes, callets are preferred as they melt uniformly, making tempering of the chocolate more effective.

White couverture is chocolate which does not contain the dark-coloured cocoa solids derived from cocoa beans. It only contains 30 per cent cocoa butter (the fatty substance derived from cocoa beans), milk and sugar. White chocolate is sweet, with a slight vanilla taste, and has a light flavour which is not too heavy or intense.

Milk couverture has added dried milk powder, along with cocoa butter, 40 per cent sweeteners and flavourings; it

contains a minimum of 10 per cent chocolate liquor and 12 per cent milk solids.

Plain couverture has a higher content of cocoa butter (60–70 per cent), which gives the chocolate more viscosity, and cocoa solids which give the chocolate its colour. It is more fluid than the white and milk varieties and is used for decorations, moulding, enrobing and flavouring.

All three chocolates are factory-tempered before being packaged. (For small quantities it is possible to melt the chocolate while keeping the tempering qualities in the chocolate. This will be discussed in more detail under tempering of chocolate.)

> **KEY TERMS**
>
> **Callets:** small, round, symmetrical pieces of couverture, which are used for melting purposes; the purchasing unit of couverture.
>
> **Tempering:** an essential process that allows chocolate to be manipulated and combined with other ingredients or made into artistic pieces; it produces a snap when bitten and has a glossy appearance.

Sugar-free couverture or diabetic couverture

Most of the top producers of couverture offer a selection of fine chocolates with no added sugar. The couverture is still produced to strict quality standards with premium grade cocoa beans, 100 per cent pure cocoa butter and natural vanilla. For example, some chocolate recipes substitute sugar with maltitol or stevia as a sweetener, creating a taste nearly identical to traditional couverture.

Furthermore, they offer the same application possibilities and can be processed in the same way as traditional chocolates. Their technical properties and behaviour are absolutely identical. The use of the denomination 'no added sugar' is permitted on these products.

Single plantation couverture

Single plantation couvertures are each made with cocoa beans from one particular country or region. Each chocolate brings you an exciting taste sensation and aromatic character and extraordinary flavour pairings. The couverture character reflects the soil, climate and environment where the cocoa beans were grown. Just like wine, the taste of these couvertures can slightly alter with each harvest.

Organic couverture

A unique selling point to customers may be the use of organic products and the guarantee of organic purity. Producers are now offering very high quality couvertures produced with cocoa beans, sugar and milk from organic agriculture, each certified by the official organisations. With such chocolate, it is possible to produce unique products from a pure and very natural base.

Hygroscopic

Hygroscopy is the ability of a substance to attract and hold water molecules from the surrounding environment. This is achieved through either absorption or adsorption with the absorbing or adsorbing substance becoming physically changed. This could be by an increase in volume, boiling point, viscosity or other physical characteristic and properties of the substance, as water molecules become suspended between the substance's molecules in the process.

If water comes into contact with melted couverture, it will immediately thicken the chocolate and restrict its use from this point. It is therefore essential to avoid contact with water and/or steam when processing or working with chocolate.

> **KEY TERM**
>
> **Hygroscopy:** the ability of a substance to attract and hold water molecules from the surrounding environment.

Cocoa

Cocoa bean

This was once called 'cocoa almond' or 'cocoa grain'. It is the seed that is found in the pods of cacao trees. After being treated, it is packed and sent to be sold on the international market. It is from this bean that cocoa butter, chocolate liquor, cocoa powder and cocoa nibs are extracted.

Cocoa nibs

These are roasted, shelled cocoa beans broken into small pieces. This is a very interesting product with an intense flavour – 100 per cent cocoa. It gives aroma, flavour and texture to many preparations, like sponge cakes, chocolate bonbons, pound cakes, muffins, ice creams, cookies and cake decorations. Care should be taken not to use excess quantities so that the balance with the other ingredients is not upset.

Chocolate liquor

This is a smooth, liquid paste. In addition to being the base for other cocoa derivatives, such as cocoa butter or cocoa powder, it can be used in all types of desserts and cakes (toffee, for example). One of its main characteristics is that it contains no sugar, which gives it a slightly bitter flavour in its pure state.

Cocoa butter

Once obtained, chocolate liquor is pressed to extract the fat (cocoa butter) and separate it from the dry extract. Cocoa butter is the 'spine' of chocolate, since its proper crystallisation determines whether chocolates (couvertures) have adequate densities and melting points. Cocoa butter should be melted at 55 °C (it begins melting at 35 °C) to achieve proper de-crystallisation.

Cocoa butter is used to coat with a spray gun (mixed with chocolate in greater or lesser quantity), for chocolate bonbon moulds, desserts, cakes and artistic pieces, or in pure form for moulds and marzipan figurines. Cocoa butter is available in various colours for decorative purposes.

Mycryo cocoa is a commercial product for use in tempering chocolate, which can save time. To use, follow the manufacturer's instructions.

Cocoa powder

Two products are extracted from pressed chocolate liquor – cocoa butter in liquid form, and dry matter, which is ground and refined to make cocoa powder. The quality of cocoa powder is a function of its finesse, its fat content, the quantity of impurities it contains, its colour and its flavour. It is very important to store cocoa powder in a dry place and in an airtight container.

The main characteristics of the cacao tree and its fruit:

- The majority of the world's cacao trees are concentrated around the equator.
- The cacao tree needs a hot, humid and rainy climate – the tropics are ideal.
- High levels of wind and sun can be damaging to the cacao tree and it must be protected from both.
- A productive tree can measure between 5 and 10 metres in height, depending on its age.
- The fruit, or 'cocoa pod', measures between 15 and 30 centimetres.
- Each cocoa pod holds approximately 30–40 seeds (cocoa beans).

Roasting
After being cleaned, the cocoa beans are roasted which develops the distinctive flavour of the cocoa bean.

Winnowing
After roasting, the beans are put through a winnowing machine which removes the outer husks or shells, leaving behind the roasted beans, now called nibs.

Milling – making cocoa liquor
The nibs are then ground into a thick liquid called chocolate liquor (this is cocoa solids suspended in cocoa butter). Despite its name, chocolate liquor contains no alcohol and has a strong unsweetened taste.

Pressing for the production of cocoa powder and cocoa butter
The next stage is to press the cocoa liquor and extract the cocoa butter. This leaves behind a solid mass which is then processed into cocoa powder.

Making the chocolate
The following ingredients are mixed together to make the three different types of chocolate:

White chocolate: made from the same ingredients as milk chocolate (cocoa butter, milk, sugar) but without the chocolate liquor. White chocolate must contain at least 20% cocoa butter and 14% total milk ingredients.

Milk chocolate: a combination of chocolate liquor, cocoa butter, sugar and milk or cream.

Plain chocolate: a combination of chocolate liquor, cocoa butter and sugar. Must contain at least 35% chocolate liquor.

Refining
The next step is to pass the chocolate through heavy rollers to form a fine flake. Additional cocoa butter is added at this stage and an emulsifying agent called lecithin. The mixture is now mixed to a paste.

Conching
The process of conching kneads the chocolate through heavy rollers which develops the flavour.

Tempering
The chocolate is now tempered ready for use, which gives it shine, snap and retraction.

Moulding
The liquid tempered chocolate is now deposited into solid block moulds or shaped into callets ready for use.

Use in the patisserie kitchen
For small quantities, melt the couverture in the microwave following the technique for microwave tempering, or fully melt to the stated temperature for the type of chocolate and then follow the tempering procedure (see below).

Process of manufacturing chocolate

Tempering chocolate

As already mentioned, cocoa butter is a vital component of chocolate, since the final result depends on its crystallisation. It determines good hardness (snap), balance, texture and shine, and it prevents excessive hardening, whitening and the formation of beads of oil on the surface.

When chocolate is melted, the cocoa butter melts and its particles separate. To achieve a perfect result, the chocolate is cooled (which re-crystallises the cocoa butter) to enable the cocoa butter and its particles to re-bond.

Tempering allows chocolate to be manipulated and combined with other ingredients or made into artistic pieces that, when re-crystallised, regain the texture and consistency of the chocolate before it was melted.

Tempering is necessary because of the high proportion of cocoa butter and other fats in the chocolate. This stabilises the fats in the chocolate to give a crisp, glossy finish when dry.

> **NOTE**
> It is essential that a thermometer is used for the tempering process and that the couverture is stirred and moved consistently to develop the beta crystals. This is known as pre-crystallising which, if performed correctly, will result in a hard (snappy), crunchy and glossy finished product. Therefore this process is a combination of monitoring temperature and movement.

The melting and working temperatures for couverture shown in Table 13.1 are a guideline. Some brands of chocolate may vary. Always check the tempering instructions on the packaging.

Table 13.1 Temperatures for tempering the three types of couverture

Initial melted temperature	Finished working temperature
Plain couverture 45°C	31–32°C
Milk couverture 40°C	30–31°C
White couverture 40°C	28–30°C

For more on tempering techniques see Section 1.4 below.

Compound chocolate

Compound chocolate substitutes two of the main ingredients found in couverture. Instead of cocoa liquor, cocoa powder is used and the cocoa butter is substituted with an oil. For general covering at less expense, chocolate compound can be melted and dipped and will generally set without any issues or complications. The taste of compound chocolate is inferior, as is the shiny appearance, snap and flavour profile.

Compound chocolate gives a crisp, hard coat. It may contain vegetable oil, hydrogenated fats, coconut and/or palm oil, and sometimes artificial chocolate flavouring. This type of chocolate does not require to be tempered.

Couverture has cocoa butter and chocolate liquor as core ingredients. This means that it requires more attention and control when melting for use. Unlike the compound coating chocolate, it cannot simply be melted down. It has to be tempered and pre-crystallised. If not, then the chocolate will bloom or may not set up properly. When it is tempered and pre-crystallised correctly, couverture sets with a glossy and shiny appearance. It also has a snap as well as a complex flavour and a finish that melts in the mouth.

1.2 Flavourings, fillings, inclusions and coatings

Chocolate products are often enhanced with additional ingredients and finishes.

Flavourings

Flavourings may be from fruit-based sources such as oils, pastes and macerated fruits. Alcohol concentrates and spices such as vanilla are also commonly used to enhance the flavour of chocolate. Herbs can be used to provide subtle and complimentary aromas and flavours to chocolate products. For further information about flavourings and their use, see Unit 309 Desserts and puddings.

Fillings

Many chocolate products, particularly filled pralines, are filled with complimentary pastes, creams and similar products. One of the most commonly utilised fillings is ganache. A simple ganache is made from an emulsion of couverture, and cream. Ganache is a very versatile product and it is often flavoured with a spirit or liqueur to give the product a base flavour, for example, cognac as a flavour for a filled chocolate praline.

Other fillings that are commonly used include lemon curd, crémeux and gianduya. Crémeux is a filling made from an emulsion of crème anglaise and couverture, and gianduya is a sweet chocolate product containing around 30 per cent hazelnut paste. As well as a filling, gianduya is also used as a flavouring in products such as creams, ice creams and mousses.

Chocolate and ginger tart with insert of mango jelly and praline in the recipes section at the end of this unit is a perfect example of a product that includes a spice-infused ganache filled with a pectin-set jelly and a praline disc.

Pectin set jelly

Inclusions

Inclusions can also be used as fillings but also, and perhaps more commonly, on the exterior surface of products. Inclusions are usually very small in size and are added to chocolate products to provide an extra dimension to the product. They can be used to enhance colour, flavour, texture or even provide an element of surprise (for example, popping candy) as well as a visual enhancement.

Inclusions

Types of inclusions include:
1 Popping candy or crackle crystal – consists of carbonated sugar crystals which crackle and pop when placed in the mouth. Used as a topping or garnish to give a surprising but pleasant effect.
2 Freeze-dried fruits – fresh fruits which have been put through a freeze-drying process, removing all the excess moisture while retaining the colour and structure. These ingredients are used to add texture and flavour to desserts, cakes and chocolates.

3 Cocoa nibs (*grue de cacao*) – dry-roasted pieces of cocoa bean. They are used to make tuiles, added to ice creams and mousses, and sprinkled over desserts.

4 Bres (Brésilienne) – made from roasted and caramelised hazelnuts, which are then crushed. It is used to flavour ice cream, in mousses or as a base to stand scooped ice cream on plates.

5 Pailleté feuilletine – a crunchy biscuit that is used as a decoration. It is made into an inclusion for mousses or flans by mixing it with equal quantities of praline paste and chocolate and then rolling out to a thin biscuit.

6 Ginger mini cubes – small cubes of sugar-coated crystallised ginger. This is an example of an inclusion that could be added to ganache-based desserts, chocolate mousses or ice creams.

7 Caramel fudge pieces – small cubes of caramel fudge, used as inclusions in iced confections.

Chocolate products for use as infusions

8 Chocolate fudge brownie pieces – used to flavour ice-creams, mousses and for inclusion in gateaux.

9 Griotines – a type of cherry. For more details on the use of Griotines see Unit 309 Desserts and puddings.

10 Crisp pearls – round biscuits coated in white, milk or plain chocolate, used as decorations and inserts.

11 Metallic/shimmer powders – brushed onto chocolate centrepieces and pralines to enhance their appearance.

12 Ready-mixed coloured cocoa butters – intended for use with spray guns. They are used to decorate and colour chocolate moulds or sprayed directly onto chocolate centrepieces. Coloured cocoa butter can also be used to create your own transfer sheets and painted into figure moulds.

13 Pâté à glace – a soft chocolate coating that comes in white, milk and plain.

14 Commercial glazes – chocolate based and give a very high gloss shine. Let the glaze down with 10 per cent stock syrup and warm in the microwave to 30 °C. When applied over mousses, the mousses can be kept frozen without losing their shine.

15 Metallic spray – used to embellish chocolate centrepieces and sprayed into praline moulds prior to coating with chocolate couverture.

16 Pure gold leaf (24 carat) – used as a decoration, especially with chocolate desserts.

17 Pure gold dust (24 carat) – can be sprinkled over dishes as a garnish.

18 Praline paste – a combination of caramelised sugar and nuts, usually hazelnuts or almonds. It is used to flavour the centres of chocolates or in desserts and ice creams.

19 Bitter cocoa powder is very dark in colour. It is used to coat truffles and dusted on desserts.

Recipe 5 (Chocolate lollipops) and recipe 8 (Convenience shells for pralines) provide good examples of how infusions are used to enhance chocolate products.

Coatings

Many chocolate products are coated to enhance the product and provide an additional texture and/or flavour. For example, truffles and coated (sometimes referred to as enrobed or dipped) chocolate pralines are coated with a fine layer of tempered couverture. Other coatings include cocoa powder, chocolate shavings (see recipe 8 – Convenience shells used for pralines), chopped nuts, praline and finely chopped dehydrated fruits.

1.3 Quality points

For flavouring purposes, couverture only requires melting and adding to the desired product. For moulding and setting purposes, the couverture needs to be 'tempered'. This is a process whereby the chocolate is taken through different temperatures to stabilise one particular chocolate crystal known as the 'beta crystal.' This crystal has all the characteristics of good tempered chocolate (snap, shine and retraction), enabling the prepared couverture to literally fall out of the mould it has been set in, giving a solid, shiny piece of chocolate with a perfect snap.

If couverture is not tempered correctly, it can develop faults. The two most common faults in prepared couverture are known as fat bloom and sugar bloom. Refer to Section 1.5 for further details.

1.4 Tempering techniques

There are three recognised tempering techniques for use with couverture:
● table top method
● injection (or seeding) method
● microwave method (not suitable for large quantities).

This couverture was tempered and spread over textured sheets to give the desired finish.

It is essential that a thermometer is used for the tempering process.

The melting and working temperatures for tempering the three types of couverture are shown in Table 13.1 (page 602).

> **PROFESSIONAL TIP**
>
> Chocolate that is to be moulded or used in decoration needs to be tempered to achieve snap, shine and retraction in the finished product.

Table top method

1 Melt carefully to the specific temperature for the type of couverture avoiding steam, moisture and over-heating. (The use of a chocolate melting tank is ideal for this.)
2 Once the couverture has reached the melting temperature, remove from the heat source.
3 Pour 70 per cent of it on to a very clean and dry marble surface or slab. Work continuously by spreading outwards and pulling back to the centre with a step palette knife until the couverture starts to thicken and the formation of the good beta-crystal is developed.
4 Quickly add the couverture back to the remaining 30 per cent, stirring continuously, dispersing and seeding the beta-crystal into the liquid chocolate until it reaches its finished working temperature (31–32 °C for plain, 30–31 °C for milk and 28–30 °C for white).
5 Check the finished temperature with a digital probe. If the chocolate is still too warm, pour a small amount once more onto the marble and repeat the steps above until the chocolate reaches the desired temperature.

Injection (or seeding) method

1 Take 30 per cent of couverture callets of the total weight being tempered.
2 Melt the remaining 70 per cent of the couverture following stages 1 and 2 of the table top method above.
3 Remove the container of melted chocolate and stand it on the table top with a folded cloth underneath (this is to prevent the chocolate from setting on the base).
4 Gradually add the remaining 30 per of the chocolate callets to the melted couverture, stirring continuously until the finished working temperature is achieved (31–32 °C for plain, 30–31 °C for milk and 28–30 °C for white).

For large-scale production a wheel tempering tank is used.

Microwave method (for small quantities)

As previously discussed, couverture is packaged already tempered.

1 Take 500 grams couverture and place into a plastic bowl.
2 Warm the chocolate in short intervals in a microwave oven set at 50 per cent until the chocolate partially melts and there are still signs of solid chocolate.
3 Remove from the microwave and continue to stir the chocolate until the solid chocolate pieces melt down. Any solid pieces are still tempered and the gentle mixing and melting will seed the newly melted chocolate.

The table top and seeding methods are generally used when a reasonably large amount of chocolate is required. This may be for producing moulded figures or a large quantity of moulded pralines, for example. The microwave method is ideal for tempering a small quantity of chocolate. This might be if a small quantity of chocolate garnishes is required, such as some chocolate spirals or run-outs.

If a very large and regular amount of tempered chocolate is required, the most effective way to have a constant supply is to invest in a tempering machine. A good quality tempering machine will help to achieve the precise temperatures that are required during the tempering process and will also provide the movement that is essential to the chocolate.

> **PROFESSIONAL TIP**
>
> ● If the working temperature is exceeded by more than 3 °C, the process will have to be repeated as the couverture will not be correctly tempered and faults will occur.
> ● If the temperature of the chocolate drops during processing, it can be gently reheated to the working temperature with a heat gun, without any detrimental effects to the characteristics of the chocolate.

- Always keep water away from melted couverture and never store in humid conditions.
- The ideal room temperature for working with chocolate is 18°C with 60 per cent humidity.
- Chocolate products should be stored in a dry place at 15–16°C and at less than 50 per cent humidity.
- Chocolate absorbs all odours and should therefore be stored well covered.
- The higher its fat content, the faster chocolate melts in the mouth.
- In tempering, it is essential to check the temperature with a thermometer and to perform the 'paper test'. This is done by dipping a piece of paper in the tempered chocolate. The tempering is optimal if, in about two minutes, it has crystallised with a flawless, uniform shine and without stains or fat drops on the surface.
- A glossy surface is a sign of good tempering.
- Two important points are to be followed to achieve good tempered couverture – correct temperature and continuous movement of the chocolate. Movement develops beta crystals in the chocolate.

1.5 Common faults

Sugar bloom

Water has a significant and, mostly, damaging affect on chocolate and chocolate products, although there are one or two exceptions. **Sugar bloom**, for example, is caused by moisture. Condensation on the surface of the chocolate or moisture in the chocolate coating causes the sugar to absorb the moisture and dissolve. Due to their minute size, it is not possible to see the sugar crystals present in chocolate with the naked eye. If water comes into contact with chocolate, it dissolves the sugar crystals on the surface. As the water dries, the dissolved sugar crystallizes and precipitates onto the surface of the chocolate. The resulting tiny sugar crystals gives the chocolate a dusty appearance.

Sugar bloom may occur in a number of ways. The most likely is as a result of direct contact with water. However, sugar bloom may occur in other, not so obvious, ways. For example, if chocolate is placed in a refrigerator where it becomes cold and then removed and placed in open air, the cold chocolate will condense moisture from the air, and the condensation will cause sugar bloom. Sugar bloom may also occur if the chocolate has been in an environment with too high a humidity.

The best way to avoid sugar bloom is to store chocolate in an area of low humidity and stable temperature so as to avoid condensation. If the chocolate is cold, such as when it has been stored in a refrigerator, it should remain covered or wrapped, if possible, and acclimatised slowly with air circulation minimised.

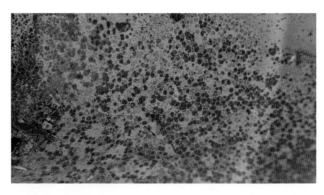

Sugar bloom

Sugar bloom is caused by:

- storage of chocolate in damp conditions
- working in humid conditions
- using hygroscopic ingredients
- products which have high liquid content packaged and stored in a warm area. (Vapour is given off which is trapped in the packaging, creating a layer of moisture on the surface of the chocolate.)

Fat Bloom

Fat bloom, unlike sugar bloom, is not always caused by a simple set of circumstances, such as the chocolate becoming wet. Fat bloom is more complicated and it is generally more difficult to discover the actual source of the problem.

Fat bloom typically appears as lighter coloured spots or areas on the chocolate. As the name suggests, the bloom is composed of fat. In this case, the fat is from the cocoa butter which is a natural by-product of the cacao bean.

When discussing the reasons for fat bloom, it is important to note that when cocoa butter hardens, it forms crystals. Some of the crystals are stable, but other crystals are not and will actually change form over time. During the tempering process, the intention is to ensure that only stable crystals form, while the chocolate hardens. Fat bloom is generally due to a lack of stability of the various crystals, usually caused by incorrect tempering of the chocolate.

Fat bloom

Fat bloom is caused by:

- poor tempering of the chocolate
- incorrect cooling methods
- covering a confectionery product that is too cold
- warm storage conditions.

1.6 Storage

It is very important to store chocolate and chocolate products in the correct environment to avoid spoilage and contamination.

To initially set chocolate products, they should be placed into a cool environment to harden. The ideal temperature for this is between 8°C and 10°C. Depending on the size of the product, this can range from 5–6 minutes for small items to 20–25 minutes for larger ones.

Beyond the initial setting, chocolate and chocolate products should be stored in a cool and dry environment, between 12°C and 18°C. They should also be stored away from odours and strong flavoured items as chocolate will quickly absorb such odours which will negatively influence the flavour of the chocolate.

Ideally, chocolate should be wrapped in protective packaging and stored in a constant temperature as fluctuations in temperature can also result in problems such as fat and/or sugar bloom. If products are to be moved from one environment to another, time should be given to allow the product time to acclimatise to the new environment/temperature. It is also necessary to avoid humid (moist) environments as this will encourage sugar bloom and condensation. It is essential that products are clearly labelled with the date of production and covered to prevent dust resting on the surface.

It is vital that all equipment is spotlessly clean and free from grease or oils and any other forms of contamination. For example, when preparing chocolate moulds for figures or filled pralines, it is necessary to use very clean and polished moulds. Correctly tempered couverture will retract slightly from shiny surfaces, leaving a highly polished and shiny surface to the finished chocolate. Any grease or oil on the mould would transfer onto the chocolate, impairing its presentation. For this reason, it is good practice to use linen gloves (supplied by many chocolate producers) when handling moulds and tempered chocolate products. This will prevent natural oils and contamination from fingerprints being transferred onto the mould and ultimately the product itself, either directly or indirectly.

It is also good practice to polish chocolate moulds using clean cotton wool to clean and polish the surface of the mould without causing damage to the mould, such as abrasions or scratches. The delicate and soft structure of cotton wool is prefect to carry out this procedure.

Producing chocolate products

2.1 Tools and equipment

A wide range of equipment is used in the specialist production of chocolate products. This ranges from dipping forks and combed scrapers though to specialist moulds and tempering machines.

Small equipment for chocolate work

Acetate provides a suitable shiny surface on which to spread chocolate. The advantage of using acetate is that it provides flexibility and can be twisted and shaped before the chocolate has set, allowing various finished shapes and styles to be produced. The recipe for Chocolate spirals (recipe 2) provides a good example of the use of acetate. The recipe for Transfer-sheet swirls (recipe 3) takes this one stage further and uses acetate transfer sheets to make products with additional décor. In this example, the transfer sheet has been coated in various pre-loaded designs using coloured cocoa butter. Practically any design and colour can be used here as manufacturers of such sheets use chocolate acetate printers to print anything a customer requests, from company logos to animated objects. It is also possible to make your own transfer sheets. If using sheets on a frequent basis, it may be worth considering the purchase of a chocolate transfer printer. If not, it is possible to produce home-made transfer sheets by brushing with coloured cocoa butter before covering with a contrasting thin layer of tempered couverture. Metallic food powders can also be used in this process to provide contrast and additional colours.

Other useful equipment:

- Dipping forks – used to dip cut and layered pralines.
- Figure or egg moulds – originally made from metal, now made from polycarbonate. Used for moulding figures in chocolate coverture.
- Hair drier – used to cool down blown sugar.
- Hot air gun – used to raise the temperature of tempered chocolate during processing.
- Leaf mould – made from silicone or metal and used to make pulled sugar leaves.
- Marble slab – used for table top tempering of melted couverture.
- Microwave oven – used for tempering chocolate, softening fondant and heating liquids for pastry dishes (e.g. heating fruit coulis before adding gelatine).
- Praline moulds – used to make filled pralines.
- Rubber gloves – used to protect the hands from high temperatures when pulling sugar. Also prevent moisture from the hands coming into contact with the sugar
- Scraper – specifically made for chocolate work. Uses include table top tempering, making chocolate copeaux and leveling the surface on praline moulds.
- Sphere moulds – made from silicone and used to make chocolate spheres.
- Stepped palette knife – used to evenly spread out chocolate during the tempering process
- Tempering couverture – this spatula has a temperature probe and gives a temperature reading when seeding the couverture.

2.2 Emulsification

It is important to consider how melted chocolate will mix and emulsify with other ingredients to produce augmented chocolate products such as ganache, creams, fillings and sauces.

Temperature is also important as mixing very hot ingredients, such as melted couverture with freshly whipped chilled cream for example, can cause issues where the chocolate can become 'grainy' and split rather than emulsifying smoothly to produce a smooth, mousse-like ganache.

There are different methods to achieve emulsification. A basic emulsion of cream and chocolate is achieved by heating cream and pouring it into a bowl of chopped or grated chocolate. Callets are ideal for this purpose as they are pre-cut or shaped and are of the same size. Chocolate callets are small pieces of couverture, similar to chocolate chips in size, but without the traditional shape of a chocolate chip. Callets are round and symmetrical with a flattened base and are perfect to use for melting purposes.

Once the cream has been added, a spatula or whisk can be used to mix the two basic ingredients to form a glossy emulsion. Depending on the use of the ganache, the ratio of chocolate to cream can vary. Generally, the more chocolate used, the thicker and more solid or dense will be the resulting ganache. Ganache in this form can be used as a coating if it is poured over a product such as a cake, for example, while it is still in its liquid form. Once cooled or chilled, the ganache will become firmer and can be used as a filling. If the chilled product is whisked further at this point, it will become lighter and fluffy, generally more mousse-like and again could be used as a filling, topping or piped for decorative purposes.

Stick blender used in emulsification

Another means to emulsify chocolate with other products is through the use of a stick-blender. In this process, melted chocolate is mixed with the other heated ingredients, such as fruit purée, crème anglaise, cream or flavourings. The stick blender is then used to emulsify the ingredients into a smooth and shiny product such as ganache or cremeaux.

It is also becoming more common in the industry to use thermostatically-controlled food processors (e.g. Thermomix) to produce ganache and similar mixed products. The beauty of using such machines is that total control of temperature and movement is possible, creating perfect results every time. In this example, a ganache can be produced whereby the lecthicin in the chocolate acts as an emulsifying agent with the cream, creating a perfect emulsion. This process somewhat takes away the traditional skills required in terms of heating, melting, mixing and monitoring the condition of ingredients through the various stages and processes. However, the assurance provided by such equipment means the chef can be confident of a perfect product and is free to concentrate on other tasks within the kitchen.

Food processor used for emulsification

2.3 Runouts

For details of how to prepare these, please see the recipes section at the end of the unit.

2.4 Truffles

For details of how to prepare these, please see the recipes section at the end of the unit.

2.5 Moulded chocolate praline

For details of how to prepare these, please see the recipes section at the end of the unit.

2.6 Moulded figures or eggs

For details of how to prepare these, please see the recipes section at the end of the unit.

2.7 Evaluation

Once chocolate products are produced, it is import to evaluate the various quality aspects that distinguish their overall quality and any opportunities for future improvement.

In terms of the processing of the couverture itself, it should have retracted slightly, particularly if moulded, and possess a very shiny and glossy exterior appearance. It should also have a crisp snap. All of these signs indicate that the couverture has been pre-crystallised and tempered accurately and correctly. It is also vital that finished products are handled very carefully to avoid smears from fingerprints, for example. Linen gloves and the use of chocolate utensils can help to avoid this.

Chocolate products should be neat and uniform and of the indicative size specified. Additional décor should be complementary and, again, uniform across the range of products. Moulded figures should have minimal sign of joins to create a high quality finish.

In terms of eating qualities, chocolate products should bring a luxurious mouth-feel during consumption. Colours, flavours and contrasts should be complementary, pure, balanced and delicate (unless a flavour is intended otherwise, for example, citrus). Textures also need to be considered. For example, a praline filling may range from a caramel to a curd or a ganache. In each example, the viscosity, smoothness and texture, as well as the flavour, plays an important role towards the overall eating quality and enjoyment of the product.

TEST YOURSELF

1 What is the difference between compound chocolate and couverture?

2 Why it necessary to 'temper' couverture?

3 What does the term 'pre-crystallising' mean?

4 How does chocolate obtain sugar bloom?

5 How does chocolate obtain fat bloom?

6 What are the ideal storage conditions for storing chocolate products?

7 Describe **three** ways in which chocolate can be emulsified with other products such as cream, fruit puree and crème anglaise.

8 Describe why it is important to form a good emulsion when making a ganache.

9 List **four** inclusions that can be used in the production of chocolate products.

10 List the steps required in the production of a tray of moulded pralines.

1 Chocolate run-outs

Preparing the dish

1 Draw two parallel lines 5 cm apart on silicone paper.
2 Take a small quantity of tempered plain couverture or melted compound chocolate.
3 Add drops of stock syrup until the chocolate slightly thickens.
4 Pipe fine line designs between the two parallel lines.
5 Allow to set, then remove using a palette knife.

2 Chocolate spirals

Preparing the dish

1 Apply tempered couverture onto an acetate strip.
2 Comb the chocolate with a grooved scraper to form parallel lines.
3 Allow the chocolate to partially set.
4 Twist the acetate and lay the strip inside a hollow drainpipe.
5 Once fully set, remove from the acetate and separate into curls.

Spread tempered couverture onto an acetate strip.

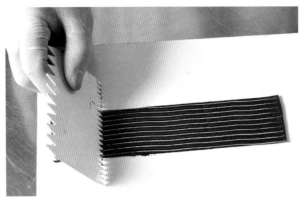

Comb into parallel lines.

Once partially set, twist the acetate and lay it inside a hollow pipe.

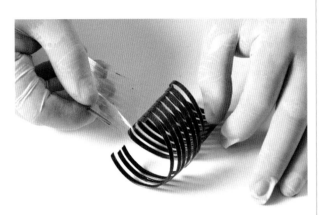

Once fully set, remove from the pipe and peel away from the acetate.

3   Transfer-sheet swirls

Preparing the dish

1   Thinly spread tempered couverture onto the reverse side of a transfer sheet.

2   Allow to partially set, then mark with a thin-bladed knife, avoiding cutting through the plastic sheet.

3   Place a sheet of silicone paper on top of the chocolate.

4   Roll up and allow to set.

5   Unfold the plastic sheet to release the chocolate swirls.

Spread tempered couverture over the acetate transfer sheet.

When partially set, mark with a knife, but do not cut through the acetate.

Cover with silicone paper, roll up and leave to set.

4 Chocolate lollipops

Preparing the dish

1 Line the base of a magnetic mould with transfer sheet.

2 Place lollipop frame on top of the transfer sheet.

3 Fill with tempered couverture and place lollipop stick in the indentation.

4 Decorate top of chocolate with crisp pearls, dried raspberries, Bres, etc.

5 When the couverture has set, remove lollipops from the mould.

6 Using transfer sheets:

7 Cut strips of transfer sheets 5 cm wide.

8 Pipe round discs of tempered couverture onto the transfer sheet.

9 Place a lollipop stick in the centre of the disc of chocolate.

10 Decorate the top of the chocolate, as in the main recipe.

11 When the couverture has set, remove from the transfer sheet.

5 Chocolate cigarettes or pencils

Preparing the dish

Two-tone coloured cigarettes or pencils:

1 Spread tempered white couverture onto a marble slab and comb using the plastic toothed edge from a cling film box.

2 When the chocolate has almost set, cover with a thin coat of tempered plain couverture.

3 Once the plain couverture has set scrape up into thin two-tone cigarettes.

Multi-coloured cigarettes or pencils:

1 Using crinkled cling film, dab different-coloured cocoa butter onto a marble slab.

2 Thinly spread with tempered white couverture.

3 Once set scrape up into white cigarettes with a multi coloured outer surface.

White or plain cigarettes or pencils:

1 Spread either white or plain tempered couverture thinly onto a marble slab.

2 When almost set scrape up into cigarettes.

PROFESSIONAL TIP
It is advisable to use a special scraper for this task.

FAULTS
If the chocolate is not at the right temperature and viscosity at the time of scraping, the cigarettes/pencils will not roll correctly. If the chocolate is too set, the chocolate will crack before it rolls. If it is too thin, it will not be able to hold its shape.

SERVING SUGGESTION
Chocolate cigarettes/pencils can be cut to the desired size and used to decorate gâteaux and cakes, on the sides or top, for example. A single cigarette/pencil is also used regularly as a garnish for a dessert, perhaps leaning on the side of a mousse or resting against a swirl or quenelle of a cream.

Spread tempered white couverture on marble, and comb through it.

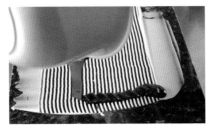

When the white couverture has almost set, pour tempered plain couverture over it.

Spread the plain couverture thinly over the white.

Clear away excess couverture from the edges.

Working at the edge of the chocolate, scrape up thin cigarettes.

6 Chocolate truffles (general-purpose ganache)

Energy	Calories	Fat	Saturated fat	Carbohydrates	Sugar	Protein	Fibre
286 kJ	69 kcal	4.7 g	2.9 g	6.1 g	5.8 g	4.6 g	0.2 g

* Using milk chocolate, unsalted butter, semi-skimmed milk and sorbitol

	60 truffles
Whipping cream	175 g
Milk	75 g
Butter	75 g
Inverted sugar (trimoline)	75 g
Powder sorbitol (optional, gives a better shelf life to the ganache)	12 g
Milk couverture	165 g
Plain couverture	333 g
Alcohol concentrate	10 g

Cooking

Boil the cream, milk, butter, inverted sugar and sorbitol to 80 °C. Remove from the heat and add the alcohol concentrate.

Partially melt both chocolates and gradually add the boiled liquid working from the centre of the chocolate forming a good elastic emulsion. Finish with a stick blender to achieve a smooth and shiny ganache.

Pour into a bowl and leave to cool and crystallise. Periodically stir the outside mixture to the centre.

When the ganache is firm, pipe out into the desired shapes.

VARIATION

This ganache can be made without the inverted sugar and sorbitol. The final product will not be quite as sweet.

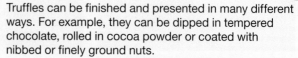

SERVING SUGGESTION

Truffles can be finished and presented in many different ways. For example, they can be dipped in tempered chocolate, rolled in cocoa powder or coated with nibbed or finely ground nuts.

7 Chocolate and ginger tart with insert of mango jelly and praline

Energy	Calories	Fat	Saturated fat	Carbohydrates	Sugar	Protein	Fibre
17606 kJ	4217 kcal	274.0 g	147.0 g	403.0 g	314.0 g	305.0 g	15.1 g

	1 × 20 cm flan
1 × 20 cm flan case, baked blind using flan case	
Chocolate and ginger ganache	
Whipping cream	200 g
Fresh ginger	22 g
Preserved ginger	22 g
Plain couverture	250 g
Unsalted butter	50 g

Cooking

Ganache:

1 Boil the cream with the finely grated ginger and allow to infuse for 10 minutes.

2 Strain, and pour onto the melted chocolate and butter. Stem blend to emulsify.

Praline disc:

	1 disc
Melted plain couverture	100 g
Praline paste	100 g
Pâte a féuilletine	100 g

1 Warm the praline paste in a microwave.

2 Mix with the melted couverture.

3 Add the pâte a féuilletine.

4 Roll out thinly between silicone paper.

5 Refrigerate until solid.

6 Cut out a disc using a flan ring that is narrower than the baked pastry case.

Mango jelly:

Mango pulp	
Caster sugar	360 g
Granulated sugar	100 g
Pectin mixture	
Pectin	30 g
Caster sugar	40 g

1 Heat the mango to 50 °C.

2 Add the caster sugar and then bring to the boil and skim.

3 Add the pectin mixture and cook to 105 °C.

4 Remove from the heat and leave to settle for 10 seconds.

5 Pour the mixture into a prepared tray. Wrap in cling film.

6 Leave to set for 4 hours.

7 Cut into a disc.

Assembly:

1 Half fill the tart case with the ganache, allow to partially set, and place on top the disc of mango jelly and praline féuilletine, cut slightly smaller than the flan case.

2 Top with the remaining ganache to fill the pastry case, refrigerate.

3 Once set, cut a wedge using a warm knife and just before service run the flame of a blow torch over the top of the ganache to give a shine. Decorate with gold leaf.

4 Serve with pear water ice.

8   Convenience shells for pralines

These shells come in various forms – spheres, squares, rectangles, rounds.

Preparing the dish

1   Fill the convenience shell with a suitable ganache and seal the top with tempered couverture.

2   Can be decorated with transfer sheets, chocolate shavings, etc.

Fill the sphere or shell with ganache, using a closing template.

Seal the top with tempered couverture.

Once set, coat in tempered couverture.

To finish, dip in cocoa powder using a dipping fork or roll in chocolate shavings (as per the main picture above).

9 Cut pralines

Energy	Calories	Fat	Saturated fat	Carbohydrates	Sugar	Protein	Fibre
546 kJ	131 kcal	8.6 g	5.3 g	12.5 g	12.0 g	10.2 g	0.4 g

* Based on 90 portions

	80–100
General purpose ganache	
Couverture	

Preparing the dish

1 Place a sheet of acetate in the bottom of a praline frame.

2 Prepare general purpose ganache and pour into the praline mould and tap to level.

3 Leave the ganache for 12 hours to crystallise and firm up.

4 Remove the ganache from the praline frame and coat one side in tempered couverture.

5 Place the couverture side down onto the guitar base and cut through the ganache.

6 Lift the slab of ganache off the guitar base, raise the wire and wipe clean.

7 Place the slab of ganache back on the guitar in the opposite direction and cut through once again giving uniform cut squares.

8 Using a dipping fork place the cut praline, chocolate-coated side, into tempered couverture.

9 Lift out of the couverture using a dipping fork and tap on the side of the container to remove excess chocolate.

10 Carefully deposit the praline onto a sheet of silicone paper.

11 The top of the pralines can be decorated by placing transfer sheets, textured sheets on top or by giving a ripple effect by using the dipping fork on the surface.

PROFESSIONAL TIP

Layered pralines can be made by pouring general purpose ganache into a frame, allowing to set, then topping with a layer of pâté de fruits. Once completely set, process in the normal manner giving a two-layered praline.

Allow the ganache to firm up in the frame, then coat the surface with tempered couverture.

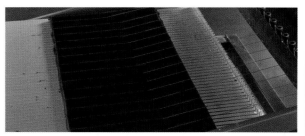

Place the couverture side down onto the guitar base and cut through the ganache.

Lift the slab of ganache off the guitar base using a take-off metal sheet.

Wipe the guitar clean, then carefully place the slab of ganache back on the guitar in the opposite direction and cut through once again, creating uniform squares.

10 Moulded pralines

Preparing the praline moulds:

1 Polish the mould with cotton wool. Never use an abrasive as any marks made on the surface of the mould will be picked up on the outer surface of the chocolate.

2 Embellish the mould with tempered coloured cocoa butter, metallic powders, different coloured tempered chocolate or just leave plain.

3 Fill the mould to the top with tempered couverture and tap the mould on the table top to raise any air bubbles to the surface of the chocolate.

4 Invert the mould upside and tap the sides with a rubber spatula, to force all the chocolate out of the mould, leaving a thin-coated cavity.

5 Using a chocolate scraper clean the surface above the chocolate coated cavities.

6 Place the mould with the open cavities coated in chocolate inverted onto a flat surface covered with silicone paper.

7 Leave the chocolate to crystallise (set).

8 Once the chocolate has set, invert the mould so that the open chocolate cavities are facing upwards.

9 Using a piping bag, fill the cavities with a ganache or any other suitable filling to just below the top of the mould.

10 Allow the ganache to harden.

11 Using a palette knife spread tempered chocolate over the ganache forming a flat smooth surface enclosing the ganache.

12 Leave the chocolate to set.

13 Once set turn the moulds upside down and tap the chocolates out of the mould.

> **PROFESSIONAL TIP**
> Always wear cotton gloves when handling chocolates to maintain the shine on the tempered couverture surface.

The following are suitable ganache recipes for filling praline moulds.

11 Mandarin ganache

Energy	Calories	Fat	Saturated fat	Carbohydrates	Sugar	Protein	Fibre
43867 kJ	10598 kcal	828.0 g	512.0 g	770.0 g	630.0 g	537.0 g	30.6 g

* Using milk chocolate, unsalted butter and tinned mandarins

Whipping cream	700 g
Glucose	50 g
Inverted sugar (trimoline)	60 g
Mandarin compound	50 g
Milk couverture	1200 g
Butter	100 g

Preparing the dish

1 Boil the cream with the glucose, the inverted sugar and the mandarin compound.

2 Pour onto the chocolate callets and mix well.

3 Cool down to room temperature.

4 Add the softened butter.

5 Pipe immediately into the praline moulds or chocolate spheres.

12 Green tea ganache

Energy	Calories	Fat	Saturated fat	Carbohydrates	Sugar	Protein	Fibre
395 kJ	95 kcal	7.2 g	4.5 g	6.2 g	5.8 g	2.3 g	1.0 g

* Using milk and dark chocolate

	60 praline shells
Whipping cream	400 g
Glucose	80 g
Green tea powder	40 g
Dark couverture	300 g
Milk couverture	400 g

Preparing the dish

1 Boil the cream, glucose and tea together and infuse for a few minutes.

2 Strain onto the melted chocolate and emulsify using a stem blender.

3 Cool down and pipe into the praline shells.

Embellish the mould with coloured, tempered cocoa butter.

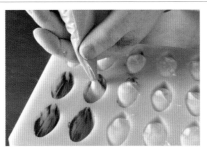

Fill to the top with tempered couverture, then tap the mould to release any air bubbles.

Invert the mould and tap it; most of the chocolate will drain out, leaving a thin coating in the cavity.

Scrape away and reserve any excess chocolate from the top of the mould.

Once the chocolate has set, fill with ganache.

Once the ganache has set, spread tempered chocolate over the top.

Scrape off any excess, leaving a flat, smooth surface.

Once completely set, tap the chocolates out of the mould.

13 Caramel and orange ganache

Energy	Calories	Fat	Saturated fat	Carbohydrates	Sugar	Protein	Fibre
410 kJ	98 kcal	6.5 g	4.0 g	9.4 g	8.9 g	6.8 g	0.7 g

* Using milk chocolate, unsalted butter, double cream and powdered glucose

	80 shells
Caster sugar	80 g
Glucose	30 g
Orange purée	325 g
Cream	200 g
Sorbitol	20 g
Butter	130 g
Plain couverture	650 g
Milk couverture	325 g

Preparing the dish

1 Prepare a dry caramel from the caster sugar and glucose.
2 Add the boiled orange purée gradually to the caramelised sugar solution.
3 Add the cream, sorbitol and butter, reboil.
4 Allow the cream to cool to 80 °C.
5 Pour onto the partially melted chocolates and emulsify using a stem blender.
6 Cool down before filling the praline shells.

14 Crémeux

Energy	Calories	Fat	Saturated fat	Carbohydrates	Sugar	Protein	Fibre
653 kJ	156 kcal	11.0 g	6.2 g	12.4 g	12.3 g	11.6 g	0.4 g

* Using semi-skimmed milk and plain chocolate

Cooking

1 Boil the milk and cream and add the mix of sugar and yolks. Cook to 84 °C. Strain and weigh 500 g.
2 Part melt the couverture and slowly pour on the warm anglaise.
3 Emulsify with a stem blender.
4 Refrigerate.
5 When firm beat to a piping consistency.

SERVING SUGGESTION

Crémeux can be served in its own right as a mousse-like dessert, although the texture is richer and smoother than a lighter and aerated mousse. However, crémeux is more often used as an accompaniment or garnish with a dessert or as a filling in a pastry product.

	15 portions
Base anglaise	
35 per cent UHT whipping cream	208 g
Milk	208 g
Egg yolk	83 g
Sugar	42 g
Crémeux	
Base anglaise	500 g
Plain couverture	220 g

15 Moulded chocolate eggs

Preparing the dish

Method 1 – single-sided mould producing half figure shapes:

1 Polish the inside of the half-figure mould with cotton wool.

2 Using a fine paint brush, highlight any areas within the mould with coloured liquid cocoa butter, white tempered couverture or plain tempered couverture.

3 Once the embellishment of the mould is set, fill with tempered liquid couverture, tap the side of the mould to release any air bubbles to the surface of the chocolate.

4 Invert the mould, once again tapping the side of the mould to remove excess chocolate, leaving a thin coating of chocolate inside the mould.

5 Using a chocolate scraper remove excess chocolate from the top of the mould and invert the mould flat side down onto a level surface lined with silicone paper.

6 Once the chocolate has set, remove the half figures from the mould.

7 Warm a flat metal tray and quickly warm the edge of one of the half figures to melt the couverture.

8 Place the other matching half figure on top and leave until fully set.

Highlight parts of the mould using coloured cocoa butter.

Fill the mould with tempered couverture and tap to release any air bubbles.

Invert the mould and empty out the excess chocolate, leaving a thin coating, then scrape any excess from the top.

Invert the mould flat side down onto a level surface lined with silicone paper.

Remove the half figures, quickly warm the edges on a flat warm tray and join them together.

Method two – fully attached figure mould producing fully shaped figures:

1 Polish the inside of the half-figure mould with cotton wool.

2 Embellish as above.

3 Join both half moulds together and clip to secure in place.

4 Fill the mould to the top with liquid tempered couverture, tap the side of the mould to release air bubbles.

5 Turn the mould upside down and tap out all the chocolate leaving a thin coating of chocolate inside the mould.

6 Using a chocolate scraper clean the base of the mould and stand on a flat surface lined with silicone paper until set. (Sometimes a second coat of chocolate may be required depending on the size of the mould.)

7 Once the couverture has fully set, remove the clips and gently remove both halves of the mould revealing a solid figure shape.

Removing a chocolate figure from a fully attached mould – the whole figure comes out, rather than a half figure.

Assessment for the Level 3 Advanced Technical Diploma in Professional Cookery

Chefs are assessed regularly in many different ways, and will receive feedback on how you are doing in different ways. This may be in the form of informal feedback and comments from customers as each meal that is served is evaluated by the customer who will consider many things about their meal experience and often provide feedback.

However, there are also more formal ways and opportunities for chefs to gain indications of how well they are doing, as well as opportunities to understand how they could improve their performance. This will occur through professional development, through education, training where you may be able to update your skills, and could also include competitions or work placements in a working kitchen environment. You may receive feedback in the following ways for example:

- Appraisal, a structured review completed with your head chef or manager, normally completed annually with a review of progress and development opportunities.
- Customer feedback that is shared with the team, for example, a letter of thanks or recommendations.
- Reports on performance following a period of work placement or experience.

When you undertake this course, you will be assessed formally at given times throughout the course.

For the Level 3 Advanced Technical Diploma the assessment takes place at given times through the development of the course, and covers a number of different tasks and assessment methods which will be formally assessed. In other words, they are set by the awarding body (for example, City and Guilds of London Institute CGLI) and the centre offering the programme. You will be marked on the quality and accuracy of your work.

For the Level 3 Advanced Technical Diploma in Professional Cookery, you will be assessed in two ways, and you must successfully complete and achieve the following assessments to gain the qualification:

1 The synoptic assignment
2 An externally marked exam.

For both assessments, the conditions and the process for formal qualifications are set to meet three main criteria:

- They must ensure the rigour of the assessment process
- They must provide fairness for all candidates, wherever they complete the assessment
- They must give confidence in relation to the outcome for both candidates and future employers

1 The synoptic assignment

The process known as 'Synoptic Assessment' is an approach where candidates have to make full use of the knowledge, understanding and skills that they have built up during the course of learning and development to tackle problems, challenges or tasks.

As part of this process, candidates will apply their full range of knowledge, skills and understanding in this assessment. The focus is on bringing together, selecting and applying learning from across the qualification rather than demonstrating achievement against units or subsets of the qualification content. They will need to ensure that they transfer their knowledge, skills and understanding to the practical situation/task, and approach the task in an independent, autonomous and confident way.

For this assignment, you will typically be asked to produce a four course taster menu to meet customer requirements and include costs for this for an event. You will be given a scenario or situation similar to one that you will be faced with as a chef in the industry and one where you will need to create multiple dishes covering a wide variety of skills. The overall assessment would be split into a number of tasks and time allocated for each of these. You will have between ten and twelve hours to complete the tasks.

Part 1 is to plan the production of the dishes: This will require you to consider the items and process you will follow to meet the brief. This could be completed two weeks before part two is attempted. Here you may need to:
- Complete a risk assessment process for health and safety, food safety, allergen information.
- Complete recipe specifications for each dish. This may include reasons or explanations for your choice. This could be linked to the brief/scenario, for instance the inclusion of specific ingredients.
- Complete a plan for producing the dishes ready for tasting, including a time plan.
- Defining a criteria for evaluating the dishes and your work, including an evaluation form to complete.

Part 2 is the practical skills element. Here, you will be expected to produce the menu items. This would be for at least two people. A six hour window for this part could be split, allowing for time to complete the preparations or mise-en-place and then a cooking and presentation phase, within the set time constraints. All dishes should be completed and evaluated against the criteria, including the completion of the relevant food safety documentation that you had in place.

Part 3 When you have reviewed the items produced, you will be required to produce a cost for the production of the event. You should take into account a range of information, such as the number of guests, the overall food costs, VAT and the recommended price per cover and overall cost: You would normally be given basic information in a scenario such as food costs to be assumed at 35% of selling price, VAT is set at 20%. You will be allowed around 2 hours to complete the assignment. Part 3 will normally be completed within a week of completing part 2.

Page 625 includes an example of a synoptic assignment and some useful tips.

There are certain conditions for the assessment process for all candidates:
- You must carry out all elements of each task on your own, under supervised conditions.
- You can complete the research and the collating of the information as part of the recipe specifications you may wish to use unsupervised.
- You are expected to work in a professional and safe manner. If you work in an unsafe manner, the assessment could be stopped.
- Work submitted in relation to all the tasks/parts should be your own.

This examination should reflect the skills and knowledge a chef at this level would be using as part of their role within a kitchen. It will therefore prove how capable you would be in the role. It is also an opportunity for you to demonstrate your higher level understanding, knowledge and skills and this may enable you to achieve a higher grade.

The assignment is externally set by the awarding organisation and although the format will be a scenario, the dishes are expected to change for each examination period. The assessment is internally marked within the setting you are completing your training and development and then externally moderated by the awarding body. This means that the awarding organisation will appoint someone to monitor elements of the assessment process. There may also be external experts (Industry representatives Chefs) as part of the assessment team. The marks awarded for each part of the Synoptic assignment would then be moderated by the Awarding Organisation before the final result is confirmed.

Marking

You will be marked against the set assessment objectives (AOs) in the specification. This will include assessing your breadth and accuracy of knowledge, understanding of concepts, and the quality of your technical skills as well as your ability to use what you have learnt in an integrated way to achieve a considered and high quality outcome. These would fall into band widths (outlined below) that enable the assessment team to consider your work and the award you could achieve.

- **Fail:** poor or limited skills and practice, elements of work not completed within time frames set, unsafe working practice for elements.
- **Pass:** some limited skills with a fair understanding of knowledge completed tasks.
- **Merit:** some good-strong skills with a good understanding of theory and practice in completing all tasks to a good level.
- **Distinction:** strong skill set across all practical areas, some creative flair and problem solving abilities, the ability to complete analysis and reflection across all tasks.

Each objective will be allocated a set number of marks and each of the assessment team will use the set criteria to judge your work and performance to ensure consistency across the assessment process. This counts for 70% of the overall mark for the qualification.

Below is an example of the type of scenario you may come across in your synoptic assignment:

SeaView Hotel has been asked to cater for a dinner for local food producers, you have been tasked with completing a tasting for four course dinner with the following menu:
- Starter: Shallow-poached fish dish, with sauce made from fish stock and garnish
- Main: Ballotine of chicken with sauce, accompanied by sessional vegetables and a potato
- Vegetarian: To include a fresh pasta dish
- Dessert: Baked chocolate tart
- Following the tasting (Practical presentation of the completed dishes) you will be expected to provide details, feedback and final costs to enable the sales team to complete the booking for the event.

Synoptic assignment: top tips

Time planning

Think about how long dishes take to cook and prepare, especially where there are a number of elements in a dish/menu, these should always be served/presented in order to ensure maximum marks are gained.
- For Ballotine of Chicken, for example, prepare the forcemeat which needs to cool before boning out the leg. Any stock to be made, garnishes or accompaniments.

- For Glazed Lemon Sole Veronique, for example, fillet the fish so that the bones can be used for the stock before preparing the garnish.
- Ravioli of squash and spinach, make the pasta and allow to rest, prepare the filling and then cool, what about the sauce
- For Baked Chocolate Tart, for example, the pastry needs to be made and blind baked before the chocolate filling which will form the tart. Presentation considerations any garnish or sauce to be offered

To write up the time plan (i.e. how you are going to use the time you have been given) if you have six hours for example, you will need to consider: whether this will be split into preparations and then cooking and finishing. Will it, for example, be two blocks of 180 minutes or a continuum with a break, therefore 360 minutes starting from where you wish to finish and work back to 000 on your first draft. You should schedule in a break time, based around the brief or instructions. Each assessment centre will be able to set the allocation of time, for instance it could be 0900-1200 then 1300-1600.

Recipe specifications

You should:

- Research your recipes for all elements of your menu.
- Consider how much time it will take to complete each element.
- Complete your food requisition and equipment list to enable you to complete the menu.

Professional practice

This area covers what you do while producing your dishes. The assessor will be looking at how you conduct yourself in the kitchen area. Throughout, there will be an expectation that your knowledge of current legislation is developed and adopted within any practical session. If at any time you are considered to be working in an unsafe manner you may be stopped and you will have to retake the assessment at a later date.

Things to consider:

- **Personal hygiene and turnout** – this will cover your dress. It is a good idea to ensure your whites are clean and pressed. Personal hygiene standards will also cover things like jewellery, and hair and nails which will need to be addressed to the expected professional standard. it will also include things like washing hands as required, and wearing the full correct uniform. You should have covered these from early stages in your training.
- **Craft skills** – use the correct craft skills that are appropriate for the dish, including knife skills; the ease of the technique used and the consistency of skills you will show are also important. For example, you will need to show that you can use the correct skill for the blanching of tomatoes.
- **Correct use of equipment** – use the correct tools or equipment safely and correctly to complete tasks. This will include use of pans and could include use of energy. For example, do you really need to have the grill on 120 minutes before you glaze a fish dish? Do you need to peel vegetables with a Chef's Knife?
- **Work methods** – ensure you work in a logical and organised manner, using appropriate time and temperature controls for food items. "Clean as You Go" is the mantra used across kitchens, and you should be aware of cross-contamination threats to your dishes.

Culinary practice and finished dishes

The final dishes you present should be completed to requirements, and linked to your recipe specification. The use and balance of ingredients should be correct to ensure that the menu flows well, and colours and flavours complement each other. The techniques you demonstrate and skills you use will show your knowledge and understanding at this level and will be well executed.

The checking of the finished dish will cover:
- Presentation
- Portioning
- flavour, texture
- seasoning
- and of course cooking.

You should consider:
- **Use of food items** – have you selected ingredients that balance in terms of texture, colour and flavours in relation to overall dish? In some settings you may be asked to consider seasonality as part of the assessment.
- **Techniques or skill** – preparation skills and techniques used for the ingredients selected, that these are completed in a logical sequence and with an understanding and awareness of portion sizes or waste from trimming and peeling. You will also need to follow legislation guidelines in relation to food safety and potential risks from contamination.
- **Cooking methods** – the correct use of these for the dish requirements and ingredients being used. You will need to consider the use of time and temperature to ensure that dishes are cooked correctly and ready for the target service time. You will also need to consider how you can maintain nutrients by using the appropriate cooking methods, for example not over cooking vegetables.
- **Monitoring cooking process** – the tasting and evaluating elements of the dish prior to finishing will cover consistency and will enable you to correct the balance of the dish through the process.
- **Presentation** – the final stage of any practical assessment is how the dish is finished to present to the potential customer. This is an opportunity to show final flair and creativity as a chef when you garnish and finish the dish. Make sure you don't allow the food to go cold while trying to get it on the plate.

2. Externally marked exam

The awarding organisation (CGLI) will set exam windows or times for centres to complete these exams. You will sit these exams under invigilated examination conditions with specific time constraints. Short answer questions will assess the breadth of knowledge. The extended response question will give candidates the opportunity to integrate and demonstrate their higher level understanding and integration through discussion, analysis and evaluation. This provides the stretch element to challenge higher achieving candidates.

The externally marked exam will make up 30% of your overall mark and will mean completing set exercises and written questions. The shorter questions will enable you to show your breadth and of understanding of the subject area across the whole content of the qualification, and will be completed towards the end of your course. This will include all units of the qualification within the examination process including:
- the current legislation including food safety and allergen requirements in relation to providing food for sale
- financial controls and costing elements within a professional kitchen environment.

- the techniques and methods used to produce dishes to meet customer requirements, from preparation, cooking and finishing dishes, linked to all the commodities in the units
- equipment and tools required, product knowledge, availability and an understanding of traditional and contemporary cooking methods.

The examination is set for 90 minutes.

All of this will be linked to the specification and assessment criteria that you will cover within your programme of study and the course you complete. You will be expected to complete an online exam paper, answer a series of short-answer or extended response style questions to test your knowledge.

The following is an example of the type of question you will be expected to answer in your exam, and the marks that will be awarded for the different answers given. The following is guidance only, but it will help you to understand what you will need to do in order to achieve the higher marks for the question.

Extended written question:

With the introduction of the requirement for food providers to provide details on allergens within dishes, what has been the effect on the responsibilities of the professional chef?

Response: (Extended written answer worth maximum 8 marks) You will need to include the following content within your answer:

- Legal requirement, what is expected and how this could be shared?
- Allergens, what commodities are covered within this what is included on the menu. Are there any substitutions that could be used (e.g. Non Dairy Ice Cream, using soya based milk or replacement product)(?)
- Operation of the kitchen and service area, link to the 5'C's:
  - **Contents**: what's in the dish or product offered
  - **Contact**: avoiding contact between products, storage procedures etc.
  - **Cross-contamination**: how products are handled and where prepared in the kitchen
  - **Cleaning**: how personal hygiene, equipment use and cleaning is completed
  - **Communication**: ensuring that all parts of the kitchen and service team involved with the dish are aware of the need of the customer in relation to allergens.

Suggested Answers for each of the mark boundaries:

Chefs need to consider the inclusion of allergens in menus, which include, Nuts, Dairy, Gluten and shellfish, the legal requirement is to make that information available to customers. (Basic response, which is correct but only factual would only gain limited Marks 1-3)

Restaurants have a legal requirement to list any of the 14 Allergens that are included in dishes on the menu. Chefs would need to consider alternatives that could be used and how to organise the kitchen to stop cross contamination. The chef is responsible for communicating details of the allergens used. (Little more depth with clear analysis in this answer mentioning communication, cross-contamination Marks 4-6)

It is a legal requirement that information about the 14 allergens be available to cover the menu/dish offered. This can be indicated on the menu, provided on separate log and/or verbal by a nominated supervisor. It would be normal to indicate any adjustments or alternatives that could be used. Customers need to be given confidence that the establishment is aware of its responsibilities so following a process as in the 5C's (Contents, Contact, Cross-contamination, Cleaning and Communication) would help in ensuring all the team are aware. The chef is the responsible person for ensuring the information is available and communicated. (Clear and detailed analysis with a range of responsibilities Marks 7-8)

Short answer question:

Give one example of a sponge and one example of a biscuit that is made using the creaming technique (worth 2 marks).

Suggested Answers:

Victoria sponge, Rich fruit cake, Lemon drizzle cake.

Viennese biscuit, Langue de chat (Cats tongues), Sable biscuits.

Other skills assessments (practical tasks and evaluations of competence)

A set of practical tasks that are set across all areas of the programme which are used as part of the formative process, this is the opportunity for you and your tutor to assess how you are progressing, these normally cover one complete dish in the first part of the course and then develop. Tutors or mentors use this process to evaluate progress being made and are part of a development programme for individual chefs, within these you should develop skills to evaluate, analyse and critique dishes against the dish specification.

Grading

The overall qualification will be grade based on individual achievement in each element of the assessment and examination process. All elements must achieve a minimum pass mark for the qualification to be awarded.

Synoptic assessment counts for 70% of the overall grade

Examination counts for 30% of the overall grade

Resits

There will be one opportunity within each academic year to sit the assignment. Candidates who fail the assignment will have one re-sit opportunity. The re-sit opportunity will be in the next academic year, and will be the assignment set for that academic year once released to centres. If the re-sit is failed, the candidate will fail the qualification.

Candidates who fail the exam at the first sitting will have one opportunity to re-sit. If the re-sit is failed the candidate will fail the qualification. For exam dates, please refer to the Assessment and Examination timetable.

Malpractice

Where a candidate is proven to have achieved a mark through malpractice then they would face penalties which could include disqualification from the assessment leading to a fail grade. Examples of malpractice include

- Plagiarism the submission of work that is not wholly your own
- Deliberate destruction or interference with another candidates' work
- Collusion with others, copying or sharing work with others to gain an advantage.

How can you prepare for the assessments

Prior planning and preparation is most important to enable you to achieve the maximum from any assessment – you cannot beat the feeling of receiving the highest mark possible!

To do that you will have to work in a methodical manner, be prepared to practise and listen to feedback as well as personally reflecting on what you have completed.

1 Read the instructions or brief that is given to you; make sure you understand what you have to do. This includes the criteria or assessment objectives you will judged or assessed on, which could include use of equipment, professional standards, use of materials, skills and techniques to be covered, time allowed, any preparations that can be completed or stages of the assessment.

2 Carry out any research, find or develop your recipes and consider which ingredients and skills you will need to include within the task.

3 Prepare and plan your time, this could be a time plan for the actual assessment. This isn't only for the day but how you may prepare before the test date.

4 Prepare a list of equipment you will require to complete the task, make sure you have available what you need.

5 Make sure you have completed risk assessments, including information with regards to ingredients being used.

6 It may be helpful to check if there is time to set out your work station before the assessment takes place.

7 Ensure that you demonstrate your level of knowledge and understanding by working to the highest level possible, covering principles, practices and legislation of professional cookery.

On the day of assessment, you should ensure you arrive at the assessment centre/kitchen in plenty of time, correctly dressed and ready to complete the task.

Tip: make sure you have a bottle of water or water available to keep yourself hydrated. It helps you to keep a clear head!

Using this book to help you to prepare for assessment

Throughout the book you will find a number of opportunities that can support your learning and development. This includes:

- Test your knowledge activities will help you prepare for assessments and questioning.
- Activities will give you a chance to look at a situation or issue around providing dishes and reviewing alternatives to meet customer expectations in terms of dishes presented for example healthy choices or dietary need
- Step-by-step guides will show processes and skills in preparation and cooking dishes.

You should also refer to the How to use this book section on page xx.

Other types of assessment

And finally, there may be other opportunities for assessment or chances to review your knowledge and skills, further education, training and development as you progress through your career. These could include competitions and skills tests.

Competitions

Many of these are set each year at local, regional, national and even international levels. These could be based on an individual skill like cutting a chicken for sauté, all the way to preparing a full menu or displaying products as culinary arts. Sponsorship is often linked to most of these with some quite prestigious development opportunities. There is also the opportunity to represent your region or country through cookery competitions both as an individual and as part of a team. The following websites give an insight into the types of competitions that could be considered:

- www.worldskillsuk.org
- www.nestle-toquedor.co.uk.

Skills test (basket of goods)

As part of the employment and application process, many employers are now using skills assessment. For some, this is a simple stage in the kitchen where you work a shift or asked to prepare a simple dish. For roles with more responsibility, you may be given what is known as a mystery basket test to assess practical skills as well as working practices prior to offering employment to an individual.

Assessments and examinations can be stressful but with good preparation you can build confidence that it is your opportunity to show what you know, how well you can perform and start you on that great career you aspire to!

Good luck with your studies!

Glossary

Aerated To introduce air into. Eggs and sugar are aerated by whisking when making a spone. An 'aerated storeroom' is one that air is allowed to circulate through

Al denté Cooked until firm, crisp and with a bite

Allergy A reaction by the immune system to certain foods or ingredients; also referred to as an allergenic reaction

Ambient storage Storing food at room temperature

Amino acid The structural units of protein

Anaphylactic shock A severe, potentially life-threatening reaction to food, which causes swelling of the throat and mouth and blocks airways

Aromats Herbs such as parsley, chervil and basil used as a flavour base; may also include vegetables such as onions and celery

Barding Laying with fat

Basting Moistening meat periodically, especially while cooking, using a liquid such as melted butter, or a sauce

Batting Tenderising using a meat mallet

Béchamel White sauce

Best before date Date coding appearing on packaged foods that are stored at room temperature and are an indication of quality. Use of the food within the date is not legally binding but it is bad practice to use foods that have exceeded this date

Binary fission The process by which bacteria divide in half and multiply.

Biological contamination Contamination caused by living organisms (for example, bacteria, toxins, viruses, yeasts, moulds and enzymes)

Bivalves Molluscs with an external hinged double shell; for example, scallops and mussels

Blanch Plunging food into boiling water for a brief time before plunging into cold water to halt the cooking process. The purpose of blanching is to soften and/or partly cook the food without colouring it

Bran Outer layers of the cereal

Broth A soup consisting of a stock or water in which meat or fish and/or vegetables have been simmered

Brown rice Any rice that has had the outer covering removed, retaining its bran

Brown stock Stock produced by browning vegetables and bones before covering in water, boiling and simmering

Bulk fermentation A proving/fermentation process in which a leavened dough will double in size

Bulk fermentation time The time it takes for dough to double in size

Callets Small, round, symmetrical pieces of couverture, which are used for melting purposes; the purchasing unit of couverture

Cellulose An insoluble substance; the main constituent of plant cell walls and of vegetable fibres such as cotton. The plant uses this for support in stems, leaves, husks of seeds and bark

Centrifuge A piece of laboratory equipment, driven by a motor, which spins liquid samples at high speed. Centrifuges work by the sedimentation principle to separate substances of greater and lesser density

Cephalopods Molluscs with a soft body and an internal shell; for example squid and octopus

Cereal The edible fruit of any grass used for food

Chemical contamination Contamination by chemical compounds used for a variety of purposes such as cleaning and disinfection

Clean in place Cleaning items where they are rather than moving them to a sink. This is used for large equipment such as mixing machines

Cleaning schedule Planned and recorded cleaning of all areas and equipment

Cleaning The removal of dirt and grease, usually with the assistance of hot water and detergent

Coagulate The transformation of liquids or moisture in meat to a solid form

Coagulation To change consistency from a fluid into a thickened mass

Coarse cut Cuts from the neck legs and forequarters; these are tougher cuts and therefore often cooked using slower methods such as braising and stewing

Coeliac disease A disease in which a person is unable to digest the protein gluten (found in wheat and other cereals). Gluten causes the person's immune system to attack its own tissues, specifically the lining of the small intestine. Symptoms include diarrhoea, bloating, abdominal pain, weight loss and malnutrition

Collagen White protein found in connective tissue

Compound chocolate Less expensive than couverture, compound chocolate contains cocoa powder instead of cocoa liquor and oil instead of cocoa butter

Connective tissue Animal tissue that binds together the fibres of meat

Consommé A completely clear broth

Contamination: when something is present in food that harms its quality, changes its taste or could cause harm, illness or an allergic reaction

Control measure Measures put in place to minimise risk

Cook/chill meals Pre-cooked foods that are rapidly chilled and packaged then held at chiller temperatures before being reheated for use.

COSHH Regulations Legal requirement for employers to control substances that are hazardous to health to prevent any possible injury to those using the substances

Couverture Chocolate which has a very high percentage of cocoa butter; used in professional kitchens and patisserie

Creaming Refers to the initial mixing of sugar and cream together using a wooden spoon or electric mixer until a smooth mixture is formed. This is often used in the production of sweet/sugar pastry

Crème renversée Egg custard mixture made using eggs, milk and cream

Critical control point (CCP) A point in a procedure or process where a hazard could occur if controls were not in place

Cross-contamination Contaminants such as pathogenic bacteria transferred from one place to another. This is frequently from raw food to cooked/high risk food

Crustaceans Shellfish with tough outer shells and flexible joints to allow quick movement; for example, crab and lobster

Cutting in the method used to describe the way in which flour is carefully folded into sponge batters

Danger zone The temperature range between 5 °C and 63 °C, at which bacteria can multiply

Deglaze To add wine stock or water to the pan in which the fish was cooked to use the sediments and flavours left in the pan

Demersal fish Live in deep water and feed from the bottom of the sea; they are almost always white fleshed fish and can be round or flat

Demi-glace Brown refined sauce

Detergent A substance that removes grease and dirt and holds them in suspension in water. It may be in the form of liquid, powder, gel or foam. Detergent will not kill pathogens

Diabetes A disease in which the body produces no insulin (type 1) or insufficient insulin (type 2) and is therefore unable to regulate the amount of sugar in the blood. If left untreated, it causes thirst, frequent urination, tiredness and other symptoms. There are different kinds of diabetes, not all of which need to be treated with insulin. Type 1 diabetes tends to occur in people under 40 and in children; type 2 diabetes is more common in overweight and older people and can sometimes be controlled by diet alone

Disinfectant Destroys pathogenic bacteria, bringing it to a safe level

Disinfection Action to bring micro-organisms to a safe level. This can be done with chemical disinfectants or heat

Due diligence Written and recorded proof that a business took all reasonable precautions to avoid food safety problems and food poisoning

Elastin Yellow protein found in connective tissue. This needs to be removed

Emulsifier Emulsifiers are made up of molecules with one water-loving (hydrophilic) and one oil-loving (hydrophobic) end. They make it possible for liquids that are normally immiscible to become finely dispersed in each other, creating a stable, homogenous, smooth emulsion

Emulsion a mixture of two or more liquids that are normally immiscible (non-mixable or un-blendable)

Endosperm middle layer of the wheat grain containing starch and protein and used to make flour

Environmental health officer (EHO) A person employed by the local authority to advise upon, inspect and enforce food safety legislation in their area. An EHO is now sometimes called an environmental health practitioner (EHP)

Enzyme Proteins that speed up the rate of a chemical reaction, causing food to ripen, deteriorate and spoil

Fat bloom Lighter colour spots of cocoa butter fat on the surface of chocolate, usually caused incorrect tempering

Fermentation Fermentation occurs when yeast and/or bacteria convert carbohydrates to carbon dioxide causing gas bubbles to form. This has a leavening (raising) effect on products such as dough when making bread

First in, first out (FIFO) A method of stock rotation that means that foods already in storage are used before new deliveries

Fixed costs Regular charges, such as labour and overheads, that do not vary according to the volume of business

Flat fish Have a flatter profile and always have white flesh because the oils are stored in the liver. They include sole, plaice, dabs, turbot, brill, bream, flounder and halibut

Food intolerance Does not involve the immune system, but it does cause a reaction to some foods

Food poisoning Illnesses of the digestive system, usually caused by consuming food or drink that has become contaminated with viruses, bacteria and/or toxins

Food spoilage Foods spoiled by the action of bacteria, moulds yeasts or enzymes. The food may smell or taste unpleasant, be sticky, slimy, dry, wrinkled or discoloured. Food spoilage is usually detectable by sight, smell or touch.

Forequarter The front section of the side of meat

Free range Animals kept in natural conditions, with freedom of movement

Gelatine A nearly transparent, glutinous substance, obtained by boiling the bones, ligaments, etc., of animals, and used in making jellies

Germ The smallest part of the grain and the part from which a new plant will shoot

Glaze a reduced stock with a concentrated flavour. Often used to enhance the flavour of soups and sauces

Grain Edible fruit of cereals

Gross profit The difference between the cost of an item and the price at which it is sold

Hazard Analysis Critical Control Point (HACCP) A system for identifying the food safety hazards within a business or procedure and putting suitable controls in place to make them safe. Details of the system must be recorded and available for inspection. All food businesses must now have a food safety system based on HACCP

Hazard Any area, activity or procedure with the potential to cause harm

Health and Safety Executive (HSE) The national independent authority for work-related health, safety and illness. It acts to reduce work-related death and serious injury across UK workplaces

Healthy carrier Someone carrying salmonella in their intestine without showing any signs of illness

Hemicellulose a polysaccharide similar to cellulose but consists of many different sugar building blocks in shorter chains than cellulose

High-ratio cakes A high ratio cake is prepared with a relatively high proportion of sugar and eggs compared to flour

High-risk food Usually ready-to-eat food that will require no further cooking before it is eaten

Hindquarter The back part of the side of meat

Hogget A sheep over a year old

Homogenise A process in the production of milk in which the fats are emulsified so that the cream does not separate

Humectant A humectant is a hygroscopic (helps to retain water) substance used to keep products moist. Honey is an example used in this way

Hydrogenated A substance that has had hydrogen gas added to it. This is used to harden oils in margarine

Hydrolysis a chemical reaction in which a compound such as sugar breaks down by reacting with water

Hydrometer Equipment used to measure density on the Baumé scale

Hygroscopy The ability of a substance to attract and hold water molecules from the surrounding environment

Immune system A system of structures and processes in the human body which defend against harmful substances and maintain good health. Occasionally the immune system in some people recognises ordinary food as harmful and a reaction occurs

Impermeable Something that liquid cannot pass through

Jus-lié Thickened gravy

Knocking back Re-kneading the dough once proved

Lactic fermentation Lactic fermentation is a biological process in which sugars (e.g. sucrose or lactose) are converted into cellular energy

Lamb A sheep under a year old

Lamination The dough that is produced by building alternating layers of dough and butter when making puff pastry, croissants or Danish pastries

Leavened Bread that has risen

Leavening The process where a substance, such as yeast or baking powder, is used to produce fermentation in a dough or batter

Liaison A mixture of egg yolks and cream that is used to thicken a sauce

Linear workflow A flow of work that allows the processing of food to be moved smoothly in one direction through the premises from the point of delivery to the point of sale or service

Long-grain rice A narrow, pointed grain that has had the full bran and most of the germ removed. Used for plain boiling and in savoury dishes

Maillard reaction The chemical reaction that occurs when heat is applied to meat, causing browning

Mandatory Something that must be done, for example rubber gloves must be worn when handling certain chemicals

Marbling White flecks of fat within meat

Marinating To tenderise with the addition of acidic sauce

Metamyoglobin Created when myoglobin is oxidised (reacts with oxygen in the air). This changes the colour of meat to dark red or brown

Mirepoix Approximately 1-cm diced carrot, onion and celery cooked in fat or oil and used as a flavour base for soups

Molluscs Shellfish with either a hinged double shell, such as mussels, or a spiral shell, such as winkles

Mutton Meat from a mature sheep

Myoglobin Pigment in the tissues which gives meat its bright red colour.

Net profit The difference between the selling price of an item and the total cost of the product (this includes food, labour and overheads)

Oily fish Are always round and, because the fish oils are dispersed through the flesh (rather than stored in the liver as in white fish), the flesh is darker. These include mackerel, salmon, sardines, trout, herrings and tuna

Osteoporosis A disease in which the density or thickness of the bones breaks down, putting them at greater risk of fracture. Exercise and good nutrition can reduce the risk of developing osteoporosis

Overheads Expenses associated with operating the business, such as rent, rates, heating, lighting, electricity, gas, maintenance and equipment

Pané Coating the fish with a light coating of seasoned flour, beaten egg then breadcrumbs

Papin An enzyme that is sometimes injected into animals before slaughter to speed up the softening of fibres and muscles

Passed A thin soup such as a consommé, served by passing through a fine muslin cloth to remove the solid particles

Pasteurisation A process where heat is applied to products such as milk and cream for a short period of time before being cooled quickly. This helps to kill harmful bacteria and extend shelf-life while maintaining flavour properties

Pathogenic bacteria Bacteria that can cause illness by infection or by the toxins they produce

Pathogens micro-organisms such as bacteria, fungi and viruses that can cause disease

Pectin A complex polysaccharide present in primary cells. Pectin is formed in the middle of plants and it helps bind cells together

Pelagic fish Live in more shallow or mid-depth waters and are usually round, oily fish such as mackerel, herrings and sardines

Physical contamination When something gets into foods that should not be there (e.g. glass, paint, plasters, hair or insects)

Potage A thick soup

Prime cut The leanest and most tender cuts of meat; these come from the hindquarters

Prove Process of allowing dough to rise and increase in size

Provenance Where food comes from, for example, where it is grown, reared, produced or finished

Purée A soup with a vegetable base that has been puréed

Raising agent A substance added to a cake or bread mixture while produces gases that give lightness to the product. Baking powder is an example of a raising agent

Refractometer Equipment used to measure sweetness on the Brix scale

Retarding dough Reducing the fermentation process of a dough by placing it in chilled conditions

Rickets A disease of the bones

RIDDOR Reporting of Injuries, Diseases and Dangerous Occurrences Regulations 2013. All injuries, diseases and dangerous occurrences happening in the workplace or because of work carried out on behalf of the employer must be reported to the Health and Safety Executive. This is a legal requirement and it is the employer's responsibility to make any such reports

Risk The possibility that someone could be harmed by a hazard

Risk assessment The process of identifying and evaluating hazards and risks and putting measures in place to control them. This process should be recorded and reviewed

Round fish Can vary greatly in size from small sardines to very large tuna. They can either have white flesh, such as bass, grouper, mullet, haddock and cod, or darker, oily flesh such as tuna, mackerel, herring, trout and salmon

Roux A soup thickened with a traditional roux of fat and flour

Rubbing in A technique where flour is rubbed into a fat to make products such as short pastry and crumbles. Using the fingertips, flour and butter are rubbed gently together until the mixture resembles fine breadcrumbs

Sanitiser A chemical with detergent and disinfecting properties; it breaks down dirt and grease and controls bacteria

Scurvy A disease that can cause bleeding gums and other symptoms

Searing Browning or colouring the outer surface of the meat, searing the surface

Septic A cut or other wound that has become infected with bacteria. It is often wet with a white or yellow appearance

Septicaemia Blood poisoning; it occurs when an infection in the bloodstream causes the body's immune system to begin attacking the body itself

Shortening Fat used in pastry making; it is made from oils and is 100 per cent fat

Shortening agent A fat used to help prevent the development of gluten strands when making pastry. This helps to make the texture of the product more crumbly

Short-grain rice A short, rounded grain with a soft texture. Suitable for sweet dishes and risotto

Spoilage bacteria Cause food to change and spoil, for example, develop a bad smell or go slimy

Spore A resistant, resting phase for some bacteria when they form protection around the essential part of the cell that can then survive boiling, freezing and disinfection

Stabiliser A substance added to food to help to preserve its structure, such as sponge cake

Sterilisation A process where heat is applied but at much higher temperatures than pasteurisation. This increases shelf-life further than pasteurisation but products treated this way tend to lose some of their flavour properties

Steriliser Can be chemical or through the action of extreme heat. It will kill all living micro-organisms

Stock rotation Managing stock by using older items before newer items, provided the older items are in sound condition and are still within use-by or best before dates

Sugar bloom A dusty appearance on the surface of chocolate caused by sugar crystals, caused by incorrect tempering

Syneresis Loss of water from a mixture.

Tempering An essential process that allows chocolate to be manipulated and combined with other ingredients or made into artistic pieces; it produces a snap when bitten and has a glossy appearance

Toxin A poison produced by some bacteria as they multiply in food or as they die in the human body

Traceability The ability to track food through the stages of production, processing and distribution

Transpiration Gradual loss of water from a plant cell

Turgid Plant cells that are in a state of maximum water content

Turgor pressure The internal pressure inside a plant cell

Turning A turn is the term used to describe the process of producing the layers in laminated pastry. Each time the paste is rolled and folded, it is referred to as a turn

Univalves Molluscs with a single spiral shell; for example, winkles or whelks

Unpassed A thin soup such as a broth which is served along with all the ingredients used

Use-by date The date by which food must be used. It must not be stored or offered for sale after this date

Variable costs Costs that vary according to the volume of business; includes food and materials costs

Vegetable stock Stock produced by simmering vegetables and herbs.

Virus Micro-organism even smaller than bacteria. It does not multiply in food but can enter the body via food where it then invades living cells

Wheatgerm Part of the wheat grain containing oil which is removed in the milling process

White stock A stock produced by blanching but not browning vegetables and bones, to give a clear stock

Index of recipes

This index lists every recipe in the book, grouped by major commodity and by type of dish.

There is a full topic index at the pack of the book.

Index

Page references in **bold** indicate key terms